AF526911

NEMATODE CONTROL IN CROPS

Edited by

M. MASHKOOR ALAM

Ph. D., D.Sc., PG Nematol, FPSI, FNSI, FBS, FBRS, FAASN

Professor of Plant Nematology
Department of Botany
Aligarh Muslim University, Aligarh, UP, India

(Formerly : Deputy Director, RAK Institute of Agriculture)

NEETA SHARMA

Ph. D., FNASc, FPSI, FES, FNRS, FBRS

Reader in Plant Pathology
Department of Botany, Lucknow University
Lucknow - 226 001, UP, India

INTERNATIONAL BOOK DISTRIBUTING COMPANY
(Publishing Division)

Published by

INTERNATIONAL BOOK DISTRIBUTING CO.

(Publishing Division)
Chaman Studio Building, 2nd Floor,
Charbagh, Lucknow 226 004 U.P. (INDIA)
Tel. : Off. : 450004, 450007, 459058 Fax : 0522-458629
E-Mail : ibdco@sancharnet.in

First Edition. 2002

Composed & Designed at :

Panacea Computers
33, Nehru Road,
Sadar Cantt. Lucknow-226 002
Phone : 481164, 480546
E-mail : prasgupt@rediffmail.com

Printed at:

Army Printing Press
33, Nehru Road,
Sadar Cantt. Lucknow-226 002
Phone : 481164, 480546
E-mail : armypress@sify.com

Dedicated to

KISHWAR SULTANA

the lady with
patience
and
affection

Preface

Our crops have been plagued with nematodes since remote past, though unnoticed by early farmers. It became obvious, only after the discovery of some very effective nematicidal chemicals during the 1940s and '50s, that growers were paying yearly taxes to these tiny but mighty pests. Plant-parasitic nematodes cause enormous losses to the farm income by way of reduced yields as well as the quality of the produce. Besides, they have pivotal role as initiators, co-operators, aggravates and synergists for many other diseases of plants.

In the quest for increasing crop productivity man has created many problems, sometimes leading to severe epidemics. The natural resistance of the wild types has been lost in many cases. Extensive use of pesticides, that sometimes accumulate in the soil and ground water, not only created ecological problems but has led to the increase in pest resistance to these chemicals. These and other problems have generated interest amongst the scientists to explore alternative approaches for combating with these noxious organisms.

The idea behind the present book is to gather information at one place on different aspects of nematode control on high value crops. The chapters have been written by recognised authorities currently involved in research in their respective fields from India as well as from U.S.A., U.K., China, Brazil and other countries. The subject matter of the book has interdisciplinary application. It has wide scope especially in plant protection sciences. We have learnt and enjoyed much while editing the book and do hope that readers will also find it useful to them.

The book in the present form could not have been published without the sincere help and co-operation of 35 contributors. We are greatly indebted to all of them. We owe much to our revered teacher Professor Abrar M. Khan, F.N.A., who initiated us into the fascinating science of plant nematology. We are also thankful to our professional colleagues and friends who have helped us during the course of preparation of the book, especially Dr. Brijesh K. Dwivedi, Dr. (Mrs.) Ghazala Parveen, Dr. Mansoor A. Siddiqui, Dr. Akhtar Haseeb and Mr. M. Arif. Sincere thanks are due to Mr. Suneel Gomber who has done excellent job in publishing the book. We also acknowledge with love the profound support given by our spouses Mrs. Kishwar Sultana and D.K. Sharma and superb children all through the period of editing of the book. They have been a constant source of joy, inspiration and honest affection.

M. Mashkoor Alam

Neeta Sharma

Contents

Chapter 1

Control of Root-knot and Cyst Nematodes Attacking Rice

Y.S. Rao[1], J.S. Prasad[2] & M.S. Panwar[1]

1. Central Rice Research Institute, Cuttack-753 006. India

2. Directorate of Rice Research, Rajendra Nagar, Hyderabad-500 030, A.P. India

INTENSIVE rice cropping dictates an ever increasing need for averting losses due to nematodes making use of nematicides, a necessary adjunct to continuous cropping. In such a system, highly effective methods of nematode control should be evolved with minimum use of chemicals per unit area. The prohibitive costs and the associated hazards in the use of soil nematicides lay special emphasis on evolving more remunerative and curative methods in infected crops and prophylactic methods for application in infested soils (Rao, 1985 ; Krishnaprasad, 1974).

ROOT-KNOT NEMATODES

The root-knot nematodes, *Meloidogyne* spp. have been established as a major constraint in the upland rice and found to cause yield losses in serious proportions. About 13 species of this genus were found to be associated with the rice culture so far throughout the world (Table 1). In India, *Meloidogyne graminicola* Golden & Birchfield is a serious pest of rice.

DISTRIBUTION

In some developing countries, root-knot nematode infection in plants of economic importance is reported to be so common and widespread that galled roots were considered normal and poor stand and low yields of agricultural crops were attributed to vague agricultural ailments like "worn-out land" or soil exhaustion (Sasser & Carter, 1985). About 13 identified species of root-knot nematode have been reported as causing root galling and yield losses in rice round the world (Table 1). The root-knot nematode is distributed in nurseries, uplands, well-drained soils, semi-deep water and deepwater soils (Rao *et al.*, 1971; Roy, 1973; Castillo *et al.*, 1976; Fortuner, 1978; Prabhu & Bedendo, 1981; Babatola, 1981; Bridge & Page, 1982; Diomonde, 1984; O'ever & Bekker, 1985; Prasad *et al.*, 1985, 1990). Among the different species, *M. graminicola is* common in most tracts.

Whereas *M. oryzae* is restricted to Indonesia and Surinam (O'ever, 1981, 1982). In India *M. graminicola* was reported from Assam, Bihar, Delhi, Gujarat, Haryana, Madhya Pradesh, Kerala, Orissa, Tripura and West Bengal and is particularly serious in Sibsagar District of Assam, Cuttack and Kalahandi districts of Orissa, Tripura, and Burdwan and West Dinajpur districts of West Bengal (Prasad *et al.*, 1987).

HOST-PARASITE RELATIONSHIP

Nematode affected plants show depletion in vigour, yellowing of leaves and curling along the midribs. The second stage juvenile nematodes cause disruption and hypertrophy of cortical cells due to their migration, movement and salivary secretions. Hyperplasia of protophloem and abnormal xylem proliferation cause swelling of stele at the sites of nematode establishment (Jena & Rao, 1973). Production of tillers is inhibited while flowering becomes poor and delayed. Infected plants showed significant reduction in N, P, K, Fe, Ca and reducing sugars in roots and shoots. Magnesium increased by 50% and manganese by 95% in roots. Total sugars, amino nitrogen and DNA increased in both roots and shoots. RNA increased by 20% in shoots and 82% in roots due to excessive hypertrophy and inhibition in protein metabolism (Rao *et al.*, 1988). Chlorophyll 'a' and 'b' fractions reduced by 20 and 28% causing chlorosis in leaves (Rao *et al.*, 1988; Swain & Prasad, 1988) and photosynthetic rate by 22.7 to 25% in susceptible rice varieties and up to 19% in resistant varieties (Swain & Prasad, 1989).

Table 1 : World distribution of *Meloidogyne* spp. on rice.

Species	Country	Reference
M. arenaria thamesi	South Africa	Van Der Linde, 1958
	Costa Rica	Lopez, 1985
M. exigua	Thailand	Kanjanasoon, 1964
	Costa Rica	Lopez, 1985
M. graminicola	Laos	Golden & Birchfield, 1986
	India	Patnaik, 1969
	Thailand	Buangsuwon *et al.*, 1971; Sontirat, 1981
	Bangladesh	Page & Bridge, 1978
M. hapla	Costa Rica	Lopez, 1985
M. incognita	Egypt	Ibrahim *et al.*, 1972
	Nigeria	Babatola, 1980
	Argentina	Sisler, 1982
	Ivory Coast	Fortuner, 1981
	5, Costa Rica	Lopez, 1985

Species	Country	Reference
M. incognita acrita	Japan	Ichinohe, 1955
	South Africa	Van Der Linde, 1956
	Costa Rica	Lopez, 1985
	Ivory Coast	Luc & Guiran, 1960
	Nigeria	Fademi, 1987
	Liberia	Lamberti *et al.*, 1988
M. javanica	South Africa	Van Der Linde, 1956
	Thailand	Leaumsang & Kanjanasoon, 1961
	Brazil	Sharma, 1981
	Comoros	Vuong, 1972
	Egypt	Ibrahim *et al.*, 1972
	Ivory Coast	Fortuner, 1981
M. lini,	China	Yang *et al.*, 1988
M. Oryzae	Indonesia	Mass *et al.*, 1978
	Surinam	O' ever & Bekker, 1985
M. graminis	Egypt	Ibrahim, 1978
M. indica	Egypt	Ibrahim, 1978
M. arenaria	Belize (North America)	Fortuner, 1981; Hollis, 1982
	Costa Rica	Lopez, 1985
M. salasi	Costa Rica	Lopez, 1984
	Panama	
Meloidogyne spp.	USA	Steiner, 1934 ; Tullis, 1934
	Brazil	Monteiro, 1968
	Taiwan	Lin, 1970
	Ghana	Addoh, 1971
	Bangladesh	Ou,1972 ; Rahman & Taylor, 1983
	Senegal	Fortuner & Merny, 1973
	Philippines	Castillo *et al.*, 1976
	Panama	Tarte *et al.*, 1978
	Thailand	Sangatian, 1978
		Sontirat, 1981
	Nigeria	Babatola, 1981
	Vietnam	Kinh *et al.*, 1982
	China	Feng *et al.*, 1980
	Burma	Yi Yi & Mya, 1981

(Prasad *et al.*, 1987)

YIELD LOSSES

Losses in grain yield ranging from 16 to 32% were reported due to infestation of *M. graminicola (*Biswas & Rao, 1971a; Chakrabarty *et al*., 1981 ; Rao & Biswas, 1973 ; Rao, 1978) *M. arenaria, M. incognita* and *M. javan*ica caused more than 60% grain losses with inoculum level of 8000 eggs and juveniles/litre of soil in upland rice areas in Nigeria (Babatola, 1984). Yield losses due to this nematode were estimated to be as high as 75% in Ivory Coast (Diomonde, 1981) and 65% in Vietnam (Kinh *et al*., 1982).

With the increase of one *M. graminicola* eggmass in the inoculum given to 10 days old plants of rice var. IR 8, a corresponding significant reduction in grain yield by 2.6% was recorded. The threshold level of *M. graminicola to* cause 10% yield losses was 1.25 eggmasses or 250 infective juveniles per seedling or direct seeded rice (Rao, 1984). Further, Rao *et al.* (1985) observed that threshold level to cause 10% loss in grain yield was 120, 250 and 600 eggs at 10, 30 and 60 days age of plants in direct seeded rice respectively. Such high levels of nematode populations prevailed in areas endemic to the nematode in India (Anon, 1985). Up to 40-70% seedling mortality was recorded in deepwater rice varieties infected with *M. graminicola* due to loss in seedling vigour (Bridge & Page, 1982; Prasad *et al.,* 1990). Pot trials with rice cv. IAC 25 inoculated with *M. javanica* showed a reduction in plant height and fresh root weight at inoculum levels of 100 nematodes/200 g soil and above and in dry weight of grain at 1,00,000 nematodes/200 g soil (Sharma, 1981). Guo *et al.* (1984) observed average yield losses of 10-20% (highest 40-50%) due to *Meloidogyne* spp. in China. Glasshouse studies revealed that grain yield was reduced by 15% in flooded soil and 9% in soil without standing water when rice plants were inoculated with *M. oryzae (*O'ever & Bekker, 1985) in Surinam. Inoculum density of *M. salasi* caused a descending linear effect on stem diameter, width of the central leaf and plant height while it caused the opposite effect on root weight under artificial inoculations with 0, 333, 666 and 999 eggs/100 ml soil in Costa Rica (Sancho *et al.*, 1987). Kinh *et al.* (1982) reported reduction in plant height by about 31-48% and rice yield by about 65% in rice var. TN 73-2 infected with *M. graminicola* (13-15 galls/plant) in Thailand.

CONTROL STRATEGIES

Resistant varieties

Golden & Birchfield (1986) found that all the varieties of rice included in their tests were susceptible to *M. graminicola.* Manser (1971) reported that more than 82 varieties including a few glutinous and semi-glutinous varieties both '*indica*' and '*japonica*' strains produced root-knots which varied in numbers. Varieties

Garem and Dumai were graded as resistant in Assam since knotted roots were absent and 11 others including Jagannath and IR 20 were rated as tolerant as they developed upto 5% galls (Roy, 1973, 1975). Based on the number of endoparasites and root-knots varieties Tkm 6, Patnai 6 and N 136 were graded as resistant (Israel & Rao, 1971).

In tolerant varieties, poor transfer cell development caused delay in the growth of the endoparasitic stages of the nematode. In resistant varieties, the nematode activity was further inhibited by cell necrosis (Jena & Rao, 1977a, b). Varieties least susceptible to *M. graminicola* were characterised by dense root hair, highly sclerotised epidermis, narrow cortex and stele, high alanine, aspartic acid and lignification in roots (Rao, 1978). The structural features offered mechanical barriers for penetration of juveniles and establishment while the chemical features retarded the development of the nematode by either antimetabolic effects or by inhibition of transfer cell development on which the nematodes depend for nutrition (Jena & Rao, 1973). That the resistant characters were heritable had been confirmed in the progenies of crosses with Tkm 6 (R) x Taichung (native) 1 (S) (Rao, 1978) and resistance is dominant over susceptibility (Swain *et al*., 1986).

Rice varieties 130^2/103/1031/F2591/5/3 and IR 20 showed high degree of resistance to *M. incognita* in Nigeria (Babatola, 1980). In Brazil, rice variety IRAT 13 was found to be moderately resistant (Sharma & Prabhu, 1981) and cultures BR 79-5276, BR 79-6257 and BR 79-32849 resistant to *M. javanica* (Sharma & Prabhu, 1983). Entries TNAU (ADT) 103, RP 1442-4-6-1-1-1, RP 1888-4259-1329-126, Akashi (Prasad *et al*., 1986a), Daya, Udaya, MW 10 and IR 36 were found to be resistant to *M. graminicola* (Swain *et al*., 1986). In Thailand, three selections from cross IR 36/RD25 (Arayarungsarit *et al*., 1985) and RD 6 were reported to show resistance to *M. graminicola* (Arayarungsarit, 1987).

Cultural practices

Eclipta alba, a common weed in rice soils was found to have nematicidal properties. Extracts of fresh roots and shoots of the weed (5g/100 ml water) killed larvae of *M. graminicola* by direct contact toxicity within 6 h. Soil amendment with fresh tissues of *E. alba* caused complete mortality of the nematode. The roots and shoots contained high phenols (0.725 and 1.720 mg/g tissue) and soils amended with the weed showed 0.114 and 0.180 mg phenols/g soil (Prasad & Rao, 1979). Analysis of phenols revealed the presence of wadalalactone, a product toxic to *M. graminicola*. Though the larvae of *M. graminicola* invaded roots of *E. alba* their survival was affected due to the toxic chemicals which are phenols or oxidation products. This toxic nature of the weed could be exploited in controlling *M. graminicola* by ploughing in the weeds, 10 to 15 days after their germination.

Potato, groundnut, blackgram and wheat grown in rotation with rice reduced *M. graminicola* infestation in the following rice crop (Rao *et al.*, 1984). *Cucumis sativus* and *Capsicum frutescens* were found to be non-hosts (Roy, 1977). Among the 18 crop plants included in a crop rotational trial, the nematode development was altogether inhibited in sweetpotato, cowpea, castor, sunflower, cauliflower, sesamum, onion and turnip while *Hibiscus cannabinus* was a non-efficient host. Brinjal, tomato, chilli and *Basella rubra* were most suitable hosts and supported high populations of *M. graminicola* (Rao *et al.*, 1984). Roy (1977) categorised *Eleucine indica* and *Digitaria adscendens* as non-hosts and french bean, jute and okra as poor hosts. He observed that decaffeinated tea waste and water hyacinth compost were also useful soil amendments in controlling *M. graminicola* (Roy, 1976).

Influence of amino acids and growth regulators

Many DL forms of amino acids inhibit root-knot development and retard growth of the nematode in roots (Overman & Woltz, 1962 ; Setty & Wheeler, 1968 ; Krishnaprasad & Rao, 1977). With soil drench (1000 ppm) treatments endoparasites were reduced in aspartic acid and increased in methionine and threonine. Giant cell development was inhibited in treatments with aspartic acid, alanine and glutamic acid treatments. Eggmasses were significantly reduced in all amino acids except in methionine and serine. In foliar sprays, endoparasitic stages and eggmasses were reduced in aspartic acid and glutamic acid and decreased in serine (Prasad & Webster, 1967).

Significant reduction of endoparasites and eggmasses were observed in treatments with IBA, NAA and MH. The number of eggs per eggmass was reduced with IAA, IBA and MH. The hormones CCC and GA were ineffective. In foliar sprays, IBA and MH had significantly reduced the endoparasites, eggmasses and eggs per eggmass (Krishnaprasad & Rao, 1977). Soil drench with GA had not increased the nematode activity in rice but plants treated with GA (Gopalakrishnan & Sircar, 1974) and glutamic acid, however, were resistant since glutamic acid accumulated in roots was proved to be antimetabolic to *M. graminicola* (Krishnaprasad & Rao, 1977). Such antimetabolic effects had been obtained with aspartic acid, glutamic acid and alanine with soil drench or foliar sprays of DL forms of the amino acids to infested crops (Krishnaprasad & Rao, 1976a, 1977).

Chemical control

Direct contact toxicity assay : Direct contact toxicity tests against infective larvae were attempted by Krishnaprasad (1974) with nematicides, fungicides and weedicides. Oxamyl or propanyl at 50 ppm and above, fensulfothion at 100 ppm and above, quinolphos, chlorpyriphos and phorate at 200 ppm were found to be effective against the nematode. There was significant reduction in the invasion particularly in treatments with 100 ppm while with 200 ppm no larvae entered roots indicating its effectiveness at par with oxamyl, fensulfothion, propanil, DBCP and quinolphos at this dose. This confirmed the earlier observations of DiSauzo (1969) that the chemical destroyed the ability of the nematodes to migrate preferentially to roots. Though the efficacy of these chemicals under direct contact studies may not show the same efficacy under field conditions, this could give a preliminary idea while selecting nematicides for the control of the nematode.

Seed soaking treatment : Over 50% reduction in the population of *M. graminicola* was obtained in 12 h seed soaking with oxamyl and fensulfothion at 1000 ppm and above. Treatments with DBCP, chlorpyriphos or carbofuran were effective at 2000 ppm. In seed soaking treatments for 24 h, oxamyl, phorate, chlorpyriphos and fensulfothion were superior to other chemicals (Krishnaprasad & Rao, 1976b). Significant reduction in number of galls, eggmasses, adults and other developing stages has been recorded in the seed soaking (at 1% for 6 hr) treatments with isofenfos and carbosulfone. Remarkably, with the fungicides, viz., thiophanate, carbendazim, mancozele and isoprothiolane treatments, a sizable population of the juveniles develop to males (Prasad *et al.*, 1986b). Ceresan dry, a mercuric fungicide @ 3 g/100 g seed was quite effective against *M. graminicola* (Sahu & Rao, 1982).

Root-dip treatment : Occurrence of root-knot nematode damage was observed not only in upland rice but also in medium and deepwater rice (Page & Bridge, 1978 ; Rahman & Taylor, 1983 ; Prasad *et al.*, 1985, 1990). As uprooting infested plants and replanting allowed continuous development of the nematode in roots (Rao, 1969) it has become necessary to evolve methods that can be prophylactic and therapeutic (Johnson & Good, 1971).

a) Post-infection root-dip treatment : Over 50% reduction in endoparasites was obtained with 100 and 200 ppm fensulfothion, 200 ppm oxamyl with 12 h root-dip. In the 24 h root-dip, 100 ppm oxamyl and 200 ppm carbofuran were effective. In the 12 h treatment, fensulfothion was superior to oxamyl, phorate or carbofuran. Mephospholan, DBCP, quinolphos and chlorpyriphos were less effective. In the 24 h treatment, there was a significant increase in the effectiveness of root-rip treatment (Krishnaprasad & Rao 1976s).

As postinfection root-dips, nemacur, phosphomidon, carbosulfone and carbofuran at 0.2% had significantly reduced the endoparasites and root-knots and were superior to root-dips with vegetable oils, viz. mustard oil, gingelly oil and sunflower oil (Prasad *et al.*, 1984).

(b) Pre-infection root-dip treatment : In 12 h treatment, oxamyl and fensulfothion at 50 ppm and above reduced endoparasites over 50% and were at par, and were significantly superior to phorate and DBCP which were at par. Doses of 100 ppm and above were also at par. In 24 h treatment, oxamyl at all doses, phorate at 50 ppm and above and fensulfothion at 100 ppm and above had reduced endoparasites by over 50% (Krishnaprasad & Rao, 1976c).

All the chemicals lost the residual toxicity from 20th day after treatment. Fensulfothion and oxamyl retained toxicity up to 15 days, phorate up to 10 days and DBCP up to 5 days following pre-inoculation root-dip treatment at 100 ppm dose. It was evident that systemic nematicides like fensulfothion, phorate or oxamyl retained residual toxicity in plants when applied as root dips (Krishnaprasad & Rao, 1976b).

Soil application of chemicals : A soil drench, to infected crops, oxamyl, fensulfothion or carbofuran at 1 kg a.i./ha resulted in complete control of both the soil and root population of *M. graminicola* (Krishnaprasad & Rao, 1976d). Carbofuran below 250 ppm had persistent toxicity and eggmass production was inhibited (Krishnaprasad & Rao, 1980). Biswas & Rao (1971b) studied the residual toxicity of metham sodium at intervals of 7, 14 and 21 days after application of nematicides. The results showed corresponding reduction in the degree of galling on the roots for 57 days following application of nematicides. Prasad *et al.* (1984) reported that soil drench with phenamiphos, phosphomidon, isofenfos, carbofuran, carbosulfone (0.02%) and *Eclipta alba* (5g root in 100 ml water) caused a significant reduction in eggmass and gall numbers. Application of carbofuran to soil in nursery and main field at the rate of 1 kg a.i./ha reduced *M. graminicola* by over 90% and resulted in increased yield of about 100% (Anon, 1985). Krishnaprasad & Rao (1982) studied therapeutic effects of oxamyl, phorate, carbofuran and fensulfothion, as soil treatments at 15 kg a.i./ha to rice plants transplanted 10 days earlier in an infested field. The pesticides reduced the endoparasitic populations, delayed the nematode development and also increased the number of males. *M. graminicola* infestation was reduced by the application of carbofuran at 3 kg a.i./ha incorporated at sowing in deepwater rice (Rahman & Taylor, 1983).

Continuous flooding or application of carbofuran or diazinon at the rate of 10 to 20 kg commercial product/ha gave good control of *M. graminicola*. Grain yields increased by 100 to 160% over untreated plots and 70 to 90% over non-diseased plots. However, at 20 days following the application of chemicals gall numbers began increasing but did not cause damage. Flooding with tide water was found to be ineffective (Kinh *et al*., 1982). Arayarungsarit (1987) treated rice varieties RD 6, Khao Dawk Mali 105 and Hahng Yi 71 known to be resistant, moderately resistant and susceptible respectively with carbofuran at the rate of 6 kg/ha ; it resulted in yield losses of 11.6, 18.2 and 32.6% respectively in the three varieties under root-knot nematode infestation.

CYST NEMATODES

The cyst nematodes are one of the most important nematode pests of agricultural crops all over the world. Among the known species of *Heterodera* five species were recorded on rice, viz., *Heterodera elachista, H. graminicola, H. oryzae, H. oryzicola* and *H. saccharin.*

DISTRIBUTION

Okada (1955) first reported the occurrence of an unnamed *Heterodera* sp. attacking upland rice in Tochigi-ken region of Japan. Cyst nematode attacking lowland rice var. Moroberekan grown at Bokakauamekro near Bouake, Ivory Coast was identified as *H. oryzae* (Luc & Brizuela, 1961) and later it was reported from Senegal also (Netscher & Perenes, 1971). Oshima (1974) described a cyst nematode from Japan as *H. elachista* which was the causative agent of successive failure of upland rice. *H. saccari* infesting sugarcane var. Pat 2878 at Jacob, Congo-Braaziaville was also found to attack rice in Senegal (Fortuner & Merny, 1973). It was recorded from Nigeria, Ivory Coast, Jamaica (Luc, 1974) and Ganta, Liberia (Volvas *et al*., 1986). In 1965, a cyst nematode identified as *H. graminophila* was found in large number on roots of barnyard grass in Louisiana, USA and host range studies proved rice cv. Saturn as an efficient host (Birchfied, 1970, 1971). Several unidentified *Heterodera* spp. attacking rice were reported from Korea (Park *et al*., 1967) and Liberia (Ou *et al*., 1975).

In India *H. oryzae* was reported from nurseries and well drained soils of Cuttack, Orissa (Rao, 1965). Later *H. oryzicola* has been recorded from Pattambi and its vicinity (Rao & Jayaprakash, 1977), Trivandrum, Anad, Alleppy, Trichur and Quilon from Kerala (Kuriyan, 1985; Raveendran *et al*., 1978; Venkitesan, 1979; Venkitesan & Charles, 1985), and Goa (Koshy *et al*., 1987), Bankura and Burdwan in West Bengal (Gupta *et al*., 1977; Rao, 1985).

YIELD LOSSES

Inoculations to rice with *H. elachista* at levels of 0, 10^2, 10^3 and 10^4 cysts per plant resulted in grain yields of 7.82, 4.37, 3.93 and 3.15 g (59.7% reduction) per plant respectively (Oda *et al.*, 1963). While inoculation of 0, 400 and 4000 larvae per rice plant resulted in grain weight of 71.2, 67.2 and 59.5 g (16.4% reduction) respectively (Shimizu, 1976). Yield reductions by 4.7, 45.4 and 56.2 per cent were recorded when inoculations were given at the rate of 1.2, 12.0 and 120.0 eggs of *H. oryzicola* per g of soil respectively (Jayaprakash & Rao, 1982). Kumari & Kuriyan (1981) reported yield losses ranging from 21 to 42 per cent due to this nematode in rice var. Triveni. Similar field trials have revealed yield losses ranging from 19 to 27 per cent (Kuriyan, 1985).

HOST-PARASITE RELATIONSHIP

Browning and chlorosis of leaves and roots, retardation in the normal functioning of roots, early flowering of plants by 10 to 13 days and partial filling of grains are the typical symptoms of infection by *H. oryzicola* (Rao & Jayaprakash, 1977). Virtually no galls developed on the roots (Jayaprakash & Rao, 1981). Vovlas *et al.* (1986) reported that the syncytia formation and disorder of the root stele structures are the main anatomical changes induced by *H. sacchari*. A significant reduction in the percentage of N, P, K and Fe occurred in roots and shoots. Leaf chlorophyll was reduced by 37.8 per cent. Calcium and sodium increased in plants to compensate for the loss of K and maintain the cation balance due to impaired photosynthesis. Total soluble sugars and proteins were reduced and starch accumulated in both roots and shoots. Soluble amino acids increased and there was no change in DNA (Jayaprakash, 1979). Impaired translocation may be responsible also for reduced IAA, cytokinin and thiamine in roots. ABA levels increased by 3.18 and 2.16 times in shoots and roots respectively similar to tungro virus infection (Rao *et al.*, 1988). Infestation by *Sclerotium rolfsii* is enhanced in the presence of the nematode but growth of the nematode is inhibited in the fungus infected rice seedlings (Jayaprakash & Rao, 1984). *H. sacchari* reproduced under both upland and simulated swamp conditions although symptoms of infection were more discernible in upland rice plants (Babatola, 1983). Hatching of eggs from cysts of *H. oryziocola* was significantly high in thiamine hydrochloride followed by rice root diffusate. Potent root diffusate was produced at 45 days age or at maximum tillering stage of plant growth. Dilution up to even 16 times was effective as a hatching agent (Jayaprakash & Rao, 1982). Unlike other cyst nematodes responding to flavionic acid (Clarke & Shepherd, 1967) where the chain length between the two terminal polarisable atoms was the factor inducing hatching. This factor for *H. oryzicola* was chained with only one polarisable atom (Jayaprakash & Rao, 1982). It was postulated (Veech, 1981) that if inactivity of pyruvic acid is the factor for dormancy in cyst nematodes, supplying cysts with thiamine hydrochloride or inorganic acids of the TCA cycle pathway should restore hatching. ATP was supplied by thiamine from rice seed and roots making the nematode pathway fully operable. Due to the above reasons, hatching was continuous and juveniles were prevalent round the year in continuously rice cropped areas in Kerala State (Rao & Jayaprakash, 1977).

CONTROL STRATEGIES

Very few varieties, viz. Lalnakanda-41, CR 143-2-2, Ratna, Hamsa, Mtu 17, Mu 4 have been found to be resistant to *H. oryzicola* (Jayaprakash & Rao, 1983). As banana (*Musa paradisiaca*) is a good host at this nematode (Venkitesan & Charles, 1985 ; Koshy *et al.*, 1987) use of banana in intercropping or rotations should be avoided. Growing any suitable crop other than banana atleast for a period of 3-4 years may bring down the nematode populations. Prasad *et al.* (1986a) reported 11 cultures resistant to *H. oryzicola*.

Rao (1986) observed that cyst nematode development can be reduced by dipping root tips in phenolics, phenyl alanine or increased by plant growth regulators and vitamins. In contact toxicity assays, the juveniles were effectively killed with oxamyl, phenamiphos, aldicarb or phorate at 250 ppm (Jayaprakash, 1979). Soaking seeds in 0.02% solutions of phenamiphos, FMC 35001, carbofuran, or phorate for 12 h reduced the cyst development and increased grain yield by 63.9 to 78.2 per cent. Soil application of aldicarb sulfone, carbofuran, phorate @ 1 kg a.i./ha at 7 and 50 days after planting has given significant reduction in the nematode population and increase in grain yield (Kuriyan, 1985). Jayaprakash (1979) tested DBCP, oxamyl, carbofuran and phenamiphos at 62.5, 125, 250 and 500 ppm against *H. oryzicola* as soil application. An effective control of over 80% was obtained with oxamyl and carbofuran followed by phenamiphos and DBCP.

REFERENCES

Addoh, P.G. (1971). The distribution and economic importance of plant parasitic nematodes in Ghana. *Ghana J. Agric. Sci.* **4**:21-32.

Anon; (1985). *In : Annual Report,* ICAR Research Complex, Shillong, Meghalaya. pp. 97-99.

Arayarungsarit, L. (1987). Yield ability of rice varieties in fields infested with root-knot nematodes. *IRR Newsl.* **12**(5):14.

Arayarungsarit, L.; Chongkid, B.; Suwanbutr, S. & Weerapat, P. (1985). Reaction of some upland rice to root-knot nematodes in rubber plantation fields. *IRR Nessl.* **10**(4):23-24.

Babatola, J.O. (1980). Reactions of some rice cultivars to the root-knot nematode, *Meloidogyne incognita. Nematropica* **10**:5-9.

Babatola, J.O. (1981). Root-knot nematode problems on rice in Nigeria. *Proc. Res. Planning Conf. on root-knot nematodes, Meloidogyne* spp. Regions IV & V, 16-20 Nov. 1981, Ibadan, Nigeria.

Babatola, J.O. (1983). Pathogenicity of *Heterodera sacchari* on rice. *Nematol. medit.* **11**:21-25.

Babatola, J.O. (1984). Rice nematode problems in Nigeria, their occurrence, distribution and pathogenesis. *Tropical Pest Management* **30**:256-265.

Birchfield, W. (1970). The biology of a new cyst nematode on grasses. *Phytopathology* **60**: 1284-1285.

Birchfield, W. (1971). Hosts and pathogenesis of a new *Heterodera* sp. on grasses. *Phytopathology* **61**: 885.

Biswas, H. & Rao, Y.S. (1971a). Studies on the nematodes of rice and rice soils. II. Influence of *Meloidogyne graminicola* Morgan & Birchfield incidence on yield of rice. Oryza **8**: 101-102.

Biswas, H. & Rao, Y.S. (1971b). Toxicological studies on nematodes with rice nematodes. *Indian Phytopath*. **24**: 159-165.

Bridge, J. & Page, S.L.J. (1982). The rice root-knot nematode, *Meloidogyne graminicola* on deepwater rice (*Oryza sativa* sub sp. *indica*). *Rev. Nematol.* **5**:225-232.

Buangsuwon, D.; Tonboon-Ek, P.; Rujirachorn, G.; Braun, A.. & Taylor, A.L. (1971). Nematodes. *In: Rice diseases and pests of Thailand*. Ministry of Agric., Thailand. pp. 61-67.

Castillo, M.B.; Arceo, M.B. & Litsinger, J.A. (1976). Nematodes in cropping patterns. V. Populations of plant parasitic nematodes associated with different land forms cropping patterns in a rainfed rice area in Itailo, Philippines. *Phytopathology* **12**:30-39.

Chakarbarty, N.K.; Kulshreshtha, J.P. & Rao, Y.S. (1971). Pest and disease problems in high yielding rice varieties and their control. *India Fmg.* **21**(7): 53-62.

Clarke, A.J. & Shepherd, A.M. (1967). Flavionic acid as a hatching agent for the potato cyst nematode, *Heteroder arostochiensis Woll. Nature* **211**:546.

Diomonde, M. (1981). Direct and combined effects of fertilizers and *Meloidogyne incognita* on upland rice. *Rev. Nematol.* **4**: 71-74.

Diomonde, M. (1984). Response of some upland rice cultivars to *Meloidogyne species*. Rev. Nematol. **7**: 57-63.

Disanzo, C.P. (1969). Some observations on the effect of carbofuran on three plant parasitic nematodes. *J. Nematol.* **1:**283.

Fademi, O.A. (1987). Resistance of some rice varieties to root-knot nematode (RKN) *Meloidogyne incognita*. *IRR Newsl.* **12(**1): 11.

Feng, Z.; Guan, Y.R. & Li, S.M. (1980). A study on root-knot nematode in rice. I. The influence of the root-knot nematode on rice growth. *J.S. China Agric. College* **1**(1):73-82.

Fortuner, R. (1978). *Meloidogyne* spp. in upland rice in Ivory Coast. pp. 8-10. *In : Proc. II IMP Res. Plang. Conf. on root-knot nematode, Meloidogyne* spp. Region IV, Feb. 20-24, 1978, Abidjan, Ivory Coast.

Fortuner, R. (1981). Les nematodes associes au rizpluvial en cote d'Ivorie. *Agronomie Troicale* **36**:70-77.

Fortuner, R. & Merny, G. (1973). Les nematodes parasites des raciness associes au riz en Basse-Casamance (Senegal) et en Gambie. *Cashiers ORSTOM., Ser. Biol.* **21**: 3-20.

Golden, A.M. & Birchfield, W. (1968). Rice root-knot nematode *(Meloidogyne graminicola)* as a new pest of rice. *Plant Dis. Reptr.* **52** : 423.

Gopalakrishnan, S. & Sircar, S.M. (1974). A comparative study of the effects of some growth regulators on the biochemistry of *Oryza sativa*. *Ann. Bot.* **38**:115-120.

Guo, Y.Y.; Zhou, N.S.; Wu, T.T. & Qio, H.L. (1984). An investigation and study on rice root-knot nematode on Hainan Island. *Guandong Agric. Sci.* No. **4** : 32-34.

Gupta, R.S.; Khera, S.& Singh, R.V. (1977). Occurrence of cyst nematode in paddy fields in West Bengal. *Newsl. Zool. Survey of India* **3:** 11.

Hollis, J.P. (1982). Summary of ecological data by country. Relationship of climate and soil characteristics to geographical distribution of *Meloidogyne* species in agricultural soils. *Intl. Meloidogyne Proj.*, North Carolina State Univ., USA. p. 20.

Ibrahim, I.K.A. (1978). The current status of root-knot nematode in northern Egypt. *Proc. Res. Plang. Conf. on root-knot nematodes* (*Meloidogyne* spp.), Jan. 29-Feb.2, 1978, Cairo, Egypt. pp. 36-37

Ibrahim, I.K.A.; Ibrahim, I.A. & Rezk, M.A. (1972). Pathogenicity of certain parasitic nematodes on rice. Alexandria *J. Agric. Res.* **20:** 175-181.

Ichinohe, M. (1955). Two species of the root-knot nematodes in Japan. *Japanese J. Appl. Zool.* **20**: 75-82.

Israel, P. & Rao, Y.S. (1971). Isolation of sources for nematode resistance in rice. *SABRO Newsl.* **3:** 7-10.

Jayaprakash, A. (1979). Pathogenesis and population dynamics of the cyst nematode, *Heterodera oryzicola n.* sp. in rice and rice soils. Ph. D. thesis, Utkal Univ., Bhubaneswar, India.

Jayaprakash, A. & Rao, Y.S. (1981). Histopathological changes in rice due to root infestation by the cyst nematode, *Heterodera oryzicola. Oryza* **18 :** 233-235.

Jayaprakash, A. & Rao, Y.S. (1982). Life history and behaviour of the cyst nematode *(Heterodera oryzicola* Rao & Jayaprakash, 1978) in rice (*Oryza sativa L.*). *Proc. Indian Acad. Sci.* (Anim. Sci.) **91:** 283-295.

Jayaprakash, A. & Rao, Y.S. (1983). Reaction of rice cultivars against the cyst nematode, *Heterodera oryzicola. Indian J. Nematol.* **13:** 117-118.

Jayaprakash, A. & Rao, Y.S. (1984). Cyst nematode, *Heterodera oryzicola* and seedling blight fungus *Sclerotium rolfsii* disease complex. *Indian J. Nematol.* **14:** 58-59.

Jena, R.N. & Rao, Y.S. (1973). Root-knot nematode resistance in rice. *In: breeding researches in Asia and Oceana.* Proc. Second General Congr. SABRAO, New Delhi. pp. 108-109.

Jena, R.N. & Rao, Y.S. (1977a). Nature of root-knot nematode resistance in rice. II. Mechanisms of resistance. *Proc. Indian Acad. Sci.* **86B:** 31-38.

Jena, R.N. & Rao, Y.S. (1977b). Histopathology of root-knot nematode infection in rice. *Proc. Indian Acad. Sci.* **86B**: 87-91.

Johnson, A.W. & Good, J.M. (1971). Nematode control advances with systemics. *Americal Nursery Manual* **13:** 30-34.

Kanjanasoon, P. (1964). Rice root-knot nematodes and their host plants. *Proc. 10th Session of Working Party on Rice Prodc. and Protc.*, Intl. Rice Comm. FAO, Manila, Philippines.

Kinh, D.; Huong, N.M. & Ut, N.V. (1982). Root-knot disease of rice in the Mekong Delta, Vietnam, *IRR Newsl.* **7**(4): 15.

Koshy, P.K.; Sosamma, V.K. & Faleiro, J.R. (1987). Occurrence of *Heterodera oryzicola* in banana in Goa. *Indian J. Nematol.* **17:** 334.

Krishnaprasad, K.S. (1974). Chemotherapy of the rice root-knot nematode. Ph. D. thesis, OUAT, Bhubaneswar, India.

Krishnaprasad, K.S. & Rao, Y.S. (1976a). Physiotherapy of the rice plant against the root-knot nematode (*Meloidogyne graminicola*). *Proc. Indian Natl. Acad. Sci*, **42**: 295-298.

Krishnaprasad, K.S. & Rao, Y.S. (1976b). Chemotherapy of the rice plant against the root-knot nematode (*Meloidogyne graminicola*) in rice. I Effect of seed soaking in nematicides. *Z. fur Pflkrank Pflschutz.* **83**: 665-668.

Krishnaprasad, K.S. & Rao, Y.S. (1976c). Chemotherapy of the rice plant against the root knot nematode (*Meloidogyne graminicola*) in rice. II. Effect of root dip treatment with nematicides in invasion and development of the nematode. *Z. fur Pflkrank. Pflschutz.* **83** : 730-735.

Krishnaprasad, K.S. & Rao, Y.S. (1976d). Chemotherapy of the rice plant against the rood-knot nematode (*Meloidogyne graminicola*) in rice. III. Evaluation of pesticides as soil drench. *Z. fur Pflkrank Pflschutz.* **83:** 736-741.

Krishnaprasad, K.S. & Rao, Y.S. (1977). Amino acids as antimetabolites against the root-knot nematode, *Melodiogyne graminicola* in rice. *Geobios* **4**: 185-187.

Krishnaprasad, K.S. & Rao, Y.S. (1980). Relative toxicity of chemicals to infective larvae of rice root-knot nematode, *Meloidogyne graminicola. Indian J. Nematol;* **10**: 221-224.

Krishnaprasad, K.S. & Rao, Y.S. (1982). Effect of a few systemic pesticides as soil treatments on the growth and development of *Meloidogyne graminicola* in rice roots. *Indian J. Nematol.* **12:** 14-21.

Kumari, U. & Kuriyan, K.J. (1981). Cyst nematode, *Heterodera oryzicola* on rice in Kerala. I. Estimation of loss in rice due to *H. oryzicola* infestation in field conditions. *Indian J. Nematol.* **11:** 106.

Kuriyan, K. (1985). *Annual Res. Progr. Reprt.* AICRP on Nematode Pests of Crops and Their Control, New Delhi.

Lamberti, F.; Tuopay, D.K.; Scrano, L.; Vito, D.I. Boiboi, J.B. & Radicci, V. (1988). *Meloidogyne* species (Nematoda : Heteroderoidea) occurring in Liberia. *FAO Plant Prot. Bull.* **36**: 69-73.

Leaumsang, P. & Kanjanasoon, P. (1961). A report on the root-knot nematode of rice plant in North Thailand. *Int. Rice Commn. Newsl.* p.1.

Lin, Y.Y. (1970). Studies on the rice root parasitic nematodes in Taiwan. *J. Agric. For. Chung-Hsing* **19:** 13-27.

Lopez, R. (1984). *Meloidogyne salasi n.* sp. (Nematoda : Meloidogynidae) a new parasite of rice (*Oryza sativa* L.) from Costa Rica and Panama. *Turrialba* **34**: 275-286.

Lopez, R. (1985). Differential plant responses, morphometrics and eletrophoretic patterns of some *Meloidogyne* species from Costa Rica and Florida, USA and the description of *Meloidogyne salasi n.* sp. *Diss. Abstr. Intl. B* (Sci. & Eng.) **45**: 2761.

Luc, M. (1974). *Heterodera sacchari* Luc & Merny 1963. *CIH description of plant parasitic nematodes.* St. Albans, Herts, UK. Set 4, No. 48.

Luc, M. & Brizuela, R.B. (1961). *Heterodera oryzae n.* sp (Nematoda : Heteroderidae) parasite due riz Cote de' Ivorie. *Nematologica* **6**: 272-279.

Luc, M. Guiran G.de (1960). Les nematodes associes aux plantes de I'quest African. Liste Pre'liminaire. *Agron. Trop. No. Gent* **15:** 434-449.

Maas, P.W.T.; Sanders, H. & Dede, J. (1978). *Meloidogyne oryzae n.* sp; (Nematoda : Meloidogynidae) infesting irrigated rice in Surinam (South America). *Nematologica* **24:** 305-311.

Manser, T.D. (1971). Notes on the rice root-knot nematode in Laos. *Plant Prot. Bull.* **19**: 135-139.

Monteiro, A.R. (1968). Ocorrencia no Brasil de importance nematoide. *Fitoparasito. O. Solo*. pp. 60-81.

Netscher, C. & Perenes, J. (1971). Etude concernant L'influence de la constitution genetique sur la longuer des larves d' *Heterodera oryzae. Nematologica* **17**: 336-346.

Oda, K.; Hoshino, M.; Yanaka, S. & Kumazawa, T. (1963). Studies on the rice cyst nematode of upland rice. III. Analysis of the injury. *Proc. Knato-Tosan. Plant Prot. Soc.* **10:**71.

O'ever, V.D.S. (1981). Root-knot nematode problems in Surinam. *Proc. 2nd Res. Plann. Conf. on Root-knot Nematodes, Meloidogyne* spp. Region I. Sept. 4-8, 1978. Sanjose, Costa Rica, Faculte de Agronomia, Univ. Panama, Panama. pp. 230-231.

O'ever, V.D.S. (1982). Research on *Meloidogyne* in Surinam. *Proc, 3rd Res. Plann. Conf. on Root-knot Nematodes, Meloidogyne* spp., Region I. Jan. 11-15, 1982. Inst. Invest, Agro. Pecuiarias de Panama, Panama. pp. 69-74.

O'ever, V.D.S. & Bekker, M.L.S. (1985). *Meloidogyne oryzae* of irrigated rice in Surinam. IRR Newsl. **10**(1):27.

Okada, T. (1955). Morphology of the upland rice nematode, *Heterodera* sp. *Appl. Zool. Ent. Ann. Meet., Japan*. pp. 11.

Oshima, Y. (1974). *Heterodera elachista n.* sp. an upland rice cyst nematode from Japan. *J. Nematol*. **4:**51-56.

Ou, S. H. (1972). *Rice Diseases* CIM, Kew, Surrey, England. pp 368.

Ou, S.H. Ling, K.G.; Kauffman, H.E. & Khush, G.S. (1975). Diseases of upland rice and their control through varietal resistance. *In : Major Research in the Upland Rice*. IRRI, Los Banos, Philippines. pp. 126-135.

Overman, A.J. & Woltz, S.S. (1962). Effects of amino acids antimetabolites, nematodes and tomato. *Proc. Fla. Stn. Hort. Soc.* **75**: 166-170.

Page, S.L.J. & Bridge, J. (1978). Plant nematodes on deepwater rice in Bangladesh (ODM). *Report on visit to Bangladesh.* June 19 to Aug. 9 Assot. Berks., U.K. Ministry of Overseas Development. p. 48.

Park, J.S.; Han, S.C. & Han, C.L. (1967). Survey on the plant parasitic nematodes in Korea. *Res. Reptr. of Office of Rural Dev.*, Suwon, Korea. **10**: 71-80.

Patnaik, N.C. (1969). Pathogenicity of *Meloidogyne graminicola* (Golden & Birchfield, 1965) in rice. *All India Nematol. Symp.* New Delhi. Aug. 21-22, 1969. p. 12.

Prabhu, A.S. & Bedendo, I.P. (1981). Root-knot nematode *Fusarium* complex in rice under dry land conditions in Brazil. *IRR Newsl.* **6**(2):5.

Prasad, J.S. & Rao, Y.S. (1979). Nematicidal properties of the weed *Eclipta alba Hassk.* (Compositae). *Rev. Parassitol.* **40**:87-90.

Prasad, J.S.; Panwar, M.S. & Rao, Y.S. (1984). Studies on the control of *Meloidogyne graminicola* in rice. *Nematol. medit.* **12:**141-144.

Prasad, J.S.; Panwar, M.S. & Rao, Y.S. (1985). Occurrence of the root-knot nematode, *Meloidogyne graminicola* in semi-deepwater rice. Curr. Sci. **54:** 387-388.

Prasad, J.S.; Panwar, M.S. & Rao, Y.S. (1986a). Reaction of some rice cultivars to the root-knot and cyst nematodes. *Oryza* **23:** 59-61.

Prasad, J.S.; Panwar, M.S. & Rao, Y.S. (1986b). Effect of seed-soaking with chemicals on the parasitic nematodes of rice. *Indian J. Nematol.* **16:**119-121.

Prasad, J.S.; Panwar, M.S. & Rao, Y.S. (1987). Nematode problems of rice in India. *Tropical Pest Management* **33:** 127-136.

Prasad, J.S.; Panwar, M.S. & Rao, Y.S. (1990). Influence of root-knot nematode infection on rice under simulated rainfed lowland conditions. *Nematol. medit.* **18:** 195-197.

Prasad, S.K. & Webster, J.M. (1967). The effects of amino acids as antimetabolites on four nematode species and their host plants. *Nematologica* **13:** 318-323.

Rahman, M.L. & Taylor, B. (1983). Nematode pests associated with deepwater rice in Bangladesh. *IRR Newsl.* **8**(3): 20-21.

Rao, V.N. (1965). Record of the occurrence of *Heterodera oryzae* Luc et Brizuela on rice in India. *Oryza* **2**: 183.

Rao, Y.S. (1969). Screening of rice varieties for their reaction to root-knot nematodes. *All India Nematol Symp.* New Delhi. Aug. 21-22, 1969. p. 59.

Rao, Y.S. (1978). Nematode problems of rice and lines of future research. ICAR Symp. on *Increasing yields of rice in Kharif,* CRRI, Cuttack, India. pp. 401-411.

Rao, Y.S. (1984). *Panel discussion on nematological problems of North Zone,* Hissar, HAU, Nematode problems of rice in submerged soils and their management (Memiogr.).

Rao, Y.S. (1985). Research on rice nematodes. *In* : *Rice in India.* ICAR Monograph (Ed. S.Y. Padmanabhan). 591-615 pp.

Rao, Y.S. (1986). Nematode problems of rice and their control. *In* : *Progress in Plant Nematology.* Prof. A.M. Khan festschrift, Aligarh Muslim University, Aligarh, India. pp. 475-502.

Rao, Y.S. & Biswas, H. (1973). Evaluation of yield losses in rice due to the root-knot nematode. *Indian J. Nematol.* **3**:74.

Rao, Y.S.; Israel, P.; Rao, Y.R.V.J. & Rao, V.N. (1971). Studies on nematodes of rice and rice soils. IV. Survey of nematodes. *Oryza* **8**: 75-82.

Rao, Y.S. & Jayaprakash, A. (1977). Leaf chlorosis due to infestation by a new cyst nematode. *IRR Newslett.* **2**:5.

Rao, Y.S.; Jayaprakash, A. & Mahanty, J. (1988). Nutritional disorders in rice due to infestation by *Heterodera oryzicola* and *Meloidogyne graminicola. Rev. Nematol.* **11**:375-380.

Rao, Y.S.; Prasad, J.S.; Padalia, C.R. & Yadav, C.A. (1984). Influence of rotation crops in rice soils on the dynamics of plant parasite nematode populations. *Biol. Agric. Hort.* **2**: 69-78.

Rao, Y.S.; Prasad, J.S. & Panwar, M.S. (1985). Nematode pests of rice. *Natl. Symp. on Impact of non-insect Pests and Predators on Food Production and Environment,* New Delhi.

Raveendran, V.; Jayaprakash, A. & Mohandas, C. (1978). Cyst nematode infestation of rice in Kerala State, India. *IRR Newsl.* **3**: 15.

Roy, A.K. (1973). Reaction of some rice cultivars to the attack of *Meloidogyne graminicola. Indian J.Nematol.* **3:** 72-73.

Roy, A.K. (1975). Studies on the resistance of rice to the attack of *Meloidogyne graminicola. Z. Pflfrunkh. Pflschutz.* **82:** 384-387.

Roy, A.K. (1976). Effect of decaffeinated tea waste and water hyacinth compost on the control of *Meloidogyne graminicola in rice. Indian J. Nematol.* **6:** 73.

Roy, A.K. (1977). Effectiveness of rotation with non-host or fallow on reducing infestation of *Meloidogyne graminicola. Indian J. Nematol.* **8:** 156-158.

Sahu, S.C. & Rao, Y.S. (1982). Seed treatment as prophylaxis against root-knot nematode in rice. *IRR Newsl.* **7:** 17.

Sancho, C.L.; Salazar, L. & Lopez, R. (1987). Effect of the initial population density on the pathogenicity of *Meloidogyne graminicola on* three rice cultivars. *Agronomia Costarricense* **11:** 233-238.

Sangatian, C. (1978). Rice varieties tolerant of root-knot nematode in rainfed areas. *IRR Newsl.* **3**(2): 7.

Sasser, J.N & Cartor, C.C. (1985). Overview of the International *Meloidogyne* Project 1975-1984. *In : Advance Treatise on Meloidogyne Vol 1 Biology* and Control. (Eds. J.N. Sasser & C.C. Carter). Coop. Publ., Deptt. of Plant Pathol., North Carolina State Univ. and US Agency Intl. Dev., Raleigh, North Carolina. p. 118.

Setty, K.G.H. & Wheeler, A.N. (1968). Growth substances in roots of tomato (*Lycopersicon esculentum Mill.*) infested with root-knot nematodes (*Meloidogyne* spp.). *Ann. appl. Biol.* **61**: 495-501.

Sharma, R.D. (1981). Pathogenicity of *Meloidogyn javanica* to rice. *In : Trabalhos apresentades* a *V Reunao Brassileira de Nematologia.* Feb. 9-13, 1981. Londoina, Brazil.

Sharma, R.D. & Prabhu, A.S. (1981). Response of some dry land rice cultivars to *Meloidogyne javanica. Brassileria de Nematologia.* Feb. 9-13, 1981. Piraciba, SP, Brazil.

Sharma, R.D. & Prabhu, A.S. (1983). Reaction of some cultivars of upland rice to the root-knot nematode *Meloidogyne javanica. In : Trabalhas appresenta-do sa VII Reuniao Brassileira de Nematologia,* Brasilia, DF, 21-25 de fevereiro de publicacao No. 7, Piraciaba, SP, Brazil, Sociedada Brasileira de nematologia. pp. 159-167.

Shimizu, K. (1976). Influence of upland rice cyst nematode, *Heterodera elachista*, on the yield of the upland-cultured paddy rice. Jap. *J. Nematol.* **6:** 1-6.

Sisler, G.M. de (1982). Summary of ecological data by country. Relationship of climate and soil characteristics to geographical distribution of *Meloidogyne* species in agricultural soils. *In :* IMP, 1982. Cooperative Publ., Deptt. Plant Pathol., North Carolina State Univ., and US Agency Intl. Dev., Raleigh, NC. p. 18.

Sontirat, S. (1981). Root-knot nematode, *Meloidogyne* sp. in Thailand. *In : Proc. 3rd Res. Plann. Conf. on Root-knot Nematodes, Meloidogyne* spp. Region VI. 20-24 July, 1981. Jakarta, Indonesia. p. 89.

Steiner, G. (1934). Root-knot and other nematodes attacking rice and some associated weeds. *Phytopathology* **24**:916-928.

Swain, B. & Prasad, J.S. (1988). Chlorophyll content in rice as influenced by the root-knot nematode. *Curr. Sci.* **57**: 895-896.

Swain, B. & Prasad, J.S. (1989). Photosynthetic rate in rice as influenced by the root-knot nematode *Meloidogyne graminicola infection. Rev. Nematol.* **12**: 431-432.

Swain, B.; Prasad, J.S. & Rao, Y.S. (1986). Reaction of some rice varieties to root-knot nematode. *IRR Newsl.* **11**(3) : 25.

Tarte, R., Osoria, J.M. & Stanzoila de M. (1978). El nematode nodular *Meloidogyne* spp. determination de especies existentes en Panama danas ocasianados d al cultivo de arroz busgueda de resistencia en tomate Industril. *Invest. Agropecu. Fac. Agron. Univ., Panama*, 1976-1977. pp. 420-445.

Tullis, E.L. (1934). The root-knot nematode on rice. *Phytopathology* **24**: 938-942.

Van der Linde, W.J. (1956). The *Meloidogyne* problem in South Africa. *Nematologica 1*: **377**-183.

Veech, J.A. (1981). Plant resistance to nematodes. *In : Plant Parasitic Nematodes* (Eds. B.M. Zuckerman & R.A. Rohde). Acad. Press. Vol. III. pp. 377-403.

Venkitesan, T.S. (1979). Note on nematode problems of upland rice in Kerala. A status paper presented in the *Task Force Meeting* held on Feb. 16-17 1979 at CRRI, Cuttack, India.

Venkitesan, T.S. & Charles, J.S.K. 1985). A note on the occurrence of *Heterodera* infection in Banana in Kerala. *Indian J. Nematol.* **15**: 279-280.

Vovlas, N.; Lamberti, F. & Tuoy, D.K. (1986). Observations on the morphology and histopathology of *Heterodera sacchari* attacking rice in Liberia. *FAO Plant Prot. Bull.* **34**: 153-156.

Vuong, H.H. (1972). Nuigibelite de *Meloidogyne javanica* surriz pluvial aux comores. *Document Ist Rech*. Agron. Madagascar 341. p. 21.

Yang, B.J.; Hu, K.J. & Xu, B.W. (1988). A new species of root-knot nematode *Meloidogyne lini n*. sp. parasitising rice *J. Yunnan Agril Univ.* **3**(1):11-17

Yi, Yi, M. & Mya, M. (1981). Report on occurrence of rice root-knot disease in Burma. *In: Proc. Agri. Res. Congr. Burma*. p. 125-143.

Chapter 2

Ufra Disease of Rice and its Control

M.L. Rahman & S.A. Miah

Plant Pathology Division, Bangladesh Rice Research Institute, Joydebpur, Gazipur-1701, Bangladesh

'UFRA' caused by the stem nematode, *Ditylenchus angustus* (Butler, 1913) Filipjev 1936, is one of the serious pests among 27 nematode species of rice listed recently (Bridge *et al.*, 1990). The disease is devastating in terms of both plant damage and yield loss in some South and South-East Asian countries (Miah & Bakr, 1977b; Catling *et al.*, 1988; Bridge *et al.*, 1990). Based on the nature and type of the symptoms and damage, it is also known as 'Dak pora', 'Akhet pet', 'Yed-Ngo' and 'Time dot San' mostly meaning burnt by lightening, twisted panicle, etc. in Bangladesh, Myanmar (Burma), Thailand and Vietnam respectively.

OCCURRENCE AND DISTRIBUTION

Ufra disease was first recorded in deep water rice fields in Naokhali district of the then East Bengal (Bangladesh) in 1913 (Butler, 1913). Then its occurrence had been noted in Burma (Seth, 1939), Uttar Pradesh of India (Singh, 1953), southern region of Thailand (Hashioka, 1963), Madagascar (Vuong, 1969) and Vietnam (Cuc & Kinh, 1981). Although its presence were reported in Malaya (Jack, 1923), Philippines (Reyes & Palo, 1956) and Malawi (Siddiqi, 1970), no further report is available from these countries. The disease is also not further reported from Thailand since its first report in 1963.

The disease predominantly occurs in low lying areas in Meghna and Ganges flood plains in Bangladesh, Ganges flood plains in India, Ayeyarwaddy in Burma and Mekong delta in Vietnam. With further intensification of rice cropping and introduction of irrigation facilities, now a days its occurrence has also been recorded from transplanted rice both *Boro* (Winter rice : October-November to April-May) and transplanted *Aman* (Autumn rice : June-July to November-December) both in low lying and under irrigated conditions in these countries (Bakr, 1977 ; Rao *et al.*, 1986a ; Cuc & Kinh, 1981). Seedlings grown in a part of ufra-infected fields in low-lying areas have also been found infected in the seedbed (Mondal & Miah, 1987a).

AREA AFFECTED AND YIELD LOSS

The area affected by ufra varies from year to year depending on the environmental conditions and cropping patterns prevailing in a particular season. Therefore yield loss assessment due to ufra on a national basis is very difficult.

In India, Singh (1953) estimated 5-25% yield loss due to ufra which may go up to 50% in a badly infected field. Rao *et al.* (1986a) estimated about 16,193 ha of *bao* rice (floating rice) and 20,000 ha of *sail* rice (transplanted aman) affected with ufra in several districts of Assam and West Bengal with 10-30% yield loss. In a recent survey Rathaiah (1988a) recorded about 2,000 ha of deepwater rice fields badly infected with ufra in several districts of Assam and West Bengal with 10-30% yield loss. In a recent survey Rathaiah (1988a) recorded about 2,000 ha of deepwater rice fields badly infected with ufra in Assam and estimated yield loss up to 75%.

Ufra disease causes 50-100% yield losses in thousands of hectares of deepwater rice and double transplanted rice in Cuu-long, Dong-Thap and An-Giang provinces in Mekong delta region of South Vietnam (Cuc & Kinh, 1981). The infested areas were about 10,000 ha in Duc-Thanh district while it was 5,000 ha in 1975 and 6,000 ha in 1980 in the Cuu-long district. Catling & Puckridge (1984) reported 60,000-10,0000 ha of rice affected with ufra specially in low lying areas of Mekong delta in 1982. Hong *et al.* (1989) reported a total of 600-1300 ha in An-Giang province and 600,000 ha in Dong-thap province affected with ufra during 1975-77 causing 100% yield loss in most cases.

Hashioka (1963) reported 500 ha of deepwater rice field affected with ufra in southern region of Thailand in 1961 and 1962 causing 20-90% yield loss.

In Myanmar (Burma) ufra attack is mostly prevalent in three divisions namely Bago, Yangon and Ayeyarwaddy causing 52% yield loss. The area affected was 1375 ha in 1984; 7,110 ha in 1985 ; and 789 ha in 1988 (Aung, 1989).

In Bangladesh ufra causes severe yield loss in deepwater rice grown in the low lying areas of southern districts. As for example, Butler (1913) estimated yield loss of about 7,400 tons of grain from 500-600 ha of deepwater rice fields in Noakhali district. He observed 100% yield loss in many parts of this district. In field experiment, Miah & Bakr (1977b) observed 40-60% or occasionally 100% yield loss due to ufra. Catling *et al.* (1979) estimated an annual yield loss of 4% (20% yield loss over 20% of the area) of total deepwater rice production in the country. At present this loss may rise up to 6-8% if damages in transplanted rice (aman and boro) are considered in addition to damage of deepwater rice. About 5,000 ha of transplanted aman rice was affected with ufra in one irrigation project area near Dhaka causing an estimated loss of about 75% of production (7,000-

8,000 tons) in 1984 (Miah & Rahman, 1985). Boro rice following aman rice is also affected in this area. It is also reported that 60-70% low lying areas covering about 200,000 ha is affected with ufra (Mondal & Miah, 1987a) where cultivation of boro rice after harvest of deepwater rice is increasing day by day. In 1990 some new areas with transplanted aman-boro cropping pattern were recorded with ufra infestation in the country. Loss of panicle density per unit area and increased grain sterility or partially filled grains are the major components of yield loss due to ufra (Cox & Rahman, 1980 ; Rao *et al.*, 1986b).

SYMPTOMS

The ufra nematode (*D. angustus*) feeds ectoparasitically on the younger tissues of unemerged leaves, leaf sheaths, peduncle and developing spikelets. Due to the feeding of the nematode, splash pattern chlorosis on the leaf sheath just above the collar appears as the diagnostic and visible symptom at the seedling to tillering stage of the crop. This symptom may appear within a week after nematode infection and may change to dark-brown stain with the advent of the disease. Sometimes leaf margins and leaf tips of the infected plants are corrugated and twisted. In severe cases, the youngest leaf is twisted and dies at the vegetative stage (Fig. 1). Emergence of several branches from the infested node is also quite common which makes the plant bushy in appearance.

Fig. 1. Ufra symptoms at the vegetative stage of rice.

At the heading stage of the crop, the panicle of the infected plant may be enclosed within the flag leaf sheath or may emerge partially or fully from it. The former type is called 'Thor ufra' while the latter 'Pucca ufra' (Butler, 1913). The first category occurs possibly in the primary infected plants while the later in the secondary and tertiary infected plants. These panicle symptoms were reclassified as ufra-1 (complete enclosure of panicle within the flag leaf sheath), ufra-2 (partial emergence), and ufra-3 (complete emergence with empty grains) in Bangladesh (Cox & Rahman, 1980). In all these categories, the grains are usually empty, malformed and wrinkled and the peduncle is shrivelled and twisted (Butler, 1913; Hashioka, 1963; Vuong, 1969; Miah, 1974; Chakrabarti *et al.*, 1985; Rathaiah & Das, 1987).

Initially the disease occurs in patches in a field where plants are usually stunted and yellowed. With the progress of the disease, the infected plants die and dry up and sometimes the whole field looks like burnt (Fig.2).

SURVIVAL AND DISEASE CYCLE

D. angustus is commonly stubble borne in nature. It usually survives in coiled state on the stubble, peduncle and in empty grains of infested plants left over in

Fig. 2. Ufra damage in rice in fields.

the field after crop harvest. Although its survival in soil and freshly harvested seed has been reported in Burma and Vietnam (Sein, 1977b; Cuc & Giang, 1982; Cuc, 1982b), transmission through these components are very minor and relatively less important. This nematode can also live in active form on the ratoons and volunteer rice and on some weeds present in the field (Miah & Bakr, 1977c; Rathaiah, 1988b). On the availability of rain, tidal or flood water, nematodes from these primary sources come out in water, resume their activity for searching the host plants and infect them. On contact with the host plant, the nematodes may pass to the space of outer leaf sheath within an hour but may take about 3 days to reach the innermost leaf sheath of 4 leaf stage of rice (Table 1). The rate of nematode invasion varies with the plant age; the older plants are relatively less infected (Rahman & Evans, 1987). Following infection, the nematodes start feeding and multiplying. They usually aggregate on the young tissues for feeding as the plant grows and proceeds to maturity. They lay eggs and reproduce on the host tissues when they feed. Butler (1913) assumed at least 3 generations in deepwater rice between the months of May-June and November but recent observations indicated that the generations could be more in a crop season. The secondary spread of the nematode from infected plant to the healthy one may take place through water, stem and leaf contact (Hashioka, 1963; Sein & Zan, 1977; Rahman & Evans, 1987) under high humid conditions (<75% R.H.).

Table 1: Number of *Ditylenchus angustus* extracted from leaf sheaths of rice cv. Habiganj Amon II after different inoculation periods.

Hours after inoculation	No. of *D. angustus* in outer to inner leaf sheaths			
	i	ii	iii	iv
1	2	2	0	0
3	2	2	2	0
12	4	2	3	0
24	4	7	3	0
48	19	43	39	0
72	3	16	7	27
96	7	73	16	31
120	4	10	3	19
144	3	9	10	24

(Source : Rahman & Evans, 1987)

As the crop proceeds to maturity the nematode accumulates around the younger tissues of glumes, inflorescences, peduncles and stem portions covered by the flag leaf sheaths where they cease feeding and coil up in a state of desiccation. Both the adult and juvenile forms, mostly J4 stage, are capable of such coiling. This desiccated nematode may survive up to 6-15 months (Butler, 1913; Miah & Bakr, 1977c). In the following crop, they again resume their activity through uncoiling on the availability of rain, tidal or flood water in the field. The period of uncoiling and motility depends on the length of desiccation period but usually takes an hour and stimulated by temperature (28-31°C) and light (Butler, 1919). However, nematodes under long desiccation period may be reactivated but may not remain infective (Butler, 1913). He also observed survival of *D. angustus* up to 4 months in flooded water without any host plant.

ALTERNATE HOSTS OF *D. ANGUSTUS*

The cultivated varieties of deepwater, irrigated or low land rices are the main hosts of *D. angustus*. The wild rice species namely *Oryza sativa* var. fatua, *O. glaberrima, O. cubensis, O. officinalis, O. meyriana, O. latifolia, O. eichingeri, O. alta, O. minuta, O. nivara, O. perennis, O. rufipogon* and *O. spontanea*, and the weed species namely *Leersia hexandra, Hygroryza aristata, Echinochloa colona* and *Sassiolepsis interrupta* have been found to be infected and to support the nematode (Hashioka, 1963 ; Voung, 1969; Miah & Bakr, 1977a; Cuc, 1982a). Vuong (1969) also reported feeding and reproduction of *D. angustus* on fungal genera of rice seeds and seedlings cultured on artificial media. This report has not been confirmed or verified elsewhere anymore. Recently, its feeding and reproduction on the fungus, *Rhizoctonia solani*, grown in autoclaved rice has been reported in Bangladesh (Bakr *et al.*, 1991).

ASSOCIATION WITH OTHER PATHOGENS

Incidence of blast (*Pyricularia oryzae*), sheath rot (*Sarocladium oryzae*) and bacterial leaf blight (*Xanthomonas campestris* PV. *oryzae*) are sometimes common in ufra-infected rice fields. It has been observed that ufra infection increased the severity of blast both in ufra susceptible and resistant variety (Table 2). From the analytical data it was found that ufra infection increases the total N content in the plant tissue which perhaps favours the blast development. Severe leaf and node blast incidence in ufra infested field has also been reported in India (Rathaiah, 1988a). Similarly results in greenhouse experiments indicated that combined effects of ufra and sheath rot (Mondal *et al.*, 1985), and ufra and bacterial leaf blight (unpublished) were the worst, i.e. cumulatively damaging both in increasing ufra severity and yield reduction compared to their individual effects. Foliar brown spots caused by *Fusarium* and *Cladosporium* species have been reported due to

Table 2 : Susceptibility of rice cultivars under different treatment combinations of ufra and blast.

Rice cultivar	Treatment*	Disease index (n=4)		N % in plant (n=3)
		Ufra	Blast	
BR 12 (susceptible to ufra)	Ufra + Blast	4.1	4.9	2.53
	Ufra	0.9	0.0	2.77
	Blast + Ufra	1.8	1.3	2.23
	Blast	0.0	3.3	1.20
	control	0.0	0.0	1.46
Rayada 16-06 (resistant to ufra)	Ufra+Blast	0.8	3.0	1.81
	Ufra	0.9	0.0	2.47
	Blast+Ufra	0.8	2.3	1.51
	Blast	0.0	3.3	0.78
	control	0.0	0.0	1.32

*The second inoculation of ufra nematode or blast fungus done after 20 days of first inoculation.

(Source : Mondal *et al.*, 1986)

secondary invasion of these fungi following nematode infection (Vuong, 1969).

FACTORS AFFECTING PARASITISM AND DISEASE DEVELOPMENT

Several factors like environmental components, number of invading nematodes per plant, age of plant at the time of infection, proportion of infested seedlings during transplanting and nutrient status in soil favour nematode parasitism and aggravate the disease severity.

Environmental components

D. angustus is a moisture loving tropical nematode and requires a relative humidity of more than 75% air temperature of about 28-30^{0}C, high rainfall and surface moisture on the host tissues for the disease development (Butler, 1919; Miah, & Bakr, 1977c; Ou, 1985; Rahman & Evans, 1987). The disease is worse in the wet years in the wet areas where annual rainfall exceeds 1.6 cm in Bangladesh (Cox & Rahman, 1980). As for example, a decline in the attack of ufra occurred in 1979 on areas where severe outbreak occurred in previous years (Cox *et al.*, 1980). The reason was the severe drought up to May and June in 1979 when the total rainfall was 1.8 cm in Joydebpur compared to 18.0 cm in 1978 during those months. The most severe disease incidence was also observed in months of high rainfall or in fields with high water level in Vietnam (Cuc & Kinh, 1981). A high disease incidence in summer rice was observed in tidal prone areas of Ganges River Island of West Bengal where crop was partially submerged for one week (Chakrabarti *et al.*, 1985).

Number of nematodes per plant

The most severe symptoms of ufra, e.g. ufra-1 (complete enclosure of panicle within the flag leaf sheath) and ufra-2 (partial emergence of the panicle) may develop with a nematode population of 10-100 per plant (Rahman & Evans, 1987).

Age of plants at nematode infection

A high incidence of ufra severity was observed when the infection by the nematode starts at earlier stages of rice (Rahman & Evans, 1987 ; Rahman & Shahjahan, 1991). Ufra incidence was less than 13% and 15% when the infection initiates 6 weeks after sowing in deepwater rice (Rahman & Evans, 1987) and 50 days after sowing in irrigated rice (Rahman & Shahjahan, 1991), respectively.

Proportion of infected seedlings at transplanting

The disease severity with the most severe panicles symptoms, e.g. ufra-1 and ufra-2 increases with the increase of diseased seedlings at transplanting (Mondal *et al.*, 1989). More than 80% plants may become infested by the time of harvest if there are 4-10% initially infected seedlings at transplanting time.

Soil nutrient

Experimental results indicate that soil application of zinc at 50 ppm potassium at 100 ppm and sulphur at 120 ppm to Zn-,K-and S-deficient soil may reduce ufra disease development with a considerable yield increase in rice (Miah *et al.*, 1984; Mondal & Miah, 1985, 1986).

CONTROL MEASURES

Although the disease had been identified in 1913, a little success has been achieved to control the disease. The information available in this aspect from different countries are discussed here.

CULTURAL CONTROL

Removal or burning of infested stubble and straw

Burning of infested stubble and straw *in situ* has long been suggested as the effective control measure for ufra (Butler, 1919 ; Hashioka, 1963 ; Miah & Bakr,

1977c). Thorough burning is essential in the practice. This requires uprooting and drying of stubble before setting fire. A little cost is involved in uprooting, still that limits the use of the practice properly. In some cases use of this practice is not possible due to standing water or plenty of soil moisture in the field after harvest. Sometimes a large portion of the stubble is taken away by the farmers for use as a cattle feed or fuel which helps in dispersal and also limits the effective burning in the field (McGeachie & Rahman, 1983). Moreover, integration of the farmers in a locality is needed to burn the stubble in each and every field. Otherwise nematodes from an unburnt field will easily spread to a burnt field at the time of flood and will equally damage the crop.

Ploughing and drying of soil

After burning, ploughing down the remaining unburnt or partially burnt stubble and straw can reduce ufra through decomposing it and exposing the nematode in moist soil where they do not survive for long (Butler, 1919) or transmit the disease (Hashioka, 1963). Nematodes which are exposed to sunshine may also die. However, several ploughing and alternate drying of soil will make the practice more effective.

Destruction of other hosts

D. angustus survives in an active state in volunteer, ratoon and wild rice, and in some weeds which also serve as a primary source of nematode infection of the new crop. Therefore, removal or destruction of these hosts will obviously help to reduce the disease incidence (Hashioka, 1963 ; Sein & Zan, 1977). Ratoon rice is a predominant source of this nematode in low land rice fields where 2-3 rice crops are grown in a year.

Escape cropping, delayed sowing and transplanting

Growing short duration (early maturing) cultivars, delayed sowing and transplanting can reduce ufra severity specially in deepwater rice (McGeachie & Rahman, 1983 ; Miah, 1984 ; Yien *et al.*, 1990). The short duration varieties mature and harvested early and thus escape post-infection damage by the nematode. 'Digha' and 'Padmapani' are such types of deepwater rice cultivars in Bangladesh (Mondal & Miah, 1987a) and India (Rathaiah & Das, 1987), respectively.

The length of the over wintering period is a critical factor in regulating the survival of *D. angustus* from one season to the next and lengthening this period may reduce primary infestation in the crop (Cox & Rahman, 1980) by reducing the live nematode population in the infected stubble. As for example less than 1% live nematodes were observed in infested panicle examined in April compared to its number in January (Cox & Rahman, 1979b). Similar results were also observed in Vietnam (Kinh, 1981). Trials in farmers' field showed that ufra severity was reduced considerably if deepwater rice is sown 2-4 weeks or transplanted 8 weeks later from its normal planting time (McGeachie & Rahman, 1983; Yien *et al.*, 1990). Yield also increases considerably with the decrease of disease severity in this practice (Table 3). It is worth while to mention that the practice of transplanting deepwater rice is increasing day by day in low lying areas following harvest of boro rice in Bangladesh.

Table 3 : Effect of delayed sowing/transplanting on ufra incidence and yield in deepwater rice cv. Hassiaman

Treatment	Ufra infestation (%) (angular transformation)	Yield (kg/ha)
Broadcast on 19.2.80	52.1	1045
Broadcast on 6.3.80	38.3	1447
Broadcast on 21.3.80	37.8	1914
Transplanted on 16.4.80	21.1	2260
S.E.	±5.0	±279.4

(Source : McGeachie & Rahman, 1983)

Crop rotation

Inclusion of non-host crops such as jute or hemp in rotation with rice for a few years in an ufra endemic area may eliminate or reduce the disease (Butler, 1919; Miah & Bakr, 1977a). This practice could be limited in the areas where flood water does not rise too fast or is not too deep for jute. Transplanted Aman-Mustard (short duration)-Boro cropping pattern in low land irrigated rice areas is reported to reduce the disease incidence (Miah & Rahman, 1985). Ufra incidence in deepwater rice (DWR) Boro-DWR pattern also appears to be decreasing compared to the DWR-non rice winter crops-DWR or DWR-fallow-DWR patterns in Bangladesh (Table 4). The reason is not clear but it is possibly due to removal of most of the stubble or straw of both DWR and Boro crops, and use of different pesticides in the boro crop in this pattern.

Leaf cutting of infested plant

Very often farmers cut leaves of luxuriant deepwater rice for cattle feed. This practice may reduce ufra incidence through eliminating nematodes living at the growing points of deepwater rice (Miah, 1984; Cox & Rahman, 1980) without inflicting yield loss (Kupkanchanakul *et al.*, 1988).

Table 4 : Ufra severity under different cropping patterns in Bangladesh.

Cropping pattern	Location/ District	Ufra severity*				
		1980	1981	1982	1983	1988
i. DWR-Fallow-DWR	Debpur	++++	++++	++++	++++	++
	Comilla					cp iii
	Motlab	++++	++++	++++	+++	++
	Comilla					cp iii
ii. DWR-non rice crop-DWR	Laksham	++	+++	++++	NV	+++
	Comilla					
	Shahrasti	++	+++	++++	+++	++
	Comilla					
iii. DWR-Boro-DWR	Hajiganj	++	++	+	+	+
	Comilla					
	Gournadi	NV	+++	++	++	+
	Barisal					

*Ufra severity: ++++ (very severe); +++ (severe); ++ (less severe); + (no visible symptom). NV (N visited) cp (cropping pattern)

DWR = Deepwater rice

(Source: BRRI Annual Internal Review Report of 19801993 and 1988; Anon. 1983)

Controlling water flow

Spreading of nematodes through water from river overflow into fields can be prevented by improved bundling of banks of rivers (Sein & Zan, 1977). Water spread of nematode is also common in irrigated rice areas where no irrigation canals or earthen canals exist. Construction of cemented canals may prevent nematode spread from field to field in such cases.

CHEMICAL CONTROL

Chemicals are relatively costly and farmers are reluctant to use it until they see the disease symptom. Farmers can recognize the ufra symptom only at the later stage of the disease when there is a little chance for its control. A considerable reduction of the disease can only be achieved if chemicals are applied either as preventive or curative measures from the very beginning of the crop in an area with known history of ufra occurrence.

Soil treatment

Soil treatment of an ufra-infested field with 100 ppm Diazinon (EC) successfully controlled the disease (Srivastava & Saxena, 1956). Application of Furadan granules (3% a.i.) at 22.42 kg/ha in soil after symptom expression recovered 61-82% plants from the disease which flowered normally (Miah & Bakr, 1977b). It also effectively reduced the disease incidence when applied at 1.0-1.5 kg a.i./ha during final land preparation for broadcasting or transplanting deepwater rice (Tables 5,6). A basal application of Furadan 3% at 24 kg a.i./ha in flooded soil in an ufra-infested deepwater rice field during mid August followed by one foliar application of Benomyl at 2.5 kg. a.i./ha in significantly increased yield by 0.87 t/ha compared to control treatment (Cox & Rahman, 1979a). Soil treatment with Fensulfothion (5 kg a.i./ha and Diazinon (15 kg a.i./ha) is also claimed to control ufra successfully (Vuong & Rodriguez, 1972).

Table 5 : Ufra incidence and yield of deepwater rice under different planting methods and Furadan treatments.

Planting method	Furadan 3% (kg a.i./ha)	Ufra incidence (%)	Yield (t/ha)
Broadcast seed	0.0	77.2	0.24
	1.5	58.3	0.44
Transplanting	0.0	52.9	0.59
	1.5	36.7	0.94

(Source : Rahman & Taylor, 1983)

Table 6 : Ufra incidence and yield under different treatment combinations of urea and Furadan in transplanted deepwater rice.

Treatment	Furadan (kg a/i/ha)	Ufra incidence (%)		Yield (t/ha)	
		Site-1	Site-2	Site-1	Site-2
Transplanting (TP)	0.0	28.7a	26.5a	1.6b	0.5b
TP + Urea	0.0	24.4a	24.5a	2.0ab	0.5b
TP + Furadan (F)	1.0	11.0b	12.7b	2.2a	1.4a
TP + urea + F	1.0	11.0b	12.0b	2.4a	1.4a

Column values with same letter do not differ significantly at 0.05 level by DMRT.
(Source : Rahman & Miah, 1989)

In a series of experiments Furadan 3% Phenamiphos 5%, Ethoprophos 10%, Agridhan 3% and Sunfuran 3% were incorporated with soil at different doses varying from 0.5 to 1.5 kg a.i./ha in irrigated rice. Transplanting of ufra-infected seedings (cv. BR11) in these treated soil showed about 80% plants apparently disease free, flowering normally and producing filled grains (Rahman & Miah,

1991). Soil application of Furadan 3%, Agridhan 3% and Sunfuran 3% at 0.50 and 0.75 kg a.i./ha was also found economically profitable in irrigated rice (Table 7).

Table 7 : Effect of soil incorporation of different nematicides to control ufra after the appearance of disease symptom in irrigated rice cv. BR11.

Nematicide/ Treatment	Dose used	Diseased Plant (%)	Yield (t/ha)	Yield difference (t/ha) over control	Benefit ratio over nematicide cost*
Control (disease)	-	69.0a	1.37a	0.00	-
Furadan 3G	0.5	20.2bc	5.37c	4.00	16.5
	0.75	13.9c	6.03a	4.66	12.9
Agridhan 3G	0.5	22.1bc	5.77bc	4.40	22.0
	0.75	20.1bc	5.77bc	4.40	14.7
Sunfuran 3G	0.5	24.7b	5.57c	4.20	19.6
	0.75	18.2bc	5.37c	4.00	12.5

Column values with the same letter do not differ significantly at 0.05 level by DMRT.

* Price of Rice, Furadan, Agridhan and Sunfuran are Tk. 5.00 72.50 60.00 and 64.00/kg respectively .Cost of other agronomic practices is same for all treatments. (1$ ₮k. 36.36).

(Source : Rahman *et al.*, 1990).

Seedbed and seedling treatment

Seedlings are infested with ufra if they are grown in an infested field or in a field nearby. Ufra may spread from field to field or from locality to locality through these seedlings. Therefore, seedbed or seedling treatment with nematicides can reduce the spread of the disease. Root soaking of ufra-infected seedlings in 7.5-10.0% solution of Miral 3G, or 2.5-10.0% solution of Tecto 40 FL for 12 hours recovered more than 80% plants (Table 8) and increased yield considerably (Mondal & Miah, 1987b). More than 80% plants also recovered from the disease when Furadan 5G or Mocap 5G at 0.75 kg a.i/ha or Tecto 40 FL at 4l/ha are applied in ufra infected seedbed 7 days before uprooting the seedlings (Rahman, 1991). Root soaking of ufra-infected seedlings in 1.5% solution of Furadan 5G or Mocap 5G for 12 hours was less effective compared to 1% solution of Tecto 40 FL in recovering of plants from the disease (Table 9). However, the yield was almost equal or more in these treatments over the untreated control. Root soaking in 1% solution of Furadan 3G for 24 hours before transplanting also reduced ufra incidence effectively in Vietnam (Rahman & Catling, 1985).

Table 8 : Effect of root-dip treatment of infected seedlings in suspension of different chemicals to control ufra in irrigated rice.

Chemical	Ufra incidence (%) under different suspension of chemicals				Yield (t/ha)
	2.5%	5.0%	7.5%	10.0%	
Furadan 3G	85.86a	85.40a	67.60a	90.83a	1.72c
Miral 3G	49.23b	50.63b	30.53b	47.10b	3.47b
Tecto 40 FL	26.43c	26.83c	29.03b	17.32c	4.86a

Column values with same letter do not differ significantly at 0.05 level by DMRT.

(Source : Mondal & Miah, 1987).

Foliar spraying

Two sprays of 1:700 dilution of Hexadrin at 15-day intervals on ufra-infested plant before flowering eliminates nematode completely (Pal, 1970). There was 82% and 58% recovery of ufra-infested plants in green house and fields tests respectively when 0.1% a.i. of Benomyl 50 w.p. was sprayed on ufra-infected plants (Miah & Bakr, 1977b). Spraying of 0.2% a.i. of Diazinon 60 EC was not effective in this test. Sein (1977c) sprayed dilutions of four chemicals thrice at

Table 9 : Effect of different methods of nematicide application to control ufra in irrigated rice cv. BR11.

Treatment/ Nematicide	Method of treatment	Dose/ha	Diseased plant (%)	% recovery of disease plant	Yield (t/ha)
Control (Diseased)	-	-	86.38	13.62	1.50
Control (Healthy)	-	-	11.00	89.00	6.25
Furadan 5G	Seedbed	0.75 kg a.i.	16.74	83.26	5.91
	Root-dip	1.5% Soln.	31.95	68.05	5.62
	Field soil	0.75 kg a.i.	15.91	84.09	6.21
Micap 5G	Seedbed	0.75 kg a.i.	17.75	82.25	6.14
	Root-dip	1.5% Soln.	29.12	70.88	5.15
	Field soil	0.75 kg a.i.	14.22	85.78	5.85
Tecto 40 FL	Seedbed	0.75 kg a.i.	17.61	82.39	5.73
	Root-dip	1.5% Soln.	10.56	89.44	6.66
	Field soil	0.75 kg a.i.	18.02	81.98	6.11

(Source : Internal Review Report for 1990 and 1991 BRRI)

weekly intervals on ufra-infected plants and achieved a good control of the disease with Furadan 3% at a costly dose (30 kg a.i./ha).

USE OF BOTANICALS

Literature on the use of botanicals to control ufra disease is not yet available. In a preliminary test of root-dip treatment of ufra-infected seedlings in the leaf extraction of *Tagetes erecta, Blumea lacera, Annonas* sp., *Andrographis paniculata, Azadirachta indica, Tamarindus indica, Hibiscus syriacus, Leucaena leucocephala, Acaia* sp., *Sapium indicum, Polygonum barbatum, Polyulthia longifolia, Sphaenoclea zeylanica* and *Eclipta alba* for 12 hours was not found effective to control the disease at Bangladesh Rice Research Institute. Transplanting of ufra-infected seedlings in the soil amended with some of these plants was also not effective to control this disease.

USE OF RESISTANCE

Butler (1919) did not find any resistant variety for ufra in the then East Bengal, now Bangladesh. Hashioka (1963) tested 5 varieties after inoculation with the ufra nematode in the coleoptiles of seedlings and reported 'Khao Tah Oo' as resistant with 42.9% stems infested. Sein (1977a) selected 'B-69-1' ('Thabaung-mee-gok') with 72% fertile tillers among 151 varieties and 71 crosses. In a screening programme in Vietnam during 1979-81, 1,197 varieties and cultivar were screened following coleoptile inoculation method in which four high yielding improved varieties namely 'IR9129-393-3-1-2', 'IR9129-169-3-2-2', 'IR9224-117-2-4-3-1' and 'IR2307-247-2-2-30' had less than 20% infestation (Kinh & Nghiem, 1982). The deepwater rice cultivars 'BKN 6986-8', 'CNL-53' and 'Jalaj' have also been recorded with 26-29% disease incidence in that country (Kinh & Phuong, 1981). Two irrigated rice 'IR36' and 'IET4094' are less susceptible to ufra and had about 40% infestation in the field in India (Chakrabarti *et al.*, 1985). 'Rayada 16-06' (deepwater rice) and *Oryza sabulata* were not found infected with ufra in an inoculation test and suggested to be a source of resistance in future breeding programmes in Bangladesh (Miah & Bakr, 1977a). Several hundred deepwater and high yielding transplanted rice varieties or lines were screened for ufra during 1981 to 1985 (Rahman & McGeachie, 1982; Rahman, 1987) following a seedling base inoculation method simulated with field conditions (Rahman, 1982). The resistant reaction of 'Rayada 16-06' was confirmed in these tests in addition to some other resistant lines of 'Rayada'. The deepwater rice lines 'CNL-319', 'Bazail-65), 'Karkati-161' and 'Gowai-50/9' were resistant to moderately resistant to ufra in this test (Rahman, 1987). Along with the continuation of national screening programme, IRRI initiated an International Ufra Screening Set (IRUSS) in 1985 for the effective screening and breeding programmes for ufra in South and South-East Asian countries specially in Bangladesh, Burma, India, and Vietnam. Screening results of IRUSS also showed 'Rayada 16-06', 'Karkati-161', 'Bazail-65', 'Gowai 50/9' and 'CNL-319' as resistant to ufra in these countries (IRRI, 1985, 1987, 1989). These entries are now being used in breeding programmes in India (Yien *et al.*, 1990), Vietnam (Hong *et al.*, 1989) and Bangladesh (Anonymous, 1990). It is hoped that ufra resistant varieties with desirable agronomic characters will be developed in near future in these countries.

CONCLUSION

Ufra is considered as one of the most important rice diseases in terms of both plant damage and yield loss in South and South-East Asian countries. Previously it was a disease of deepwater rice but now a days it is also spreading to irrigated rice. Thus this disease is becoming a potential threat to rice production in this region. Thorough burning of crop residues is still considered as the best method of controlling this disease since its report in 1913 by Butler. Delayed sowing or transplanting and use of short duration deepwater rice cultivars are the recent approaches to control it. Use of nematicides in deepwater rice field is relatively less feasible and profitable compared to irrigated rice. Resistant varieties are yet to be developed with desirable agronomic traits when its use will be the most effective measure to control this disease.

REFERENCES

Anonymous (1983). BRRI *Annual Report for 1983*, BRRI, Gazipur, Bangladesh. pp. 138-139.

Anonymous (1990). *Technical Report of Deepwater Rice Pest Management Project for 1989-1990*. BRRI, Gazipur, Bangladesh (Mimeo). p. 40.

Aung, T. (1989). *Nematode pests of deepwater rice in Myanmar (Country report)*. A report presented at the Int. Training Workshop on Nematode Pests of Deepwater Rice (Sept. 16-28, 1989). BRRI, Joydebpur, Gazipur, Bangladesh. p. 3.

Bakr, M.A. (1977). Occurrence of ufra disease in transplanted rice. *Int. Rice Res. Newsl.* (IRRN).**3**(3): 16.

Bakr, M.A.; Fakir, G.A. & Miah, S.A. (1991). Gnotobiotic culture of rice stem nematode (Abst. no. 21). *In: Proc. 4th Biennial Conf. of Bangladesh Phytopathol. Soc.* (6-7 Feb. 1991) BRRI, Joydebpur, Bangladesh.

Bridge, J.; Luc, M. & Plowright, R.A. (1990) Nematode parasites of rice. *In : Plant Parasitic Nematodes in Subtropical and Tropical Agriculture* (Eds.- M. Luc, R.A. Sikora & J. Bridge). CAB International, U.K. pp. 69-108.

Butler, E.J. (1913). Diseases of rice. *Agr. Res. Inst. Pusa Bull.* **34** B: 1-27.

Butler, E.J. (1919). The rice worm *(Tylenchus angustus) and its control. Bot. Series* **10**(1):1-37.

Catling, H.D., Cox, P.G.; Islam, Z. & Rahman, L. (1979). Two destructive pests of deepwater rice : Yellow stem borer and Ufra. The ADAB News **6**(8). 16-21.

Catling, H.D.; Islam, Z. & Rahman, L. (1988). Status of pests and disease in Asian deepwater rice. *In: Proc. 1987 Int. Deepwater Rice Workshop*. IRRI, Philippines. pp. 525-537.

Catling, H.D. & Pukridge, D.W. (1984). Report on a visit to Vietnam, October, 1983. *Deepwater Rice:* 3-4.

Chakrabarti, H.S.; Nayak, D.K. & Pal, A. (1985). Ufra incidence in summer rice in West Bengal. *IRRN* **10**(1): 15-16.

Cox, P.G. & Rahman, L. (1979a). Synergy between Benomyl and Carbofuran in the control of ufra. *IRRN* **4**(4): 11.

Cox, P.G. & Rahman, L. (1979b). The overwinter decay of *Ditylenchus angustus. IRRN* **4**(5):4.

Cox, P.G. & Rahman, L. (1980). Effects of ufra disease on yield loss of deepwater rice in Bangladesh. *Tropical Pest Management* **26**(4): 410-415.

Cox, P.G.; Rahman, L. & Hannan, M.A. (1980). 1979 setback for ufra. *IRRN* **5**(1): 10-11.

Cuc, N.T.T. (1982a). New weed host of rice stem nematode identified in Vietnam. *IRRN* **7**(3): 15.

Cuc, N.T.T. & (1982b). Field soil as a source of rice stem nematode. *IRRN* **7**(4): 15.

Cuc. N.T.T. & Giang, L.T. (1982). Relative humidity and nematode number and survival in rice seeds. *IRRN* **7**(4): 14.

Cuc, N.T.T. & Kinh, D.N. (1981). Rice stem nematode disease in Vietnam. *IRRN* **6**(6): 14-15.

Hashioka, Y. (1963). The nematode *Ditylenchus angustus* in Thailand. *FAO Plant. Prot. Bull.* **11**(5): 97-102.

Hong, L.T.T.; Cuc, N.T.T.; Kien, D.V.; Minh, N.N.; Nha, L.V.; Truc, N.T. & Tuan, P.V. (1989). *Rice resistance to ufra in Vietnam*. A paper presented at the Int. Training Workshop on Nematode Pests of Deepwater Rice (16-28 Sept., 1989). BRRI, Gazipur, Bangladesh.

IRRI (1985). *The First International Rice Ufra Screening Set. IRRI*, Philippines.

IRRI (1987). *The Third International Rice Ufra Screening Set. IRRI*, Philippines

IRRI (1989). *The Fifth International Rice Ufra Screening Set. IRRI*, Philippines

Jack, H.W. (1923). Rice in Malaya. *Malaysian Agril. J.* **11**: 103-169.

Kinh, D.N. (1981). Survival of *Ditylenchus angustus* in diseased stubble. *IRRN* **6**(6): 13.

Kinh, D.N. & Nghiem, N.T. (1982). Reaction of rice varieties to stem nematode in Vietnam. *IRRN* **7**(3): 6-7.

Kinh. D.N. & Phuong, C.D. (1981). Reaction of some deepwater rice varieties to *Ditylenchus angustus. IRRN* **6**(6): 6-7.

Kupkanchanakul, T.;Vergara, B.S. & Robles, A.T. (1988). Herbage production in a rice crop under natural deepwater conditions. *In: Proc. 1987 Int. Deepwater Rice Workshop*. IRRI, Manila, Philippines. pp. 517-523.

McGeachie, I. & Rahman, L. (1983). Ufra disease : A review and a new approach to control. *Tropical Pest Management* **29** : 325-332.

Miah, S.A. (1974). Disease problems of deepwater rice in Bangladesh. *In: Proc. of the Inst. Seminar on Deepwater Rice* (August, 1974). BRRI, Joydebpur, Bangladesh. pp. 199-204.

Miah, S.A. (1984). Problems and progress of research of ufra disease of deepwater rice in Bangladesh. *Int. Rice Commission Newsl.* **33**(2): 35-38.

Miah, S.A. & Bakr, M.A. (1977a). Sources of resistance of ufra disease of rice in Bangladesh. *IRRN* **2**(15): 8.

Miah, S.A. & Bakr, M.A. (1977b). Chemical control of rice. *PANS* **23**(4): 412-413.

Miah, S.A. & Bakr, M.A. (1977c). Nematode diseases of rice *In: Lit. Rev. of Insect Pests and Diseases of Rice*, BRRI, Bangladesh. pp. 125-129.

Miah, S.A. & Rahman, M.L. (1985). Severe ufra outbreak in transplanted rice in Bangladesh. *IRRN* **10**(3): 24.

Miah, S.A.; Shahjahan, A.K.M.; Sharma, N.R.; Shah, A.L. & Bhuiyan, N.I. (1984). Interaction of zinc deficiency and ufra disease on rice in Bangladesh. *Bangladesh J. Agric.* **9**(2): 61-67.

Mondal, A.H.; Akanda, S.I. & Miah, S.A. (1985). Interaction between ufra and sheath rot disease of rice. *In: Proc. 10th Ann. Bangladesh Sci. Conf.* (March 22-27, 1985).

Mondal, A.H. & Miah, S.A. (1985). Effect of potash on the severity of ufra disease of rice. *Bangladesh J. Plant Pathol.* **1**(1): 67.

Mondal, A.H. & Miah, S.A. (1986). Sulphur reduces rice stem nematode pathogenicity. *In: Proc. 11th Aun. Bangladesh Sci. Conf. Sec.* 1: p. 11.

Mondal, A.H. & Miah, S.A. (1987a). Ufra problem in low lying areas of Bangladesh. *IRRN* **12** (4): 29-30.

Mondal, A.H. & Miah, S.A. (1987b). Control of ufra of rice by seedling treatment. *Bangladesh J. Plant Pathol.* **3**(1-2): 25-30.

Mondal, A.H.; Rahman L.; Ahmed, H.U. & Miah, S.A. (1986). The causes of increasing blast susceptibility of ufra-infected rice plants. *Bangladesh J. Agric.* **11**(4): 77-79.

Mondal, A.H.; Rahman, M.L. & Miah, S.A. (1989). Yield loss due to transplanting of ufra-infected seedlings. *Bangladesh J. Bot.* **18**(1): 67-72.

Ou, S.H. (1985). Diseases caused by nematodes. *In: Rice diseases* (2nd edition). The Commonwealth Mycol. Inst., Kew, Surrey, England. pp. 337-364.

Pal, A.K. (1970). Survey on nematode infection of paddy and successful treatment with hexadrin. *In: Proc. Indian Sci. Cong. Assoc.* **57**: p. 496.

Rahman, L. & Catling, H.D. (1985). *Report on a visit to Vietnam to review research on ufra nematode* (Sept., 1985). p. 8.

Rahman, L. & McGeachie, I. (1982). Screening for resistance to ufra (*Ditylenchus angustus*) in deepwater rice. *In: Proc. 1981 Int. Deepwater Rice Workshop*. IRRI, Manila, Philippines. pp. 459-466.

Rahman, M.L. (1982). Source for ufra resistance in deepwater rice. *IRRN* **7**(5): 12-13.

Rahman, M.L. (1987). Source of ufra resistant deepwater rice. *IRRN* **12**(1): 8.

Rahman, M.L. (1991). Curative control of ufra through seedbed, root-dip and soil treatments with nematicides. *In: Proc. 4th Biennial Conf. of Bangladesh Phytopathol. Soc*. BRRI, Gazipur, Bangladesh (Feb. 6-7, 1991).

Rahman, M.L. & Evans, A.A.F. (1987). Studies on host-parasite relationship of rice stem nematode, *Ditylenchus angustus* (Nematoda: Tylenchida) on rice (*Oryza sativa* L.). *Nematologica* **33**: 451-459.

Rahman, M.L. & Miah S.A. (1989). Effect of seedling establishment methods and Carbofuran application to control ufra disease (*Ditylenchus angustus*) in deepwater rice. *Bangladesh J. Plant Pathol.* (In press).

Rahman, M.L. & Miah S.A. (1991). Nematicides for control of rice ufra disease. *In: Proc. Int. Bot. Conf.* (Jan. 10-12,1991). Dhaka, Bangladesh.

Rahman, M.L.; Mondal, A.H. & Miah, S.A. (1990). Nematicides for control of ufra disease in transplanted rice. *Bangladesh J. Bot.* (In press).

Rahman, M.L. & Shahjahan, A.K.M. (1991). Relationship of ufra infection at different age of plants on disease severity and yield of rice. *In: Proc. 4th Biennial Conf. of Bangladesh Phytopathol. Soc.*, January, 6-7, 1991). BARI, Joydebpur, Bangladesh.

Rahman, M.L. & Taylor, B. (1983). Nematode pests associated with deepwater rice in Bangladesh. *IRRN* **8**(5): 20-21.

Rao, Y.S.; Prasad, J.S. & Panwar, M.S. (1986a). Stem nematode (*Ditylenchus angustus*) a potential pest of rice in Assam and West Bengal, India. *Int. Nematol. Network Newsl.* **3**(4): 24-26.

Rao, Y.S.; Prasad, J.S. & Panwar, M.S. (1986b). Nematode problems in Rice: Crop losses, Symptomatology and Management. *In: Plant Parasitic Nematodes of India : Problems and Progress* (Eds.— G. Swarup & D.R. Dasgupta). Indian Agril. Res. Inst., New Delhi. pp. 279-299.

Rathaiah, Y. (1988a). Disease of deepwater rice in Jorhat district, Assam, India. *IRRN* **14**(4): 32-33.

Rathaiah, Y. (1988b). Volunteer rice, host for diseases in DWR. *Deepwater Rice* **14**: 1-2.

Rathaiah, Y. & Das, G.R. (1987). Ufra threatens deepwater rice in Majuli, Assam, *IRRN* **12**(4): 29.

Reyes, G.M. & Palo, A.V. (1956). Nematode disease of rice. *Aranata J. of Agric.* **3**(3): 72-77.

Sein, T. (1977a). Varietal resistance to ufra disease in Burma. *IRRN* **2**(2): 3.

Sein, T. (1977b). Seed-borne infection and ufra disease. *IRRN* **2**(2): 6.

Sein, T. (1977c). Testing some pesticides against ufra disease. *IRRN* **2**(2): 6.

Sein T. & Zan, K. (1977). Ufra disease spread by water flow. *IRRN* **2**(2): 5.

Seth, L.N. (1939). *Report of the Mycologist, Burma*, for the year ending 31st March 1939. p. 6.

Siddiqi, M.R. (1970). White Tip and Ufra Disease of Rice. Plant Pathology Advisory Circular No. 4 Ministry of Agriculture & Natural Resources, Zomba, Malawi. p. 4.

Singh, B. (1953). Some important diseases of paddy. *Agric. & Animal Husbandry* **3**(10-12): 27-30.

Srivastava, A.S. & Saxena, H.P. (1956). Effect of diazinon on paddy nematodes. *In: Proc. of the 43rd Indian Sci. Cong. New Delhi*. p. 286.

Vuong, H.H. (1969). The occurrence in Madagascar of the rice nematodes. Aphelenchoides besseyi and Ditylenchus angustus. *In : Nematodes of Tropical Crops* (Ed.J.E. Peachy). Technical Communication No. 40. Commonwealth Bureau of Helminthology, St. Albans, England. pp. 274-288.

Vuong, H.H. & Rodriguez, H. (1972). II. Result of trials for chemical control of nematode parasites of rice in Madagascar: *Aphelenchoides besseyi* Christie, 1942, *Ditylenchus angustus* (Butler, 1913) Filipjev, 1936. *Document Institute de Recherches Agronomiques a Madagascar No.* **335**. p. 25.

Yien, B.R.; Medhi, B.N. & Bora, L.C. (1990). *A technical report of strengthening research on integrated pest management for deepwater rice farming systems (phase II)*. DWR project document no. 5, Regional Agril. Res. Stn. Assam Agril-Univ., Assam. p. 160.

Chapter 3

Control of Rice Root Nematodes, *Hirschmanniella* Species

Qaiser H. Baqri
Desert Regional Station, Zoological Survey of India, Pali Road, Jhalamand, Jodhpur-342 005, India.

WITH the rise in demand of food supplies due to increase in human population, our endeavour is to increase the food production through mechanised cultivation, improved agronomical and cultural practices, development of high yielding and resistant varieties, management of weeds and pests, etc. Rice is one of the most popular cereal foods in the world. The most commonly cultivated species of rice, *Oryza sativa* has thousands of varieties, which can be grown under different climatic conditions, waters and soils (from temperate to tropical climate, upland to deep undrained swamps, fresh to saline waters, and sandy-loam to clayey soils) because of the wide differences in their genetic characteristics. As a result, rice is grown in about 140 million hectares of area in Asia, Africa, Australia, Europe, North and South America. However, the major rice producing countries are China, India, Bangladesh, Burma, Pakistan, Japan, Philippines, Thailand, etc. India has produced about 79 million tonnes rice on 37 million hectares in 1990. Previously rice was mainly grown and considered as main food in eastern, some central and western, and southern states of India. Now rice is becoming an important popular staple food and widely cultivated crop in states like Punjab, Haryana and western Uttar Pradesh which were supposed to be wheat eating and wheat growing areas. Unfortunately the average yield per hectare is still significantly lower than Japan and other developed countries despite India being the third largest scientific force in the world. In view of this fact, it is necessary for us to improve the average yield. It assumes added importance as living conditions of farmers, depending on paddy crop as their only source of livelihood, can be improved. Accordingly, during the last two decades, efforts have been made to improve the yield, including research on pest management.

IMPORTANCE OF NEMATODES

Like other crops, the rice (paddy) is attacked by all kinds of pests, e.g. bacteria, viruses, fungi, insects, nematodes, molluscs, birds and rodents. The importance of nematodes as rice pathogens was recognised after Butler (1913) reported ufra disease caused by *Ditylenchus angustus* (stem nematode) from the then East Bengal, India (now Bangladesh). Only after a couple of years, a new disease of rice was described, caused by an undescribed nematode *(= Aphelenchoides besseyi)* from Japan, under the name "black grain disease." Later, the disease was named as "white tip disease". Since then many other nematode species have been recorded from paddies. Jairajpuri & Baqri (1991) have provided consolidated information on important nematode species and the problems in rice. They have listed more than 150 nematode species under different ecological and systematic groups, with special reference to India. Though majority of these species not parasitic in nature, some of them are serious pests of rice, viz. *Ditylenchus angustus, Hirschmanniella* spp., *Heterodera* spp., *Melodiogyne* spp., *Criconemella* spp., *Aphelenchoides besseyi*, etc.

According to Hollis & Keoboonrueng (1984), "when we consider rice nematodes on world basis, we are concerned first with species of *Hirschmanniella*. Despite the severity of ufra and white tip diseases, *Hirschmanniella* species are considered more important because these have been reported from all the rice growing countries whereas the incidences of the above mentioned diseases are endemic. The host-parasite relationship of *Hirschmanniella* spp. with rice is so strong that the former are generally called as "rice root nematodes".

Van Breda de Hann (1902), while describing *Tylenchus oryzae* (now *Hirschmanniella oryzae*), had suspected that the species was causing injuries to the rice plants, but he did not provide any experimental evidence. Van der Vecht & Bergman (1952) were the first to observe through experiments that the heavily infested roots with *H. oryzae* (= Radopholus oryzae) adversely affected rice plants. Perhaps the economic importance of rice root nematodes could not be acknowledged earlier because they do not cause any specific above ground symptoms or disease like ufra or white-tip. Though several species of *Hirschmanniella* have been reported from the rhizosphere of rice, only seven have been considered important all over the world (Sher, 1968; Hollis & Keoboonrueng, 1984; Jairajpuri & Baqri 1991). In India, Prasad (1964) was first to suspect the possibility of *H. oryzae* as a bottleneck in the rice production. Mathur & Prasad (1971) provided important information on the distribution of *H. oryzae*, *H. mucronata* and *H. gracilis*. The first one was found to be widely distributed species in 12 states of India. Now the various reports confirm, though many *Hirschmanniella* spp. have been recorded from the country, that these three species are the key/potential pests causing 12-70% loss in rice yield (Panda & Rao, 1969; Mathur & Prasad, 1972; Ahmad *et al*., 1984; Rao & Panda, 1970).

CONTROL STRATEGIES

It is in the above that important information on the control of rice root nematodes have been gathered in the present write up specially in India. These can also be helpful in the neighbouring countries (Rao, 1988). During the last two decades, a large number of research papers have appeared on the control of rice root nematodes through nematicides, soil amendments, antagonistic or trap plants, resistant varieties, bio-control agents, etc. Most of the papers, however, have reported the achievements with chemical treatments. Since it is beyond the scope of this paper to report all the findings, only the important ones are being discussed here.

NEMATICIDES

Prasad *et al.* (1986a) have noted that the seed soaking treatment in 1% solution of carbofuran, carbosulfone and isofenfos for six hours checks the high build up of *H. mucronata* population in roots.

Pre-sowing treatment of seedbeds and pre-planting root-dip treatment of seedlings with different chemicals have recorded significant reduction in the nematode population and improvement in the length and weight of rice seedlings and plants. The negative correlation between the nematode population and plant growth resulted in the increase of rice grain yield. Mathur & Prasad (1973) found that the treatment of nursery bed with DD @ 400 kg/ha decreased *H. oryzae* population and increased the weight and length of the seedlings. Many reports confirm that the application of carbofuran at 1-1.25 kg a.i./ha in the seedbed reduced the population of rice root nematodes and increased the grain yield (Jonathan & Velyutham, 1984; Rao *et al.*, 1986, etc.). The Hissar Centre of All India Coordinated Research Project on Nematode Pests has reported that the treatment of seedbeds with carbofuran and aldicarb @ 1 kg a.i./ha has resulted in the significant increase in length and weight of seedling than in control during 1981 and 1982. Carbofuran was, however, found more effective than aldicarb. Muthukrishnan *et al.* (1977) achieved good results against *H. oryzae* when they dipped the roots of seedlings in 0.1% solution of carbofuran 3G for 24 hours. Pre-planting root-dip treatment in 0.2% solution of carbosulfone, isofenfos and carbofuran for six hours has been reported to increase the shoot and root weight of rice plants and the grain yield 45.8%, 30.7% and 24.8%, respectively, over the control (Prasad *et al.*, 1986b). The application of these nematicides in root-dip treatment also reduced *H. mucronata* population significantly. Balasubramanian & Palanisamy (1983) have recorded significant increase in grain yield when chlorophyriphos was used at 0.2% concentration for 20 min. as root-dip against *H. gracilis*. The solution of triazophos, phosphomidon, phenamiphos and BPMc at 0.2% concentration have also been found effective in the root-dip treatment against *Hirschmanniella* species.

The results of some trials reveal that higher grain yield may be obtained when the seedlings grown in nematicide treated seedbed are exposed to the root-dips before transplantation, i.e. seedbed treatment followed by root-dip treatment. Venkitesan *et al.* (1980) noted remarkable increase in the grain yield over the control when the seedbeds were treated with DBCP @ 30 1/ha followed by pre-planting root-dip treatment of seedlings with 0.2% solution of phosphomidon. The treated plots were found to produce 85% more grain yield than the check plots. A significant reduction in the rice root nematode population was also recorded. Subramanian & Velayutham (1983) treated the nursery bed with carbofuran at 1 kg a.i./ha at sowing time and found *H. oryzae* population at low level. After 30 days the seedlings were uprooted and roots dipped in 0.2% solution of phenamiphos or carbosulfan before transplantation. The combined treatment reduced the nematode population and increased the rice yield. Similar results have been obtained after the seedbeds were treated with metham sodium at 250 l/ha and the seedlings were exposed to 0.2% solution of phenamiphos or FMC 35001 for six hours.

Muthukrishnan *et al.* (1977) tested the efficacy of diazinon 5G, fensulphothion 5G, aldicarb 5G, carbofuran 3G, phenamiphos 5G and phorate 10G against *H. oryzae* and observed that the application of these nematicides in the split doses @ 1.5kg a.i./ha, first at planting time and thereafter on 40th and 60th day, increased rice yield significantly. The application of same chemicals @ 1kg a.i./ha on 30th day after planting was also found to be effective. In an other trial, they (l.c.) noted that the pre-plant application of DBCP 60EC at 38 l/ha and the post-plant application on 30th day at 18 l/ha in the irrigation water reduced *H. oryzae* population.

The soil treatment of nursery beds followed by two treatments of the main field with carbofuran @ 1 kg a.i./ha reduced the rice root nematode population significantly in the soil as well as in the roots. The seedbed should be treated a day before seed sowing while the main field on 7th and 50th day after transplanting the seedlings. The increase in the grain yield has been noted from 12.5 to 42.8% over the untreated plots (Ahmed *et al.*, 1984 ; Prasad & Rao, 1984). The same treatments, however, have increased the grain yield by 51.2% at Bhubaneswar Centre of Coordinated Project on Nematode Pests.

OTHER METHODS

Despite the good results achieved through the application of chemicals in the control of *Hirschmanniella* species, other methods like soil amendments, bio-control, crop rotation, and resistant varieties, etc. are preferred because these methods not only reduce the cost but also protect the environment from pollution. Moreover, the health hazards can also be avoided.

Modification of soil environment with organic and inorganic substances has proved to be very effective in reducing the population of rice root nematodes (Rao *et al.*, 1984a). The application of fresh plant materials of *Azola* sp. and *Ipomea* sp. has been reported to reduce *H. mucronata* population and improve the plant weight (Prasad & Rao, 1983). Mathur & Prasad (1973) observed that the application of neem and mustard cakes at 120-240 kg/ha reduced 68.90% *H. oryzae* population. Higher doses of 'til' and groundnut seed cakes were also found effective. Various reports confirm that the application of ammonium sulphate, water hyacinth, urea, calcium, ammonium nitrate and sulfur coated urea @ 60 kg N/ha as basal dose reduces the *Hirschmanniella* population (Prasad & Rao, 1983; Routaray & Das, 1985; etc.). Baqri *et al.* (1988) have found that incorporation of fresh chopped leaves of water hyacinth @ 19 kg/16 sp ft microplot, so as to get 60 kg N/ha, at the time of main field preparation (before seedling transplantation) brings down the soil and the root populations of *H. gracilis* (39.56% reduction in soil and 36.62% in roots). This reduction resulted in the increase of total weight of plant by 19.5-38% and the grain yield by 27-29% over the control. They (l.c.) also noted that the application of ammonium sulphate and calcium ammonium nitrate @ 60 kg/ha as basal dose reduced the rice root nematode population and increased the grain yield 10-17% over the control.

A large number of rice varieties have been tested to find out the resistance against rice root nematodes in India. Israel & Rao (1971), Ramakrishnan *et al.* (1984) and others have found only TKM 9 as resistant and the following as fairly resistant R. RP 155-128-1, CR. 142-3-2, CR. 141-6058-1-35, Ptb. 27, Annapurna, CR. 130-203, CR. 44-140-2-1051, CR. 320, CR. 294-548, Mtu. 28, CR. 52, N. 36, W.113, Tinpakhia, and CR. 44-32. The fairly resistant varieties contained one or a few as against 100 specimens of rice root nematodes per gram roots in susceptible varieties.

Antagonistic or trap plants have also been found as a tool in the management of rice root nematodes in other countries. Inter-cropping with *Sesbania rostrata* has been confirmed as a plant trap for *H. oryzae* and *H. spinicaudata* (Pariselle, 1987; Rinaudo *et al.*, 1988). *Sphenoclea zeylanica*, a weed in rice fields, has been reported controlling 95% rice root nematode population due to some toxic substance found in its roots. Nematode trapping fungi and sulphate reducing bacteria also bring down *Hirschmanniella population* (Jacq & Fortuner, 1979).

In India, the cultivation of some non-host crops like groundnut, jute, wheat, potato or blackgram has been found effective in controlling rice root nematode (Rao *et al.*, 1984b).

Muthukrishnan *et al.* (1977) have recommended to keep the infested fields fallow for one year because the *Hirschmanniella* species may survive without host up to one year. As a result, the population declines gradually with the increase in the fallow period after six months. Since *Hirschmanniella* species may survive on grasses and sedges, the weeds should also be removed from the field.

REFERENCES

Ahmad, N.; Das, P.K. & Baqri, Q.H. (1984). Evaluation of yield losses in rice due to *Hirschmanniella gracilis* (de Man, 1980) Luc & Goodey, 1963 (Tylenchida : Nematoda) at Hooghly (West Bengal). *Bull. Zool. Surv. India* **5**: 85-91.

Balasubramanian, P. & Palanisamy, S. (1983). Evaluation of bare root dip with chemicals for the control of *Hirschmanniella gracilis* on rice. *Indian J. Nematol.* **13**: 215-217.

Baqri, Q.H.; Dey, S. & Ghosh, S. (1988). Effect of different sources of nitrogen on the management of *Hirschmanniella gracilis* (de Man 1880) Luc & Goodey, 1963 (Tylenchida : Nematoda) associated with paddy crop. *Indian J. Helminth (n.s.)* **4**: 12-20.

Butler, E.L. (1913). Ufra disease of rice. *Agric. J. India* **8**: 205-220.

Hollis, J.P. & Keoboonrueng, S. (1984). Nematode parasites of ride. *In : Plant and Insect Nematodes* (Ed.W.R. Nickle). Marcel Dekker, Inc, NY, Basel. pp. 95-146.

Israel, P. & Rao, Y.S. (1971). Isolation of sources for nematode resistance in rice. *SABRAO Newsl.* **III**: 7-10.

Jacq, V.A. & Fortuner, R. (1979). Biological control of rice nematodes using sulphate reducing bacteria. *Revue Nematol,* **2**: 41-50.

Jairajpuri, M.S. & Baqri, Q.H. (1991). *Nematode pests of rice.* Oxford & IBH Publ., New Delhi, India p. 66.

Jonathan, E.I. & Velayutham, B. (1984). Nursery application of carbofuran for control of rice root nematode. *Int. Rice. Res. Newsl.* **9** : 27.

Mathur, V.K. & Prasad, S.K. (1971). Occurrence and distribution of *Hirschmanniella oryzae* in the Indian union with description of *H. mangaloriensis* sp. n. *Indian J. Nematol.* **1**: 220-226.

Mathur, V.K. & Prasad, S.K. (1972). Role of the rice root nematode, *Hirschmanniella oryzae* in rice culture. *Indian J. Nematol.* **2** : 158-168.

Mathur, V.K. & Prasad, S.K. (1973). Control of *Hirschmanniella oryzae* associated with paddy. *Indian J. Nematol.* **3** : 54-60.

Muthukrishnan. T.S.; Rajendran, G.; Ramamurthy, V.V. & Chandrasekran, J. (1977). Pathogenicity and control of *Hirschmanniella oryzae. Indian J. Nematol.* **7**: 8-16.

Panda, M. & Rao, Y.S. (1969). Evaluation of losses caused by incidence of *Hirschmanniella mucronata* Das in rice (abstract). *All India Nem. Symp.*, New Delhi, IARI, Aug. 2-22.

Pariselle, A. (1987). Comportement du nematode *Hirschmanniella oryzae* (van Breda de Haan) dans les racines de la legumineuse *Sesbania rostrata* Brem. *Revue Nematol.* **10**:333-336.

Prasad, J.S.; Panwar, M.S. & Rao, Y.S. (1986a). Effect of seed-soaking with chemicals on the parasitic nematodes of rice. *Indian J. Nematol.* **16**: 119-121.

Prasad, J.S.; Panwar, M.S. & Rao, Y.S. (1986b). Chemical control of the root nematode (*Hirschmanniella mucronata*) in rice. *Beitrage Trop. Landwirtsch. Veterinarmed.* **24:** 65-69.

Prasad, J.S. & Rao, Y.S. (1983). Influence of different sources of nitrogen fertilisation to rice crop on the build-up of the root nematode. *Agric. Sci. Digest* **3**: 190-192.

Prasad, J.S. & Rao, Y.S. (1984). The status of root nematodes (*Hirschmanniella* spp.) *Beitrage Trop. Landwirtsch. Veterinarmed.* **22**: 281-284.

Prasad, S.K. (1964). Plant parasitic nematodes. *In. Entomology in India.* Silver Jubilee Number, Entomological Society of India.pp. 397-400.

Ramakrishnan, S.; Varadharajan, G. & Sutharsan, P.D. (1984). TKM-9 is resistant to rice root-root nematode. *Int. Rice Res. Newsl.* **9**: 20.

Rao, Y.S. (1988). Current research on cyst and root-knot nematodes of rice in India. *In: Advances in Plant Nematology.* Proc. US-Pak Intl. Workshop on Plant Nema. (1986). pp. 205-227.

Rao, Y.S. & Panda, M. (1970). Study of plant parasitic nematodes affecting rice production in the vicinity of Cuttack (Orissa), India *Final Tech. Rept.* (US-PL480 Project). Central Rice Research Institute.

Rao, Y.S.; Prasad, J.S. & Panwar, M.S. (1984a). Influence of fertiliser application in rice on the incidence of plant parasitic nematodes. *Environ. Ecol.* **2**: 201-208.

Rao, Y.S.; Prasad, J.S. & Panwar, M.S. (1986). Nematode problems in rice: Crop losses, Symptomology and Management. *In: Plant Parasitic Nematodes of India: Problem & Progress* (Eds.—G. Swarup & D.R. Dasgupta). pp. 279-299.

Rao, Y.S.; Yadava, C.P. & Padalia, C.R. (1984b). Influence of rotation crops in rice soils on the dynamics of plant parasitic nematode population. *Biol. Agric. Hortcult.* **2**: 69-78.

Rinaudo, G.; Alazard, D.& Mondiongui, A. (1988). Stem modulating legumes as green manure for rice in West Africa. *In: Green manure in rice farming.* IRRI, Manila, Philippines. pp. 97-701.

Routaray, B.N. & Das, S.N. (1985). Effect of organic and inorganic sources of nitrogen controlling *Hirschmanniella mucronata* on rice. *J. Indian Bot. Soc.* **64**: 390-392.

Sher, S.A. (1968). Revision of the genus *Hirschmanniella* Luc & Goodey, 1963 (Nematoda : Tylenchoidea). *Nematologica* **4**: 243-275.

Subramanian, P. & Velayutham, B. (1983). Control of rice root nematode. *Int. Rice. Res. Newsl.* **8**: 15-16.

Van Breda de Haan, J. (1902). Een aaltjes - ziekte der rijst ''omo mentek" of ''omo bambang" Voorlooping rapport. *Mededeel's Lands Plantentuin* (Buitenzorg) **53**: 1-65.

Van Der Vecht, J. & Bergman, B.H.H. (1952). Studies on the nematode Radopholus oryzae (van Breda de Haan) Thorne and its influence on the growth of rice plant. *Pemberitaan Balai Besar Penjelidikan Pertanian, Bogor* No. 131:82.

Venkitesan, T.S.; Charles, J.S. & Nair, V.R. (1980). Checking infection of rice root dips to increase yield. *Int. Rice Res. Newsl.* **5**: 9-10.

Chapter 4

Control of Molya Disease of Wheat and Barley

D.S. Bhatti* & R.S. Dahiya

Department of Nematology, Haryana Agricultural University, Hisar-125 004, India

CEREALS are the main staple food crops throughout the world. Wheat and barley, being the major cereal crops of India, serve as a direct source of food for the people and livestock and are the rich sources of protein, carbohydrate and energy. In our country major areas under wheat cultivation are confined to Punjab, Haryana, Rajasthan, Delhi, Uttar Pradesh, Madhya Pradesh, Bihar and Gujarat states which produce nearly 87 per cent of India's wheat. India today ranks fourth among the wheat growing countries of the world in terms of both area and production. In India, wheat and barley are grown in 2,30,74,000 and 13,78,000 hectares producing 4,68,65,000 and 19,52,000 tonnes, respectively (Anon, 1986). Average yield in India is very low in comparison to the developed countries of the world. This is due to the poor crop management of which pests and diseases constitute an important part.

Wheat and barley crops are attacked by a large number of pests which cause heavy crop losses. The yield losses are due to pests and pathogens such as insects, nematodes, fungi, bacteria, viruses, etc. In recent years losses, due to nematodes have been abundantly demonstrated. Among various genera of phytonematodes causing economic losses the cereal cyst nematode, *Heterodera avenae* Woll, is considered economically the most destructive nematode and cause 'molya' disease of wheat and barley.

HISTORY

Cereal cyst nematode, *Heterodera avenae* was first observed as a parasite of cereals in Germany by Kuhn (1874), who considered it a biological form of beet nematode, *Heterodera schachtii* Schmidt. Mortensen *et al.* (1908) recognised it as a distinct form and first used the name *H. schachtii* var. *avenae* for cereal infecting nematode. Wollenweber (1924) demonstrated clearly the morphologi-

* Present address : 435-Defence Colony, Hisar (Haryana), India.

cal and physiological differences by which cereal cyst nematode could be separated from sugar beet cyst nematode and named it as *H. avenae*. Schmidt (1930) observed that larvae from oats were larger than those from sugarbeets and designated as *H. schachtii* sub sp. major Franklin (1940) raised the sub species to species rank and designated as *H. major*. Nematologists call *H. major* or *H. avenae* in different countries for this pest. Franklin *et al.* (1959) reviewed the nomenclature and proposed the adoption of *H. avena*.

DISTRIBUTION

Cereal cyst nematode, *Heterodera avenae* was first reported by Kuhn in 1874 in Germany. However, later on it was reported from Holland, Denmark, Sweden, England, West Germany, U.S.S.R., Norway, Australia, Canada, Scotland, Tunisia, Italy, Japan, Israel, Belgium, Peru, India, Poland, France, Eire, Spain, Portugal, Northern Ireland, Switzerland, Greece, Yugoslavia, Bulgaria, Czechoslovakia, USA, New Zealand, Pakistan, Iraq, Egypt, Libya (Meagher, 1977; Maqbool, 1981; Ibrahim *et al.*, 1986).

In India, Prasad *et al.* (1959) first reported this nematode from Sikar district of Rajasthan. Later on it was recorded from Alwar, Ajmer, Bhilwara, Chittorgarh, Churu, Jaipur, Jhunjhunu, Marwar, Pali, Sirohi, Jaipur, Udaipur districts (Koshy & Swarup, 1971; Mathur *et al.*, 1975a). In Punjab, it is present in Faridkot, Ferozepur, Hoshiarpur, Jallundhar, Kapurthala, Ludhiana, Patiala and Sangrur districts (Koshy & Swarup, 1971; Chhabra, 1973; Singh *et al.*, 1990). Bhatti *et al.* (1980a) reported it from Mahendergarh, Gurgaon, Rohtak and Bhiwani districts of Haryana. Later on, Bajaj & Walia (1985) recorded its infestation in Ambala district. Authors have also noticed its presence in Faridabad, Sirsa and Hisar districts of Haryana. *H. avenae* infestation was recorded in Kangra district of Himachal Pradesh (Koshy & Swarup, 1971) and Jammu and Leh of Jammu and Kashmir state (Singh *et al.,* 1976; Khan *et al.,* 1990). Recently it has been reported from Union Territory of Delhi (Swarup *et al.,* 1982), and Ghaziabad, Aligarh, Badaun, Bulandshahar, Sikandrabad, Ballia, Jalaun, Lalitpur of Uttar Pradesh (Swarup *et al.* 1982; Siddiqui *et al.*, 1986; Husain *et al.*, 1989).

LOSSES

'Molya' is one of the serious diseases of wheat and barley in India and abroad. From abroad, as early as 1953, Stapel estimated an average annual loss of 50,000,000 Kroner in Denmark. Dixon (1965) found that ten eggs or larvae/g soil resulted in losses of 190.5 and 76.2 kg/ha for spring wheat and winter barley, respectively. Hansen (1985) estimated losses of 30,000 to 75,000 tons of barley in Denmark. From USSR, Tikhonova & Smirnov (1969a) found that losses in spring wheat grown in successive year in infested field were 18-20, 28-

30 and 66% in first, second and third year, respectively. Ishamaev (1974) estimated yield loss up to 60 per cent. From Australia Meagher (1972) reported 73.89 per cent loss in wheat yield. Brown (1981b) estimated annual losses in grain yield in wheat alone to be Aus. $ 72 m.

In India, lot of work has been done on loss estimation. Van Berkum & Seshadri (1970) estimated losses worth $ 8 m in wheat and barley in Rajasthan state alone. In another study, annual losses worth rupees 4 crore in wheat and 3 crore in barley in Rajasthan state have been reported by Seshadri & Dasgupta (1980). Mukhopadhyaya *et al.* (1972) reported that 10 cysts/kg soil caused 10 per cent loss in wheat and barley yield, the loss reached 64 per cent with 1250 cysts/kg soil. Handa *et al.* (1978b) estimated yield loss of 41.6 and 54.9 per cent in 1976-77 and 66.6 and 70.2 per cent in 1977-78 in wheat and barley crops, respectively.

HOST-PARASITE RELATIONSHIP

Upon hatching from cysts in November, second-stage juveniles penetrate the roots usually in the meristematic region behind the root cap and other parts of root. The juveniles move through the cortex both intercellularly and intracellularly and establish for feeding in the stellar region and lie parallel to the stele. The cells around the head of nematode begin to enlarge and their protoplasm becomes dense. Dissolution of cell wall and merging of protoplasm results in the formation of elongated multinucleate specialised cells called syncytia with thickened outer boundaries (Williams, 1969). The cells involved in the formation of syncytia include those of endodermis, phloem and protoxylem. With further development of different larval stages, disorganisation of cortex, endodermis, pericycle and central vascular bundles takes place. Johnson & Fushtey (1966) observed the formation of syncytium within 6 days after inoculation in Garry's cultivar of oat. Syncytial formation was found within stele only and outward development appeared to be restricted by the endodermis. Cotten (1970) observed the initiation of syncytium in resistant barley cv. KVL 91, but subsequently these cells become vacuolated with less dense granular cytoplasm than in susceptible cultivar.

The ultimate effect of host-parasite relationship is the development of disease symptoms. Root system becomes bushy, bunchy and shallow due to root proliferation and has slight swelling near the root tip. Plant remains yellow, stunted with reduced tillering (Fig. 1).

ECOLOGY

Cereal cyst nematode has been reported to prefer open textured well drained soil. Meagher (1970) working with Australian populations observed a marked reduc-

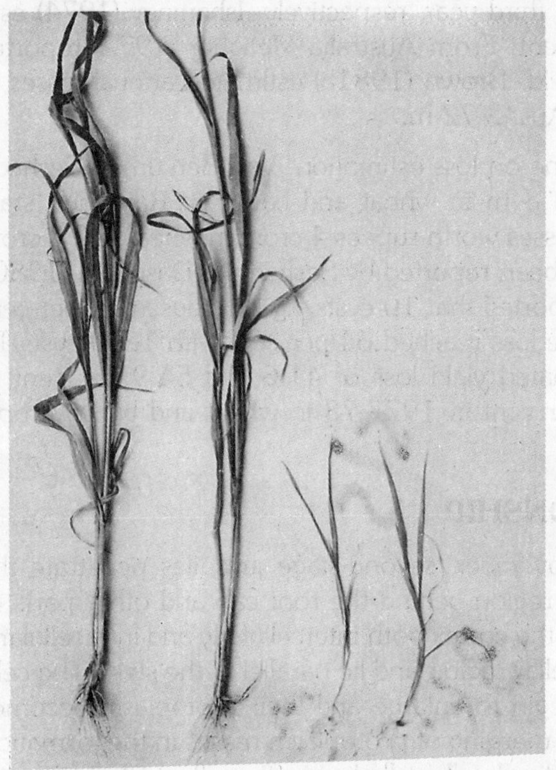

Fig. 1. Wheat plants infected with cereal cyst nematode, *Heterodera avenae* (infected on right, healthy on left).

tion in egg viability in wet soil (24.0% soil moisture) after one month and almost complete loss of viability after 5 months, while egg viability in dry soil (3.4% soil moisture) remained very high even after 7 months. In India, Mathur *et al.* (1974) recovered viable larvae from cysts stored at 10.0% relative humidity (RH) up to 3 months and 6 months or more at 32.3 and higher RH and no hatch at 0% RH. Similarly, reduction in egg viability was more pronounced in wet soil than dry soil.

Temperature also plays an important role for nematode survival. Winslow (1956) observed that storage at high temperature reduced cyst viability. Viability of one year old cysts appeared to be adversely affected when stored at room temperature at 15-40°C (Swarup & Singh, 1961). Mathur *et al.* (1974) did not find any change in cyst viability even after 15 months when stored at 0°C, and at 30°C complete loss of viability was noted after 6 months. While Rao & Dhawan (1988) indicated that encysted eggs can be conveniently stored for at least 6 months in dry soil at 30°C without losing the viability and infectivity.

Winslow (1955) showed that larval hatching is almost totally suppressed from dried cyst. Hesling (1956) reported that after drying the cysts in incubator for 4 hours, hatch from eggs was greatly decreased and after 8 hours hatching was totally inhibited. Duggan (1960), however, found that eggs survived in air-dried soil for 6 months and after longer period of drying the encysted eggs lose their viability and no larval hatch took place. In Australia, by contrast eggs within cysts have been observed to survive for 13 months of storage in air-dried soil (Meagher, 1970) and 5.5 years in dry atmosphere (Meagher, 1974). In Rajasthan (India) cysts can withstand 2 years dry storage without soil and yet hatch freely when placed in water (Mathur *et al.*, 1975b). Cysts stored with soil survived for longer periods than cysts stored without soil (Mathur, 1969; Mathur *et al.*, 1974).

Winslow (1955) observed that hatch was greater at temperature fluctuating at 18-20°C than at constant temperature of 25°C. The optimum temperature for hatch of Canadian and Australian population is 15°C (Fushtey & Johnson, 1966; Bayner & Fisher, 1971), while that of Dutch population is about 17°C (Mathur, 1969). Maximum emergence in an Indian population took place at 20-22°C and fluctuating temperature between 17 and 23°C in laboratory (Swarup & Gill, 1972; Mathur *et al.*, 1975b). Bhatti & Malhan (1982) observed differential temperature optima for larval emergence from new generation cysts (10-15°C) and old cyst (18°C). Larval emergence from white, yellow and light brown cysts occurred at 10°C but not at 26-30°C. Hatching did not take place from dark brown cysts at any of the temperatures (Rajan & Swarup, 1985).

Degree of aeration also greatly influence the larval emergence. Rate of hatch under partially submerged cysts in water was more than under the completely submerged cysts (Swarup & Gill, 1972). Mathur *et al.* (1975b) also observed maximum larval emergence from well aerated cysts than relatively poor aerated ones.

Gair (1965) found *H. avenae* in all types of soil but Meagher (1972) reported that the nematode was predominant in soils with good physical structure and not in heavy textured soil. In India, it is found in all soil types but more in light soils (Koshy & Swarup, 1971; Bhatti & Malhan, 1981; Mathur *et al.*, 1981).

Based on soil pH of 635 samples in Eire, it was concluded that there was an upward trend of infestation level with pH above 6.0, but the relationship did not extend to pH below 6.0 (Duggan, 1963). However, in England, it was concluded that pH had no noticeable effect on soil population (Fiddler & Bevan, 1963). Swarup & Gill (1972) in India found that larval emergence takes place over a wide range of pH (3.5-11.5) with a maximum hatching at 8.5 and minimum at 11.5.

Root leachings from one-week old barley seedlings have been observed to stimulate hatching (Swarup & Gill, 1972). Jones (1973) has reported that hatching is influenced by root diffusates from cereals or grasses in soil during spring. Leaching from hosts (wheat, barley, oat and rye) and non-hosts (carrot, gram, fenugreek, mustard and marigold) failed to stimulate larval hatching (Winslow, 1955; Mathur *et al.*,1975b).

CONTROL STRATEGIES

Cultural Control

These methods are aimed to adopt crop husbandry practices so as to minimise the effects of nematodes. Following cultural practices have been attempted to control *H. avenae*.

Crop rotation

There is much scope for adopting such practices due to highly host specific nature of the nematode. Kort (1959) has suggested avoiding wheat and barley for at least two years in heavily infested fields and rotation with non-host crops such as alfalfa and beets. Meagher & Rooney (1966), however, found that improved wheat yields occurred when a wheat-legume rotation was used. Tikhonova & Smirnov (1969b) reported that crop rotation with pea and subsequently winter rye sharply reduced nematode's population. Handa *et al.* (1975b) found that rotation with non-hosts and keeping the field fallow during rabi season reduced the cyst population by about 50% and 75% in one and two years, respectively as against population build-up up to 250 to 335% if rotation or fallowing is not practised. Bachthaler & Hien (1976) observed that maize rotation reduced the cyst population to 50 per cent. Singh (1985) demonstrated that non-host crops reduced the nematode population significantly. The resistant barley when included in rotation, reduced population by 57-58.3% and mustard reduced it by 50%. Under field condition resistant barley cv. Rajkiran reduced the cyst population up to 84.8% (Singh & Yadav, 1986). Growing of *sarson, toria, raya* and *taramira* for one season resulted in 87-100% reduction in cyst population (Singh *et al.*, 1987).

Summer ploughing

Handa *et al.* (1975a) indicated a reduction in the cyst population, under conditions where soil was kept exposed, by ploughing, to direct sun heat during May-June. Bhatti & Dabur (1982a) studied the effect of summer ploughings at 10 days interval during May-June. Yield increases over the control were 2.7, 5.6 and 11.2 per cent for 1, 2 and 3 ploughings, respectively and cyst populations were 9.4, 9.0, 8.9 and 5.5 eggs and larvae/g soil in 0, 1, 2 and 3 ploughings, respectively. Mathur *et al.* (1987) also studied the effect of 1-5 deep summer ploughings at 7-10 days intervals in May-June. Population reduction and yield increase was found to be directly proportional to the number of ploughings. Cyst population reduction of 9.3-42.4 per cent and increase in wheat yield of 4.4-97.5 per cent were recorded.

Trap cropping

Bhatti & Dabur (1982b) studied the effect of wheat cv. 306 as a trap crop and reported that final cyst populations in aldicarb and trap crop were decreased by 7.5 and 9.4 per cent, respectively, and respective yield increases were 9.0 and 13.5 per cent.

Mixed cropping

Handa *et al.* (1980a, 1989) studied the effect of mixed sowing of resistant barley cv Rajkiran with susceptible wheat cv. Kalyan Sona/susceptible barley cv. RD 103 (50:50). The yield was significantly increased and cyst population decreased in mixed sowing compared with susceptible wheat.

Additional irrigation

Additional post-harvest irrigation did not influence the nematode population but wheat yield increased to 36.2 per cent (Bhatti *et al.* 1980c). Handa *et al.* (1989) showed significant yield increase in wheat and barley crops by changing irrigation frequency but nematode population also increased.

Time of sowing

Studies in Australia show that crops sown early in late autumn are less damaged than those sown late in the season (Meagher, 1982). Bhatti *et al.* (1980b) found that yield of wheat sown early in November in 'molya'-infested field was more than late sown crop.

Soil amendment

Bhatti & Dhawan (1980) observed the reduction in cyst populations with the addition of crushed seeds of carrot (@ 1, 2 or 4% w/w) and of coriander (@ 4%) in a pot experiment. The number of tillers and grain weight increased with all levels of crushed seeds of carrot and coriander. Sharma *et al.* (1981) demonstrated that mustard cake amendment @ 1 to 5% (w/w) significantly reduced cyst population and increased number of tillers and shoot weight. Dahiya *et al.* (1986) found 52.85 and 48.66 per cent reduction in cyst population with the addition of *subabool* and sunhemp leaves each @ 60 g/kg soil, respectively and significant increase in shoot weight, number of tillers and earhead length under screenhouse condition.

HOST RESISTANCE

It is an effective, economical and environmentally safe means of reducing crop losses. The use of plant varieties resistant to nematode may be the only method of this kind of control available.

Wheat

The first source of resistance in wheat was found in cv. Loros (C13779) in Denmark by Nielsen (1966). It was later shown to be resistant in Australia (O'Brien &

Fisher, 1974). Then resistance was reported in varieties like AUS 10894, AUS 11577, AUS 90248, Festiguay, Katyil and Cocomba Brown (1974), McLeod (1976), O'Brien & Fisher (1979), O' Brien *et al.* (1980), Cook & McLeod (1980), Brown (1985), Brown & Young (1982), and Archer *et al.* (1984) from Australia. Sanchez *et al.* (1985) reported wheat cv. Pane resistant in Spain. A single dominant gene is reported to be responsible for resistance in Loros spring wheat (Nielsen, 1966; Cook & McLeod, 1980).

In India, Sikora *et al.* (1972) identified three USDA wheat selections, PI 183868 from Turkey and PI 185205 and PI 185207 from Portugal, to be highly resistant. Singh & Yadav (1978) found HD 2257, HD 2265, WH 220, WH 250, WL 922, WL 2199, WG 736, Raj 1409, Raj 1470, Raj 1470, Raj 1487 free from infection. Varieties like PV 18, Sonora 64 and Lerma Rojo (Swarup & Gill, 1969), Agra Local, NP 718, Sonora 64 (Joshi *et al.* 1970), Kalyan Sona and PV 18 (Mukhopadhyaya *et al.*, 1972), HD 1981, HD 1999, AI-8-1 and CC 62 (Seshadri & Sethi, 1978), WL 711, WH 147, Raj 3037, HD 2285 (Dhawan & Nagesh, 1987) were found as tolerant.

Dahiya & Bhatti (1983b) found varieties WH 147, WH 157, WH 283, WH 291, C 306, Kalyan Sona, Sonalika, HD 2009 and HD 2204 susceptible to varying degrees. Khurma & Sakhuja (1986) found PBW 34, Raj 1556, DWL 5023, HD 4579 and Mexicali 75 varieties of macroni wheat as susceptible. About 5,000 wheat varieties/lines have been screened at HAU, Hisar during 1974-1990 but none has shown resistance. Katyil, a resistant variety in Australia was found susceptible to Indian populations of the nematode (Dhawan, 1988).

Barley

The search for resistance was apparently initiated by Nilsson-Ehle (1920) in Sweden and reported resistance in Primus, Swawnhals and Chevalaier. Later on, resistant varieties Drost, Fero and Kron were bred in Denmark by Bogh (1952). Andersen (1959) reported resistance in PL 191, No. 14 and No. 291. Cultivars 746534 and 747475 were recorded resistant to both the races 1 and 2 in Denmark (Anon., 1980). From Sweden, Lundin (1974) reported cv. Ansgar resistant to *H. avenae*.

From Australia, varieties Morocco (CI 3902), Marocaine (CI 18334) and Martin 403-2 (Brown, 1974), Marocaine 079 and Sabarlis (Ellis & Brown, 1976; Brown, 1985; McLeod *et al.*, 1986), Morocco (O'Brien & Fisher, 1979), Morocco, Nile, CI 8147 and Athenais (Sparrow & Dube, 1981) were recorded resistant.

Sanchez *et al.* (1985) recorded cv. Sekag resistant in Central Spain. Soetopol (1985) recorded Morocco and Martin resistant to Australian pathotype in Indonesia. Benedicte, Kara, Nemax were found resistant by Ewertson & Lundin (1985) and Ewertson (1986) in Denmark, Sweden and Germany, respectively.

In India, Bhatnagar *et al.* (1965) found Prior, AQ 1127, AQ 253, Research and AQ 853 cultivars of barley (*Hordeum distichon*) resistant under field condition. Bhatti *et al.* (1976) found BP 263, BP 264 and C 164 resistant where ratio of male to female was 106:1, 86:1 and 109:1, respectively. Large number of varieties/lines have been screened by all the cooperating centres during 1966-1976 and resistance was reported in cv. Morocco, CI-3902 and Martin 403-2 (Seshadri & Sethi, 1978). Handa *et al.* (1980b) reported resistance in 62 out of 1300 genotypes tested. Resistance was also reported in DL 349, DL 375, DL 376 and DL 379 (Dhawan & Sethi, 1983). Yadav *et al.* (1987) reported resistance in cv. C164, BH 75 and Rajkiran.

BIOLOGICAL CONTROL

Exploitation of natural enemies in reducing the nematode populations is receiving an increased attention at present, although parasitization of eggs and larvae was reported by Kuhn as early as 1877. A number of fungi and other organisms like *Mycocentrospora acerina* var. *castelnaudariensis, Verticillium chlamydosporium, Scopulariopsis acremonium, Paeciloymces lilacinus, Acremonium roseum, Nematophthora gynophila, Catenaria auxilaris, Nematoctonus haplocladus* etc. have been isolated from cysts and larvae (Tribe & Cayrol, 1982; Kerry *et al.*, 1982; Kerry & Andersen, 1983, Jonowiczk & Wronkowska, 1984). Kondokoa & Tikhonova (1981) recorded reduction in the populations of *H. avenae* to an extent of 76.7 to 95.9 per cent by *Catenaria anguillulae* and *Cylindrocarpon radicicola.* Similarly Kerry (1982) recorded about 60-70 per cent mortality of eggs and females in the presence of *N. gynophila* and *V. chlamydosporium.* Kerry *et al.* (1984) reported that some isolates of *V. chlamydosporium* reduced *H. avenae* population up to 80%.

In India, Jain (1979) recorded *Cephalosporium* sp., *Torula* sp., *Phialophora* sp. *Fusarium equiseti, Pencillium spinulosum* etc. from cyst wall, eggs and larvae. *Cylindrocarpon radicicola* was found to be most effective in destroying cysts, eggs and larvae in a minimum time. Gill & Swarup (1977) also found reduced nematode reproduction with a combination of *Fusarium moniliforme* and *Helminthosporium gramineum* on barley. Chowdhry & Dhawan (1984) recorded a parasitic fungus, *Catenaria vermicola* in the eggs and juveniles for the first time from India. Dhawan (1985) observed that infection in healthy juveniles took place only thorough zoospores and life cycle of fungus is completed in 2-3 days. Chowdhry & Kaushal (1984) isolated *Cylindrocarpon uniseptatum* from unhatched eggs of *H. avenae.* Sharma & Swarup (1988) isolated *Allomyces anomalus, Catenaria auxiliary, C. vermicola, Glomus fasciculatum, Nematophthora* sp., *Olpidium* sp. *Pythium* sp., 'Black Yeast' and *Lagenidiaceous* sp. from females, cysts, eggs and larvae. *G. fasciculatum* and *Pasteuria penetrans* were found to be most destructive against *H. avenae.* Bhattacharya & Swarup (1988) further studied the effect of *P. penetrans* on the nematode. The life cycle of bacterium completed in 49 days at 10-17°C. Amongst application methods direct mixing of bacterial spore in infested soil was found to be most effective against nematode population. Maximum cyst reduction was obtained when spore infested soil was incubated at 30°C before application.

NEMATICIDAL CHEMICALS

Control of phytonematodes by the use of chemicals is recommended for successful crop production when other methods are inadequate to suppress the nematode sufficiently. Their use is dependent upon the cost/benefit ratio.

Lot of work has been done in India and abroad, using various soil fumigants and systemics. Williams & Salt (1970) studied the effect of methyl bromide, chloropicrin, DD, formalin and mercury salt on *H. avenae* in Britain. All the sterilants except mercury decreased number of *H. avenae* and increased crop yield. Kyrou (1970) reported that methyl bromide at the rate of 40 kg/1000 sq m improved plant growth, increased yield and markedly decreased cyst population on wheat. Swarup *et al.* (1976) found that yield increase in barley was 6-14 times with DBCP at 3 gal/acre. The best return was obtained with 40 gal/acre of DD where 2800 lbs/acre of wheat and 3070 lbs/acre of barley was produced. Swarup *et al.* (1976) later concluded that DD and DBCP were best chemicals against 'molya' disease of wheat. Seshadri (1971) found that pre-plant application of DD @ 400 kg/ha resulted in 3 and 13 folds increase in yield of wheat and barley, respectively and significantly reduced the nematode population. Mukhopadhyaya *et al.* (1971) reported that DD @ 500 l/ha proved effective in reducing cyst population and increasing grain yield to 253% over control. Bhatti & Dalal (1972) found the increase in wheat yield by 128.5 per cent following DBCP application @ 30 l/ha in irrigation water and cost/benefit ratio was 1:1.45. Jain & Seghal (1975) reported DD @ 450 l/ha and DBCP @ 45 l/ha as pre-plant treatment gave better yield and nematode control over lower doses. Yadav & Singh (1975) observed 10 quintal higher wheat yield and 81.8 per cent population reduction over control by DBCP @ 7.5 gal/ha. Swarup *et al.* (1976) found that best nematode control and optimum yield increases in wheat and barley were obtained by DD @ 300 l/ha or DBCP @ 45 l/ha. Singh & Yadav (1978) found DBCP @ 45 l/ha most superior over others by increasing 95 per cent yield and reducing 75 per cent cyst population. Meagher *et al.* (1978) found that methyl bromide-chloropicrin @ 45 kg/ha increased grain yield up to 323 per cent. Brown (1981a) reported that EDB and DBCP applied to wheat gave yield increases up to 0.9 t/ha and reduction in cyst numbers. Bhatti *et al.* (1981) evaluated DBCP @ 45 l/ha and obtained 12.9

to 95.3% and 8.9-83.9% increase in yield, respectively and decrease in nematode population to 0.0-4.8 eggs and larvae/g soil.

Regarding systemic nematicides, Brown *et al.* (1970) observed that broadcasting of aldicarb 10G @ 11 kg a.i./ha provided best control and more than trebled wheat yield. Brown (1973) reported that aldicarb 10G at 9 kg/ha as broadcast and 2.2 kg/ha in rows completely prevented cyst formation on wheat. Bhatti *et. al.* (1981) recommended aldicarb 10G @ 1.5-2.0 kg a.i/ha for the control of *H. avenae* in infested fields and cost benefit ratios were 1:4 to 11. Handa & Mathur (1982) found carbofuran and aldicarb @ 2 kg a.i/ha significantly better than counter and cytrolane in pots while in fields 2 kg of these chemicals significantly increased grain yield than control and lowered cyst population. Dahiya & Bhatti (1983a) evaluated aldicarb 10G @ 1.5-2.0 kg a.i./ha in naturally infested field at different locations and maximum yield increase in wheat was 1250 per cent and cost benefit ratio was 1:2.67 to 5.99. Mathur *et al.* (1986) tested aldicarb, carbofuran, phorate, terbufos, isofenphos and mephosfolan under field conditions. All, except phorate, were found effective in reducing cyst population and improving crop yield. Drill application was found to be superior to broadcast and aldicarb was found to be most economical. Singh *et al.* (1988) studied the time of application of aldicarb and carbofuran @ 2 kg at sowing, 28 days and 43 days after sowing on nematode control in wheat. Mean yield increased by 24.83, 7.62 and 5.04 per cent and cyst population reduced by 64.39, 56.55 and 56.66 per cent respectively. Singh *et al.* (1988) studied the efficacy of cytolane, phorate, terbufos and aldicarb in field. Minimum cysts (5/250 ml of soil) were observed with phorate @ 2 kg and minimum number of eggs and larvae/cyst were recorded with cytrolane followed by phorate, terbufos and aldicarb. Phorate @ 2 kg resulted 13.2 per cent increase in grain yield over control thereby indicating the potentiality of phorate for managing *H. avenae* in field.

Sakhuja & Singh (1990) found that sebufos @ 1.5 and 3.0 kg a.i./ha reduced population by 88.8% and 97% respectively. Phorate, diazinon and carbofuran @ 1.5 kg reduced the population by 69.2, 50.7 and 73.1%, while at 3 kg a.i./ha by 81.9, 95.6 and 75% respectively. Maximum increase (119.7%) in grain weight occurred with diazinon @ 3 kg/ha followed by Naemin 37.5 kg/ha (88.3%), diazinon @ 1.5 kg/ha (82.7%) and carbofuran @ 3 kg/ha (62.7%).

Gill & Swarup (1979) reported inhibition of egg hatch of H. avenae with DD at 400 I and DBCP at 20 I/ha in a pot experiment. Sakhuja *et al.* (1981) reported significant reduction in emergence of H. avenae with aldicarb sulfone and carbofuran each @ 20, 40 and 80 ug/ml. Carbofuran was found to be more inhibitory than aldicarb sulfone. Sultan & Singh (1984) found that Nemacur @ 20, 40 80 adn 160 ppm significantly inhibited cyst hatching and more so on higher doses and suggested temporary inhibition on transfer of cysts in water.

Kaushal & Dhawan (1986) exposed the second-stage juveniles to 125, 250 and 500 ppm and nemacur, BPMC and FMC and found maximum (81%) larval mortality in 500 ppm of nemacur after 4 days of exposure while larval emergence was minimum at all doses of nemacur and BPMC at 500 ppm. They also found the stimulatory effect of BPMC at 250 ppm and FMC at all doses. Kaushal & Seshadri (1986b) observed that larval penetration was significantly affected with carbofuran at 2.0 and 4.0 kg a.i./ha and cyst multiplication was negligible. Sultan & Sakhuja (1986) concluded the carbamates do not kill the eggs but reduce larval emergence temporarily.

Singh & Yadav (1978) reported reduction in nematode multiplication by seed treatments with carbofuran and aldicarb sulfone at 2.0%, the reduction being to the extent of 93% in carbofuran. Kaushal & Seshadri (1986a) recorded significant reduction in cyst number with aldicarb, aldicarb sulfone, carbofuran @ 2.5 and 5.0% fensulfothion @ 2.0% and isofenphos @ 1.0 to 5% as seed dressers.

INTEGRATED CONTROL

Integrating two or more practices at a time is aimed at achieving greater efficiency and economy in control. Handa *et. al.* (1978a) recorded an increase in grain yield from 18.29 to 126.74 and 35.01 to 105.23 per cent in wheat and barley, respectively when nitrogen was combined with carbofuran and aldicarb. Sakhuja *et al.* (1978) also reported that DBCP @ 45 l/ha alongwith 150 kg/ha of nitrogen was found most effective in reducing build-up of cysts and recorded a grain yield of 49.33 g/ha as compared to 27.5 in control. They also observed the inhibition of germination of wheat at 300 kg/ha of nitrogen. Bhatti *et al.* (1980b) studied the effect of sowing time and nematicide and found that aldicarb @ 2 kg a.i./ha with early November sowing gave better yield and 84.6 per cent decrease in cyst population. Rovira & Simon (1982) obtained increase in wheat yield by applying nematicides with seed, growing resistant cultivars, alternate legume with wheat in crop rotation and growing wheat with minimum tillage. Singh & Yadav (1986) recorded higher wheat yield and greater reduction in nematode population over control when organic amendments were combined with nitrogen fertilizer. Handa *et al.* (1989) combined aldicarb @ 2 kg a.i./ha with susceptible wheat/barley, resistant barley, and resistant barley + susceptible wheat in a field experiment and found significantly higher yield while final population decreased significantly over the untreated control.

CONCLUSION

Wheat and barley are major cereals crops of India. Among various pathogens responsible for low production of yield, cereal cyst nematode, *Heterodera avenae*, the causal agent of 'molya' disease, is most severe and destructive. This nematode was first reported from Rajasthan in 1959. After a lapse of 32 years, it could only be reported from seven states including Union Territory of Delhi in light and heavy soils. Therefore, systematic extensive/intensive survey is required at block level for proper mapping of its distribution.

Studies on loss estimation in different climatic zones at multilocations are necessary to find out damage threshold levels. There are many abiotic and biotic factors operating in the soil which ultimately affect population dynamics of the nematodes and the crop losses are dependent upon the initial population densities. Information on host parasite relationship is scanty. Therefore, study may be undertaken in different crops/varieties, soil types and at different population levels.

Due to high cost of chemicals, research on cultural and other methods of control should get prime attention in future research programmes. Successful use of cultural practices can be made only when a complete knowledge of the nature of pathogen in relation to soil environment is available. Efforts must continue for finding out suitable resistant sources so that it can be utilized by crop breeders to breed suitable varieties which are widely acceptable to the growers of different agroclimatic zones. Phytotherapy, as an alternative to nematicides should also get attention for spot/patch application under field conditions since this nematode is not distributed homogeneously in the field. Although biological control, so far, is not practicable on field scale in India, we should make our efforts to find out endemic organisms for controlling the nematode.

REFERENCES

Andersen, S. (!959). Resistance of barley to various populations of the cereal root eelworm (*Heterodera major*). *Nematologica* **4**: 91-98.

Anon. (1980). Report No. 34 on the work of the Agricultural Cereal Breeding Station, Denmark. *Pl. Br. abstr.* **51**: 10239-41.

Anon. (1986). *Statistical Abstract, India*, 1986. p. 685.

Archer, M.J.; Griffiths, J.B. & Pye, D.L. (1984). Cocamba, a new wheat variety resistant to cereal cyst nematode. *A note, Dep. Agric. Victoria*. p. 2.

Bachthaler, G. & Hien, L. (1976). Results of long-term cereal rotations with varying proportions of maize. *Soil & Fert*. **40**: 3191.

Bajaj, H.K. & Walia, R.K. (1985). A new report on the occurrence of cereal cyst nematode, *Heterodera avenae* Woll. 1924 in Ambala district (Haryana). *Geobios New reports* **4**: 169-170.

Bayner, R.J. & Fisher, J.M. (1971). Effect of temperature on hatching of eggs of *Heterodera avenae. Nematologica* **17**: 519-534.

Bhatnagar, M.P.; Sanghi, A.K. & Sharma, S.K. (1965). Resistance of barley varieties to the cyst forming nematode, *Heterodera avenae. Indian J. Genetics & Pl. Br.* **25** : 381-383.

Bhattacharya, D. & Swarup, G (1988). *Pasteuria penefrans*, a pathogen of the genus *Heterodera*, its effect on nematode biology and control. *Indian J. Nematol*. **18**: 61-70.

Bhatti, D.S. & Dabur, K.R. (1982a). Effect of summer ploughings on *Heterodera avenae* population and wheat yield. *J. Nematol.* **14**: 429.

Bhatti, D.S. & Dabur, K.R. (1982b). Use of wheat as a trap crop for the control of *Heterodera avenae. J Nematol.* **14**: 429.

Bhatti, D.S.; Dahiya, R.S.; Dalal, M.R. & Dhawan, S.C. (1976). Resistant barley varieties for the control of *Heterodera avenae. Curr. Sci.* **45**: 678.

Bhatti, D.S. Dahiya, R.S.; Gupta, D.C. & Malhan, I. (1980a). Plant parasitic nematodes associated with various crops in Haryana. *Haryana agric. Univ. J. Res.* **10**: 413-414.

Bhatti, D.S. & Dalal, M.R. (1972). Yield responses in wheat to nematicidal treatments in a soil heavily infested with wheat cyst nematode. *Haryana agric. Univ. J. Res.* **2**: 231-232.

Bhatti, D.S.; Dalal, M.R.; Dahiya, R.S. & Dhawan, S.C. (1980b). Integrated control of *Heterodera avenae* in wheat. *J. Nematol.* **12**: 214-215.

Bhatti, D.S.; Dalal, M.R.; Dahiya, R.S. & Malhan, I. (1981) Chemical control of *Heterodera avenae* in wheat. *J. Nematol.* **13**: 433.

Bhatti, D.S.; Dalal, M.R. & Malhan, I. (1980c). Effect of additional irrigation on population of *Heterodera avenae* and wheat yield. *Haryana Agric. Univ. J. Res.* **10**: 571-572.

Bhatti, D.S. & Dhawan, S.C. (1980). Effect of crushed seeds of carrot and coriander on wheat plant growth and multiplication of *Heterodera avenae. Haryana Agric. Univ. J. Res.* **10**: 419-420.

Bhatti, D.S. & Malhan, I. (1981). Effect of soil texture on the development of *Heterodera avenae* population. *Indian J. Nematol.* **11**: 98.

Bhatti, D.S. & Malhan, I. (1982). Effect of temperature on larval emergence from cysts of *Heterodera avenae. Indian J. Nematol.* **12**: 73-78.

Bogh, H. (1952). Pajbjergfondens Forsogs-ogforadling sarbejde 1952. Borkop 1952.

Brown, R.H. (1973). Chemical control of the cereal cyst nematode (*Heterodera avenae*) — a comparison of methods and rates of application of two systemic nematicides. *Aust. J. Exp. Agric. Anim. Husb.* **13**: 587-592.

Brown, R.H. (1974). Further studies on the Victorian biotype of cereal cyst nematode (*H. avenae*). *Aust. J. Exp. Agric. Anim. Husb.* **14**: 394-398.

Brown, R.H. (1981a). Low volume in row application of fumigant nematicides, at seedling, for the control of cereal cyst nematode *Heterodera avenae* in wheat. *Pl. Res. Inst. Burnley Vict. Australia* **3121**: 40-41.

Brown, R.H. (1981b). Nematode diseases. *In: Economic importance and biology of cereal root diseases in Australia*. Report to Plant Pathology Subcommittee of Standing Committee on Agriculture.

Brown, R.H. (1985). The selection of management strategies for controlling nematodes in cereals. *Agric. ecosystems & Env.* **12**: 371-388.

Brown, R.H.; Meagher, J.W. & McSwain, N.K. (1970). Chemical control of the cereal cyst nematode (*Heterodera avenae*) in the Victorian Mallee. *Aust. J. Exp. Agric. Anim. Husb.* **10**: 172-173.

Brown, R.H. & Young, R.M. (1982). Katyil, a new wheat resistant to cereal cyst nematode. *A note, Dep. Agric. Victoria, Agdex.* 112/38, p. 2.

Chhabra, H.K. (1973). Distribution of *Heterodera avenae* Wollenweber, cereal cyst nematode in Punjab. *Curr. Sci.* **42**: 441.

Chowdhary, P.N. & Dhawan, S.C. (1984). *Catenaria vermicola* in *Heterodera avenae*, nematode — a new record. *Curr. Sci.* **53**: 96.

Chowdhary, P.N. & Kaushal, K.K. (1984). *Cylindrocarpon uniseptatum* sp. Nov.— a new fungus from India. *Curr. Sci.* **53**: 205-207.

Cook, R. & McLeod, R.W. (1980). Resistance in wheat to *Heterodera avenae* in Australia and Britain. *Nematologica*. **28**: 274-277.

Cotten, J. (1970). Some aspects of breeding for cereal root eelworm (*H. avenae*) resistance in barley. *Proc. IX Int. Nematol. Symp*. 231-237.

Dahiya, R.S. & Bhatti, D.S. (1983a). Field evaluation of aldicarb for the control of *Heterodera avenae. III Nematol. Symp., HPKVV, Solan, India*. pp. 39-40.

Dahiya, R.S. & Bhatti, D.S. (1983b). Relative susceptibility of wheat varieties to *Heterodera avenae. III Nematol. Symp., HPKVV, Solan, India*. p. 26.

Dahiya, R.S.; Dutt, R. & Gupta, D.C. (1986). Effect of subabool and sunnhemp leaves on cereal cyst nematode, *Heterodera avenae*. (Woll, 1924) in wheat. Haryana Agric. Univ. J. Res. **16**: 389-391.

Dhawan, S.C. (1985). Mode of infection and life cycle of *Catenaria vermicola* infecting juveniles of *Heterodera avenae. Indian Phytopath.* **38**: 380-381.

Dhawan, S.C. (1988). 'Katyil' a resistant wheat cultivar, susceptible to Indian population of cereal cyst nematode, *Heterodera avenae. Indian J. Nematol.* **18**: 140.

Dhawan S.C. & Nagesh, M. (1987). Tolerance in wheat against cereal cyst nematode, *Heterodera avenae. J. Nematol.* **17**: 336-338.

Dhawan, S.C. & Sethi, C.L. (1983). Resistance in barley to the cereal cyst nematode (*Heterodera avenae* Woll). *Indian J. Nematol.* **13**: 235-237.

Dixon, G.M. (1965). The effect of cereal root eelworm (*Heterodera avenae* Woll.) on spring sown cereals. *VIII Int. Nematol. Symp. Antibes*. p. 26.

Duggan, J.J. (1960). Effect of soil drying on the viability of *Heterodera major* cyst. *Nature* **185**: 534-555.

Duggan, J.J. (1963). Relationship between cereal root eelworm, *Heterodera avenae* (Wollenweber, 1924) infestation and pH value of soil. *Irish J. Agric. Res.* **25**: 105-110.

Eillis, S.E. & Brown, J.A.M. (1976). The reactions of resistant cultivars of barley to some Australian populations of *Heterodera avenae. Nematologica* **22**: 87-93.

Ewertson, G. (1986). Kara—the nematode resistant spring barley best worth growing. *Helm. Abstr.* **55**: 105.

Ewertson, G. & Lundin, P. (1985). Benedict, Kara and Nemex—three cereal cyst nematode and powdery mildew resistant spring barley varieties. *Agri. Hortique Genetica* **43**: 29-42.

Fidler, J.H. & Bevan, W.J. (1963). Some soil factors influencing the density of cereal root eelworm (*Heterodera avenae* Woll) populations and their damage to oat crop. *Nematologica* **9**: 412-420.

Franklin, M.T. (1940). On the specific status of the so-called biological strains of *Heterodera schachtii* Schmidt. *J. Helminth.* **18**: 193-208.

Franklin, M.T.; Thorne, G. & Oostenbrink, M. (1959). Proposal to stabilise the scientific name of the cereal-root eelworm (Class Nematoda) : Z.N. (S.) 375, *Bull. Zool. Nomencl.* **17**: 76-85.

Fushtey, S.G. & Johnson, P.W. (1966). The biology of the oat cyst nematode *Heterodera avenae* in Canada. 1. The effect of temperature on the hatchability of cysts and emergence of larvae. *Nematologica* **12**: 313-320.

Gair, R. (1965). Cereal root eelworm. *In* : Plant Nematology (J.F. Southey, Ed.) pp. 199-221. Tech. Bull (2nd Ed.). Minist. Agric. Fish., HMSO, London.

Gill, J.S. & Swarup, G. (1977). Effect of interaction between *Heterodera avenae* Woll. 1924, *Fusarium moniliforme* and *Helminthosporium gramineum* on barley plants and nematode reproduction. *Indian J. Nematol.* **7**: 42-45.

Gill, J.S. & Swarup, G. (1979). Efficacy of some nematicides on the viability of cyst of *Heterodera avenae* Woll. 1924 and plant growth. *Indian J. Plant Protec.* **5**: 202-204.

Handa, D.K. & Mathur, B.N. (1982). Control of cereal cyst nematode (*Heterodera avenae*) of barley through systemic nematicides. *Nematologica* **28**: 150.

Handa, D.K.; Mathur, R.L. & Mathur, B.N. (1975a). Studies on the effect of deep summer ploughings on the cereal cyst nematode (*Heterodera avenae*) and yield of wheat and barley. *Indian J. Mycol. & Pl. Path*; **5**: 18.

Handa, D.K.; Mathur, R.L. & Mathur, B.N. (1975b). The effect of crop rotation and nematicides on the cereal cyst Nematode *Heterodera avenae* and yield of wheat and barley. *Indian J. Mycol. & Pl. Path.* **5**: 20-21.

Handa, D.K. Mathur, R.L. & Mathur, B.N. (1978a). Studies on the integrated control of molya disease of wheat and barley caused by *Heterodera avenae* in Rajasthan; *Indian J. Mycol. & Pl. Path,* **8**: 70.

Handa, D.K.; Mathur, R.L., Mathur, B.N. & Sharma, G.L. (1978b). Survey and assessment of losses in wheat and barley by 'molya' disease caused by *Heterodera avenae* in the light soils of Rajasthan. *Indian J. Mycol. & Pl. Path*; **8**: 83.

Handa, D.K.; Mathur, B.N.; Sharma, G.L. & Yadav, B.D. (1980a). Effect of mixed cropping of susceptible wheat and resistant barley on the control of molya disease. *Indian J. Mycol. & Pl. Path.* **10**: vii-viii.

Handa, D.K.; Mathur, R.L., Yadav, B.D. & Sharma, G.L. (1980b). Evaluation of barley varieties for resistance to molya disease incited by *Heterodera avenae, Indian J. Mycol. & Pl. Path.* **10**: ix.

Handa, D.K.; Mathur, B.N.; Yadav, B.D. & Sharma, G.L. (1989). Cultural practices and their effects on *Heterodera avenae* and yield of cereals in Rajasthan, India. *Indian J. Nematol.* **19**: 250-253.

Hansen, L.M. (1985). Incidence of *Heterodera avenae* in spring barley fields in Denmark. *In* : Cyst Nematodes (F. Lamberti & C.E. Taylor, Eds.) Plenum Press, New York, London. p. 291.

Hesling, J.J. (1956). Some observations on *Heterodera major. Nematologica* **1**: 56-63.

Husain, S.I.; Siddiqui, Z.A. & Siddiqui, M.R. (1989). Prevalence and geographical distribution of cyst forming nematodes in Uttar Pradesh, India. *Indian J. Nematol.* **19**: 108-114.

Ibrahim, I.K.A.; Rezk, A. & Ibrahim, A.A.M. (1986). Occurrence of cyst nematodes *Heterodera avenae* and *H. rosii* in northern Egypt. *J. Nematol.* **18**: 614.

Ishaniaev, A.M. (1974). The oat cyst nematode on spring wheat and its control. *Ufa, USSR, Akademiya Nauk USSR, Bashkir Filial Institute Biologic.* 115-119.

Jain, R.K. (1979). Occurrence of nematophagous fungi and their mode of action on cereal cyst nematode (*Heterodera avenae*). Thesis *Abstracts* **5** : 88-89.

Jain, R.K. & Sehgal, S.P. (1975). Efficacy of nematicides in controlling the cereal cyst nematode, *Heterodera avenae* in wheat in light sandy soils in Rajasthan. *Indian J. Mycol. & Pl. Path.* **5**: 19.

Johnson, P.W. & Fushtey, S.G. (1966). The biology of the oat cyst nematode *Heterodera avenae* in Canada. II. Nematode development and related anatomical changes in roots of oats and corn. *Nematologica* **12**: 630-636.

Jones, F.G.W. (1973). Nematodes and cereals. *Agric. Develop. Advisory Service Quarterly Review* **6**: 20-35.

Jonowiczk, M & Wronkowska, H. (1984). Fungi parasitising *Heterodera avenae. Ochrona Roslin.* **28**: 5-6.

Joshi, L.M.; Renpro, B.L.; Saari, E.E.; Wilcoxson, R.D. & Raychaudri, S.P. (1970). Diseases of wheat in India other than rusts and smuts. *Plant Dis. Reptr*. **54**: 594-597.

Kaushal, K.K. & Dhawan, S.C. (1986). Evaluation of chemicals against *Heterodera avenae. Nat.* Conf. Plant Parasitic Nematodes of India. p. 98.

Kaushal, K.K. & Seshadri, A.R. (1986a). Evaluation of chemicals as seed dressers against *Heterodera avenae* on wheat. *Indian J. Nematol*. **16**: 140.

Kaushal, K.K. & Seshadri, A.R. (1986b). Effect of chemical treatment in soil on emergence, penetration and development of *Heterodera avenae* on wheat. *Indian J. Nematol*, **16**: 171-174.

Kerry, B.R. (1982). The decline of *Heterodera avenae* populations. *Bull. EPPO*. **12**: 491-496.

Kerry, B.R. & Andersen, S. (1983). *Nematophthora gynophila* and *Verticillium chlamydosporium*, fungal parasites of cyst nematodes, occur frequently in Swedish soils infested by cereal cyst nematodes. Vaxtaskydsnotiser **47**(5/6): 79-80.

Kerry, B.R.; Crump, D.H. & Mullen, L.A. (1982). Studies on the cereal cyst nematode, *Heterodera avenae* under continuous cereals, 1975-1978. II. Fungal parasitism of nematode females and eggs. *Ann. appl. Biol*. **100**: 489-499.

Kerry, B.R.; Simon, A. & Rovira, A.D. (1984). Observations on the introduction of *Verticillium chamydosporium* and other parasitic fungi into soil for the control of the cereal cyst nematode *Heterodera avenae. Ann. appl. Biol*. **105**: 509-516.

Khan, E.; Gaur, H.S. & Lalitha, Y. (1990). Occurrence of cereal cyst nematode, *Heterodera avenae* at high altitude in Leh, India. *Intl. Nematol. Network Newsl*. **7**(1):3.

Khurma, U.R. & Sakhuja, P.K. (1986). Parasitism of cereal cyst nematode on macaroni wheat. *Indian J. Agric. Sci*. **56**: 681.

Kondokova, E.I. & Tikhonova, L.V. (1981). The effect of parasitic fungi on the infection of wheat with *Heterodera avenae* Woll. *Byulleten Vseuyuznogo Institute Gel' mintologii im.K.1. Skryabina*. No. 31: 33-35.

Kort, J.J. (1959). Some observations on soil populations of cereal root eelworm, *Heterodera avenae* (*H. major*) under the influence of various crops in sandy soils. *T. Pl. Ziekten* **65**: 1-4.

Koshy, P.K. & Swarup, G. (1971). Distribution of *Heterodera avenae, H. zeae, H. cajani* and Anguina tritici in India. *Indian J. Nematol*. **1**: 107-111.

Kuhn, J. (1874). Uber das vor Kommen Von Rubennematoden an den wurzehdea Halmfruchte. *Landw. Jbr*. **3**: 47-50.

Kuhn. J. (1877). Vorlaufiger bericht uber die bisherigan Ergebnisse der Seit dem Johre 1875 in Aftrage des verline fur Ruberqucher-Industrie ausqegiihrten versuche zur Emitteling der ursacho der Rubenmudiqui des Bondens und zur Er Forschung der Natur der Nematoden. *Z. Ver. Ruben. Ind. Deut. Reich*. (Ohne Band) : 452-457.

Kyou, N.C. (1970). Chemical control of cereal root eelworm, *Heterodera avenae*, in field soil. *Plant Dis. Reptr*. **54**: 134-136.

Lundin, P. (1974). What is the situation with the nematodes in your field. *Weilbulls Arsbok*. 20-22.

Maqbool, M.A. (1981). Occurrence of root-knot and cyst nematodes in Pakistan. *Nematol. Medit.* **9**: 211-212.

Mathur, B.N. (1969). Studies on the cereal root eelworm (*Heterodera avenae* Woll). with special reference to Molya disease of wheat and barley in Rajasthan. Ph.D. Thesis, Univ. Rajasthan, Jaipur. p. 233.

Mathur, B.N.; Arya, H.C.; Handa, D.K. & Mathur, R.L. (1974). Effect of moisture and temperature on the survival of cyst contents. *Indian J. Mycol. Pl. Path.* **4**: 151-156.

Mathur, B.N.; Arya, H.C.; Handa, D.K. & Mathur, R.L. (1975a). On the symptoms and distribution of molya disease of wheat and barley caused by *Heterodera avenae* in the light soils of Rajasthan. *Indian J. Mycol. Pl. Path* **5**:91-94.

Mathur, B.N.; Arya, H.C.; Handa, D.K. & Mathur, R.L. (1975b). The biology of cereal cyst nematode, *Heterodera avenae* in India. II. Factors affecting the emergence of larval from cysts. *Indian J. Mycol. Pl. Path.* **5**: 149-155.

Mathur, B.N.; Arya, H.C.; Handa, D.K. & Mathur, R.L. (1981). The biology of cereal cyst nematode *Heterodera avenae* in India. III. Factors affecting population level and damage to host crops. *Indian J. Mycol. Pl. Pathol.* **11**: 5-13.

Mathur, B.N.; Handa, D.K. & Swarup, G. (1987). Effect of deep summer ploughings on the cereal cyst nematode, *Heterodera avenae* and yield of wheat in Rajasthan, India. *Indian J. Nematol.* **17**: 292-295.

Mathur, B.N.; Handa, D.K.; Swarup, G.; Sethi, C.L.; Sharma, G.L. & Yadav, B.D. (1986). On the loss estimation and chemical control of molya disease of wheat caused by *Heterodera avenae* in India. *Indian J. Nematol.* **16**: 152.

McLeod, R.W. (1976). Sources of resistance to *Heterodera avenae* Woll. In New South Wales. *Proc. Linnean Society of N.S.W.* **100**(3): 195-201.

McLeod, R.W.; Wong, P.T.W. & Southwell, R.J. (1986). Biology and control of cereal cyst nematode in Northern New South Wales. *Aust. J. Expt. Agric.* **26**: 375-381.

Meagher, J.W. (1970). Seasonal fluctuations in number of larvae of the cereal cyst nematode, *Heterodera avenae* and *Pratylenchus minyus* and *Tylenchorhynchus brevidens* in soil. *Nematologica* **16**: 333-346.

Meagher, J.W. (1972). Cereal cyst nematode (*Heterodera avenae* Woll.). Studies on ecology and control in Victoria. Tech. Bull. 24 dept. Agric. Vict. p. 50.

Meagher J.W. (1974). Cryptobiosis of the cereal cyst nematode (*Heterodera avenae*) and effects of temperature and relative humidity on survival of eggs in storage. *Nematologica* **20**: 323-335.

Meagher, J.W. (1977). World dissemination of the cereal cyst nematode (*Heterodera avenae*) and its potential as a pathogen of wheat. *J. Nematol.* **9**: 9-15.

Meagher, J.W. (1982). Yield losses caused by *Heterodera avenae* in cereal crops grown in a Mediterranean climate. *Bull EPPO* **12**: 325-331.

Meagher, J.W.; Brown, R.H. & Rovira, A.D. (1978). The effects of cereal cyst nematode (*Heterodera avenae*) and *Rhizoctonia solani* on the growth and yield of wheat. *Aust. J. Agric. Res.* **29**: 1127-1137.

Meagher, J.W. & Rooney. D.R. (1966). The effect of crop rotations in the Victorian Wimmera on the cereal cyst nematode (*Heterodea avenae*) nitrogen fertility and wheat yield. *Aust. J. Exp. Agric. Anim. Husb.* **6**: 425-431.

Mortensen. M.L.; Rostrup, S. & Kilpin Ravn, F. (1908). Oversight over Landbrugsplanteznes Sygdomme i 1907. *Todsskr. Landbr. Planteavl.* **15**: 145-158.

Mukhopadhyaya, M.C.; Dalal, M.R. & Kharub, S.S. (1971). Effects of nematicides on the cyst nematode *Heterodera avenae* and wheat yield. *Haryana Agric. Univ. J. Res.* **1**: 36-38.

Mukhopadhyaya, M.C.; Dalal, M.R. & Saran, S. & Kharub, S.S. (1972). Studies on the molya disease of wheat and barley. *Indian J. Nematol.* **2**: 11-20.

Nielsen, C.H. (1966). Onter suchwagen uber die vererbung der Resistemz gengan die Getreidenmatiden (*H. avenae*) beim weizen. *Nematologica* **12**: 575-578.

Nilsson-Ehle, H. (1920). Uber Resostenz qeqen *Heterodera schachtii* be gewssen Gersten sortenihre verez bungs welse and Bedeutung furdie Praxis. Hereditas **1**: 1-34.

O'Brien, P.C. & Fisher, J.M. (1974). Resistance within wheat, barley and oat cultivars to *Heterodera avenae* in South Australia. *Aust. J. Exp. Agric. Anim. Husb.* **14**: 399-404.

O'Brien, P.C. & Fisher, J.M. (1979). Reactions of cereals of populations of *Heterodera avenae* in South Australia. *Nematologica* **25**: 261-267.

O'Brien, P.C.; Fisher, J.M. & Rathjen, A.J. (1980). Inheritance of resistance in two wheat cultivars to an Australian population of *Heterodera avenae. Nematologica* **26**: 69-74.

Prasad, N.; Mathur, R.L. & Sehgal, S.P. (1959). Molya disease of wheat and barley in Rajasthan. *Curr. Sci.* **28**: 453.

Rajan & Swarup, G. (1985). Dormancy in cereal cyst nematicide, *Heterodera avenae. Indian J. Nematol.* **15**: 118-120.

Rao, M.S. & Dhawan, S.C. (1988). A suitable method of storage of *Heterodera avenae* cysts for use in laboratory experiments. *Indian J. Nematol.* **18**: 84-88.

Rovira. A.D. & Simon, A. (1982). Integrated control of *Heterodera avenae. Bull EPPO* **12**: 517-523.

Sukhuja, P.K. & Singh, I. (1990). Comparative efficacy of some nematicides and Naemin against cereal cyst nematode infesting wheat. *Pl. Dis. Res.* **5**: 237-238.

Sakhuja, P.K.; Singh, I. & Sharma, S.K. (1978). Effect of nitrogen alone and in combination with some nematicides on the population of cereal cyst nematode, *Heterodera avenae* and yield of wheat. *Indian J. Nematol.* **8**: 159-162.

Sakhuja, P.K.; Singh, I. & Sharma, S.K. (1981). Effect of some nematicides and fungicides on the emergence of *Heterodera avenae. Indian J. Nematol.* **11**: 219-221.

Sanchez, A.; Scristan, J.C. & Bello, A. (1985). Investigation of cereal varieties resistant to the nematode *Heterodera avenae* Woll. Anales de Edofologiay Agrobiologia **44**(1/2): 169-173.

Schmidt, O. (1930). Sind Ruben-und Hafer-Nematoden identisch. *Arch. Pb Bau* **3**: 420-464.

Seshadri, A.R. (1971). Recent research on control of plant parasitic nematodes in India. Proc. Symp. Epidemiology Forecasting & Control of plant diseases in India. *Nat. Sci. Acad*. p. 15.

Seshadri. A.R. & Sethi, C.L. (1978). Nematode problems. *In : Wheat Research in India (1966-76).* ICAR, New Delhi, pp. 168-187.

Seshadri, A.R. & Dasgupta, D.R. (1980). Nematodes. *In: Hand Book of Agriculture*. ICAR, New Delhi. pp. 551-557.

Sharma, R. & Swarup, G. (1988). Pathology of cyst nematodes. Malhotra Publishing House, New Delhi, p. 87.

Sharma, S.K.; Sakhuja, P.K. & Singh, I. (1981). Effect of mustard cake on the hatching and cyst population of *Heterodera avenae* on wheat. Indian J Nematol. **11**: 11-14.

Siddiqui, Z.A. Siddiqui, M.R. & Husain S.I. (1986). First report on the occurrence of *Heterodera avenae* Wollenweber, 1924 in Uttar Pradesh, India. *Int. Nematol. Network Newsl.* **3**: 11.

Sikora, H.A.; Koshy, P.K. & Malek, R.B. (1972). Evaluation of wheat selections for resistance to the cereal cyst nematode. *Indian J. Nematol.* **2**: 81-82.

Singh, H. (1985). Effect of crop rotation on the population of cereal cyst nematode. *Indian J. Nematol.* **15**: 263-264.

Singh, H & Yadav, B.S. (1978). Control of molya disease of wheat caused by *Heterodera avenae. Indian J. Mycol. & Pl. Path.* **8**: 73.

Singh, H. & Yadav, B.S. (1986). Effect of integrated control on wheat yield and population of *Heterodera avenae. Nat. Conf. Plant parasitic nematodes of India — Problems and Progress* p. 97.

Singh, I.; Sakhuja, P.K. & Sharma, S.K. (1987). Influence of some crucifers on *Heterodera avenae* — the cyst nematode. *Plant Dis. Res.* **2**: 93-94.

Singh, I.; Sakhuja, P.K. & Sharma, S.K. (1988). Studies on the time of application of nematicides against cyst nematode. *Heterodera avenae. Plant Dis. Res.* **3**: 144-145.

Singh, I.; Sakhuja, P.K.; Sharma, S.K. & Sharma, N.K. (1990). Distribution of cereal cyst nematode in Punjab. *Pl. Dis. Res.* **5**: 207-208.

Singh, I.; Sharma, N.K. & Chohan, J.S. (1976). Incidence of cereal cyst nematode, *Heterodera avenae* Woll. in Jammu Kashmir state. *Food Fmg. & Agric.* **8**: 20.

Soetopo, I. (1985). Genetic analysis of the resistance in barley (*Hordeum vulgare* L.) to the cereal cyst nematode (*Heterodera avenae* Woll.). Part 1. The Cultivar reaction Australian Patho. Type. *Agrivita* (185/86): 8-9.

Sparrow, D.H.B. & Dube, A.J. (1981): Breeding barley cultivars resistant to cereal cyst nematode in Australia. *Helm. Abstr.* **53**: 135.

Stapel, C. (1953). Plantesygdommens og skadedyrenes okonomiske betydning i landbrouget. *Tidsskr. Londokon; yr. 1953*: 229-244.

Sultan, M.S. & Sakhuja, P.K. (1986). Emergence of *Heterodera avenae* larvae as influence by some systemic chemicals. *Indian J. Nematol.* **16**: 264-265.

Sultan M.S. & Singh. I. (1984). Effect of nemacur on the hatching of cereal cyst nematode, *Heterodera avenae. Indian J. Nematol.* **14**: 50.

Swarup, G & Gill, J.S. (1969). Varietal response of wheat and barley to infestation by *Heterodera avenae.* All Indian Nematol. Symp. held at IARI, New Delhi, pp. 64-65.

Swarup, G. & Gill, J.S. (1972). Factors influencing emergence of larvae from cysts of *Heterodera avenae* Wollen w. 1924. *Indian J. Expl. Biol.* **10** : 219-223.

Swarup, G; Mathur, R.L.; Seshadri, A.R.; Raski, D.J. & Mathur, B.N. (1976). Response of wheat and barley to soil fumigation by DD and DBCP against molya disease caused by *Heterodera avenae. Indian J. Nematol.* **6**: 150-155.

Swarup, G.; Sethi, C.L.; Kaushal, K.K. & Siya Nand (1982). Distribution of *Heterodera avenae*, the causal organism of molya disease of wheat and barley in India. *Curr. Sci.* **51**: 896-897.

Swarup, G. & Singh, K. (1961). Molya disease of wheat and barley in Rajasthan. *Indian Phytopath.* **14**: 127-133.

Tikhonova, L.V. & Smirnov. P.I. (1969a). Economic losses due to *Heterodera avenae* on spring wheat. *Trudy uses. Inst. Gel' mint* **15**: 305-306.

Tikhonova, L.V. & Smirnov P.I (1969b). The role of crop rotation in the reduction of the oat nematode population *Heterodera avenae. Bull. uses, Inst. Gel' mint. K.I. Skryabinan No.* **2** pp. 102-103.

Tribe, H.T. & Cayrol, J.C. (1982). Mycocentrospora scerina var. *Casteinaudariensis* var. Nov. in eggs of the cyst nematode *Heterodera avenae. Trans. British Mycol. Soc.* **78**: 368-370.

Van Berkum, J.A. & Seshadri, A.R. (1970). Some important nematode problems in India. Xth Int. Nematol. Symp. Pescara, Italy. pp. 136-137.

Williams, T.D. (1969). The effects of formation, nabam, irrigation and nitrogen on *Heterodera avenae* Woll., Ophiobolus gramins Sacc. and the growth of spring wheat. *Ann. appl. Biol.* **64**: 325-334.

Williams, T.D. & Salt, G.A. (1970). The effects of soil sterilants on the cereal cyst nematode (*Heterodera avenae* Woll), take all (*Ophiobolus graminis* Sacc.) and yields of spring wheat and barley. *Ann. appl. Biol.* **66**: 329-338.

Winslow, R.D. (1955). The hatching responses of some root eelworms of the genus *Heterodera. Ann. appl.Biol.* **43**: 19-36.

Winslow R.D. (1956). Seasonal variations in the hatching responses of the potato-root eelworm, *Heterodera restochiensis* Wollenweber, and related species. *J. Helminth.* **30**: 157-164.

Wollenweber, H. (1924). Zur Kenntnis der Kartobbel-Heterideren. III. *Landw. Ztg.* No. **12**: 100-101.

Yadav, B.S. & Singh, H. (1975). Effect of certain chemicals for the control of molya disease of wheat caused by *Heterodera avenae. Indian J. Mycol. & Pl. Path.* **5**: 17.

Yadav, R.; Behl, R.K.; Singh, D. & Singh, A. (1987). Inheritance of resistance against *Heterodera avenae* in barley. *Crop Improv.* **12**: 188-190.

Chapter 5

Control of Earcockle Disease of Wheat

I.J. Paruthi.

Department of Nematology, Haryana Agricultural University, Hisar-125 004, India

WITH the green revolution ushered in mid sixties, wheat has become a commercially viable crop, apart from basic need as an important component of human food. This crop is vulnerable to a large number of pests and pathogens which cause a considerable reduction in its yield. Diseases even in a mild form can cause economical and social upheavals, especially in land where food surpluses are marginal. According to FAO publication (Stapley & Gaynes, 1969), 35 per cent of wheat crop which could theoretically be harvested, is lost as a result of pests and diseases. Amongst various disease causing agents, the nematodes play an important role and wheat seedgall nematode, *Anguina tritici* unfortunately is still considered an important contributing agent towards the total disease pest loss spectrum in developing countries like India. This nematode alone causes earcockle disease (locally known as 'gegla', 'mamni' or 'sehun') and yellow ear rot (locally known as 'tundu' or 'tanan') in association with a bacterium, *Clavibacter tritici*. Both these diseases had been reported from all wheat growing areas of the world but due to adoption of modern cleaning methods, now a days these are not the problems in developed countries. Koshy & Swarup (1971) reported widespread occurrence of *A. tritici* in India. In Haryana, the occurrence of this nematode has been reported throughout the state by Paruthi & Bhatti (1985a).

HOST RANGE

Wheat is generally considered as the main host of this nematode but some other crops like rye, emmer, spelt, barley, triticale, etc. have also been reported to be attacked by this nematode. Paruthi & Bhatti (1987 a, b) tested host differential response of this nematode on some cereal crops and found that wheat and triticale as good hosts.

ECONOMIC IMPORTANCE

This nematode causes a considerable quantitative loss in wheat yield due to non-formation of the grains in the diseased portion (Paruthi & Gupta, 1986). However, the magnitude of the loss depends upon the number of diseased earheads (Paruthi & Bhatti, 1985a). Average incidence of the disease and the losses caused due to them has been reported to be 1-3 per cent. But on individual farms, it has been reported as high as 30%.

Contamination of wheat grains with the galls changes the colour of the flour from creamy white to dull whitish depending upon the quantity of the galls present in the grain mixture and reduces the market price considerably. Unlike other species of *Anguina*, particularly *A. agrostis*, this nematode does not produce any toxins (Midha & Swarup, 1974). However, Singh (1978) has mentioned the toxic effect of galls, if consumed in large quantities.

DISEASE SYMPTOMATOLOGY

Symptomatology of earcockle and 'tundu' disease has recently been reviewed by Paruthi & Bhatti (1990), and Paruthi (1992). This nematode produces typical above ground symptoms on the plants. Initially the diseased plants show basal swelling of the stem, crinkling, curling, and twisting of leaves with increased tillering and stunted growth (Gupta & Swarup, 1968). Such symptoms on leaves are more pronounced in early stage of the plant which start disappearing at the later stage. Expression of such symptoms depends upon the level of nematode infestation.

Fig. 1 Ear of wheat i) healthy (left), ii) infected with *Anguina tritici* (right). Symptoms are also visible on leaf. (Courtesy : M.M. Alam, Aligarh, India)

Typical cockled earheads (Fig. 1) are smaller or broader than the healthy ones with or without awns on them (Gupta & Swarup, 1968). Some ears though look normal are sometimes also diseased (Gupta &

Swarup, 1968). Some ears though look normal are sometimes also diseased (Gupta, 1968; Paruthi & Bhatti, 1987a). Normally the galls are formed in place of grains, but these have also been reported to be formed on some other non-floral organs like leaves, awns, glumes, lemma and palea (Byars, 1920; Gupta & Swarup, 1968; Paruthi & Bhatti, 1985 b). The galls on such organs are however considered of minor importance because of their rare occurrence and less gall content (Christie, 1959). In addition to these symptoms, Paruthi & Gupta (1985) reported two new abnormalities in wheat ears due to *A. tritici*, i.e. sterility in wheat ears and elongation of distance between first and second spikelet.

Yellow ear rot or 'tundu' : Since this disease is caused by the association of *A. tritici* with the bacterium *Clavibacter tritici*, hence the early symptoms of this disease (like basal swelling of stem, crinkling, curling and twisting of the leaves, increased tillering and stunted growth, etc.) are similar to that of earcockle disease. At the later stage of the plant growth, bright yellow bacterial slime can be seen on the abortive ears and leaves in a humid atmosphere which becomes hard and brittle in dry weather. Such diseased culms either die in early stage or grow until heading. In the latter case, the earhead is narrower and shorter with the grains partially or completely replaced by the bacterial mass. In partially attacked earheads, some wrinkled grains of poor quality are formed (Paruthi *et al.*, 1987). In addition to these symptoms, narrow white flecks (Vasudeva & Hingorani, 1952a, b) and some temporary local lesions (Sabet, 1954) on the upper leaf surface have also been reported.

CONTROL STRATEGIES

Use of healthy and clean seeds

Since both the diseases (earcockle and yellow ear rot) are seed borne in nature, use of healthy and clean (gall free) seeds is the best and cheapest method to eradicate this nematode pest, which can be achieved by either of the following practices :

Fanning

Cleaning the seed by this practice has been recommended by Cobb (c.f. Coleman & Regan, 1918) and is commonly used by the farmers in China (Chu, 1945). This method is not effective in removing the gall from wheat. About 45 per cent of the galls remained in wheat even if the fan revolves at 850 revolutions per minute (Coleman & Regan, 1918). Chu (1945) reported that fanning resulted in complete removal of the galls only in 6 out of 357 samples examined by him.

Sieving/Screening

This method is based on the difference in size between the nematode galls and sound wheat kernels. The galls generally being smaller in size than the seeds, fall down and healthy seeds remain on the sieve. Complete removal of the galls is not achieved by this method, because some large sized compound galls are retained on the sieve along with the seeds and some small sized grains also pass through the pores of the seed alongwith the galls (Chu, 1945; Paruthi, 1982). Paruthi & Bhatti (in press) reported that sieving the infested seed through 3 mm size removed 77.5 to 91.8 per cent galls and resulted 4.8 to 13.8 per cent reduction in seed weight.

Winnowing

Winnowing of contaminated seed by 'chhaj' (winnowing basket) in an open area also separates the cockles from the seed. The removal of the galls by this method depends upon the velocity of the wind. Jat *et al.* (1986) reported that none of the fractions of the samples cleaned by winnowing basket was completely free from cockles. The cleaned fraction, though had more than 93% pure seed but they also contained 0.3%, 0.39% and 0.51% inert matter in the fractions obtained from the samples containing 0.5%, 1.0% and 2.0% cockles, respectively.

Water Floatation

Putting the contaminated seed in a drum containing water is very effective practice of removing the cockles from the grains. The galls being lighter in weight than the healthy seeds, will float on the surface of water which can easily be removed by some sieve. According to Paruthi (1982) immersion of contaminated wheat seed in water for 5-15 minutes resulted in 90.0 to 93.3 per cent removal of the galls. Chu (1945) reported that the galls floated on the surface of water for about 5 minutes. Consequently, this method is only effective if the operator can remove the galls right after the contaminated seed is poured into the water. It is very effective recommended practice in Indian conditions. However, it is not generally advocated in China due to the problem of drying the moist grains particularly in Southern China where wheat crop occurs during autumn rainy season and non-availability of sufficient amount of water required for this purpose in mountainous districts (Chu, 1945).

Salt Sedimentation

This method was used for the first time to remove the cockles by Byars (1920). Various workers have used different concentrations of salt ranging from 10-30 per cent (Singh *et al.*, 1953; Jain *et al.*, 1956; Paruthi, 1982; Jat *et al.*, 1986). Addition of salt increases the specific gravity of water in which the galls remain floating for longer period. This method, though more effective than water floatation (Paruthi & Bhatti, 1983), but is somewhat more cumbersome because the seeds treated with salt are required to be washed several times before sowing to avoid the injurious effect of salt on germination of seed (Jain *et al.*, 1956).

SEED TREATMENT

Hot Water Treatment

Putting the contaminated wheat seeds in hot water at 54-56°C for 10-12 minutes (Marcinowski, 1910), 122°F for 30 minutes (Limber, 1938) and 30°C for 4 h has been found effective in destroying the larvae of *A. tritici* in seed galls. Byars (1919) and Chu (1945) advocated pre-soaking of galls in cold water before putting them in hot water. The larvae in pre-soaked galls were destroyed in less time than dry galls when subjected to hot water treatment (Chu, 1945).

Fumigation

Romascu (1972) tested some fumigants against *A. tritici* on stored wheat seeds and found that methyl bromide applied at the rate of 40-80 g/m^2 seed for 24 hours caused higher mortalities of *A. tritici* than the other treatments.

PLANT RESISTANCE

Use of resistant variety is another economical method of controlling this nematode. A limited success achieved in this direction is summarised in Table 1.

Table 1: Resistance in wheat against *Anguina tritici*.

Resistant lines/varieties	Country	Reference
Kanred	U.S.A.	Leukel (1929)
Novokrymka and Co-operatorka	USSR	Meyer (1939)
PC171, PC633, PC690, PC876 Hsuchow resistant No.2	China	Chu (1945)
Conche, Funone, Bison, Dica, Marquis Megurele	Rumania	Romascu (1972)
Acodde	Yugoslavia	Tesic (1969)
India 66, Saberbeg, Genotypes 342, 346, 347 and 352 derived from Saberbeg x Mesipak irradiated with 400 r Nf in the F_2 generation	Iraq	Saleh & Fattah (1989, 1990)
Nainari, Pictic-2, V5009, V5040, Zamindar and local white	Pakistan	Amin (1970); Shahina *et al*. (1989)
E650, HP4530, Karchia 65, Moti, N1747-9, NP888, CPAN 1676, HD2177, Raj 1774	India	Singh & Khanna (1974); Alam (1983); Verma & Yadav (1986)

CULTURAL PRACTICES

Sowing of non-host crop for about 1-2 years in the infested areas has been reported to reduce the nematode infestation to a great extent (Carne, 1926; Leukel, 1957). In tropical and sub-tropical countries like India and China, where the temperature in summer goes very high, the nematodes left in the soil do not seem to play an important role in disease causation in the next cropping season (Chu, 1945; Paruthi, 1982).

Sukul (1970) found decreased incidence of disease caused by *A. tritici* when wheat was grown alongwith *Polypogon hydropipea*. Marcinowski (1910) advised catch crop method for removal of *A. tritici* in wheat. Singh *et al.* (1953) advocated removal and burning of affected plants.

ORGANIC AMENDMENTS

Aqueous extracts of *Anagralis arvensis* (Nene & Thapliyal, 1966), *Erigeron linifolius* (Nene & Kumar, 1967), *Azadirachta indica* (Qamar *et al.*, 1989) have been reported to be effective against *A. tritici*. Nematicidal activity of oils of 3 *Cymbopogon* grasses against *A. tritici* has been reported by Sangwan *et al.* (1985). Kumari *et al.* (1986) reported that methanolic extracts of leaves, stems and buds of *Datura stramonium, Ipomoea carnea, Tagetes patula* and Lawsonia alba caused 75-100 per cent mortality of *A. tritici* at a concentration of 1:5. Malik *et al.* (1987b) reported nematicidal activity of essential oils, its isolates and extracts of root and stem of *Xanthium strumarium* against this nematode. Malik *et al.* (1987a) reported the effectiveness of citral, graniol, methanol, cyclohexanol, their carbamates, acetates, butanoates, benzoates, cyclohexanone, against *A. tritici, Meloidogyne javanica* and *Tylenchulus semipenetrans* under *in vitro* conditions. Malik *et al.* (1989) tested the nematicidal activity of various substituted phenols, phenoxy acetic acid esters and hydrazides against *A. tritici, M. javanica* and *Heterodera cajani*. Phenols with electron donating substituents in general and chlorosubstituents in particular and corresponding substituted phenoxy acetic acid esters were found to show good nematicidal activity while hydrazides were largely inactive. Sangwan *et al.* (1990) reported the toxic effect of essential oils of *Syzygium aromaticum* and its major constituents eugenol, linalol, and geranol. Gokte & Swarup (1988b) observed a significant reduction in larval penetration of *A. tritici* when wheat seeds were treated with neem powder @ 600-800 mg/100 g seed.

BIOLOGICAL CONTROL

Very little is known about biological control of *A. tritici*. Dhawan *et al.* (1979) observed a nematophagous fungus, *Arthrobotrys oligospora* parasitising the larvae of *A. tritici* under laboratory conditions. Gokte & Swarup (1985, 1988a) reported some larvicidal bacterial isolates from the nematode galls. Gokte & Swarup (1988b) tested the performance of such five bacterial isolates individually and in combination against *A. tritici* under pot conditions. All the isolates caused reduction in per cent penetration of the larvae. However, better results were obtained by *B. subtilis* and *B. pumilus*. Akhtar & Aqil (1986) reported that application of *Arachnitus* sp. culture @ 2 kg/acre to soil significantly reduced earcockle disease in wheat. Galper *et al.* (1991) reported that the culture filtrate of the fungus *Cunninghamella elegans* showed some response against *A. tritici*. Bilgrami & Jairajpuri (1989) reported that second stage *A. tritici* larvae were most susceptible to predation by *Mononchoides longicaudatus* and *M. fortidens*.

CHEMICAL CONTROL

Very little information is available on the chemical control of *A. tritici*. Srivastava & Katiyar (1957) reported that three applications of diazinone @ 0.15% reduced the infection of *A. tritici* from 30.6 to 2.48%. Singh & Prasad (1972) reported that spray of ethyl parathion @ 0.05% was found to be the most effective; and completely eliminated the galls of *A. tritici*.

CONCLUSION

Anguina tritici, though the rarest in advanced countries, is still an important troublesome pest of wheat in underdeveloped and developing countries like India. This nematode alone is responsible for earcockle disease but in association with a bacterium (*Clevibacter tritici*) causes yellow ear rot or 'tundu' disease. In both the diseases, there is no grain formation in the diseased portion which causes a heavy loss in the yield. Although the average incidence of these diseases is about 1-3% but on individual farms their occurrence may be as high as 30% depending upon the level of seed contamination. The diseases being seed borne in nature can easily be managed by the use of gall-free healthy seeds as has been done in advanced countries. Although the galls can easily be distinguished from the healthy grains by their colour and appearance yet their occurrence is normally ignored by the farmers because of very low level of incidence in the grains. There is a great responsibility on the part of agriculture field staff and extension agencies to educate the farmers to use healthy (gall free) seeds either obtained from some seed certification agencies or cleaning it (by water or salt sedimentation or sieving, etc.) before sowing.

Once the diseases have appeared in the field it becomes difficult to manage them. However, there is a great need to pursue research in evolving suitable control technologies as a curative measure for this troublesome pest.

REFERENCES

Akhtar, C.M. & Aqil, T. (1986). Biology and control of earcockle of wheat. *J. Agric. Res.* **24**(4): 237.

Alam, M.M. (1983). Reaction of certain varieties of wheat to the seed galls nematode. *Anguina tritici. Indian J. Plant Pathol.* **1**(2): 210-211.

Amin, M. (1970). Relative resistance of 25 varieties of wheat against earcockle of wheat (*Anguina tritici*). *In* : Pakistan Science Conference (21st-22nd) Rajshahi, for Part III, Proceedings, Lahore, Pakistan : Association for Advancement of Science : A 80-A 82.

Bilgrami, A.L. & Jairajpuri, M.S. (1989). Resistance of prey to predation and strike rate of the predators *Mononchoides longicaudatus* and *M. fortidens* (Nematoda : Diplogasterida). *Revue de Nematologie* **12**(1): 45-49.

Byars, L.P. (1919). The eelworm disease of wheat and its control. *Frms. Bull. 1041, U.S. Dep. Agric.*

Byars, L.P. (1920). The nematode disease of wheat caused by *Tylenchus tritici*. *Frms, Bull. 842, U.S. Dep. Agric.*

Carne, W.M. (1926) Earcockle (*Tylenchus tritici*) and a bacterial disease (*Pseudomonas tritici)* of wheat. *J. Dep. Agric. West Aust. (Ser. 2)*: 508-513.

Christie, J.R. (1959). Plant Nematodes, their bionomics and control. *Florida Agric. Exp.* St. p. 256.

Chu, V.M. (1945). The prevalence of wheat nematode in China and its control. *Phytopathology* **35**: 288-295.

Coleman, D.A. & Regan, S.A. (1918). Nematode galls as a factor in the marketing and milling of wheat. *Frms. Bull. 734. U.S. Dept. Agric.*

Dhawan, S.C. Jain, R.K. & Swarup, G. (1979). A nematode trapping fungus of *Anguina tritici* larvae. *Indian J. Nematol.* **9**: 180-183.

Galper, S.; Cohn, E.; Spiegel, Y. & Chet, I. (1991). A collagenolytic fungus *Chunninghamella elegans* for biological control of plant-parasitic nematodes. *J. Nematol.* **23**(3): 269-274.

Gokte, N. & Swarup, G. (1985). Bacteria as potential biocides against *Anguina tritici*. *Indian J. Nematol.* **15**: 266-267.

Gokte, N. & Swarup, G. (1988a). On the association of the bacteria with the larvae and galls of *Anguina tritici. Indian J. Nematol.* **18**: 313-318.

Gokte, N. & Swarup, G. (1988b). Efficacy of some bacterial isolates as seed treatment on penetration of *Anguina tritici* larvae to the growing point of wheat seedlings and viability of the larvae in the galls formed. *Indian J. Nematol.* **18**: 368-369.

Gupta, P. & Swarup, G. (1968). On the earcockle and yellow ear rot diseases of wheat. I. Symptoms and histopathology. *Indian Phytopath.* **21**: 318-323.

Jain, J.S.; Sinha, R.P. & Mathur, R.S. (1956). Controlling earcockle of wheat by floating seed in salt solution. *Agric. Anim. Husb. India* **5**: 3.

Jat, R.R.; Pathak, V.N. & Verma, O.P. (1986). Removal of earcockle from contaminated wheat seeds. *Seed Sci. Technol.* **14**: 237-243.

Koshy, P.K. & Swarup, G. (1971). Distribution of *Heterodera avenae, H. zeae, H. cajani* and *Anguina tritici* in India. *Indian J. Nematol.* **1**: 106-111.

Kumari, R.; Verma, K.K.; Dhandsa, K.S. & Bhatti, D.S. (1986). *Datura, Ipomoea, Tagetes,* and *Lawsonia* as control of *Tylenchulus semipenetrans* and *Anguina tritici. Indian J. Nematol.* **16**: 236-240.

Leukel, R.W. (1929). The nematode disease of wheat and rye. *U.S. Dep. Agric. Bull. Frms. 1607* p. 12.

Leukel, R.W. (1957). Nematode disease of wheat and rye. *U.S. Dep. Agric. Bull. 1607 (Revised)*. p. 16.

Limber, D.P. (1938). Notes on hot water treatment of *Anguina tritici* galls on wheat and comparison of an Indian and Chinese collection by use of weight criteria. *Proc. Helminth. Soc. Wash.* **5**: 20-23.

Malik, M.S.; Pal, V.; Sangwan, N.K.; Dhindsa, K.S.; Verma, K.K. & Bhatti, D.S. (1989). Nematicidal efficacy of substituted phenols, phenoxy acetic acid esters and hydrazides : a structure activity relationship study. *Nematologica* **35**(3): 366-370.

Malik, M.S.; Sangwan, N.K.; Dhindsa, K.S. & Bhatti, D.S. (1987a). Nematicidal activity of extracts of *Xanthium strumarium. Pesticides* **21**(10): 19-20.

Malik, M.S.; Sangwan, N.K.; Dhindsa, K.S.; Verma, K.K. & Bhatti, D.S. (1987b). Nematicidal efficacy of some monoterpenes and related derivatives *Pesticides* **21**: 30-32.

Marcinowski, K. (1910). Parasitisch and semiparasitisch an Pflanzen lebende *Nematoden. Arb. K. Biol. Anstalt Landu. Forstw.* **7**: 1-192.

Mayer, A.A. (1939). The wheat nematode, *Anguillulina tritici* in the Crimea. Leningrad Academy of Agricultural Sciences. Institute for Plant Protection. *Collected Works on Nematode of Agricultural Crops* (Ed.—E.S. Kiryanova). Moscow and Leningrad. pp. 160-167.

Midha, S.K. & Swarup, G. (1974). Studies on the wheat seed gall nematode caused by *Anguina tritici. Indian J. Nematol.* **4**: 53-63.

Nene, Y.L. & Kumar, K. (1967). Nematostatic properties of *Erigeron linifolius* extracts. *Indian Phytopath*. **20**: 172-174.

Nene, Y.L. & Thapliyal, P.L. (1966). Nematicidal properties of *Anagralis arvensis* L. extracts. *Indian Phytopath*. **19**: 26-29.

Paruthi, I.J. (1982). Studies on *Anguina tritici* in wheat and barley. Ph. D. Thesis, Haryana Agricultural University, Hisar, India.

Paruthi, I.J. (1992). Nematode pests of wheat and barley *Anguina tritici. In: Nematode Pests of Crops* (Eds.—D.S. Bhatti & R.K. Walia) C.B.S. Publishers, New Delhi : pp. 14-26.

Paruthi, I.J. & Bhatti, D.S. (1983). Studies on *Anguina tritici* in wheat and barley. *III All India Nematol. Symp.* Solan, India. p. 50.

Paruthi, I.J. & Bhatti, D.S. (1985a). Estimation of loss in yield and incidence of *Anguina tritici* on wheat in Haryana. *Int. Nematol. Network. Newsl.* **2**(3): 13-16.

Paruthi, I.J. & Bhatti, D.S. (1985b). Comparative morphology and contents of galls caused by *Anguina tritici* : Nematoda (Anguinidae) on various parts of wheat plants. *Nematologica* **33**: 243-245.

Paruthi, I.J. & Bhatti, D.S. (1987a). Host differential responses of some cereals to *Anguina tritici. Int. Nematol. Network Newsl.* **4**(3): 8-10.

Paruthi, I.J. & Bhatti, D.S. (1987b). Variation in earcockle symptoms on wheat ears caused by *Anguina tritici. Nematologica* **33**: 360-365.

Paruthi, I.J. & Bhatti, D.S. (1990). Wheat seed gall nematode, *Anguina tritici* — a review. *Curr. Nematol.* **1**(2): 107-136.

Paruthi, I.J. & Gupta, D.C. (1985). Two unrecorded abnormalities in wheat ears infested with *Anguina tritici. Nematologica* **32**: 357-360.

Paruthi, I.J. & Gupta, D.C. (1986). Control of earcockle and yellow ear rot diseases of wheat. *Indian Fmg.* **30**(4): 32-35.

Paruthi, I.J.; Singh, M. & Gupta, D.C. (1987). Quantitative and qualitative losses in wheat (*Triticum aestivum*) grains due to earcockle and 'tundu'. *Seed Res.* **15**: 83-76.

Qamar, F.; Saeed, M.; Khan, A. & Seema, N. (1989). Control of anguinosis of wheat using plant extracts. *Pakistan J. Sci. and Indus. Res.* **32**(4): 260-261.

Romascu, E. (1972). Contribution to the study of chemical control of *Anguina tritici* Steinb. by seed dressings. *Anable Institutili de Cercolari Pentru Protectia Plantelor* **9**: 253-259.

Sabet, K.A. (1954). Pathological relationship between host and parasite in yellow slime disease of wheat. *Bull. Fac. Agri. Cairo. Univ.* **43**: 10.

Saleh, H.M. & Fattah, F.A. (1989). Reaction of wheat genotypes to infection by *Anguina tritici. Revue de Nematologie* **12**(4): 429-431.

Saleh, H.M. & Fattah, F.A. (1990). Studies on wheat seed gall nematode. *Nematol. Medit.* **18**: 59-62.

Sangwan, N.K.; Varma, K.K. & Dhindsa, K.S. (1990). Nematicidal activity of some essential plant oils. *Pesticide Science* **28**(3): 331-335.

Sangwan, N.K.; Varma, K.K.; Malik, M.S. & Dhindsa, K.S. (1985). Nematicidal activity of essential oils of Cymbopogon grasses. *Nematologica* **31**: 93-99.

Shahina, F.; Maqbool, M.A. & Ghaffar, A. (1989). Screening of some wheat varieties for resistance against *Anguina tritici*. *Pak. J. Nematol.* **7**: 61-65.

Singh, B.; Singh, J. & Mathur, S.C. (1953). Earcockle or sehun disease of wheat. *Agric. Anim. Husb. India* **3**(10): 7-9.

Singh, D.V. & Khanna, B.M. (1974). Resistance of wheat varieties to earcockle disease (*Anguina tritici* Steinb.) Filipjev. *Indian J. Farm. Sci.* **2**: 102.

Singh, I. & Prasad, S.K. (1972). Effects of some pesticides and growth regulators on *Anguina tritici* in wheat. *Indian J. Ent.* **34**: 77-79.

Singh, R.S. (1978). *Plant Diseases* (4th edn.) Oxford & IBH Publishing Co., New Delhi, India. p. 564.

Srivastava, A.S. & Katiyar, K.P. (1957). Field scale trials on effect of diazinon, salicyalic acid and benzoic acid on wheat nematode. *Proc. 44th Indian Sci. Congr. Part III.* p. 301.

Stapley, J.S. & Gaynes, F.C.H. (1969). World crop production Vol. I. London, ILIFFE Books Ltd. p. 270.

Sukul, N.C. (1970). Inhibition of nematode infestation of wheat seedlings by *Polygonium hydropiper. Nature* **226**: 771-772.

Tesic, T. (1969). A contribution to the investigation of the resistance of wheat varieties to wheat nematode. *Savremen Polijofr Novi sad* **17**(11/12): 541-543.

Varma, M.K. & Yadav, B.S. (1986). Studies on *Anguina tritici*. III. Screening of *I. aestivum* wheat cultivars to earcockle disease caused by *Anguina tritici. Nat. Conf. Plant Parasitic Nematodes of India : Problem and Progress* (December 17-20, 1986). p. 106.

Vasudeva, R.S. & Hingorani, M.K. (1952a). 'Tanan' or 'tundu' disease of wheat. *Indian Fmg.* **2**(3): 14-19.

Vasudeva, R.S. & Hingorani. M.K. (1952b). Bacterial disease of wheat caused by *Corynebacterium tritici* (Hutchinson) Bergeyal. *Phytopathology* **42**: 191-193.

Chapter 6

Control of Phytophagous Nematodes Attacking Potato

K.S. Krishna Prasad
Central Potato Research Station, Muthorai-643 004, Nilgiris, India

A native of Andes mountains in South America, potato (*Solanum tuberosum*) was used as food by the native Mayas, Incas and Red Indians. It was brought to Europe by the Spaniards in the later half of the 16th century and became popular in entire Europe. It was introduced into India in about 1615 and its cultivation seems to have picked up after another 150 years. Basically, the crop grows in the cooler regions of the world and is more suited to the temperate climatic conditions. However, the crop adapts to the warmer climatic conditions and is presently grown in about 130 countries with an annual production of about 310 million metric tonnes occupying fourth place in importance after rice, wheat and maize. The crop has a great potential for exploitation to meet the food energy for the 21st century, particularly in the developing countries as it can provide nutritionally balanced diet. Our country, best among the developing ones, grows potato now in about 0.97 million hectares, standing in fourth place in potato area after CIS countries, China and Poland. Its production in India is about 15.8 million metric tonnes and stands fifth in position after U.S.A.

The estimated world potato production should have been about 480 million tonnes, but for the diseases, pests, nematode parasites, weeds and post-harvest handling, which account for 35% loss (Anon., 1975). The disease and nematode parasites account for a major share, i.e. about 25 per cent amounting to 120 million tonnes, both the groups sharing equally. The nematode parasites are known to infect potatoes for more than a century and now about 140 species are reported to be associated with this crop throughout the world. In India, Thirumalachar (1951) recorded the root-knot nematode infestation for the first time on potato tubers from Simla hills. This was exactly 50 years after Barber had recorded the root-knot nematode infestation on tea in the country. Later, the report of occurrence of potato cyst nematodes from the Nilgiri hills by Jones in 1961 boosted the nematological research activities throughout the country not only in potato but also in other crops. The dawn of this decade saw nearly 90 different species of nematodes belonging to 38 genera associated with potato culture in the country. Among these, the potato cyst nematodes and the root-knot nematodes are the major nematode problems of the crop (Gill, 1974) in India and around the world which have been discussed in detail.

CROP LOSSES

The damages caused by parasitic nematodes have been recognised as major obstacles in potato production particularly in developing countries (Sasser & Freckman, 1987). In U.S.A. alone an estimated 2 million tonnes are lost annually due to nematodes which amount to about ten per cent of total potato produced in that country. In United Kingdom it has been estimated that the potato cyst nematode alone accounted for a loss of 20 million pounds (Hague, 1979). In South Africa potato losses due to root-knot nematodes was estimated to be $ 1250/ha during 1960's while in South Carolina the losses amounted to $ 2500 (Koen, 1966; Sitterly & Fassuliotis, 1965). Though there are no authentic records of crop loss estimates in India one can visualise the extent of losses due to nematodes, since the crop is growing in diversified agroclimatic zones which is congenial for the nematodes also. The root-knot nematodes alone may cause losses ranging between apparently negligible (Sanwal, 1979) to heavy as have been observed in Mahasu area of Himachal Pradesh (CPRI, 1963) and Hassan in Karnataka where the farmers could not harvest even as much seed as planted by them (Gill & Krishnananda, 1977). An initial level of 2 root-knot larvae per gram of soil resulted in 100 per cent tuber infection with an overall yield reduction of 42.5 per cent (CPRI, 1978) indicating more of qualitative yield reduction. Similarly, the losses due to potato cyst nematodes also depended on the initial inoculum levels (Seshadri, 1978), wherein up to 200 per cent increased potato yields were obtained in pesticidal treated plots decreasing the nematode populations by 75 per cent. The detailed studies conducted during 1984 to 1985 at CPRS, Oatacamund have shown that the avoidable yield loss in susceptible potato due to cyst nematodes was 99.5 to 99.8 per cent in summer and autumn crops.

ROOT-KNOT NEMATODES

The root-knot nematodes, causing typical root galls are the most widely distributed parasitic nematodes of crop plants. These are prevalent in most parts of the world, especially in the sub-temperate, sub-tropical and tropical regions affecting almost all agricultural crops including potatoes. Its infection on potato was recorded in 1889 by Neal from Florida, U.S.A. At that time it was designated as *Anguillula arenaria*. This nematode was also termed as *Heterodera radicicola* till 1932, when all root-knot causing nematodes were grouped as *H. marioni*. Later, the genus *Meloidogyne* was established by Chitwood in 1949 by describing four most common and widely distributed *M. incognita, M. javanica, M. hapla* and *M. arenaria*. Now about 65 species of *Meloidogyne* have been described throughout the world of which 10 species are reported on potato. Among these the infection of *M. incognita* and *M. javanica* on potato is more damaging as they are able to infect potato tubers in addition to roots. This infection causes warty out-growths, which are typical in potato (Fig. 1) that decreases the marketable value of the produce in addition to quantitative losses.

DISTRIBUTION

The root-knot nematodes have been recorded from all the potato growing countries of the world and are the most important pests of commercial crops next only to potato cyst nematode. The most important species on a world basis is *M. incognita*, followed by *M. javanica* and *M. arenaria* distributed in the warmer climates while *M. hapla* occurs in cooler zones. In India, Thirumalachar (1951) observed scab like warts on potato tubers from Simla during 1950 for the first time and since then it has been recorded from all the potato growing regions of the country. The dominant root-knot species affecting potato both in hills and plains is *M. incognita* while *M. javanica* infestation is confined to mid hills and plains. The infestation of *M. hapla* on potato roots was later recorded from Simla, Kufri, hilly regions of Kashmir and the Nilgiri hills in South India.

Detailed surveys undertaken in about 225 villages of eleven districts of Himachal Pradesh and eight hilly districts of Uttar Pradesh indicated that root-knot nematodes were prevalent on potato in a total of 153 villages. The determination of species by perennial pattern of root-knot females indicated that *Meloidogyne incognita* occurred in 116 villages either alone or in combination with other species. *M. javanica* was least prevalent occurring only in 10 villages of Himachal Pradesh situated at 1100 metres and below where potatoes were generally grown during October-February-March. The seeds were obtained from Daulakuan in Himachal Pradesh and Jallandhar in Punjab where *M. javanica* has been reported earlier. The infestation of *M. hapla* was observed in 59 villages situated at 2150 to 2650 metres above mean sea level and its prevalence could not be determined by routine root observations and soil simple processing but required growing of tomato as an indicator plant in the laboratory. The studies also revealed that these nematodes were prevalent at higher intensities wherever brown rot disease was a major problem indicating a synergistic association between the two organisms.

BIOLOGY AND ECOLOGY

Generally the aerial parts of potato do not express symptoms of root-knot nematode disease, however, heavily infected plants are stunted with yellow leaves. The typical galls on potato roots are small and often go unnoticed but wart like struc-

Fig. 1. Root-knot nematode infection on potato tubers.

tures formed on tubers are very prominent. The female nematodes which are sedentary in nature deposit about 300 to 400 eggs into a gelatinous matrix which is usually found adhering to the root galls. (Fig.2) The eggs hatch readily and at the time of tuber formation the larvae usually enter tubers since the root system would have started decaying. Healthy looking tubers from nematode sick plots also developed typical root-knot symptoms possibly because the larvae would have entered the tubers before suberisation at harvest and developed during storage. At Simla, hilly region in North Western India, *M. incognita* completed its life cycle in 25-30 days during summer (April-September) while in winter (October-March) it took about 65 to 100 days. This has been mainly due to the summer temperatures which were congenial for nematode development, i.e./11 to 19°C (minimum) and 17 to 29°C (maximum). At Ranchi another hilly region in Eastern India it took 28-35 days in rainy season (June-July) and 50-56 days in winter (November-March) months to complete one generation. However, in plains of Bihar the life cycle was completed in 35 to 45 days during November to February.

In the hilly regions two generations of nematode are completed by the time tuber development takes place and hence tuber infestation is invariably observed even in low infested fields. Whereas in plains, tuber infestation could be low mainly because the crop duration is short and the fresh roots are preferred by the newly emerging larvae. Absence of tuber infestation generally leads to the conclusion that the root-knot nematodes may be absent from a locality although they may be present on the roots. Further, hot summers (up to 42-43°C) in North Indian plains reduce the initial soil populations.

The eggs and larvae survived for more than 100 days even in the absence of hosts during summer months in Simla hills. This could be the reason for higher initial inoculum which could build-up for subsequent tuber infestations. The root-knot nematode infection has been found to predispose potato plants for infection by other organisms and it is suspected that these nematodes may be helping in the spread of brown rot disease in hilly areas of Himachal Pradesh and Uttar Pradesh and also in Peninsular India particularly the plateau regions of Karnataka.

CONTROL STRATEGIES

Complete eradication of the root-knot nematode from an infested area is not possible as it has a very wide host range. Further, the nematode in having an

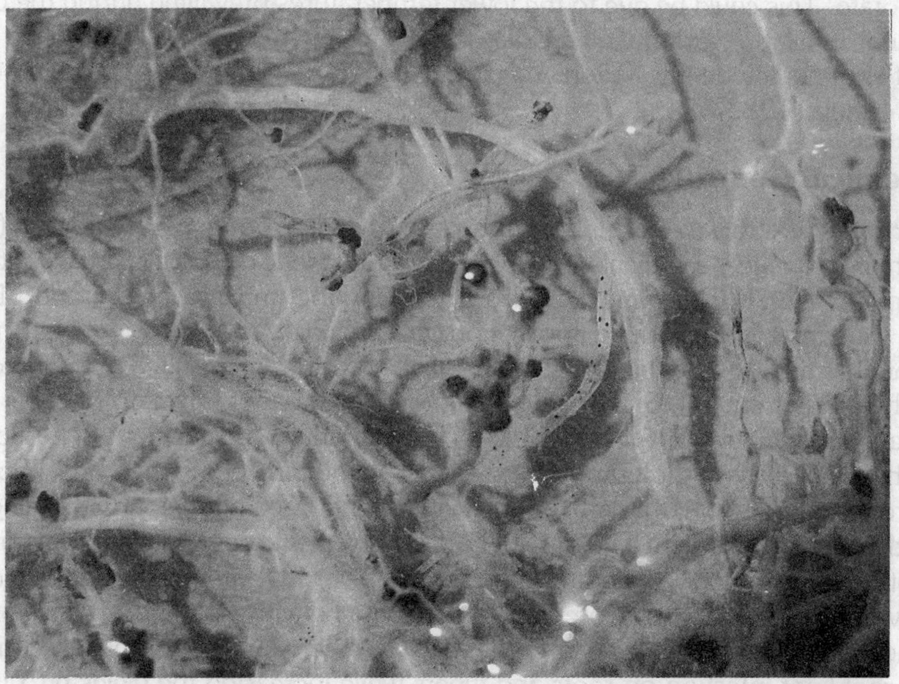

Fig. 2. Egg masses of root-knot nematode adhering to potato roots.

endoparasitic life cycle makes it very difficult to attain maximum control as any of the measures used to eradicate it should not affect the host. The following management strategies are adopted to cope up with the root-knot nematode problem of potato under Indian conditions.

Cultural Practices

Potato rotated with non-hosts such as maize, wheat coupled with summer fallow after two or three deep ploughings in north western plains reduced root-knot nematode infection. Similarly in north western hills crop rotation of potato with French beans, leafy vegetables, maize and wheat helped in minimising the nematode damage and increasing the productivity of the crop (Raj & Nirula, 1969). Growing trap crop such as French marigold, *Targets* patula in alternative rows with potato reduced larval populations in the soil, number of galls and tuber infestation while the yields increased up to 123% over control. Similarly, growing *Cassia* species and incorporation of saw dust at 26 q/ha also helped in reducing tuber infestations (CPRI, 1972, 1982).

At Jalandhar early planting of spring crop in first week of January at late planting of autumn crop in 2nd and 3rd week of October reduced the root-knot infection in potato. This could be due to the lower temperatures prevailing during the crop period. Similarly, early planted potatoes in 3rd or 4th week of March had less root-knot nematode infestation and yielded better than late planted crop in Simla hills (Ahuja, 1983).

Chemical Control

During 1960-65 several halogenated hydrocarbons such as DD (1-3 Dichloropropane + 1-2 dichloropropene), EDB (Ethylene dibromide), MBr (Methyle bromide), DBCP (Dibromochloropropane), Dorlone (mixture of DD and EDB), etc. were tested. Preplanting application of DD @ 200 l/ha, EDB @ 90 kg/ ha at Jalandhar and 180 kg/ha at Simla and DBCP at 20 l/ha resulted in an increase in the yields from 10 to 105% with little or no root-knot infection of roots and tubers (Raj & Nirula, 1969; CPRI 1976). Later, granular formulations of pesticides such as aldicarb, carbofuran, ethoprop, fensulfothion, nemaphos and phorate were tested at 1 to 10 kg a.i./ha individually and in combination with halogenated hydrocarbons (CRPI, 1971, 1972). These pesticides reduced infection of roots and tubers but were not as efficient as DD or EDB in increasing the yields. However, combination of preplant soil treatment with DD and application of aldicarb, nemaphos, phorate or DBCP 50 days after planting completely controlled root-knot infection on roots and tubers while increasing the yields by 116 to 140% at Simla (CPRI, 1973).

The studies conducted at Simla, Deesa and Ranchi under All India Coordinated Potato Improvement Project have indicated that application of carbofuran @ 3 kg a.i/ha or aldicarb @ 2 kg a.i./ha or ethoprop @ 10 kg a.i./ha increased the yields by 18 to 23% while accounting for 80% reduction in tuber infestation. However, the efficiency of these pesticides increased when applied in equal split doses once at planting and once at earthing time (Krishna Prasad, 1986).

Breeding for Resistance

Work on breeding potato varieties resistant to *M. incognita* was initiated at CPRI Simla in 1961. Out of 101 hybrid cultures tested in a nematode infested field one line 'HC-294' a cross between 'Kufri Red x (Gladstone x Taborky)' was found to be resistant to root-knot nematode (Khanna, 1966). Further tests confirmed that only few larvae invaded the roots of this selection and the giant cells were not well developed. The resistance to root-knot nematode in this variety was inferred to be polygenic in nature. Later, another selection 'HC 115' from a cross of 'HB 829 x Kufri Red' was also reported to be resistant to *M. incognita*. Both these lines were tolerant to heat, drought and frost. However, crosses involving 'HC-294' did not possess any resistance to root-knot nematode (CPRI, 1978, 1979) indicating that the resistance was controlled by recessive genes (Krishna Prasad, 1984).

Further, screening of accessions of tuber bearing wild *Solanum* species indicated that a high degree of resistance was available in *S. spegazinii* and *S. vernei* which were used as parental material for producing several crosses. During testing of commercial varieties and advanced cultures low infestation was observed in several lines of *S. tuberosum andigena* (Deshraj & Gill, 1983). The post-inoculation development and the reproduction of root-knot nematode was hindered to a great extent in several advanced cultures such as 'QB/D-64-12' 'Kufri Dewa', and 'QB-B-92-4' as compared to its development in 'Kufri Jyoti' (CPRI, 1985). The availability of resistant material in various tuber bearing *Solanum* species indicate that a thorough breeding and hybridization programme could be envisaged in the near future since at least three races of *M. incognita* are prevalent in India.

Ingegrated Approach

By practice, it has been observed that a single method of nematode control is uneconomical and it has been realised that proper blending of one or more method has always been economical and effective in achieving better nematode management. This sort of approach should be aimed especially in potato particularly under Indian conditions since we have to keep in mind the various agro-ecological factors into account. Adopting any single method of nematode control is bound to effect the ecological balance and hence various factors have to be carefully considered before advocating nematode management practices. However, by judicious application of above methods it is not difficult to manage root-knot nematode in potato for achieving higher production.

CYST NEMATODES

The potato cyst nematode, is considered as one of the major pests of potato and is popularly called as the golden cyst nematode. An average loss of about nine per cent of global potato is accounted to these nematodes amounting to about 45 million tonnes (Krishna Prasad, 1988). The ability of this nematode to build-up to damageable levels in the short span of 5-6 years, substantial yield reductions in potato, lack of inexpensive nematicides for soil treatment capable of providing adequate level of control under field conditions, the relative ease with which the cysts are dispersed with soil adhering to the seed tubers and the long persistence of eggs within the cysts in the absence of the host makes this nematode as probably the most important pest of many cultivated crops.

This nematode was first noticed by Julius Kuhn in 1881 at Rostoch in Germany which thought to be a strain of sugarbeet cyst nematode, *Heterodera schachtii*. Differences between these two nematodes were later observed and it was recognised as an independent species, *H. rostochiensis*. Further developments in breeding potato for resistance to these nematodes, separation of round cyst nematodes as genus *Globodera* and detailed studies on chromogenesis and morphology of potato cyst nematode variants, made it clear that there are two species (Stone, 1973). The populations with white or cream coloured females were designated as *G. pallida* while those with golden yellow coloured females being retained as *G. rostochiensis*. Although the females of these two species could be differentiated initially by their colour at developing stage, it is difficult to separate them once they become brown cysts at which stage cannot be easily distinguished by other *Globodera* species (Mulvey & Stone, 1976).

Most of the countries thought to be free of potato cyst nematode have regulations governing the import of produce that might carry soil containing cysts. Several countries such as U.S.A., Canada and India have internal quarantine regulations to prevent further spread of the nematode. The classical example of regulations followed was in U.S.A. Regular surveys were being conducted in all potato growing regions of the country to locate the infested areas. The regulating measures strictly followed were avoiding growing of potatoes and imposing restrictions on the sale and transportation of top soil or potato products from infested areas. By 1968, about 4,700 hectares of infested agricultural land in Nassau country in Long Island was permanently converted for real estate developments to avoid spread of this nematode. Such strict regulatory measures have reduced the distribution of the pest in other potato growing regions of U.S.A.

DISTRIBUTION

Andes mountains of South America seems to be the original home for the cyst nematodes that might have originated alongwith their host potato (Stone, 1985). The cysts must have got introduced into Europe with the breeding material brought for late blight disease resistance in 1850's and later spread throughout the world with the improved potatoes developed in Europe at that time. So, Europe has been considered as the secondary distribution centre (Evans & Stone, 1977) and at present these nematodes are prevalent in about 60 countries. The yellow cyst nematode *G. rostochiensis* is more widespread being reported from 58 countries while *G. pallida* is prevalent in about 27 countries. The latter species is able to adopt itself under subtropical climatic conditions and shows a wide genetic variability in its original home Andes mountains.

In India, the nematode was detected by Jones in 1961 from a field at Vijayanagaram State Farm in Ootacamund of Nilgiri hills in Tamil Nadu. The nematode was probably introduced from Britain since these fields contained European weeds. Later their occurrence was noticed in Kodaikanal hills also. Detailed surveys conducted earlier in other major potato growing areas of Assam, Karnataka, Himachal Pradesh, Punjab and Uttar Pradesh had indicated that this nematode was absent in these states confirming its prevalence only in Tamil Nadu State. Realising the potential danger of these nematodes to potato cultivation in the country, the Destructive Insect Pest Act 1919 was amended by the Tamil Nadu Government in 1971 to ensure strict checking of potato for marketing from infested fields. However, recently the presence of potato cyst nematodes have been recorded from neighbouring states of Karnataka and Kerala. Though the cysts observed at Karnataka were non-viable, the nematode pose as a major quarantine problem and calls for reviewing the domestic quarantine. The latest surveys at Nilgiri hills have shown that the potato cyst nematodes are prevalent in all the 28 major potato growing localities with varying intensities.

BIOLOGY AND ECOLOGY

The low nematode infestations are not easily detected and are noticed when small patches of poorly growing plants appear in field after the nematode populations build-up to appreciable levels. The root system of such plants shows mustard seed sized cyst sticking to it (Fig.3) and closer examination reveals yellow or white spherical females which later turn to brown colour (Fig.4.) The hatching of eggs in the cysts are initiated by root diffusates of potato or the members of the family Solanaceae. There are reports that the eggs may remain viable up to 15-

20 years in the absence of root diffusates of suitable plants. It takes about 35 to 40 days for completion of one life cycle (Logiswaran & Menon, 1974) and as the cysts undergo a dormancy period, usually one generation is completed in one crop season. However, there are evidences of a second generation of *G. rostochiensis* being completed if a second crop of potato is taken up in the same soil as observed at Ootacamund (CPRI, 1987). The multiplication rate ranges from 4 to 13 times depending on the environmental conditions such as high altitude with mild climatic conditions (CPRI, 1988). In Nilgiris foothills there was an indication of very low build-up of potato cyst nematode. The parasitism in these parasites are very much advanced and the populations exhibit variations so as to break the resistance of potato (Howard, 1977). Such variants are termed as pathotypes and at Nilgiris the pathotypes Pa_2 of *G. pallida* and Ro_1 of *G. rostochiensis* are prevalent (Krishna Prasad, 1988). There are also indications that other pathotypes of both species could be prevalent in this region as some resistant potato lines show nematode infections (CPRI, 1988).

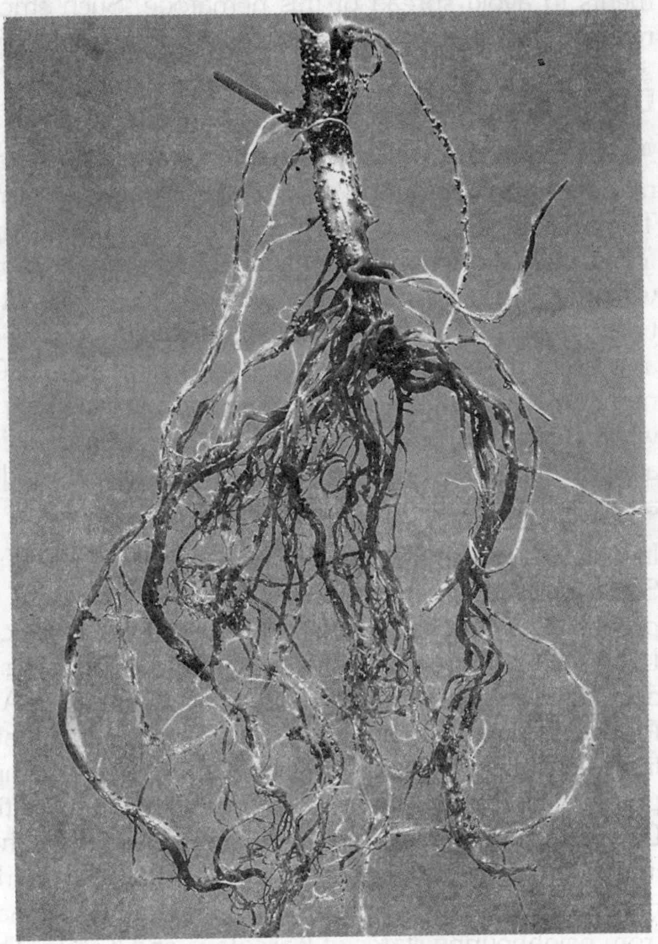

Fig. 3. Nematode cysts attached to potato tubers.

CONTROL STRATEGIES

The potato cyst nematodes are the most successful plant-parasitic nematodes exhibiting a highly specialised survival mechanism. They are restricted to a limited

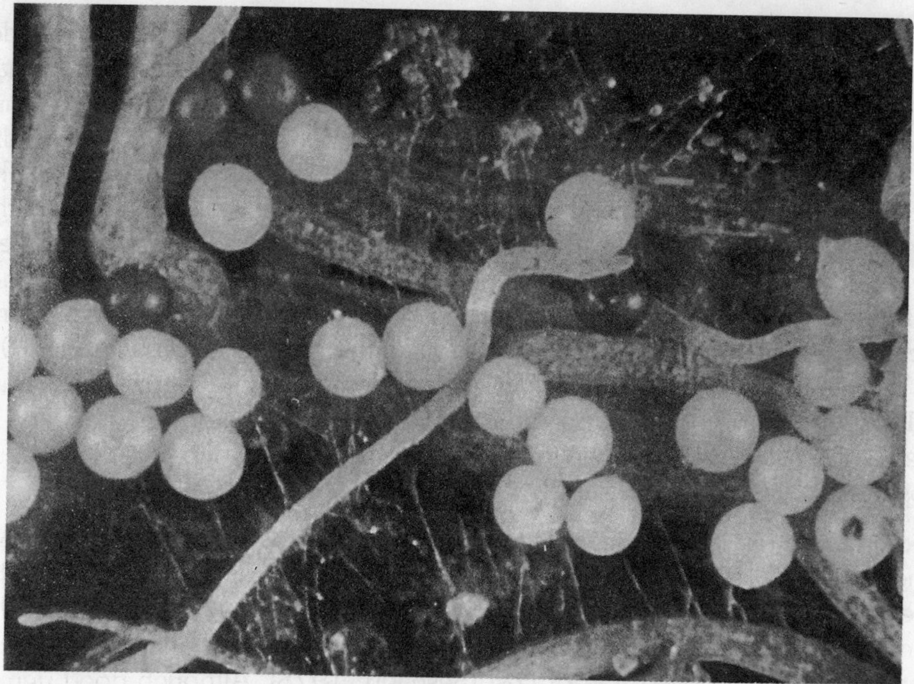

Fig. 4. White and brown cysts on potato roots.

host range in the family Solanaceae and have got themselves distributed along with their host adjusting to the surrounding variations. The experience has shown that they cannot be completely eradicated once they establish in a locality but they can be managed by adopting several plant protection strategies.

Cultural Practices

Growing non-host crops and following crop rotation at least for one year with any non-solanaceous vegetables such as beetroot, cabbage, carrot, cauliflower, French bean, garlic, radish, turnip, etc. during autumn season brings down the cyst populations to a great extent, and thus the management of cyst nematode becomes more feasible (Seshadri, 1970). The rotational sequence of potato-French bean-pea-potato in a four-year rotation had decreased the cyst populations by 98 to 99% with 90% increased yields in potato (Menon & Thangaraju, 1973). Growing of vegetables and resistant potato in a long-term rotation trial brought down the cyst propagule population up to 92% and was effectively used for nematode management at Ootacamund (CPRI, 1989, 1990).

Chemical Control

Several nematicidal experiments were conducted during 1964 to 1967 under Golden Nematode Scheme using DD, DBCP, Nemaphos, etc. These studies had showed DD was the best by reducing the nematode populations up to 60% while increasing the yields by 110% at a dose of 500 l/ha (Menon *et al.*, 1969; Seshadri, 1970). In another trial where DD was applied to an area of 1.96 ha at 1000 l/ha in two equal doses the yields had increased by 200 per cent over control and the net returns per ha was Rs. 7,400 during 1967 ($ 1 = Rs. 30). However, the yield increase due to pesticides depended upon the level of field infestations. Further experiments conducted under Indo-German Nilgiris Development Project during 1968 to 1970 with granular pesticides showed that fensulfothion (Dasanit 10% G) at 30 kg a.i./ha almost completely prevented nematode infestation (Bhaskaran, 1971). A lower dose of 15 kg a.i./ha for next 2-3 seasons and 7.5 kg a.i./ha in subsequent seasons was effective in keeping the nematode under check to obtain almost double yields even in heavily infested fields (Menon *et al.*, 1974).

Based on these results, a large scale chemical control programme was carried out during 1971 to 1975 under Indo-German Nilgiris Development Project. The entire infected area of about 2000 ha in Nilgiri hills at that time was covered. The treatment was made mandatory under Tamil Nadu Pest Act and the pesticide Dasanit 10G was supplied free of cost (Seshadri, 1978). Although good nematode control (75%) was achieved, the complete eradication was not possible. Several field trials conducted with systemic pesticides such as aldicarb, carbofuran, ethoprop, fensulfothion, phorate, phenamiphos and quinalphos have shown that aldicarb or carbofuranan 2 kg a.i./ha as spot treatment to be the most economical dose with ensured reasonable level of nematode control while increasing the yields significantly (CPRI, 1983, 1985).

Breeding for Resistance

Research work on breeding for cyst nematode resistance began at CPRS, Ootacamund in 1968 (Hari Kishore *et al.*, 1969). Initially, 41 clones of 34 accessions belonging to 25 *Solanum* species were screened and the results indicated that high degree of resistance was found in S. *ajanhurri, S. bulbocastanum, S. gandarlassi*, and *S. tuberosum andigena* while *S. vernei* and *S. spegazzinii* were highly resistant (CPRI, 1974). A large hybrid seedling population was produced using resistant *S. tuberosum andigena* clones, *S. multidissectum* (selection No. 3246) obtained from Scottish Plant Breeding Station, Edinburgh, U.K. and some commercial varieties as parental material. Testing of these hybrids had indicated 3 hybrids possessing high degree of resistance to cyst nematodes (CPRI, 1973). However, further studies and screening showed that genotypes reported resistant earlier proved to be susceptible since there existed two nematode species and several pathotypes occurred within them (Howard, 1977).

Later, high degree of resistance to several populations of cyst nematode was exhibited in *S. vernei* clone VTn[2] 62-33-3 obtained from Netherlands (Thangaraju & Sundaram, 1981). However, this hybrid was highly susceptible to late blight disease and hence, 'Kufri Jyoti' a late blight resistant commercial variety was used as female parent with this clone to obtain several genotypes (CPRI, 1985). Further screening resulted in the release of one of the progenies as 'Kufri Swarna' in 1985 which combined resistance to both the species of potato cyst nematodes and the late blight disease. The variety is also drought resistant and has shorter dormancy which is advantageous to Nilgiris region where potato is taken up mainly as a rainfed crop almost throughout the year and thus covers about thirty per cent of potato area. At present, three hybrids possessing combined resistance to cyst nematodes and late blight disease are under advanced trial stages of which one of the most promising hybrid 'D 79-56' is under adoptive research trial in farmers' fields (CPRI, 1990).

Integrated Management

The experience has shown that potato cyst nematodes cannot be completely eradicated once they establish in a locality. They have to be managed by adopting several plant protection strategies. Now the problem is being managed in Nilgiris by chemical treatments, crop rotations and utilizing the available sources of resistance in tuber bearing *Solanum* species. Initially, several halogenated hydrocarbons were used as fumigants. Due to the hazardous nature and difficulties in application of these pesticides they were slowly replaced by systemic granular pesticides. The escalating costs of these pesticides, associated residual problems and slow build-up of nematode populations to unmanageable levels have made chemical treatments as uneconomical at several places. However, the restriction of the parasite only to a selected host range has helped in the management of the problem to a great extent. Further the Indian populations containing pathotypes Ro_1 of *G. rostochiensis* and Pa_2 of *G. pallida* can be managed by *S. vernei* source as it combines resistance to both these populations.

OTHER NEMATODES

The potato tuber worm *Ditylenchus destructor* was reported from Shillong in 1961 (CPRI, 1963). The nematode was recovered on tubers which had shown small grayish cracks with whitish glistening superficial tissues. Fortunately, this nematode has not been encountered anywhere else in India excepting in imported tubers and thus has been considered as one of the quarantine problems (NBPGR, 1980). Several other plant-parasitic nematodes such as lesion, stunt, spiral and reniform nematodes are being constantly encountered during surveys of potato fields. However, not much information was available on their role as pathogens of potato. However, the presence of virus transmitting nematode genera such as *Longidorus, Trichodorus* and *Xiphinema* with potato cultures may later become a constraint in disease free seed production (Nagaich & Vashisth, 1983).

Recently, the pathogenicity of *Quinisulcius capitatus*, a stunt nematode frequently occurring in hill tracts, was established on potato variety 'Kufri Jyoti' at Simla (Krishna Prasad & Sharma, 1985). The nematode build-up ranged from 5 to 8 times at initial inoculum levels of 10 to 1000 at 45 days. Concomitant to the nematode build-up, plant characters (such as shoot length, fresh and dry weights of shoot and root) reduced which affected the tuber production. The tuber reduction ranged from 14 to 29% by weight which was related to the reduction of dry weights of plant parts and was negatively correlated with nematode build-up index. The spiral nematode, *Helicotylenchus dihystera* was also pathogenic to potato accounting for 9-27 per cent yield reductions in 90 days with 2-4 times build-up (Krishna Prasad, 1990). Both these nematodes are commonly occurring in major potato growing belts of potato in Himachal Pradesh and could be potential pests on a long run. Other parasitic nematodes constantly found in potato soils such as pin nematode (*Paratylenchus* species), reniform nematode (*Rotylenchulus reniformis*) and lesion nematode (*Pratylenchus coffeae*) are already established pests of other crops and hence may also pose as plant protection problem but certainly not to the extent as either root-knot nematodes or cyst nematodes.

REFERENCES

Ahuja, S. (1983). Studies on the variability in infection of root-knot nematode *Meloidogyne incognita* on potato. IIIrd Nematology Symp., H.P.K.V.V., Solan (May 24-26). p. 4.

Anonymous (1975). Crop loss estimation in potato. *FAO Plant Protec. Bull.* **23**(3-4): 1-4.

Bhaskaran, A.R. (1971). Control of the potato golden nematode (*Heterodera rostochiensis* Woll.) campaign in the Nilgiris. *Madras Agric. J.* **58**(9-10): 800-808.

Central Potato Research Institute (1963). *Scientific Report 1961 and 1962* : 108-111.

Central Potato Research Institute (1971). *Annual Scientific Report 1966* : 34, 119-131.

Central Potato Research Institute (1972). *Annual Scientific Report 1967* : 25-26, 28, 73-79.

Central Potato Research Institute (1973). *Annual Scientific Report 1968* : 87-101.

Central Potato Research Institute (1974). *Scientific Report 1971-73* : 18, 43-45, 175-180.

Central Potato Research Institute (1976). *Annual Scientific Report 1975* : 14-24, 174-181.

Central Potato Research Institute (1978). *Annual Scientific Report 1977* : 14, 92-99.

Central Potato Research Institute (1979). *Annual Scientific Report 1978* : 10-11, 22-24, 124-129.

Central Potato Research Institute (1982). *Annual Scientific Report 1981* : 1-2, 10-11, 103-108.

Central Potato Research Institute (1983). *Annual Scientific Report 1982* : 24-25, 155-160.

Central Potato Research Institute (1985). *Annual Scientific Report 1983* : 131-133.

Central Potato Research Institute (1987). *Annual Scientific Report 1986* : 120-125.

Central Potato Research Institute (1988). *Annual Scientific Report 1987* : 150-152.

Central Potato Research Institute (1989). *Annual Scientific Report 1988* : 146-153.

Central Potato Research Institute (1990). *Annual Scientific Report 1989-90* : 110-114.

Chitwood, B.G. (1949). Root-knot nematodes - Part I. A revision of the genus *Meloidogyne Goeldi*, 1887. *Proc. Helminth. Soc. Wash.* **16**(2): 90-104.

Desh Raj & Gill, J.S. (1983). Reaction of some potato varieties to root-knot nematode *Meloidogyne incognita.* IIIrd Nematol. Symp. HPKVV, Solan (May 24-26) : 27-28.

Evans, K. & Stone, A.R. (1977). A review of the distribution and biology of potato cyst nematode. *PANS* **23**(12) : 178-189.

Gill, J.S. (1974). Advances made in the potato nematology in India. *JIPA* **1**(1-2): 51-55.

Gill, J.S. & Krishnananda, N. (1977). Plant nematodes and potato cultivation. *In : Recent technology in potato improvement and production* (Ed.—B.B. Nagaich). CPRI, Simla: 277-289.

Hague, N.G.M. (1979). Nematodes — the unseen enemy. *World Crops 31*(3): 2-17.

Hari Kishore; Nirula, K.K.; Menon, P.P.V. & Sharma, A.K. (1969). Breeding potato varieties for resistance to golden nematode. All India Nematol. Symp., IARI, New Delhi (Aug. 21-22): 58-59.

Howard, H.W. (1977). Report on the experiments at the Potato Experimental and Trial Centre, Ootacamund to test resistance of material of Plant Breeding Institute Cambridge to Nilgiri populations of potato cyst nematodes. p. 17.

Jones, F.G.W. (1961). Potato root eelworm *Heterodera rostochiensis* Woll. in India. *Curr. Sci.* **30**(5): 187.

Khanna, M.L. (1966). Multiple resistance to heat, drought, frost, charcoal, rot and nematode in commercial potato. *Curr. Sci.* **35** : 494-496.

Koen, H. (1966). Potato cyst nematode crop losses. *Nematologica* **12**: 109-112.

Krishna Prasad, K.S. (1984). Nematode problems of potato and their management. *In* : Panel Discussion on Nematological Problems of Economic Importance in North Zone, Haryana Agricultural University, Hisar (20-21 September).

Krishna Prasad, K.S. (1986). Potato Nematodes. *In* : *Plant Parasitic Nematodes of India* (Ed.—Gopal Swarup & D.R. Das Gupta). Indian Agricultural Research Institute, New Delhi. pp. 350-370.

Krishna Prasad, K.S. (1988). Nematology in 2000 AD as I see it in relation to potato cyst nematode. National Symp. IARI, New Delhi (28th May, 1988). pp. 26-29.

Krishna Prasad, K.S. (1990). *Helicotylenchus dihystera*, a potential pest of potato. *J. Indian Potato Assoc.* **17**(3-4): 204-205.

Krishna Prasad, K.S. & Sharma, R.K. (1985). Parasitic nematodes associated with potato and the pathogenicity of *Quinisulcius capitatus* on variety Kufri Jyoti. *J. Indian Potato Assoc.* **12**(1-2): 56-62.

Kuhn, J. (1881). Die Ergebnisse der versuche zur Ermittelung der Ursache der Rubenmudigkeit und zur Erforschung der Natur der Nematoden. *Ber. Physiol. Lab. Landw. Inst. Univ. Halle* **3**: 1-153.

Logiswaran, G. & Menon, P.P.V. (1974). Observations on the host range of *Heterodera rostochiensis* in Nilgiris, Tamil Nadu State, India. *Bull. Hort. Dept* : 36-39.

Menon. P.P.V.; Logiswaran, G.; Seshadri, A.R. & Thangaraju, D. (1969). Observations on the control of the golden nematode (*H. rostochiensis*) of potatoes in the Nilgiris, All India Nematol. Symp., IARI, New Delhi (Aug. 21-22): 48-50.

Menon P.P.V. & Thangaraju, D. (1973). Effect of crop rotation on the population of potato golden nematode (*Heterodera rostochiensis*) in the Nilgiris. *Indian J. Agric. Sci.* **43**: 304-306.

Menon, P.P.V.; Vijayaragavan, S.; Logiswaran, G.; Thangaraju, D.; Sadasivan, M. & Dakshina Murthy, A. (1974). Observations on the large scale chemical control of the potato golden nematode (*Heterodera rostochiensis*) in the Nilgiris, Tamil Nadu State. *Bull. Hort. Dep.*: 27-32.

Mulvey, R.H. & Stone, A.R. (1976). Description of *Punctodera metadorensis n.* gen. *n.* sp. with lists of genetic diagnosis of *Globodera, Heterodera* and *Sarisodera. Chandian J. Zool.* **54**: 772-785.

Nagaich, B.B. & Vashisth, K.S. (1983). Transmission of viruses and mycoplasmas by vectors. *In: Potato-Production, Storage and Utilization* (Ed.: B.B. Nagaich). CPRI, Simla : 317-333.

National Bureau of Plant Genetic Resources (1980). *Scientific Manograph No. 2 of NBPGR, New Delhi*. p. 99.

Neal, J.C. (1889). The root-knot disease of the peach, orange and other plants in Florida due to the work of Anguillula. *Bull. U.S. Bur. Ent.* **20**: 31.

Raj, B.T. & Nirula K.K. (1969). Trials on the eradication of root-knot nematodes from potato fields with DD. All India Nematol. Symp., IARI, New Delhi (Aug. 21-22).

Sanwal, K.C. (1979). Nematode pests of potato. *In : Int. Symp. on post-harvest technology and utilisation of potato*. (Ed.—Hari Kishore). Int. Potato Centre, Lima, Peru. pp. 197-207.

Sasser, J.N. & Freckman, D.W. (1987). A world perspective on Nematology. *In : Vistas on Nematology* (Eds.—J.A. Veech & D.W. Dickson). Society of Nematologists, U.S.A. pp. 7-14.

Seshadri, A.R. (1970). The golden nematode problem (*Heterodera rostochensis*) in India. Int. Nematol. Symp. Pescara, European Society of Nematologists, (8-13th September): 141-143.

Seshadri, A.R. (1978). Chemical control of potato nematodes. *In : Developments in the control of nematode pests of potato*. Report of the 2nd Nematode Planning Conference, Lima, Peru (Nov. 13-17): 173-187.

Sitterly, W.R. & Fassuliotis, G. (1965). Potato in South Carolina due to cotton root-knot nematode *Meloidogyne acrita. Plant Dis. Reptr.* **49**: 723.

Stone, A.R. (1973). *Heterodera pallida n* sp. (Nematoda: Heteroderidae) a second species of potato cyst nematode. *Nematologica* **18**: 591-606.

Stone, A.R. (1985). Coevolution of potato cyst nematodes and their hosts. Implication for pathotypes and resistance. EPPO Bulletin 15: 131-137.

Thangaraju, D. & Sundaram, R. (1981). Studies on *Solanum vernei* 62-33-3 and hybrids of *S. vernei* 62-33-3 x K. Jyoti for resistance against potato cyst nematodes in Nilgiris. Nematol. Symp., Coimbatore. *Indian J. Nematol.* **11**(1): 123.

Thirumalachar, M.J. (1951). Root-knot nematode on potato tubers in Simla. *Curr. Sci.* 20-104.

Chapter 7

Nematode Diseases of Pulses and their Control

Renu Sharma[1] & S.B. Sharma[2]

1. Central Plant Protection Training Institute, Rajendranagar, Hyderabad-500 030, India

2. International Crops Research Institute for the Semi-Arid Tropics (ICRISAT), Patancheru-502 324, India

PULSES are one of the essential components of the diets of people in India. They provide necessary protein supplement to the largely cereal-based diets. These crops are very beneficial not only in context of human diet but also in enriching the agricultural soil by fixing atmospheric nitrogen. India is a major pulse growing country and it ranks first in pulse production in Asia. Pulses are cultivated over an area of 23 m ha and average yield ranges from 500 to 600 kilograms. Madhya Pradesh, Maharashtra, Orissa, Rajasthan and Uttar Pradesh are the important pulse producing states and account for around 68 per cent of the total pulse production in India. Chickpea and pigeonpea are the two most important pulse crops and together they account for nearly 45 per cent of the area and 60 per cent of the total production of pulses. Green gram and black gram are also important pulses and account for an area of more than 20 and 17 per cent of the total production. Horse gram and *Lathyrus* also occupy a sizable area but are localized in distribution. Other important pulse crops are cowpea and field beans.

Production of pulses in India has remained more or less static over the last two decades and with a gradual increase in human population the per capita availability of pulses has declined. Reasons for low productivity are many and diverse. Plant-parasitic nematodes are one of the biotic factors that adversely effect the productivity of pulse crops in India. Several species of plant-parasitic nematodes are associated (Table 1) with different pulse crops (Alam *et al.*, 1978, 1979; Sitaramaiah, 1984; Sharma, 1985; Gill & Singh, 1989). Many of these nematode species invade the roots, induce physiological disorders resulting in reduced

Submitted as Journal Article No. 1243 by the International Crops Research Institute for the Semi-Arid Tropics (ICRISAT), Patancheru-502 324, India.

plant vigour and low crop yields of poor quality. The nematodes also suppress the *Rhizobium* root modulation and nitrogen fixation.

Major Nematode Problems

The root-knot nematodes (*Meloidogyne incognita* and *M. javanica*). the pigeonpea cyst nematode (*Heterodera cajani*) and the reniform nematode (*Rotylenchulus reniformis*) attack different pulse crops in India. Populations of these nematodes are associated with poor plant growth of different pulse crops in many states. These nematodes are of national economic importance on pulse crops in India. Attempts have not been made to establish losses on either a state wide or national basis. Yield loss estimates in nematicide trials and best-guess estimates indicate that grain and biomass losses attributable to these nematodes may range from negligible to significant, depending on the population densities of these nematodes (Table 2). Besides many other plant-parasitic nematodes are associated with pulses in India.

Table 1: Plant-parasitic nematodes associated with pigeonpea, chick-pea, greengram, blackgram and cowpea in India.

Nematode species	Pigeon-pea	Chickpea	Green gram	Black gram	Cowpea
Helicotylenchus abunnammai			0		0
dihystera	0	0	0	0	0
erythrinae					0
elegans	0				
indicus	0	0			
retusus	0	0			
sharafati	0	0			
Hemicriconemoides cocophillus	0	0			
Heterodera cajani	*	0	*	0	*
Hirschmanniella mucronata	0	0			
Horyzae	0	0			
Hoplolaimus indicus	0	0	0	0	0
dimorphicus		0			
seinhorsti	*				
Lonqidorus siddiqi	0				
elongatus					0
Longidorus brevicaudatus				0	
Macroposthonia ornata	0	0	0	0	
Meloidogyne arenaria		0			
javanica	*	*	*	0	*
incognita	*	*	*	0	*

Nematode species	Pigeon pea	Chickpea	Green gram	Black gram	Cowpea
Paratrichodorus mirzai	0				0
Pratylenchus thornei		0		0	
zeae	0	0			
mulchandi		0			
Radopholus similis					0
Rotylenchulus reniformis	*	0	*	*	*
Scutellonema erectum	0				0
Telotylenchus tonkiensis			0		
Tylenchorhynchus dubius					0
divittatus	0		0	0	
vulgaris	0	0	0	0	0
martini				0	
goffarti	0				0
phaseoli	0				0
mashoodi	0	0			0
brassicae			0		
brevilineatus		0			

0 = associated, * =cause significant reduction in crop growth under field conditions.

Table 2 : Nematode-caused losses in yields of different pulses in different crop-loss assessment trials.

Crop	Nematode	Loss (%)	Reference
Black gram	*Meloidogyne javanica*	39.0	Handa, 1990
Chickpea	*M. incognita*	40.07	Upadhyay & Dwivedi, 1987
French beans	*Meloidogyne* sp.	43.0	Reddy, 1986
Green gram	*Heterodera cajani*	86.2	Saxena & Reddy, 1987
Pea	*Meloidogyne* sp.	20.0-33.3	Reddy, 1985; Upadhyay & Dwivedi, 1987
Pigeonpea	*H. cajani*	30.1	Saxena & Reddy, 1987

Root-knot nematodes, *Meloidogyne* species first reported in India in 1901 are sedentary endo-parasites and are characterized by the galling of roots. They complete their life cycle in 21-28 days, and have many generations in a crop season. Of the 11 species reported in India, *M. incognita* and *M. javanica* attack a wide range of pulse crops (Sitaramaiah, 1984; Sikora & Greco, 1990; Gill & Singh, 1989). Control of these nematodes generally results in a three to fourfold yield increase (Sethi & Gaur, 1986). Affected fields are characterized by patches of poorly growing plants. These patches may expand each year. Symptoms advance with crop age and are more severe during drought stress periods and/or in low fertility soil. Usually symptoms on above ground plant parts are detectable when crop yield losses exceed 15-20%.

The reniform nematode, first recorded in India in 1959, has a very wide host range. The nematode feeds in the cortex, pericycle, endodermis, and phloem parenchyma of the hosts and completes its life cycle in 24-30 days. It is a semiendoparasite of the roots and has multiple generations in each crop season. The preadult stage is the infective stage of the life cycle. Yield loss estimates for this nematode are not available in India; however, it is reasonable to assume that significant losses occur. Control of the nematode significantly increases crop vigour and yield.

The cyst nematode, *Heterodera cajani*, was reported on pigeonpea in 1967. Presence of pearl-like white females of cyst nematode on the roots is a characteristic symptom of nematode infection. *H. cajani* is the only species of cyst forming nematodes that attacks pulse crops in India. The nematode completes its life cycle in 3 weeks at 25-30°C and reproduces rapidly, having multiple generations during a crop season. The nematode is widely distributed and is the most important nematode pest of pigeonpea and other pulses (Sharma & McDonald, 1990). It significantly reduces plant vigour and seed yield.

Damage thresholds

Information on the relationship between performance of crop and population densities of plant-parasitic nematodes is essential for making valid nematode management decisions. The damage threshold levels (Table 3), largely based on glasshouse pot experiments of *M. incognita* and *M. javanica*, *R. reniformis* and *H. cajani*, range from 0.1 to 2.0 juveniles per g or cm^3 of soil for different pulses (Srivastava *et al.*, 1974; Sharma & Sethi, 1975; Gupta, 1979; Gupta & Yadav, 1979; Nath *et al.*, 1979; Panda & Seshadri, 1979; Raut, 1980; Raut & Sethi, 1980; Mishra & Gaur, 1981, 1981a; Mani & Sethi, 1984; Pathak *et al.*, 1985; Mahapatra & Padhi, 1986; Zaki & Bhatti, 1986; Padhi & Mishra, 1987; Mohanty & Pathi, 1987; Dwivedi & Upadhyay, 1988; Ahmad & Husain, 1988; Patel & Thakar, 1988; Tiyagi & Alam, 1986, 1987, 1988, 1989; Tiyagi *et al.*, 1989).

Interrelationships with other microorganisms

H. cajani, R. reniformis, M. javanica and *M. incognita* reduce the number of *Rhizobium* nodules on different pulse crops (Table 4). Species of *Glomus, Fusarium, Rhizoctonia*, and *Sclerotium* reduce the population densities of the

Table 3 : Damage thresholds (nematode per cm³ or per g of soil) of *Rotylenchulus reniformis, Meloidogyne javanica, M. incognita* and *Heterodera cajani* on pulses in India.

	R. reniformis	*M. javanica*	*M. incognita*	*H. cajani*
Black gram	1.0-2.0	0.5	1.0	
Chickpea	>0.5	0.2-2.0	0.2-2.0	
Cowpea	<1.0	2.0	0.2	0.1-10
French bean	1.0	0.1		
Green gram	<0.1-1.0	0.1-2.0	2.0	
Lathyrus			1.0	
Mothbean	1.0		1.0	0.1
Pigeonpea	0.2-2.0	2.0	0.2-2.0	0.2-1.0
Soybean		0.1		

nematodes. The nematodes and fungi together reduce the plant growth more than that by either of them alone. The root-knot nematodes modify the resistance of resistant chickpea and pigeonpea to fusarium wilt (Table 4).

MANAGEMENT OPTIONS

Nematode-caused diseases can be managed by judicious use of an integrated combination of chemical and non-chemical methods. The use of nematicides (dichloropropane dichloropropene, dibromochloropropane, high doses of aldicarb, and carbofuran) is effective but the use is restricted because of high costs, non-

Table 4 : Interrelationship of plant-parasitic nematodes with other soil microorganisms on pulses in India.

Crop	Nematode	Microorganisms	Interaction	Reference
Cowpea	*M. incognita*	*Rhizoctonia solani* +*Rhizobium*	Nematode and fungus reduced the number of nodules	Varshney *et al.*, 1987
	M. incognita	*Rhizobium*	Nematode reduced the number of nodules and nitrogen content of plants	Sharma & Sethi, 1976; Moni & Nair, 1986
	M. javanica	*R. solani*	Fungus reduced nematode reproduction and hatching	Kanwar *et al.*, 1988; Singh *et al.*. 1986

Crop	Nematode	Microorganisms	Interaction	Reference
	Heterodra cajani	*Rhizoctonia bataticola*	Fungus reduced nematode reproduction	Walia & Gupta 1986
	H. cajani	*Rhizobium*	Nematode reduced the number of nodules	Sharma & Sethi, 1976
	H. cajani	*R. solani*	Fungus reduced nematode population	Walia & Gupta, 1986a
	H. cajani	*Glomus fasciculatum G. epiqques*	Fungus reduced nematode penetration in roots	Jain & Sethi, 197; Jain & Sethi, 1988
	M.incognita + *R. reniformis*	Rhizobium and R. solani	Root-knot and fungus caused severe damage to unbacterized plants	Khan & Husain, 1988
Green gram	*M. incognita*	*Rhizobium*	Nematode reduced number of nodules and chlorophyll content of leaves	Chahal & Chahal 1987, 1987a, 1988, 1989
Black gram	*M. incognita*	*Rhizobium* strains	Nematode reduced leghaemoglobin, bacterial content, number of nodules, and dry shoot and root weights	Chahal *et al.*, 1988
	M. incognita + *R. reniformis*	*Rhizobium*	Both the nematodes reduced number of nodules	Mishra & Gaur, 1981a
French beans	*M. incognita*	*Rhizobium*	Reduced nitrogen fixation	Singh & Reddy, 1981
Soybean	*M. incognita*	*Fusarium* sp. + *Pythium* sp.	Caused maximum damage to plant when 1 g pythium mycelium/pot+2500 nematode larvae inoculated	Sharma, 1990
Lentil	*M. javanica*	*Macrophomina phaseolina*	Fungus reduced the nematode reproduction	Tiyagi *et al.*, 1988
	R. reniformis	*M. phaseolina*	Fungus reduced the nematode reproduction	Anver *et al.*, 1991

Crop	Nematode	Microorganisms	Interaction	Reference
Pea	*M. incognita*	*Rhizobium*	Reduction in number of nodules and leghaemoglobin content	Chahal & Singh, 1984; Sharma & Tiyagi, 1988
	M. incognita	*R. solani* Pea Mosaic Virus	Synergistic reduction of nodulation and plant growth. Fungus multiplication and galling	Husain *et al.*, 1985
Chick-pea	*M. javanica*	*Fusarium oxysporum* f. sp. *ciceri*	Both synergistically reduce plant growth; nematode reduced modulation; fungus on prior inoculation inhibited nematodes.	Goel & Gupta, 1986
	M. javanica	*Sclerotium rolfsii*	Additive pathogenic effect on plant growth. Low nematode population in presence of fungus	Mukhtar & Khan 1989
	M. javanica	*E. oxysporum* f. sp. *ciceri*	Nemaode breaks wilt resistance in chickpea variety "avrodhi"	Upadhyay & Dwivedi, 1987a
	M. javanica	*R. bataticola*	More damage to plant when both inoculated together. Root gall reduced by fungus	Goel & Gupta, 1987
	M. incognita	*F. oxysporum* f. sp. *ciceri* and *F. soloni*	Additive adverse effect of combined inocula of all three pathogens; Nematode favourably predisposes plants to fungal attack; *F. solani* antagonistic to nematodes.	Mani & Sethi, 1987
	M. incognita	*Fusarium* sp.	Nematode breaks wilt resistance. 50% mortality in presence of nematodes	Thakar *et al.*, 1986
	M. incognita	*F. oxysporum* f.sp. ciceri	Presence of nematode enhanced wilt incidence. Additive adverse effect on plant growth	Patel *et al.*,1987; Kumar *et al.*, 1988

Crop	Nematode	Microorganisms	Interaction	Reference
Pigeon-pea	*H. cajani*	*Fusarium udum*	Additive damage to plants by both organisms. Highest incidence of wilt (93.6% plants) when nematode inoculated 3 weeks before fungus & 100% plants wilted within 25 days of simultaneous inoculation	Edward & Singh, 1979; Hasan, 1984; Sharma & Nene, 1989
	R. reniformis	*F. udum*	Fungus enhanced the damage caused by nematode alone. Nematode induced early wilting in the fungus susceptible plants. Fungus reduced nematode population in roots	Sharma & Nene, 1990
	Meloidogyne spp. *(M. incognita +M. javanica)*	*F. udum*	Nematodes more damaging than fungus. Nematode induced 55% wilting in wilt resistant plants	Sharma & Nene, 1990
	M. incognita	*R. reniformis*	Reniform nematode suppressed gall formation and final population of the root-knot nematode	Pathak *et al.*, 1985

availability, and possibility of health hazards associated with application. Perhaps that is why nematicides form a very small portion (<2.5%) of the total pesticide market (Premachandran, 1986). Many crop management practices such as summer ploughing, fallowing, and planting of non-host crops usually reduce the crop damage associated with plant-parasitic nematodes.

Summer ploughing and soil solarization

Two to three deep ploughings during the hot summer months expose nematodes to high temperatures, leading to their desiccation. The efficacy of summer ploughing is enhanced by polyethylene mulching (solarization) that traps and retains solar heat for a longer duration. Active nematodes are very susceptible to solar heating. The pigeonpea cyst nematode and the reniform nematode are effectively controlled by this method (Sharma & Nene, 1990a). Summer ploughing in combination with summer fallowing is more effective.

Organic amendments

The non-edible seed-cakes of *neem, karanj, mahua*, castor, mustard, sesamum are useful in management of nematodes on pulses (Table 5). These oilseed-cakes suppress nematode populations (Khan & Husain, 1988a; Ahmed *et al*., 1990). These amendments are in general beneficial for soil and deserve more attention by scientists who are attempting to develop effective and economical nematode management systems. The synergistic effect of oilseed-cake amendments and various pesticides are useful in integrated crop management programmes (Sundaram & Velayutham, 1988).

Delayed sowing

Late planting of chickpea in November reduces root-knot nematode damage and increases chickpea yield in Northern India (Gaur *et al*., 1979).

Table 5 : Effects of different organic amendments on population densities of plant-parasitic nematodes in different pulses in India.

Crop	Nematode	Organic amendment	Reference
French beans	*R. reniformis*	Neem cake 2000 kg/ha Karanj cake 1000-2000 kg/ha Prawn exoskeleton 2000 kg/ha	Padhi & Mishra, 1987a,
Cowpea	*H. cajani*	Shialkanta, *bhang, datura, kaner* aqueous extract of 100 g leaves at 5-100% conc./100 second-stage juveniles for 1-48 hrs	Vijayalakshmi *et al*., 1989
	M. incognita	2 g neem oil-cake/10 g seeds 2 g groundnut oil-cake/10 g seeds 2 g neem leaf powder/10 g seeds	Khan & Husain, 1988a
		Neem cake	Jain & Hasan, 1986
Garden bean	*R. reniformis* *Helicotylenchus dihystera*	Neem cake 250-500 kg/ha	Sundaram & Velayutham, 1988

Crop	Nematode	Organic amendment	Reference
Green gram	*Helicotylenchus* sp. *Tylenchorhynchus* sp. *Pratylenchus* sp.	(120-180 kg N equivalent cakes) neem, mahua, mustard, groundnut, sesamum, cotton 1500-3600 kg/ha	Misra & Prasad, 1974
Chickpea	*M. javanica*	Aqeuous extract 40 g fresh neem leaves/kg soil	Kaliram & Gupta, 1980
		Neem leaves 20000 kg/ha Datura leaves 20000 kg/ha	Gupta & Kaliram, 1981
		Neem leaves 20000-40000 kg/ha	Kaliram & Gupta, 1981
		Aqueous extract 40 g Datura leaves/kg soil	Kaliram & Gupta, 1982

Host-plant resistance

Use of resistant varieties is an economical and effective way to manage nematode-caused damage. Major research emphasis has been on the evaluation of germplasm lines in pots mainly against root-knot nematodes. More than 400 lines of chickpea and pigeonpea have been identified as probable sources of resistance to the root-knot nematode (Sharma, 1987). Many genotypes with resistance to reniform, root-knot and cyst nematodes have been identified in pulse crops (Table 6). Research on breeding varieties for nematode resistance has been limited. However, some information is available on the nature of resistance. Some amino acids: arginine, leucine/norleucine and threonine were absent in resistant varieties of mung (Hussaini & Seshadri, 1976). In cowpea resistance is attributed to high magnesium (Mg) and low iron (Fe) concentrations in plants (Sharma & Sethi, 1976a). The resistant plants allow less larval penetration and development; majority of penetrated juveniles develop into males. In roots of resistant lines the cortex layer is thick, scleroids are present in cortex, and there are fewer giant cells (Singh *et al.*, 1984). Anver & Alam (1989) and Alam *et al.* (1991) have observed that loss in plant weight, number of pods per plant, chlorophyll contents of leaves and bulk density of stem parts of pigeonpea was directly correlated with the multiplication of the nematodes (*M. incognita, R. reniformis*), the highest being in susceptible varieties with minimum or no change in resistant varieties.

Bio-control

A large number of bacteria and fungi have been reported as potential bio-control agents for *H. cajani* and *M. incognita* (Sharma & Swarup, 1988). *Paecilomyces lilacinus* reduces the damage caused by *R. reniformis* or *M. incognita* or *R. solani* (Khan & Husain, 1988c). However, more field research is required. Several plants including mustard, African marigold, asparagus, crotolaria and sesamum reduce population densities of plant-parasitic nematodes. Alam & Saxena (1986) studied the effect of preceding crop on the population-growth pattern of *Meloidogyne incognita* and *Tylenchorhynchus brassicae* on chickpea cv. RS-6. Growing cluster bean and kochia prior to chickpea were less favourable to *M. incognita* and zinnia, green gram and eggplant to *T. brassicae.*

Nematicides

Aldicarb, carbofuran and phenamiphos (1-4 kg a.i./ha) effectively reduce nemato

Table 6 : Sources of resistance to plant-parasitic nematodes in different pulses.

Crop	Nematode	Sources of resistance	Reference
Cowpea	*M. incognita*	RS-9, Iron	Mathur *et al.*, 1974
		Barsati mutant	Sharma & Sethi, 1976a
		HFC 719	Phukan *et al.*, 1983
		C-152, IHR 29-5	Singh & Reddy, 1982
		82-1-B	Darekar & Patil 1981
		IC 8444, IC-20626, IC-44671	Das *et al.*, 1988
	M. incognita, M. javanica	*Barsati mutant* 6-1B, F5 (M x iron) 2-22-2, S-18, 19 and 20, F4 (M x 126)-1-34-4, CM-11-1, F6 (CB_2x910)-4-10-3, C-152, GC-162, TVU-1190, TVV-201-1D, P-46, P-309 (85R), P-370, P-426 (126R), P-427, P-435, P-910, P-1171, P-1262, P-1265, P-1325, P-1346, P-1371, P-1399, Victor, new selection and iron	Patel *et al.* 1977
	M. javanica	HFC-12, HFC-311, C-28	Gupta, 1979
	R. reniformis,	V-16 RC-48 Pusa phalguni IC 9611 S-488, EC-4213A, RC-8	Thakar & Patel 1984 Gaur, 1986 Makadia *et al.* 1987 Nandwana, 1985 Khan & Husain, 1988 b
	H. cajani	Sel-A, Sel-1-2, SEl-1-4, 28-8-2, P33-2-1, P 85-2 (Eq) NP2	Sharma & Sethi, 1976 a

Crop	Nematode	Sources of resistance	Reference
Green gram	*M.incognita*	T 44	
		53 P, 61, 218 P 330 68 P81, 91P,	Mathur *et al.*, 1974
		131-68, 244-P, 368-68, 47 g Pb2	Hussaini & Seshadri, 1976
		g-4, 243P, 362-68, NP 18, RS-4,	Nandwana, 1985
		160 P 68	
		ML 80	
	R. reniformis	T-44, Ml-3, ML-5, ML-13, K-851,	Routaray *et al.*, 1986
		Hyd-12-4 ML-80, ML-62, TT 8E,	Patel *et al.*, 1989
		Pusa 103, PDM-14, IC 8924, IC 11312	
		ML 56, Pusa 105	Chavda *et al.*, 1988
	H. cajani	FM1, MC 131,	Devi & Gupta, 1987
Black gram,	*R. reniformis,*	Ratnapur-1, UG 201, UG 135,	Routaray *et al.*, 1986
	M. javanica	DDU 5, UN 81-7, PDU-10, VGG 2	Handa, 1990
	M. cajani		
		Type 9, PDU 1, PDU 6, Pant U-19-MR	Devi & Gupta 1987
	M. incognita	T-9	Thakar & Patel, 1985
Pea	*M. incognita*	JP1 69-1	Ravichandra *et al.*, 1987
Soybean	*M. javanica*	PK 628, JS-75, 20-3, MACS No. 57, 64, 92, PK-660, DS-74, S-20	Kundu & Mishra, 1985
Horse gram	*R. reniformis*	BGM-1 Kuruvai, K-42, DP-1-1236 and Kalarahang,	Nayak *et al.*, 1987
	M. incognita	Kalarahang, BGM-1-1-8-3, K-42, Kuruvai	Routaray *et al.*, 1987

de population densities (Table 7) on different pulse crops (Singh & Reddy, 1982a; Reddy & Singh, 1983; Verma & Gupta, 1986; Patel & Thakar, 1986; Meher *et al.*, 1985; Padhi & Mishra, 1987a). Seed treatment with aldicarb, carbofuran, phenamiphos and thimet in the range of 1-6% w/w have been tried against the root-knot species and the Reniform nematode (Table 8). More work is required in the efficacy of seed treatments. This is a potentially effective and economical treatment for nematode control and requires much more attention.

CONCLUSION

In view of several plant-parasitic nematode species associated with pulse crops there are enough reasons to suspect that nematode diseases are one of the important factors that adversely affect the production of pulses in India. Most of the research information on pulse nematodes is based on pot trials and laboratory

Table 7 : Chemical control of plant-parasitic nematodes on pulses.

Crop	Nematode	Chemical	Dose (kg a.i. per ha)	Reference
Cowpea	*M. incognita*	Aldicarb	0.5-1.5	Singh & Reddy, 1982a
		Carbofuran		
		Ethoprop		
		Phenamiphos		
	M. javanica	Aldicarb	1.0	Reddy & Singh, 1983
	T. vulgaris		0.5 1.0	Verma & Gupta, 1986
	T. mashoodi			
	Helicotylenchus indicus	Mephosfolan	1.0-4.0	Prasad *et al.*, 1982
	H. delhiensis			
	T. vulgaris			
Green gram	*Meloidogyne*	Aldicarb	1.5	Yien *et al.*, 1977
		DBCP	60L	Sakhuja & Singh, 1980
	R. reniformis	Phenamiphos	4.0	Patel & Thakar, 1986
Pea	*M. incognita*	Phenamiphos	1.0	Reddy, 1985
	M. incognita	Carbofuran	3.0	Meher *et al.*, 1985
	R. reniformis			
French beans	*R. reniformis*	Carbofuran	4.0	Padhi & Mishra, 1987a
	M. incognita	Aldicarb	1.0	Reddy & Singh, 1983
Chickpea	*M.javanica*	Aldicarb		Gupta & Kaliram, 1981
		Carbofuran	1.5	
		Phorate	3.0	
		Aldicarb	1.5-3.0	Kaliram & Gupta, 1981
		Carbofuran		
		Phorate	3.0-6.0	
	Meloidogyne spp.	Fensulfothian	10.0,4.0	Reddy, 1975
		aldicarb & methomyl	8.0	
Pigeonpea	*H.cajani*	Carbofuran	4.5	Gunasekaran *et al.*, 1976
		Fensulfothion	1.0-2.0	
	M. incognita + R. reniformis	Aldicarb & Fensulfothian	2.0	Jaiswal *et al.*, 1987
		Gamma BHC	1.0	

Table 8 : Seed treatment with different chemicals to control plant-parasitic nematodes on pulses.

Crop	Nematode	Chemical and dose		Reference
Cowpea	*M. incognita*	Phenamiphos	0.25	- Reddy, 1984
		Aldicarb Carbofuran	1.00%	
	M. incognita *R. reniformis* *R. solani*	Aldicarb (5 g/m ml) Bavistin (0.2 g/5 ml)	5 ml slurry/10 g seeds	Khan & Husain, 1988a
Black gram	*M. javanica*	Thimet 6%		Gupta *et al.*, 1987
Green gram	*M. javanica*	Carbofuran Bendiocarb Phenamiphos	2-4% 4%	Kaushik & Bajaj, 1981
	M. javanica	Carbofuran Aldicarb	1, 2, 4%	Paruthi & Gupta, 1985
	M. incognita	Carbofuran	1%-3%	Darekar *et al.*, 1990
Peas	*M. incognita*	Phenamiphos Aldicarb, carbofuran	1%	Reddy, 1984
French beans	*M. incognita*	Phenamiphos Aldicarb Carbofuran	1%	Reddy, 1984
	M. incognita	Carbofuran Buphos Triazophos	125, 250 ug/ml	Hong & Sethi, 1988
	M. incognita	Carbofuran Aldoxycarb Carbosulfan Triazophos	0.5, 1 & 2%	Hong & Sethi, 1988a
Chichpea	*M. incognita*	Benlate	0.25%	Mani & Sethi, 1984a
	F. oxysporum f. sp. *ciceri* *F. solani*	Carbofuran	2%	
	M. javanica	Aldicarb sulfone Carbofura Bendiocarb Phenamiphos	1-4%	Kaushik & Bajaj, 1981
Pigeonpea	*H. cajani*	Carbofuran Aldicarb, Phorate & Fensulfothion	0.5 - 2.0%	Zaki & Bhatti, 1986a
	M. incognita	Carbofuran Benfuracarb	2%	Mishra, 1986
	Pratylenchus thornei	Aldicarb Carbofuran Fensulfothion Phorate	1.2%	Walia & Seshadri, 1985

experiments and there is an urgent need to move towards on-farm field trials. The adverse effects of parasitic nematodes on vegetative growth of crops and on the damage thresholds have been studied mainly in glasshouse pot experiments. How much loss does a given nematode pest causes in a district, state, or country basis? We do not know. Some work on crop loss estimates has already been initiated but much additional crop loss information is needed. In addition to finding out the direct losses, the nematologists should divulge the indirect losses being caused by nematodes in association with other pathogens. Emphasis has been on the evaluation of different concentrations of insecticides, nematicides and fungicides as seed treatments. Many of the chemicals (particularly the nematicides) are not available and the formulations used are not developed for seed treatment. Published information on seed treatment is apparently of academic interest ; however, testing of biocides that are registered for seed treatment for control of other pests and diseases may provide useful information. Many promising sources of resistance to the root-knot nematodes and to the reniform nematodes have been identified in chickpea and pigeonpea at different nematology centres in India (Sharma, 1987; Darekar & Jagdale, 1987; Chavda *et al.*, 1988). Reactions of these promising sources have not been confirmed at different hot spot locations, and there is a need to conduct multilocation field screening of the identified sources of resistance before using them in the breeding programmes. One of the important areas of research that has been hitherto unexplored is cropping system based management of nematode problems. There are evidences in literature that cropping systems can modify the incidence and intensity of nematode diseases. Studies on influence of traditional cropping systems on nematode populations will be extremely helpful in identifying cropping sequences that can control the nematode caused diseases. A perusal of literature indicates that there are apparently no priorities assigned to work on nematode problems of different pulses in India. On the basis of area under cultivation, the nematode diseases of chickpea, pigeonpea, urdbean, mungbean and horsegram require more attention than other pulses. The first two crops are most important in the post-rainy (*rabi*) and rainy (*kharif*) seasons. It is clear that more research emphasis should be placed on pulse diseases caused by nematodes in India.

REFERENCES

Ahmed, A.; Tiyagi, S.A. & Alam, M.M. (1990). Efficacy of oilseed cakes and nematicides against root-knot nematode attacking black gram. *Sci. Khyber* **3**: 165-170.

Ahmed, S. & Husain, S.I. (1988). Effect of root-knot nematode on qualitative and quantitative characters of chickpea. *Intl. Nematol. Network Newsl.* **5**: 12-13.

Alam, M.M.; Anver, S. & Yadav, A. (1991). Effect of *Meloidogyne incognita* on plant growth and bulk-density of plant residues of pigeonpea *Cajanus cajan*. *Afro-Asian J. Nematol.* **1**: 73-76.

Alam, M.M. & Saxena, S.K. (1986). Effect of preceding crop on the population-growth pattern of *Meloidogyne incognita* and *Tylenchorchynchus brassicae* on chickpea cv. RS-6. *Intl. Chickpea Newsl. No.* **14**: 33-34.

Alam, M.M.; Saxena, S.K. & Khan, A.M. (1978). Suitability of crops to certain ecto-parasitic nematodes. *Acta Bot. Indica* **6** (Suppl.): 205-208.

Alam, M.M.; Saxena, S.K. & Khan, A.M. (1979). Suitability of crops to the reniform, stunt and root-knot nematodes. *In* : *Recent researches in plant sciences* (Ed.—S.S. Bir). Kalyani Publishers, New Delhi. pp. 630-634.

Anver, S. & Alam, M.M. (1989). Effect of root-knot and reniform nematodes on plant growth and bulk-density of plant residues of pigeonpea. *Biol. Wastes* **30**: 245-250.

Anver, S.; Tiyagi, S.A.; Yadav, A. & Alam M.M. (1991). Interaction between *Rotylenchulus reniformis* and *Macrophomina phaseolina* on lentil. *Pak. J Nematol.* **9**: 127-130.

Chahal, P.P.K. & Chahal, V.P.S. (1987). Adverse effect of *Meloidogyne incognita* on the functioning of nodules of mungbean (*Vigna radiata*). *Nematol. medit.* **15**: 13-19.

Chahal, P.P.K. & Chahal, V.P.S. (1987a). Interrelationships between different strains of *Rhizobium*, root-knot nematode and moong (*Vigna radiata*). *Zentralblatt fuer Mikrobiologie* **142**: 129-132.

Chahal, P.P.K. & Chahal, V.P.S. (1988). Interaction of *Rhizobium* and *Meloidogyne incognita* on symbiotic nitrogen fixation in mungbean under different concentrations of iron. *J.Pl. Nutrition* **11**: 821-828.

Chahal, P.P.K. & Chahal, V.P.S. (1989). Effect of *Meloidogyne incognita* on leghaemoglobin, bacteroides and nitrogenase activity of nodules produced by *Rhizobium* spp. on mungbean (*Vigna radiata*). *Pak. J. Nematol.* **7**: 21-25.

Chahal, P.P.K. & Singh, I. (1984). Effect of population density of *Meloidogyne incognita* on pea in association with *Rhizobium leguminosarum*. *J. Res. Punjab Agric.* Univ. **21**: 311-315.

Chahal, V.P.S.; Chahal, P.P.K. & Kapoor, S. (1988). Effect of *Meloidogyne incognita* on symbiosis between *Rhizobium* and mash (*Vigna mungo*). *Crop Improvement* **15**: 95-97.

Chavda, J.C.; Patel, B.A. & Patel, D.J. (1988). Screening of pigeonpea lines to *Rotylenchulus reniformis*. *Intl. Nematol. Network Newsl.* **5**: 28-29.

Darekar, K.S. & Jagdale, G.B. (1987). Screening of chickpea varieties against root-knot nematode, *Meloidogyne incognita Curr. Res. Reptr. Mahatma Phule Agric. Univ.* **3**: 92-93.

Darekar, K.S.; Mhase, N.L. & Shelke, S.S. (1990). Effects of green gram seed treatment with certain nematicides on root-knot nematode and crop yields *Int. Nematol. Network Newsl.* **7**: 4-5.

Darekar, K.S. & Patil, B.D. (1981). Reaction of some cowpea varieties to root-knot nematode. *Indian J. Nematol.* **11** : 83-84.

Das, P.; Sarmah, D.K. & Phukan, P.N. (1988). Reaction of some pulse varieties to root-knot nematode, *Meloidogyne incognita. Indian J. Nematol* **18**: 116.

Devi, L.S. & Gupta, P. (1987). Evaluation of green gram and black gram varieties against cyst nematode, *Heterodera cajani. Indian J. Nematol.* **17** : 341.

Dwivedi, K. & Upadhyay, K.D. (1988). Pathogenicity of the root-knot nematode *Meloidogyne incognita* to French bean. *Intl. Nematol. Network Newsl.* **5**: 20-21.

Edward, J.C. & Singh, K.P. (1979). Interaction between *Heterodera cajani* and *Fusarium udum* on pigeonpea. *Alld. Farmer* **50**: 23-24.

Gaur, H.S. (1986). Invasion and reproduction of the reniform nematode on some cultivars of cowpea and green gram. *Intl. Nematol. Network Newsl.* **3**: 8-10.

Gaur, H.S.; Mishra, S.D. & Sud, V.C. (1979). Effect of date of sowing on the relation between the population density of the root-knot nematode, *Meloidogyne incognita* and the growth of three varieties of chickpea *Cicer arietinum. Indian J. Nematol.* **9**: 152-159.

Gill, J.S. & Singh, R.V. (1989). Nematodes associated with pulse crops and their control. *In* : *Progress in Plant Nematology* (Eds.—S.K. Saxena *et al.*). CBS Publishers & Distributors Pvt. Ltd., Delhi, India. pp. 503-514.

Goel, S.R. & Gupta, D.C. (1986). Interaction of *Meloidogyne javanica* and *Fusarium oxysporum* f. sp. *ciceri* on chickpea. *Indian Phytopath.* **39**: 112-114.

Goel, S.R. & Gupta D.C. (1986a). Interaction of *Meloidogyne javanica* and *Rhizoctonia bataticola* on chickpea (*Cicer arietinum* L.). *Indian J. Nematol.* **16**: 133-134.

Gunasekaran, C.R.; Muthukrishnan, T.S. & Rajendran, G. (1976). Evaluation of chemicals for controlling pigeonpea cyst nematode *Heterodera cajani* in red gram. *Madras Agri. J.* **63**: 382-383.

Gupta, D.C. (1979). Studies on the pathogenicity and relative susceptibility of some varieties of cowpea (*Vigna unguiculata* (L) Walp.) against *Meloidogyne javanica. Forage Res.* **5**: 141-145.

Gupta, D.C. & Kaliram (1981). Studies on the control of *Meloidogyne javanica* infecting chickpea in different types of soil. *Indian J. Nematol.* **11**: 77-80.

Gupta, D.C.; Verma, K.K. & Paruthi, I.J. (1987). Studies on root-knot nematode, *Meloidogyne javanica* on black gram, *Vigna mungo. Indian J. Nematol.* **17**: 177-179.

Gupta, D.C. & Yadav, B.S. (1979). Studies on the pathogenicity of reniform nematode, *Rotylenchulus reniformis* to urd, *Vigna mungo* (L.) Wilezek. *Indian J. Nematol.* **9**: 48-50.

Handa, D.K. (1990). Reaction of urdbean cultivars and losses caused by *Meloidogyne javanica. Intl. Nematol. Network Newsl.* **7**:20-21.

Hasan, A. (1984). Synergism between *Heterodera cajani* and *Fusarium udum* attacking *Cajanus cajan. Nematol. Medit.* **12**: 159-162.

Hong. L.T.T. & Sethi, C.L. (1988). Effect of nematicidal seed soaking on penetration and development of *Meloidogyne incognita* in French bean. *Indian J. Nematol.* **18**: 212-215.

Hong, L.T.T. & Sethi, C.L. (1988a). Effect of nematicidal seed treatment on *Meloidogyne incognita* in French beans (*Phaseolus vulgaris*). *Indian J. Nematol.* **18**: 219-222.

Husain, S.I.; Khan, T.A. & Jabri, M.R. (1985). Studies on root-knot, root-rot and pea mosaic virus complex of *Pisum sativum. Nematol. medit.* **13**: 103-109.

Hussaini, S.S. & Seshadri, A.R. (1976). Resistance in some mung bean (*Vigna radiata*) varieties and breeding lines to the root-knot nematode, *Meloidogyne incognita. Indian J. Nematol.* **6**: 131-137.

Jain, R.K. & Hasan, N. (1986). Efficacy of "neem cake" on fodder production, photosynthetic pigments and nematodes associated with oats and its residual effect on cowpea. *Indian J. Nematol.* **16**: 98-100.

Jain, R.K. & Sethi, C.L. (1987). Pathogenicity of *Heterodera cajani* on cowpea as influenced by the presence of VAM fungi, *Glomus fasciculatum* or *G. epigaeus*. *Indian J. Nematol*. **17** : 165-170.

Jain, R.K. & Sethi, C.L. (1988). Influence of endomycorrhizal fungi *Glomus fasciculatum* and *G. epigaeus* on penetration and development of *Heterodera caajani* on cowpea. *Indian J. Nematol.* **18**: 89-93.

Jaiswal, B.K.; Nath. R.P.; Haider, M.G. & Pathak, K.N. (1987). Efficacy of some pesticides on *Meloidogyne incognita* and *Rotylenchulus reniformis* infesting pigeonpea. *Indian J. Nematol.* **17**: 60-61.

Kaliram & Gupta, D.C. (1980). A note on the efficacy of fresh neem leaf extract in the control of *Meloidogyne javanica* infecting chickpea (*Cicer arietinum*). *Indian J. Nematol.* **10**: 96-98.

Kaliram & Gupta, D.C. (1981). Studies on the control of *Meloidogyne javanica* infected chickpea (*Cicer arietinum*). *Haryana Agric. Univ. J. Res.* **11**: 77-81.

Kaliram & Gupta, D.C. (1982). Efficacy of plant leaves, nematicides and fertilizers alone and in combination against *Meloidogyne javanica* infecting chickpea (*Cicer arietinum* L.). *Indian J. Nematol.* **12**: 221-225.

Kanwar, R.S.; Gupta, D.C. & Walia, K.K. (1988). Effect of nematode inoculation at different intervals and soil types on interaction between *Meloidogyne javanica* and *Rhizocotonia solani* on cowpea. *Indian J. Nematol.* **18**: 112-113.

Kaushik, H.D. & Bajaj, H.K. (1981). Control of root-knot nematode *Meloidogyne javanica* infesting mung bean and gram by seed treatment *Haryana Agric. Univ. J. Res.* **11**: 106-108.

Khan, T.A. & Husain, S.I. (1988). Effect of individual, concomitant and sequential inoculations of *Rhizobium, Rotylenchulus reniformis, Meloidogyne incognita* and *Rhizoctonia solani* on cowpea plant growth, disease development and nematode multiplication. *Indian J. Nematol.* **18**: 232-238.

Khan, T.A. & Husain, S.I. (1988a). Studies on the efficacy of seed treatments with pesticides, oil-cakes, neem-leaf and culture filtrate of *Paecilomyces lilacinus* for the control of diseases caused by the presence of *Rotylenchulus reniformis, Meloidogyne incognita* and *Rhizoctonia solani* either individually or concomitantly on cowpea. *Indian J. Nematol.* **18**: 192-198.

Khan, T.A. & Husain, S.I. (1988b). Response of cowpea cultivars to *Rotylenchulus reniformis*. *Nematropica* **18**: 159-162.

Khan, T.A. & Husain, S.I. (1988c). Studies on the efficacy of *Paecilomyces lilicinus* as bio-control agent against a disease complex caused by the interaction of *Rotylenchulus reniformis*, *Meloidogyne incognita* and *Rhizoctonia solani* on cowpea. *Nematol. medit.* **16**: 229-231.

Kumar, R.; Ahmad, S. & Saxena, S.K. (1988). Disease complex in chickpea involving *Meloidogyne incognita* and *Fusarium oxysporum*. *Intl. Nematol. Network Newsl.* **5**: 12-14.

Kundu, G.G. & Mishra, S.D. (1985). Field evaluation of some soybean germplasm for resistance to stem fly, *Melanagromyza sojae* (Zehnt.), root-knot nematode, *Meloidogyne incognita* and yellow mosaic virus. *Indian J. Ent.* **47**: 230-232.

Mahapatra, B.C. & Padhi, N.N. (1986). Pathogenicity and control of *Rotylenchulus reniformis* on *Cicer arietinum*. *Nematol. Medit.* **14**: 287-290.

Makadia, B.M.; Patel, D.J.; Jogani, D.K. & Shah, H.M. (1987). Reactions of root-knot nematode resistant and susceptible cowpea varieties to reniform nematode, *Rotylenchulus reniformis*. Gujarat *Agric. Univ. Res. J.* **12**: 62-64.

Mani, A. & Sethi, C.L. (1984). Plant growth of chickpea as influenced by initial inoculum levels of *Meloidogyne incognita. Indian J. Nematol.* **14**: 41-44.

Mani, A. & Sethi, C.L. (1984a). Influence of seed treatment on seedling emergence of chickpea in presence of *Meloidogyne incognita, Fusarium oxysporum* f. sp. *ciceri* and *F. solani. Indian J. Nematol.* **14** : 68-69.

Mani, A. & Sethi, C.L. (1987). Interaction of root-knot nematode, *Meloidogyne incognita* with *Fusarium oxysporum* f. sp. *ciceri* and *F. solani* on chickpea. *Indian J. Nematol.* **17**: 1-6.

Mathur, R.L.; Handa, D.K. & Mathur, B.N. (1974). Resistance of varieties of cowpea, green gram and guar to root-knot nematode (*Meloidogyne javanica* & *M. incognita*). *Indian J. Mycol. Pl. Pathol.* **3**: 181-183.

Mehar, H.C.; Agnihotri, N.P. & Sethi, C.L. (1985). Persistence of carbofuran in soil and pea crop and its bioefficacy against soil nematodes. *Indian J. Nematol*. **15**: 155-158.

Mishra, S.D. (1986). Control of root-knot nematode, *Meloidogyne incognita* infesting pieonpea by nematicidal seed treatment. *In: Proc. Nat. Symp. New Dimensions Parasitol.* (Eds.: V.N. Capoor, S.L. Mishra & S.K. Malhotra). Univ. Alld. pp. 34-36.

Mishra, S.D. & Gaur, H.S. (1981). Pathogenicity of *Meloidogyne incognita* and *Rotylenchulus reniformis* to moth bean. *Vigna aconitifolius* Jacq. *Indian J. Nematol*. **11**: 228-230.

Mishra, S.D. & Gaur, H.S. (1981a). Effect of individual and concomitant inoculation with *Meloidogyne incognita* and *Rotylenchulus reniformis* on growth of black gram (*Vigna mungo*). *Indian J. Nematol.* **11**: 25-28.

Mishra, S.D. & Prasad, S.K.(1974). Effect of soil amendments on nematodes and crop yields. I. Oilseed cakes, organic matter, inorganic fertilizers and growth regulators on nematodes associated with wheat and their residual effect on mung. *Indian J. Nematol*. **4**: 1-19.

Mohanty, G.B. & Padhi, N.N. (1987). Pathogenic effects of reniform nematode of varying levels of inoculation on pigeonpea. *Intl. Nematol. Network Newsl.* **4**: 15-16.

Moni, D.N. & Nair, S.K. (1986). Effect of root-knot nematode infestation on nodulation by *Rhizobium* in cowpea *Agri. Res. J. Kerala.* **24**: 225-228.

Mukhtar, J. & Khan, A.A. (1989). Disease complex in chickpea involving *Meloidogyne javanica* and *Sclerotium rolfsii. Intl. Nematol. Network Newsl.* **6**: 31-32.

Nandwana, R.P. (1985). Evaluation of cowpea varieties for resistance to the root-knot nematode, *Meloidogyne incognita* (Kofoid & White, 1919) Chitwood, 1949 on cowpea. *Indian J. Nematol.* **15**: 277.

Nath, R.P.; Banerjee, A.K.; Haider, M.G. & Sinha, B.K. (1979). Studies on nematodes of pulse crops in India. I. Pathogenicity of *Meloidogyne incognita* on gram. *Indian Phytopath*. **32**: 28-31.

Nayak, D.K.; Routaray, B.N. & Das, S.N. (1987). Relative susceptibility of some horse gram cultivars against reniform nematode, *Rotylenchulus reniformis. Intl. Nematol. Network Newsl.* **4**: 19-20.

Padhi, N.N. & Mishra, R.P. (1987). Pathologic reactions of French bean to reniform nematode at varying levels of inoculation. *Indian J. Nematol*. **17**: 128-129.

Padhi, N.N. & Mishra, R.P. (1987a). Control of *Rotylenchulus reniformis* on French bean (*Phaseolus vulgaris*). *Indian J. Nematol*. **17**: 130-131.

Panda, M. & Seshadri, A.R. (1979). Effects of initial inoculum levels of *Rotylenchulus reniformis* on different growth attributes of cowpea, mungbean and urd. *Indian J. Nematol.* **9**: 54-60.

Paruthi, T.J. Gupta, D.C. (1985). Chemical control of root-knot nematode, *Meloidogyne javanica* in mung by seed coating alone and in combination with soil drenching. (Abs.) *Indian J. Nematol.* **15**: 270.

Patel, B.A.; Chavda, J.C. & Patel, D.J. (1989). Screening of certain green gram lines against reniform nematode, *Rotylenchulus reniformis. Intl. Nematol. Network Newsl.* **6**: 25-26.

Patel, G.J.; Shah, H.M. & Patel, D.J. (1977). Screening of cowpea cultivars against root-knot nematodes. *Indian J. Nematol.* **7**: 169-170.

Patel, H.R.; Thakar, N.A.; Patel, B.K. & Patel, C.C. (1987). Interaction between *Meloidogyne incognita* and *Fusarium oxysporum* f. sp. *ciceri* on chickpea variety Chaffy. *Indian J. Nematol.* **17**: 124.

Patel, O.S. & Thakar, N.A. (1986). Efficacy of some systemic chemicals in control of *Rotylenchulus reniformis* infecting mungbean. *Indian J. Nematol.* **16** : 281-282.

Patel, O.S. & Thakar, N.A. (1988). Pathogenicity of the reniform nematode on a susceptible mungbean variety Gujarat-1. *Indian J. Nematol.* **18**: 248-251.

Pathak, K.N.; Nath, R.P. & Haider, M.G. (1985). Effect of initial inoculum levels of *Meloidosgyne incognita* and *Rotylenchulus reniformis* on pigeonpea and their interrelationships. *Indian J. Nematol.* **15**: 177-179.

Phukan, P.N.; Das, P. & Bora, B.C. (1983). Reaction of some cowpea cultivars to root-knot nematode (*Meloidogyne incognita*). *J. Res. Asam Agri. Univ.* **3**: 97-99.

Prasad, D.; Singh, R.P.; Srivastava, K.P.; Agnihotri, N.P. & Sethi, C.L. (1982). Evaluation of mephosfolan for nematode and insect control of okra, cowpea and tomato. *Indian J. Nematol.* **12**: 277-284.

Premachandran, D. (1986). Newer approaches to nematode control. *In: Plant-parasitic nematodes of India: Problems and Progress* (Eds. G. Swarup: & D.R. Dasgupta). Indian Agricultural Research Institute, New Delhi. pp. 446-457.

Raut, S.P. (1980). Effect of initial inoculum levels of *Meloidogyne incognita* on plant growth and rhizobial nodulation of mungbean *Indian Phytopath.* **33**: 351-353.

Raut, S.P. & Sethi, C.L. (1980). Studies on the pathogenicity of *Meloidogyne incognita* on soybean. *Indian J. Nematol.* **10**: 166-174.

Ravichandra, N.G.; Krishnappa, K.; Saifulla, M. & Anilkumar, T.B. (1987). Evaluation of field peas varieties for resistance to root-knot nematode, *Meloidogyne incognita. Intl. Nematol. Network Newsl.* **4**: 9-11.

Reddy, D.D.R. (1975). Pathogenicity and control of root-knot nematode (*Meloidogyne* spp.) infecting chickpea. *Mysore J. Agri. Sci.* **9**: 434-439.

Reddy, P.P. (1984). Efficacy of seed treatment with three nematicides for the control of *Meloidogyne incognita* infecting cowpea, French bean and peas. *Indian J. Nematol.* **14**: 39-40.

Reddy, P.P. (1985). Estimation of crop losses in peas due to *Meloidogyne incognita. Indian J. Nematol. Network Newsl.* **15**: *226.*

Reddy, P.P. (1986). Analysis of crop losses in certain vegetables due to *Metoidogyne incognita. Intl. Nematol Network Newsl.* **3**: 3-5.

Reddy, P.P. & Singh, D.B. (1983). Chemical control of *Meloidogyne incognita* on selected crops. *Nematol. Medit.* **11**: 197-198.

Routaray, B.N.; Nayak, D.K. & Das, S.N. (1987). Resistance of horse gram varieties against *Meloidogyne incognita. Indian J. Nematol.* **17**: 333.

Routaray, B.N.; Sahoo, H. & Das, S.N. (1986). Evaluation of green gram and black gram varieties against reniform nematode, *Rotylenchulus reniformis. Indian J. Nematol.* **16**: 27-29.

Sakhuja, P.K. & Singh, I. (1980). Effect of nematicidal treatments against *Meloidogyne incognita* on mung (Phaseolus aureus) and on bacterial nodulation. *Nematol. medit.* **8**: 201-203.

Saxena, R. & Reddy, D.D.R. (1987). Crop losses in pigeonpea and mungbean by pigeonpea cyst nematode, *Heterodera cajani. Indian J. Nematol.* **17**: 91-94.

Sethi, C.L. & Gaur, H.S. (1986). Nematode management — An overview. *In* : *Plant Parasitic Nematodes of India: Problems and Progress.* (Eds.: G. Swarup & D.R. Dasgupta). Indian Agricultural Research Institute, New Delhi. pp. 424-445.

Sharma, N. (1990). A disease complex of soybean involving nematode *Meloidogyne incognita* and soil inhabiting fungi *Fusarium* sp. and *Pythium* sp. *Intl. Nematol. Network Newsl.* **7**: 17-19.

Sharma, N.K. & Sethi, C.L. (1975). Effects of initial inoculum levels of *Meloidogyne incognita* and *Heterodera cajani* on cowpea and on their population development. *Indian J. Nematol.* **5**: 148-154.

Sharma, N.K. & Sethi, C.L. (1976). Interrelationship between *Meloidogyne incognita*, *Heterodera cajani* and *Rhizobium* sp. on cowpea (*Vigna sinensis* L.) Savi. *Indian J. Nematol.* **6**: 117-123.

Sharma, N.K. & Sethi. C.L. (1976a). Reaction of certain cowpea varieties to *Meloidogyne incognita* and *Heterodera cajani. Indian J. Nematol* **6**: 99-102.

Sharma, R. & Swarup, G. (1988). Pathology of cyst nematodes. Malhotra Publishing House, New Delhi.

Sharma, R.K. & Tiagi, B. (1988). Effect of *Meloidogyne incognita* infestation on leghaemoglobin content of pea nodules. *Indian J. Nematol.* **18**: 362.

Sharma, S.B. (1985). *A world list of nematode pathogens associated with chickpea, groundnut, pearl millet, pigeonpea and sorghum*. ICRISAT Pulse Pathology Progress Report 42. International Crops Research Institute for Semi-Arid Tropics, Patancheru, India.

Sharma, S.B. (1987). Nematology, report of work. *Progress Report* 1: International Crops Research Institute for the Semi-Arid Tropics, Patancheru, India.

Sharma, S.B. & McDonald, D. (1990). Global status of nematode problems of groundnut, pigeonpea, chickpea, sorghum and pearl millet and suggestions for future work. *Crop Protection* **9**: 435-458.

Sharma, S.B. & Nene, Y.L. (1989). Interrelationship between *Heterodera cajani* and *Fusarium udum* in pigeonpea. *Nematropica* **19**: 21-28.

Sharma, S.B. & Nene, Y.L. (1990). Effect of *Fusarium udum*—alone and in combination with *Rotylenchulus reniformis* or *Meloidogyne* spp. on wilt incidence, growth of pigeonpea and multiplication of nematodes. *Intl. J. Trop. Pl. Dis.* **8**: 95-101.

Sharma, S.B. & Nene, Y.L. (1990a). Effect of soil solarization on nematodes parasitic to chickpea and pigeonpea. *J. Nematol.* **22** S: 658-664.

Sikora, R.A. & Greco, N. (1990). Nematode parasites of food legumes. *In: Plant Parasitic Nematodes in Subtropical and Tropical Agriculture.* (Eds.—M. Luc, R.A. Sikora & J. Bridge). CAB International, U.K. pp. 181-235.

Singh, D.B. & Reddy, P.P. (1981). Influence of *Meloidogyne incognita* infestation on rhizobium nodule formation in French bean. *Nematol. Medit.* **9**: 1-5.

Singh, D.B. & Reddy, P.P. (1982). Cultivars of cowpea resistant to root-knot nematode in India. *Trop. Pest Manag.* **28**: 185-186.

Singh, D.B. & Reddy, P.P. (1982a). Chemical control of *Meloidogyne incognita* infecting cowpea. *Indian J. Nematol.* **12**: 196-197.

Singh, D.B. & Reddy, P.P. & Shyam Sundar, J. (1984). Histological, histopathological and histochemical investigations on root-knot nematode resistant and susceptible lines of cowpea. *Nematol. Medit.* **12**: 213-219.

Singh, R.V.; Gupta, D.C. & Walia, K.K. (1986). Effect of *Rhizocotonia solani* culture filtrate on hatching of *Meloidogyne javanica* larvae. *Indian Phytopath*, **39**: 624-625.

Sitaramaiah, K. (1984). *Plant parasitic nematodes of India.* Today & Tomarrow's Printers & Publishers, New Delhi.

Srivastava, A.S.; Upadhyay, K.D. & Singh, G. (1974). Effect of root-knot nematode, *Meloidogyne javanica* on gram crop. *Indian J. Nematol.* **4**: 248-251.

Sundaram, R. & Velayutham, B. (1988). Relative efficacy of some insecticides and neem cake in the control of *Rotylenchulus reniformis* and *Helicotylenchus dihystera* affecting garden bean. *Indian J. Nematol.* **18**: 329-331.

Thakar, N.A. & Patel, C.C. (1984). Screening of few cowpea varieties against reniform nematode, *Rotylenchulus reniformis* Linford & Oliveira, 1940. *Indian J. Nematol.* **14**: 204.

Thakar, N.A. & Patel, C.C. (1985). Reaction of some varieties of mungbean, black gram and okra to root-knot nematodes, *Meloidogyne incognita* and *M. javanica. Indian J. Nematol.* **15**: 102-103.

Thakar, N.A.; Patel, H.R.; Patel, B.K. & Patel, C.C. (1986). Breaking of resistance to *Fusarium* wilt in resistant chickpea by root-knot nematodes. *Madras Agri. J.* **73**: 410-411.

Tiyagi, S.A. & Alam, M.M. (1986). Root-knot development and plant growth of chickpea in relation to inoculum density of *Meloidogyne incognita. Intl. Chickpea Newsl. No.* **14**: 30-32.

Tiyagi, S.A. & Alam, M.M. (1987). Pathogenicity of *Rotylenchulus reniformis* to two cultivars of chickpea. *Intl. Nematol. Network Newsl.* **4**(1): 19-21.

Tiyagi, S.A. & Alam, M.M. (1988). Pathogenicity of the root-knot nematode, *Meloidogyne incognita* to mungbean *Intl. Nematol. Network Newsl.* **5**(3): 22-24.

Tiyagi, S.A. &s Alam, M.M. (1989). Effect of different inoculum densities of stunt nematode, *Tylenchorhynchus brassicae* on the growth of chickpea. *Intl. Chickpea Newsl. No.* **20**: 14-15.

Tiyagi, S.A.; Anver, S. & Alam, M.M. (1989). Studies on the pathogenicity of root-knot and reniform nematodes on pegeonpea. *Intl. Nematol. Network Newsl.* **6**(4): 12-15.

Tiyagi, S.A.; Zaidi, S.B.I. & Alam, M.M. (1988). Interaction between *Meloidogyne javanica* and *Macrophomina phaseolina* on lentil. *Nematol. Medit.* **16**: 221-222.

Upadhyay, K.D. & Dwivedi, K. (1987). Analysis of crop losses in pea and gram due to *Meloidogyne incognita. Intl. Nematol. Network Newsl.* **4**: 6-7.

Upadhyay, K.D. & Dwivedi, K. (1987a). Root-knot nematode, *Meloidogyne javanica* breaks wilt resistance in chickpea variety "Avrodhi". *Curr. Sci.* **56** : 915-916.

Varshney, V.P.; Khan, A.M. & Saxena, S.K. (1987). Interaction between *Meloidogyne incognita, Rhizoctonia solani* and *Rhizobium* species on cowpea. *Intl. Nematol. Network Newsl.* **4**: 11-14.

Verma, K.K. & Gupta, D.C. (1986). Studies on the effect of NPK fertilizers singly and in combination with pesticides on cowpea (*Vigna unguiculata* (L.) Walp) infected with *Meloidogyne javanica*. *Indian J. Nematol.* **16**: 51-55.

Vijayalakshmi, M.; Mishra, S.D.; Haque, M.M. & Goswami, B.K. (1989). Nematicidal efficacy of some wild plants against pigeonpea cyst nematode, *Heterodera cajani*. *Intl. Nematol. Network Newsl.* **6**: 21-24.

Walia, K.K. & Gupta, D.C. (1986). Effect of fungus, *Rhizoctonia bataticola* on the population development of pigeonpea cyst nematode, *Heterodera cajani* on cowpea (*Vigna unguiculata*). *Indian J. Nematol.* **16**: 131-132.

Walia, K.K. & Gupta, D.C. (1986a). Antagonism between *Heterodera cajani* and *Rhizoctonia solani* on cowpea (*Vigna unguiculata* (L.) Walp.) *Indian J. Nematol.* **16**: 41-43.

Walia, R.K. & Seshadri, A.R. (1985). Chemical control of *Pratylenchus thornei* on chickpea through seed treatment. *Intl. Chickpea Newsl.* **13**: 32-34.

Yien, B.R.; Singh, H. & Chaabra, H.K. (1977). Effect of pesticides and fertilizers singly and in combination on the root-knot nematode infesting mung. *Indian J. Nematol.* **7**: 117-122.

Zaki, F.A. & Bhatti, D.S. (1986). Pathogenicity of pigeonpea cyst nematode, *Heterodera cajani*, Koshy, 1967 on some pulse crops. *Indian J. Nematol.* **16**: 30-35.

Zaki, F.A. & Bhatti, D.S. (1986a). Control of pigeonpea cyst nematode, *Heterodera cajani* Koshy, 1967 by chemical seed treatment. *Indian J. Nematol.* **16**: 106-107.

Chapter 8

Nematode Problems of Millets and their Control

I.J. Paruthi & H.K. Bajaj
Department of Nematology, CCS Haryana Agricultural University, Hisar-125 004, India

MILLETS are warm weather cereals with small grains and form the staple diet for the millions of people in Asia and Africa. They are utilised both as grains and forage crops and give some returns even in rainfed areas. Millets are classified as major and minor millets on the basis of their grain size, production and utility:

Major Millets

a) Sorghum (Giant millet): *Sorghum bicolor* (L.) Moench. and *S. vulgare* Pers.
b) Pearl millet: *Pennisetum typhoids* (Burn.) Stapf & Hubb.

Minor Millets

a) Finger millet: *Eleusine coracana* (L.) Gaertn.
b) Foxtail millet: *Setaria italica* (L.) P. Beauv.
c) Common millet (Little millet): *Pacnium miliare* Lamk.
d) Proso millet: *P. miliaceum* L.
e) Kodo millet: *Paspalum scrobiculatum* L.
f) Barnyard millet: *Echinochloa colonum* L. (Link)

In India major millets (sorghum and pearl millets) are chiefly grown in the states of Gujarat, Haryana, Madhya Pradesh, Maharashtra, Rajasthan and Uttar Pradesh. These crops are subjected to number of insect pest and pathogens. All these millet crops have also been reported to be infected with various nematode species. Most of the studies are confined to nematode associations with a crop. However, a few reports on some other aspects like biology, life history, pathogenicity, interaction with other organisms, screening of germplasms for resistance, etc. are also available particularly on sorghum, pearl millet and finger millet (*ragi*). The important nematode problems on these crops, with special reference to the reports from India, are briefly discussed hereunder.

SORGHUM

There are about 100 species of plant-parasitic nematodes reported to be associated with this crop. Cyst nematodes (*Heterodera* spp.), root-knot nematodes (*Meloidogyne* spp.), stunt nematodes (*Tylenchorhynchus* spp.) and lesion nematodes (*Pratylenchus* spp.) are the main nematode problems in the field.

CYST NEMATODES (*Heterodera* spp.)

Sorghum cyst nematode (*Heterodera sorghi*) and corn cyst nematode (*H. zeae*) have been recorded parasitising the roots of *Sorghum vulgare* under field conditions. Larvae of cereal cyst nematode (*H. avenae*) nut grass cyst nematode (*H.mothi*) and *H. delvii* have also been observed to penetrate the roots of this crop under experimental conditions but they either fail to develop further or form only a few cysts (cf. Sharma & Swarup, 1984). *H. sorghi* is more widely distributed and is considered as a serious problem on this crop in certain parts of India.

Heterodera sorghi

H. sorghi first recorded from village Raispur (district Ghaziabad, U.P., India) on the roots of *Sorghum vulgare* by Jain *et al.* (1982) is now known to occur in the states of Andhra Pradesh, Delhi, Haryana, Himachal Pradesh, Jammu and Kashmir.

Host range: Sorghum (*S. vulgare*) is the host of this nematode under field conditions. Barnyard millet, finger millet, Italian millet, kodo millet, pearl millet, wheat, barley and nut grass, etc. have also been reported to be attacked by this nematode under experimental conditions (Sharma & Swarup), 1984; Srivastava & Sethi, 1987).

Biology. Emergence of larvae from the cysts can take place at 20-35°C, with optimum being 28°C (Sharma & Swarup, 1984; Sakhuja & Singh, 1985; Srivastava, 1985). Temperature lower than 20°C and higher than 35°C adversely affects the emergence of larvae from the cysts. Larval penetration starts within 24 h and increases with passage of time till 5th day. Second, third and fourth moults occur on 4, 11 and 14 days after inoculation, respectively. Adult females are encountered on 19th day and females with egg sac attached to the posterior portion are formed on 20th day. Cysts with second stage larvae are reached by 26th day. The life cycle is completed in 20-30 days (Sharma & Swarup, 1984; Sakhuja & Singh, 1985; Srivastava, 1985). The average number of eggs and larvae from the cysts ranges between 155-322.8 depending upon the cultivar (Srivastava & Sethi, 1988). The nematode completes several generations on a single crop (Anon., 1986). Multiplication of this nematode is better in clay loam than sandy soil (Mittal & Dhawan, 1989).

Host-parasite relationship: There are no clear cut symptoms caused by this nematode on above ground parts of the plants. The infested plants are normally poor in growth with thin yellow and narrow leaves and thin stalk (Sakhuja & Singh, 1985). The manifestation of such symptoms depends upon the number of nematode larvae present in the soil and an inoculum level of 500 larvae/plant is sufficient to cause significant reduction in plant growth parameters (Mittal, 1988).

Control : Sorghum vars. 296-B and CK 608 are found to be moderately resistant to *H. sorghi* (Srivastava & Sethi, 1988). Seed dressing with phenamiphos or aldicarb @ 1-2 per cent a.i. w/w can reduce the population of *H. sorghi* (Anon., 1986). *Glomus fasciculatum, Pasteuria penetrans* are considered to be most potent biocontrol agents against *H. sorghi* (Sharma & Swarup, 1986).

Heterodera zeae

H. zeae is mainly a parasite of roots of *Zeae mays* (maize). It has also been reported infecting the roots of sorghum in U.P. (Singh & Edward, 1979). Studies on chemical composition of various plant parts in *H. zeae* infested sorghum plants by Singh *et al.* (1979) indicate that there is an increase in concentration of organic matter, ash and calcium whereas potassium remains unchanged in the leaves and slightly decreases in the affected roots. The concentration of other constituents like crude fat, crude fibre, gross energy, total and reducing sugar (analysis only in leaves) is decreased in the affected plants. However, the number of amino acids and their quality is decreased in the roots and leaves of affected plants.

ROOT-KNOT NEMATODES (*Meloidogyne* spp.)

Root-knot nematodes (*Meloidogyne* spp.) are known to parasitize sorghum in different parts of the world. However, *M. incognita* and *M. naasi* are known to cause more damage. Race 3 (Taylor & Sasser, 1978) of cotton root-knot nematode M. incognita is a problem in sorghum.

Meloidogyne incognita

Host range: It is a polyphagous pest and also infects other millets besides sorghum.

Host-parasite-relationship: Root-knot nematodes are basically parasites of roots. The formation of galls on the roots is the main symptom which are initially small in size and in later stage may coalesce to form large galls. The above ground symptoms are those of slow debility of roots in its function of nutrients are water uptake and translocation. The plants may be dwarfed, yellowish with smaller foliage and may show nutrient deficiency symptoms (Orr, 1967; Dasgupta & Gaur, 1986). Orr (1967) reported 15 per cent reduction of dry weight of grain sorghum infested with *M. incognita*. Babatola & Idowu (1990) reported reduction in growth, vigour and yield of the diseased sorghum plants. Anatomical response of grain sorghum to *M. incognita* was studied in detail by Orr (1967) and Orr & Morey (1974, 1978). The nematode causes syncytia in the cortex or stele of lateral roots. Giant cells develop either singly with few nuclei or in groups with many nuclei. The pericycle and endodermis are interrupted at the site of nematode invasion. Large areas of these tissues are absent for one third of the circumference of stele and extend 1.5 mm longitudinally along the root. In areas where pericycle and endodermis are absent, the parenchyma of cortex extends to the vascular elements and normal xyelem surrounding giant cells extends into region of cortex. Root-knot galls appear on the sorghum roots as elongate swellings, discrete knots or swellings with root proliferation.

Control: Several workers have screened various varieties of sorghum against *Meloidogyne* spp. Funk's hybrid 8 and Haskel Harris (Johnson *et al.*, 1977), hybrid Funk Brand and Pioneer Brand B 185 (McSorley *et al.*, 1987), Ironian local (Babatola & Idowu, 1990), E 5764 m, E 6250, E 6256, E 6263 and E 6487 (Anon., 1975) have shown resistance to *Meloidogyne* spp. Dunn (1988) has advocated the use of Temik 15G (aldicarb), Furadan 15G (carbofuran) and counter 15G (tebufos) for control of *Meloidogyne* spp. infesting sorghum in Florida (U.S.A.).

Meloidogyne naasi

This polyphagous nematode is very common in U.S.A. and other European countries. Michael *et al.* (1973 a,b) demonstrated that this species contained at least 5 host races out of which race 5 from Kansas is the only race to reproduce on sorghum.

Host range: It is a polyphagous pest. Ediz & Dickerson (1976) reported *Hordeum vulgare, Secale cereale* and *Sorghastrum nutans* as good hosts of race 5 of *M. naasi* besides *S. bicolor*.

Biology : Second stage larvae of this nematode penetrate the roots just behind the root cap and then move through the tissues inter and intra-cellularly. The optimum soil temperature for its development is 26°C. Completion of life cycle takes place in 34, 36 and 40 days at 26, 30 and 34°C respectively (Ediz & Dickerson, 1976).

Host-parasite relationship : Infected plants generally are shorter in height with smaller diameter and pale green yellowish leaves and have knots on the roots. Giant cells are initiated either in procambial region or in the young phloem when the giant cells develop in cortex, corresponding areas of the vascular system do not have an endodermis, pericycle of phloem fibre. All the nematodes developing in cortical sites are males (Ediz & Dickerson, 1976).

Control : Not much information is available on control aspects of this nematode. Ediz & Dickerson (1976) reported that use of corn (a poor host) and soybean (non-host) in cropping sequences keeps the population of this nematode below economic damaging level.

LESION NEMATODES (*Pratylenchus* spp.)

Four species of *Pratylenchus* (*P. zeae, P. thornei, P. brachyurus* and *P. hexincisus*) have been reported parasitising sorghum roots. Out of these, *P. zeae* is of economic importance on sorghum.

Partylenchus zeae

Host range : This nematode is polyphagous. Sudheer & Sivakumar (1983) compared the reproduction of *P. zeae* on maize, sorghum finger millet and foxtail millet. Maize and sorghum were found to be its good hosts. The reproduction factors ($R_f = P_f/P_i$) were 6.4 and 5.9 on maize and sorghum respectively while it ranged from 1.6 to 2.4 on the other millets.

Host-parasite relationship : This nematode is migratory endo-parasite and forms lesions on the roots through which several parasitic fungi and bacteria may enter. The affected plants are normally stunted with yellowing of leaves. Bee-Rodriguez & Ayala (1977b) reported that 1500 larvae of *P. zeae* suppress the top and root growth of *S. bicolor*.

Pathogenesis of primary roots of sorghum seedlings by *P. zeae* results in a decreased H CN P (Hydrocyanic potential) of infected plants. However, such response is not observed in the lateral roots which may be due to low concentration of dhurin (Starr *et al.*, 1984) in lateral roots (3-5 g/kg dry weight) than primary roots (25 g/kg dry weight).

Association with other organisms : Bee-Rodriguez & Ayala (1977a) reported that combination of *P. zeae* with *Curvularia* species, *Fusarium moniliforme, Rhizoctonia solani* or *Macrophoma* species retard top growth as do *F. moniliforme* and *R. solani* alone. Root growth is suppressed by all nematode-fungus combinations. Greatest damage however results from *P. zeae* with *Curvularia* species. Severe root necrosis is caused by *P. zeae* alone and in combination with *R. solani* and by *Curvularia* species and *F. moniliforme. P. zeae* and *Qunisulcius acutus* are predominant species of nematodes on sorghum in U.S.A. Curazema-Teran & Trevathan (1985) studied the combined effect of these nematodes on sorghum and also reported that reproduction of *P. zeae* is significantly less when combined with *Q. acutus*.

Control : Application of Furadan or Temik @ 1 kg a.i./ha has been found effective in suppressing the population of *P. zeae* in sorghum (Boast, 1984; Laxmanan, 1986). Dean *et al.* (!957) found that ethylene dibromide (W-85) @ 7.5 U.S. gallons/acre is effective in controlling *P. zeae* on sorghum. VAM (vesicular-arbuscular mycorhizae) fungi can be used to manage the population of *P. zeae* on sorghum (Jain & Hasan, 1986a).

OTHER NEMATODES

Besides cyst, root-knot and lesion nematodes several others like *Helicotylenchus dihystera, Longidorus africanus, L. elongatus, Quinisulcius acutus, Belonolaimus longicaudatus, Hoplolaimus indicus, Tylenchorhynchus mashhoodi* and *T. vulgaris* are frequently encountered in the rhizosphere of sorghum and may be involved in crop damage.

Jain & Hasan (1986b) recorded high population of *H. dihystera* from the rhizosphere of forage sorghum from U.P., India. Pathogenicity trials conducted by them revealed that 10,000 nemas per plant cause significant reduction in plant growth parameters of sorghum. They also screened 21 varieties of sorghum against *H. dihystera* and found Etawah-2 and JHS 822 to be the least susceptible. Hunt *et al.* (1973) recorded a marked reduction in population of *Helicotylenchus* sp. on *S. bicolor* in U.S.A. by use of compost at the rate of 8, 16 and 32 tonnes/ha.

Longidorus africanus and *L. elongatus* are commonly found in the rhizosphere of sorghum in Sudan and Canada. The affected plants normally remain stunted and chlorotic with shortened, thick deformed roots. The grain yield is reduced by 41.4% at an inoculum level of 65 *L. elongatus*/100 g soil (Semokina, 1975). High reproduction of *L. africanus* has been reported at 80-85^{0}C (Lamberti, 1965).

Quinisulcus acutus is a predominant nematode in sorghum in Mississippi state of U.S.A. Rapid increase in nematode population occurs in the sorghum rhizosphere (Cuarezema-Teran & Trevethan, 1985). The affected roots are generally smaller in size and show lesions and discolouration.

PEARL MILLET

Among major millets, pearl millet, *Pennisetum typhoides* Stapf & Hubbard (*H. bajra*) occupies an important position as an staple food in India. With the introduction of hybrids, the diseases and pest problems have increased considerably. Rot-knot nematodes (*Meloidogyne incognita* and *M. javanica*), stunt nematode (*Tylenchorhynchus vulgaris*) and root-lesion nematode (*Pratylenchus* sp.) have received more attention as pests of pearl millet.

ROOT-KNOT NEMATODES (*Meloidogyne* spp.)

Occurrence of root-knot nematodes, *M. incognita* and *M. javanica* on pearl millet was first recorded from Rhodesia (Collin, 1938; Martin, 1955). In India, Handa *et al.* (1971) first recorded infestation of pearl millet roots with *M. javanica* from Rajasthan. The infestation is widespread in state of Gujarat where pearl millet is cultivated as a rotational crop with tobacco (Vaishnav, 1977). These nematodes have got a wide host range (Vaishnav & Sethi, 1977).

Biology : Life cycle of *M. incognita* on pearl millet is completed in 27 days from second to second stage juveniles at 22-32°C (Vaishnav, 1977).

Host-parasite relationship : Underground and above ground symptoms on pearl millet are similar to those described for sorghum. The galls formed are regular and more or less rounded with a diameter of 1-2 mm and generally show conspicuous swellings only on one side of the roots. Histopathological changes induced are typical of other graminaceous hosts like sorghum (Vaishnav, 1977). Pathogenicity trials conducted by Vaishnav & Sethi (1978) gave a conclusive evidence of destructive potential of *M. incognita* on pearl millet. They reported significant reduction in plant growth parameters at the above 1,000 larvae per plant. The plants inoculated with 1,000 and 10,000 larvae exhibited varying degree of chlorosis and stunting.

Association with other organisms : Vaishnav & Sethi (1979) noticed that coinoculation of fungus, *Sclerospora graminicola* and nematode advance the appearance of downy mildew disease of pearl millet by 12-13 days. More incidence of diseases (83-96%) occurs by 16-20th day after seed germination in the presence of *M. incognita*. Nematode and fungus interact synergistically in adversely affecting the plant growth (Vaishnav & Sethi, 1979). Combined effect of *M. incognita* with *Tylenchorhynchus vulgaris* on reduction of plant growth of pearl millet (hybrid BK 560) is additive although *T. vulgaris* exhibits antagonistic relationship towards *M. incognita* (Vaishna & Sethi, 1978).

Control : Application of DD mixture (11.2 kg a.i./ha), EDB (33.6 kg a.i./ha), phenamiphos (6.7 kg a.i./ha) and aldicarb (6.7 kg a.i./ha) reduces the population of *M. incognita* on millet (Johnson & Burton, 1977). Soil application of DD at 400 l/ha, aldicarb 2 kg a.i./ha, DBCP at 30 and 40 l a.i./ha is highly effective in controlling root-knot nematodes in pearl millet (Vaishnav, 1977). Johnson *et al.*

(1977) reported pearl millet variety Gahi 3 as resistant to *M. incognita, M. javanica* and *M. hapla*. Studies from Gujarat have shown that varieties 5054 A, BJ 104, GHB 27, Inbred 355 and Inbred 358 are resistant to mixed populations of *M. incognita* and *M. javanica* (Anon., 1990).

STUNT NEMATODE (*Tylenchorhynchus vulgaris*)

This nematode occurs most frequently in sandy soils of A.P., Bihar, Delhi, Haryana, H.P., Punjab, Rajasthan and U.P. states of India (Upadhyaya & Swarup, 1976).

Host range : This species is apparently polyphagous. Vaishnav (1977) reported that the nematode reproduces well on all the 17 varieties of 10 plant species tested by him.

Biology : The nematode completes its life cycle in 15 days at 24-35°C on pearl millet (Vaishnav, 1977).

Host-parasite relationship: The reduction in plant growth parameters in pearl millet was observed at and above 1000 nemas per plant. The rate of multiplication of nematodes was highest (133.9 times) in the treatments with inoculum level of 10 nematodes and lowest (10.6 times) at 10,000 nematodes per plant (Vaishnav, 1977).

Control : *T. vulgaris* can be successfully controlled by soil application of DD at 400 l/ha carbofuran and aldicarb @ 2kg a.i./ha and DBCP @ 30 and 40 l a.i./ha (Vaishnav, 1977).

LESION NEMATODES (*Pratylenchus* spp.)

Amongst various species of lesion nematodes, *P. mulchandi* and *P. zeae* are common on pearl millet in India.

Pratylenchus mulchandi

This nematode was reported for the first time from the roots of *Pennisetum typhoides* from Rajasthan, India by Nandakumar & Khera (1971).

Biology : The nematode normally prefers to feed near the root tip between meristematic region and region of differentiation. The life cycle is completed in 24-36 days at room temperature (25-30°C): 6-10 days for eggs to hatch, 15-20 days for second stage larvae to reach maturity, 3-6 days from maturation to egg laying. Reproduction is apparently by parthenogenesis (Nandkumar & Khera, 1973).

Host-parasite relationship : Epidermal and cortical tissues are affected by the invasion of this nematode (Nandakumar *et al.*, 1970a, b).

Association with other organisms: *P. mulchandi* alone or in combination with fungi or bacteria isolated from the rhizosphere of pearl millet do not cause any pathological symptoms on pearl millet (Nandakumar *et al.*, 1969a, b).

LANCE NEMATODE (*Hoplolaimus indicus*)

Hoplolaimus indicus is also commonly found associated with pearl millet in Rajasthan, India (Nandakumar & Khera, 1971).

Host range : It is a polyphagous pest. Among millets sorghum, pearl millet and little millet have been reported as hosts of this nematode (Dasgupta & Seshadri, 1969; Raman & Das, 1969).

Biology : Nandakumar & Khera (1973) conducted studies on the feeding behaviour of *H. indicus* on pearl millet. The nematode prefers to feed on cortical tissues. They feed as semi-endoparasite embedding their anterior part of body deep into cortical tissues. Dasgupta *et al.* (1969, 1970) reported that prior feeding is necessary for oviposition and development of successive post-embryonic stages on pearl millet. The life cycle from egg to egg stage is completed in 27-36 days at 28-32°C. Oviposition at various pH ranges showed that almost equal number of eggs are deposited at pH 4.5-7.0 and no eggs are deposited at pH 2.0 or 10.0.

Association with other organisms : *H. indicus* when combined with various seed borne fungi like Alternaria alternata, Cochiliobolus lunatus and Fusarium semitechum enhances the harmful effects on the germination and seedling emergence of pearl millet (Nath *et al.*, 1986).

OTHER NEMATODES

Tikyani & Khera (1968) reported common occurrence of *Telotylenchus loofi* in the samples from millets from Rajasthan, India. In about 30% of the samples the population *T. loofi* was 4 times as high as in rest of the samples. Another species *T. housei* has been reported to cause poor plant growth of millets in U.S.A. Tikyani & Khera (1970) reported that *T. loofi* fails to survive during hottest months of May and June but is capable of surviving for 9 months without host.

MINOR MILLETS

Among minor millets, finger millet *Eleusine coracana* alone accounts for more than 40 per cent area and 60 per cent production of small millets. In India this crop is widely grown in A.P., Bihar, Gujarat, Karnataka, M.P., Maharashtra, Orissa and U.P. Only little information is availab e on the nematode problems on other millet crops (foxtail millet, common millet, barnyard millet and kodo millet).

Srivastava & Swarup (1975) reported *Setaria italica* as a good host of *Heterodera zeae*. Lal & Mathur 1988) recorded the occurrence of *Aphelenchoides besseyi* in the seeds of *S. italica* from India. Padhi & Das 1986) recorded high population of *Helicotylenchus abunaamai* on *Panicum milliare* from Orissa, India. These millets are not severely damaged by the nematodes. Nevertheless, they may be playing very important role in the ecology and epidemiology of these nematode diseases by acting as an alternate/reservoir hosts or as agents of dispersal of nematode pests of crops of economic importance.

CONCLUSION

Although certain millets, particularly sorghum, pearl, millet and ragi are cultivated in many parts of the world but not much attention has been paid to study detailed aspects of diseases caused by plant-parasitic nematodes. Since sorghum and millets meet the food requirements of poor people of the world and of cattle, there is an urgent need to concentrate our future lines of work in identification of major nematode problems of these crops and their management. It is also important to assess the role of these millets in the disease development, spread and quarantine of nematode pests of important agricultural crops. Information generated on nematode behaviour, biology, host-parasite relationship, etc. can also be exploited for the inclusion of these small grain crops in the cropping sequences in nematode management strategies.

REFERENCES

Anonymous (1975). East African Agriculture and Forestry Research Organization. Annual Report, 1974: 128-133.

Anonymous (1986). Two decades of Nematological Research and Training. Division of Nematology, I.A.R.I., New Delhi. pp. 26-27.

Anonymous (1990). Reaction of various crop germplasms to plant parasitic nematodes. Deptt. of Nematology, B.A. College of Agriculture, Gujarat Agricultural University, Anand Campus, Anand-388 110, India.

Babatola, J.O. & Idowu, K.A. (1990). Responses of some sorghum cultivars to the root-knot nematode, *Meloidogyne incognita* in *Nigeria. Nematol. Medit.* **18**: 65-66.

Bee-Rodriguez, D. & Ayala, A. (1977a). Interaction of *Pratylenchus zeae* with four soil fungi on sorghum. *J. Agric. Univ. Puerto Rico* **61**: 501-506.

Bee-Rodriguez, D. & Ayala, A (1977b). Nematodes associated with sorghum in Puerto Rico. *Nematropica* **7**: 16-20.

Bost, S.C. (1984). Evaluation of nematodes for control of lesion and lance nematodes on grain sorghum. *Fungicide and Nematocide Tests. American Phytopathol. Soc.* (1985) **40**: 98.

Collin, J.C. (1938). Nematode investigations. *In*: Report of tobacco research board for the year ending December, 1937, *Rhodesia Agric. J.* **35**: 424.

Cuarezma-Teran, J.A. & Trevathan, L.E. (1985). Effects of *Pratylenchus zeae* and *Quinisulcius acutus* alone and in combination on sorghum. *J. Nematol.* **17**: 169-174.

Dasgupta, D.R. & Gaur, H.S. (1986). The root-knot nematodes, *Meloidogyne* spp. In India. *In : Plant Parasitic Nematodes of India: Problems & Progress* (Eds.—G. Swarup & D.R. Dasgupta). I.A.R.I., New Delhi, India. pp. 139-171.

Dasgupta, D.R. & Seshadri, A.R. (1969). Embryology and life history studies on the lance nematode, *Hoplolaimus indicus* Sher. *Ist. All India Nematology Symposium, New Delhi*., p. 31.

Dasgupta, D.R.; Nand, S. & Seshadri, A.R. (1970). Culturing, embryology and life history studies on the lance nematode, *Hoplolaimus indicus. Nematologica* **16**: 235-248.

Dunn, R.A. (1988). Sorghum and nematodes. *Nematology Plant Protection Pointer* No. 20: p. 2.

Ediz S.A. & Dickerson, O.J. (1976). Life cycle, pathogenicity, histopathology, and host range of race 5 of the barley root-knot nematode. *J. Nematol.* **8**: 228-232.

Handa, D.K.; Mathur, B.N. & Bhargava, L.P. (1971). Occurrence of root-knot on pearl millet. *Indian J. Nematol.* **1**: 24.

Hunt, P.G.; Hortenstine, C.C. & Smart, G.C. Jr. (1973). Response of plant parasitic nematode populations to composted municipal refuse. *J. Environ. Quality* **2**: 264-266.

Jain, K.K.; Sethi, C.L.; Swarup, G. & Srivastava, A.N. (1982). *Heterodera sorghi* sp. *n.*, a new cyst forming nematode parasitising sorghum. *Revue de Nematol.* **5**: 201-204.

Jain, R.K. & Hasan, N. (1986a). Association of vesicular-arbuscular mycorhizal (VAM) fungi and plant parasitic nematodes with forage sorghum (*Sorghum bicolor* L.). *Sorghum Newsl.* **29**: 84.

Jain, R.K. & Hasan, N. (1986b). Relative susceptibility of some selected forage sorghum (*Sorghum bicolor* L.) varieties to *Helicotyleenchus dishystera. National Conference Plant Parasitic Nematodes of India: Problems & Progress. I.A.R.I., New Delhi* (Dec. 17-20, 1986). p. 107.

Johnson, A.W. & Burton, G.W. (1977). Influence of nematicides and yield of sorghum Sudan grass hybrids and millets. *Plant Dis. Reptr.* **61**: 1013-1017.

Johnson, A.W.; Burton, G.W. & Wright, W.C. (1977). Reaction of sorghum Sudan grass hybrids and pearl millet to three species of *Meloidogyne. J. Nematol.* **9**: 352-353.

Lal, A. & Mathur, V.K. (1988). Record of *Aphelenchoides besseyi* on *Setaria italiaca. Indian J. Nematol.* **18**: 131-132.

Lamberti, F. (1965). Pathogenicity of *Longidorus africanus* on selected field crops. *Plant Dis. Reptr.* **53**: 421-424.

Martin, G.C., (1955). Plant and soil nematodes of the Federation of Rhodesia and Nyasaland: Preliminary investigations nematodes catalogued under hosts and associated plants. *Rhodesia Agric. J.* **52**: 345-361.

McSorley, R.; Lamberts, M.L.; Parrado, J.L. & Reymolds, J.S. (1987). Reaction of sorghum cultivars and other cover crops to two races of *Meloidogyne incognita. Proc. Soil & Cros Sci. Soc. Florida* **46**: 141-143.

Michell, R.E.; Malek, R.B.; Taylor, D.P. & Edwards, D.I. (1973a). Races of barley root-knot nematode, *Meloidogyne naasi*. I. Characterisation by host preference. *J. Nematol.* **5**: 41-44.

Michell, R.E.; Malek, R.B.; Taylor, D.P. & Edwards, D.I. (1973b). Races of barley root-knot nematode, *Meloidogyne naasi*. II. Development rates. *J. Nematol.* **5**: 44-46.

Mittal, A. (1988). Studies on the biology of sorghum cyst nematode. M.Sc Thesis. Div. of Nematology, I.A.R.I., New Delhi.

Mittal, A. & Dhawan, S.C. (1989). Effect of soil texture on sorghum cyst nematode, *Heterodera sorghi* infesting sorghum. *Indian J. Nematol.* **19**: 186-188.

Nandakumar, C. & Khera, S. (1971). Plant parasitic nematodes from the rhizosphere of pearl millet in Rajasthan. *Indian J. Helminthol.* **22**: 136-138.

Nandakumar, C. & Khera, S. (1973). *In vitro* feeding habits of *Patylenchus mulchandi* and *Hoplolaimus indicus* on pearl millet roots. *Indian J. Nematol.* **3**: 138-142.

Nandakumar, C.; Khera, S. & Bhatnagar, G.C. (1969a). Studies on pathogenicity of *Pratylenchus muli* on pearl millet. *Labdev J. Sci. Technol.* **7**B: 175-176.

Nandakumar, C.; Khera, S. & Bhatnagar, G.C. (1969b). Studies on the susceptibility of pearl millet to *Heterodera avenae. All Indian Nematol. Symp., New Delhi.* p. 65.

Nandakumar, C.; Khera, S. & Bhatnagar, G.C. (1970a). *In vitro* studies on the host range of *Pratylenchus mulchandi. Indian Phytopath.* **22**: 521-522.

Nandakumar, C.; Khera, S. & Bhatnagar G.C. (1970b). On the histopathology of *Pennisetum typhoides* infected with a species of *Pratylenchus. Proc. Indian Sci. Assoc.* **57**: 551.

Nath, R.P.; Dasgupta, N.C. & Pathak, K.N. (1986). Combined effect of some seed borne fungi and *Hoplolaimus indicus* on germination and seedling growth of some millets. *National Conf. Plant Parasitic Nematodes of India: Problems & Progress, I.A.R.I., New Delhi (Dec. 17-20, 1986)*: 52-53.

Orr, C.C. (1967). Observations on cotton root-knot nematode in grain sorghum in West Texas. *Plant Dic. Reptr.* **51** : 29.

Orr, C.C. & Morey, E.D. (1974). Anatomial response of grains sorghum roots to root-knot nematode. *J. Nematol.* **6**: 148.

Orr, C.C. & Morey, E.D. (1978). Anatomical response of grain sorghum roots to *Meloidogyne incognita acrita*. *J. Nematol.* **10**: 48-53.

Padhi, N.N. & Das, S.N. (1986). Biology of *Helicotylenchus abunaamai. Indian J. Nematol.* **16**: 141-148.

Ramana, K.V. & Das, S.N. (1969). Pathogenicity of *Hoploaimus indicus* Sher, 1963 on ragi. *Ist All India Nematol. Symp*. pp. 1-2.

Sakhuja, P.K. & Singh, I. (1985). A note on the occurrence of *Heterodera sorghi* in Punjab state. *Indian J. Nematol*, **15**: 250.

Semokina, A.I. (1975). The effect of *Longidorus elongatus* (de Man, 1876) Thorne of Swanger, 1936 on the crop of millet. Cheliyabinsk, USSR; *Chelyabinskii Gosudarstvennyi Pedagogicheskii Institute, Voprosy Zoologii* No. **4**: 3-8.

Sharma, R. & Swarup, G. (1986). Potential microorganisms associated with cyst nematodes in soil. II. Pathology. *National Conf. Plant Parasitic Nematodes of India: Problems & Progress*, I.A.R.I., New Delhi (Dec. 17-20, 1986). p. 100.

Sharma, S.B. & Swarup, G. (1984). *Cyst Forming Nematodes of India*. Cosmo Publications, New Delhi, India. p. 152.

Singh, K.P.; Bind, B.N. & Edward, J.C. (1979). Effect of *Heterodera zeae* on the growth of *Sorghum vulgare* and on the rhizosphere microflora. *Indian Phytopath*. **32**: 167.

Singh, K.P. & Edward, J.C. (1979). Effect of *Heterodera zeae* on growth of *Sorghum vulgare* on the rhizosphere microflora. *Indian J. Mycol. Plant Pathol.* **8**: 167.

Srivastava, A.N. (1985). On the biology of *Heterodera sorghi. Indian J. Nematol.* **15**: 292.

Srivastava, A.N. & Sethi, C.L. (1987). Additional host records of *Heterodera sorghi. Indian J. Nematol.* **17**: 342-343.

Srivastava, A.N. & Sethi, C.L. (1988). Response of sorghum cultivars to sorghum cyst nematode, *Heterodera sorghi. Indian J. Nematol.* **18**: 397-398.

Srivastava, A.N. & Swarup, G. (1975). Preliminary studies on some graminaceous plants for their susceptibility of *Heterodera zeae* Koshy *et al.*, 1970. *Indian J. Nematol.* **5**: 257-259.

Starr, J.L., Newton, R.J. & Miller, F.R. (1984). Presence of dhurrin in sorghum root tissue and the effect of pathogenesis of hydrogen cyanide potential. *Crop. Sci.* **24**: 739-742.

Sudheer, M.J. & Sivakumar, C.V. (1983). Host range of *Pratylenchus zeae* Graham, 1951. *Third All India Nematol. Symp.* H.P.K.V.V. (May 24-26, 1983). p. 4.

Taylor, A.L. & Sasser, J.N. (1978) Biology, Identification and Control of Root-knot Nematodes (*Meloidogyne* spp.) A co-operative publication of Department of Plant Pathology, North Carolina State University and United States Agency for International Development. p. 111.

Tikyani, M.C. & Khera, S. (1968). A survey of plant parasitic nematodes around roots of great millet (*Sorghum vulgare*) from Rajasthan, India. *Plant Dis. Reptr.* **52**: 396-398.

Tikyani, M.C. & Khera, S. (1970). Studies on the longevity of *Telotylenchus loofi* in the absence of host plant. *Indian J. Helminthol.* **21**: 168-176.

Upadhyay, K.D. & Swarup, G. (1976). Reaction of some maize varieties against *Tylenchorhynchus vulgaris. Indian J. Nematol.* **6**: 105.

Vaishnav, M.U. (1977). Studies on plant parasitic nematodes associated with bajra (*Pennisetum typhoides*). Ph. D. Thesis, I.A.R.I. New Delhi.

Vaishnav, M.U. & Sethi, C.L. (1977). Reaction of some graminaceous plants to *Meloidogyne incognita* and *Tylenchorhynchus vulgaris. Indian J. Nematol.* **7**: 176-177.

Vaishnav, M.U. & Sethi, C.L. (1978). Pathogenicity of *M. incognita* and *Tylenchorhynchus vulgaris* on bajra and their interrelationships. *Indian J. Nematol.* **8**: 1-8.

Vaishnav, M.U. & Sethi, C.L. (1979). Interaction of root-knot and stunt nematodes with *Sclerospora graminicola* on bajra. *Indian Phytopath.* **31**: 497-500.

Chapter 9

Nematodes Pathogenic to Tea and their Control

Nalini C. Gnanapragasam
Tea Research Institute of Sri Lanka, St. Coombs, Talawakelle, Sri Lanka

TEA is one of the most popular beverage crop grown in more than 50 countries in the temperate, sub-tropical and tropical regions of the world, at latitudes ranging from 27°C to 43°N and altitudes ranging from mean sea level up to 2300 m, in varying soil types. For optimum growth tea needs well drained soils (pH 4.5-5.5) and reasonably well distributed rainfall, totalling not less than 1,000 mm per annum.

Amongst abundant number of pests attacking the tea plant, parasitic-nematodes rank as important pests limiting growth and productivity.

HISTORY

The first plant-parasitic nematode to be recorded as a pest of tea was the burrowing nematode *Radopholus similis*, in Java in 1899 (Zimmermann, 1899). This was followed by the detection of the *Meloidogyne* sp. in tea seedlings in South India (Barber, 1901). In Sri Lanka, the first nematode pest of tea to be detected was *Meloidogyne* sp. (Stuart Light, 1928). Since then several plant-parasitic nematodes have been encountered in the tea soils collected from different tea growing areas of the world. Amongst them the species that are known or suspected to be pathogenic to tea include, *Pratylenchus loosi, P. brachyurus, Radopholus similis, Meloidogyne incognita, M. javanica, M. hapla, M. arenaria, M. thamesi, M. brevicauda, Hemicriconemoides kanayensis, Rotylenchulus reniformis, Rotylenchus* sp., *Helicotylenchus* sp., *Helicotylenchus erythrinae, H. dihystera, Hoploaimus* sp., *Paratylenchus curvitatus, Xiphinema* sp. (Sivapalan & Gnanapragasam, 1991).

DISTRIBUTION

Detailed studies carried out in Sri Lanka on the effect of environment on the distribution of nematodes in the different agro-ecological regions of the tea areas indicated variation in soil temperature, rainfall intensity and distribution, soil moisture and soil type to be the dominant environmental factors determining the distribution of the different species of plant-parasitic nematodes.

Plant-parasitic nematodes such as *Pratylenchus loosi* and *M. brevicauda* are found in cooler soil temperature zones whilst *R. similis* and *R. reniformis* favour warmer temperature. *Radopholus similis* has been found to thrive well in warmer areas receiving high and uniform rainfall whilst *R. reniformis* could survive in drier areas as well. However, reduction in rainfall below 100 mm during any period of the year brought down a drastic reduction in population of *R. reniformis* (Gnanapragasam, 1988). *Meloidogyne* sp. (other than *M. brevicauda*), *Paratylenchus* sp. and *Helicotylenchus* sp., on the other hand, seem to be well adapted to different climatic conditions and found in all the tea growing areas of the world. In China, optimum soil temperature for pest incidence by this species of nematode has been reported to be 20°-C30°C in soils with 20% moisture (Rong *et al.*, 1984). In Sri Lanka, however, since tea has been planted by means of vegetative propagation, the incidence and damage due to *Meloidogyne* sp. has been very rare. This may be because all soils used for nursery in most tea growing areas are routinely fumigated and this being a pest of only young tea, does not get a chance of infesting the plants.

Soil type and texture have also a significant influence on the reproductive rate and population build-up of parasitic nematodes. Earlier studies have revealed greatest build-up of *P. loosi* and largest amount of damage to tea in clayey loam, whilst the lowest was in gravelly and sandy soils (Sivapalan, 1972). Recent trials carried out in Sri Lanka for *R. similis*, on the other hand, has shown highest build-up of population in sandy, gravelly and loamy soils. There was hardly any nematode build-up in the heavier clayey soil (Gnanapragasam, 1990). *P. brachyurus* has been reported to be prevalent in sandy-loams and it fails to survive in clayey soils (Basu, 1968).

No detailed studies have yet been undertaken in other countries, to correlate the influence of different environmental factors on the distribution and pathogenicity of other species of nematodes.

HOST-PARASITE RELATIONSHIP AND SYMPTOMS OF DAMAGE

All species of nematodes that are pathogenic to tea are root feeders that damage the root system, thereby seriously impeding the uptake of soil nutrients and water, as well as destroying the other normal functions of the roots. This results in nutrient imbalance, reduction in the rate of photosynthesis, development of deficiency symptoms and a significant lowering of the general vigour of the plant. As in other crops, the weakened tea bushes become susceptible to other forms of stress including drought, attack by other pathogens, etc.

Fig. 1. A young tea plant infested with *Pratylenchus loosi* showing stunted growth with premature flowering and fruiting, as well as poor root system (left) and healthy uninfested plant of the same age (right)

The infested bushes show above ground symptoms in the form of unthrifty tea with weak maintenance foliage (Figs. 1,2). The leaves appear pale yellowish and dull mainly due to nutritional imbalances caused by altered rate of uptake of essential nutrients. The heavily infested plant show tendency for premature flowering and fruiting. The infested plants are some times found to have small shoots and internodes and the leaves at the plucking table remain as tufts with few buds giving rise to pluckable shoots (Hutchinson, 1960).

Examination of the roots of such infested plants show typical marked reduction in the feeder root system. The feeder roots of plants infested with *P. lossi* appear brownish in colour and dark brown necrotic patches appear on both the feeder roots, as well as the storage roots (Fig.3). Peeling pieces of the bark or cortex from the junction between the dying and healthy tissues and examination of the undersurfaces of the peeled strips would reveal large numbers of the nematodes.

Fig. 2 A tea field showing weak patchy growth of mature plants (7 years) infested with *Radopholus similis*.

P. brachyurus attacks mainly the feeder roots of young tea and occasionally the tap roots forming reddish brown lesions. *Radopholus similis* also forms brownish lesions mostly on the feeder roots (Fig.4) and the infested roots appear dried-up. The ectoparasite *H. kanayensis* feeds only on feeder roots, resulting in sloughing-off of the cortex and discolouration of the stele. The roots infested with *M. brevicauda* are gnarled and knotted due to the formation of characteristic galls. *R. reniformis* attacks the finer roots attached to the feeder roots giving a clipped off appearance.

In N.E. India, *Xiphinema* sp. has been found to feed on the root tips of the feeder roots of nursery plants resulting in slight swellings of the affected roots tips (Basu, 1967).

DECLINE IN YIELD AND DAMAGE THRESHOLD LEVEL

In Sri Lanka, crop loss in tea infested with *P. loosi* has been earlier estimated to average between 225-350 kg made tea/ha/year (Visser, 1959; Sivapalan, 1972). Based on figures obtained from a questionnaire, Sasser estimated the average

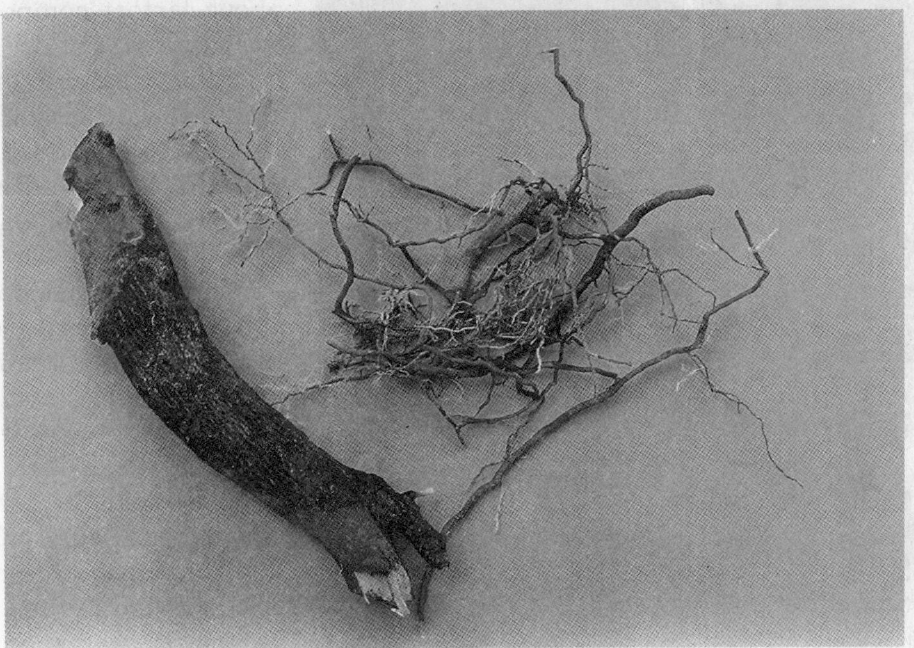

Fig. 3 Brown necrotic patches on storage roots (left) and feeder roots (right) due to damage caused by *Pratylenchus loosi.*

crop loss on a worldwide scale to be 8.2% (Sasser, 1989). In many tea growing countries in-depth studies on nematodes have yet not been carried out, and thus the extent of injury cannot be yet correctly established. Moreover, although nematodes by themselves can bring about significant loss in crop, under field condition, it is not possible to assess with any precision the actual loss incurred, as this is very much dependent on the interaction of several factors including:

a) The particular species of nematode.

b) Population density of the relevant species of nematode.

c) Prevailing climatic condition.

d) Type of cultivar (resistant/tolerant/susceptible var. of tea).

e) Age and vigour of the tea plant.

f) Type of soil in which tea is growing.

g) Agronomic inputs.

h) Attack by other pathogens, etc.

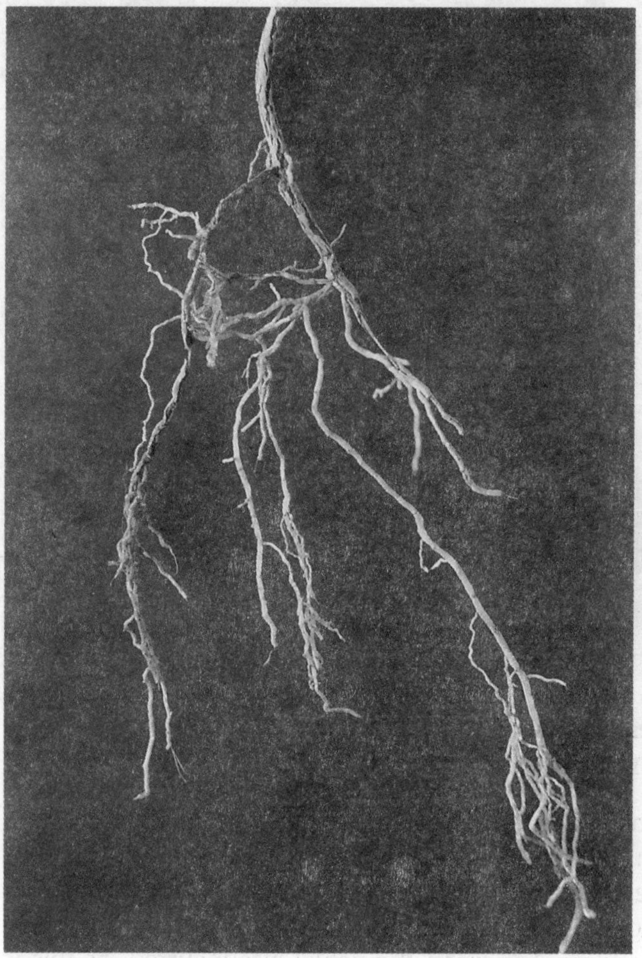

Fig. 4. Necrotic patches on feeder roots of tea caused by the burrowing nematode, *Radopholus similis*.

Since the interaction of the above factors compound the expression of damage symptoms, studies to determine damage threshold levels have to be carried out under controlled environmental condition in specific soils, inoculated with the particular species of nematode. Such studies to determine the damage threshold level have been carried out so far only for two species of nematodes, *P. loosi* and *R. similis*. In Sri Lanka, under green-house condition, the damage threshold level of *P. loosi* to tea was estimated to be 40 nemas/ 100 g soil at 24°C in a clayey loam soil (Gnanapragasam & Manuelpillai, 1984) and that for *R. similis* to be 28 nemas/100 g soil at 25°C grown in the same type of soil as above (Gnanapragasam & Herath, 1989).

CONTROL OF NEMATODES

In most countries, the pathogenic nematodes detected in the different tea growing areas cause damage to only young tea. These include *P. brachyurus, R. reniformis* and the commoner species of *Meloidogyne, M. incognita, M. arenaria, M. javanica, M. hapla* and *M. thamesi*. As such treatment at the nursery stage would be adequate to bring about moderate control, to produce healthy plants suitable to be planted in the field.

However, in those countries where nematode species that are pathogenic to both young and mature tea are encountered, a more holistic integrated approach becomes warranted to help mitigate their effects on the growth and productivity of tea. The pathogenic species damaging both young and mature tea reported from different tea growing areas of the world include : *Pratylenchus loosi, Radopholus similis, Hemicriconemoides kanayensis* and *Meloidogyne brevicauda.*

CONTROL OF NEMATODES IN THE NURSERIES

Tea nurseries are the focal point of spread of infestations to areas hitherto uninfested, as thousands of plants propagated from each nursery are distributed throughout the plantations. Furthermore, since plants from commercial nurseries are also distributed to different locations they would be a source of spreading infestation rapidly to uninfested locations. It is therefore, necessary to take precautionary measures to prevent contamination and infestation of nursery plants. Care should also be taken to avoid transplanting of infested plants. All infested plants in the nurseries should be destroyed by burning. This is of special importance with infestations of nematode species that are also pathogenic to mature tea.

Since all nematodes pathogenic to tea are root feeders and tea is propagated by vegetative means or by seeds, it is easy to prevent infestation in the nurseries by ensuring that the plants are grown in soil free of nematodes. In the case of root-knot nematodes, moderate control in the nursery to produce healthy plant is adequate, as the plants develop complete resistance by about 9-15 months (Gnanapragasam *et al.*, 1989). On the other hand, in the case of nematodes which attack both young and mature tea, more drastic control measures are necessary to totally eradicate nematodes.

Physical Control

Physical control is confined exclusively to heating of nursery soils. Nursery soils spread over galvanized sheets to a depth of 4 cm and heated to 60-62°C for 5 min. have been found to effectively suppress populations of *Meloidogyne* sp. in nurseries (Rao, 1976; Basu, 1978). However, this method of treatment is not practical for large nurseries.

Controlling nematodes by heat solarization is another method that could be adopted. Since complete eradication of nematodes is not possible, this method could also be used only for controlling those species of nematodes attacking young tea such as root-knot nematodes (other than *M. brevicauda*), *P. brachyurus*, etc.

Chemical Control

In countries like India, granular nematicides are added to soils planted to young nursery plants (Rao, 1976; Basu, 1979; Basu & Gope, 1985). Since granular nematicides can not completely eradicate nematodes, this method of control in the nurseries is also suitable to control only those species attacking young tea.

In order to achieve complete eradication of nematodes in the nurseries, it is essential to routinely fumigate all nursery soils collected from nematode prone areas with potent fumigants such as MBr at the rate of 0.4 kg/2.83 cu.m. of soil or Dazomet 98% (Basamid) at the rate of 0.5 kg/2.83 cu.m of soil (Sivapalan *et al.*, 1980).

Cultural Control

Despite routine fumigation, many nurseries have been observed to be contaminated with nematodes pathogenic to tea. This has been traced to many factors (Gnanapragasam, 1989) such as :

a) improper method of fumigation, including fumigation during wet weather, using incorrect dosage, inadequate waiting periods and the use of damaged plastic covers;

b) soil brought to the nursery prior to fumigation is heaped in close proximity to fumigated soil or nursery bags containing uninfested plants;

c) use of agricultural implements which are contaminated with nematodes;

d) use of irrigation water contaminated with nematodes;

e) presence of alternate hosts in the nursery;

f) movement of water and soil from infested fields into the nursery site, etc.

a), (b) and (c) can be avoided by proper supervision of the nursery operations, by trained personnel.

Nurseries are often irrigated with ravine water that flows through infested fields. Plant-parasitic nematodes can survive for varying periods of time in water and get carried over long distances with the movement of water and soil. In order to circumvent this problem, it is essential to use clean source of water, such as from a well or water collected directly from springs. If this is not possible, water flowing through infested field should be allowed to sediment in settling tanks for over 48 hours (Gnanapragasam & Jebamalai, 1982).

The movement of soil and water into the nursery site from adjacent infested areas could be avoided by having a deep drain around the nursery and also by planting marigold (*Tagets* sp.) *Eragrostis curvula or Veti-veria zizanoids* (savendra) along the edges of these drains. These plants act as trap crops and thereby significantly reduce movement of nematodes into the protected area.

Irrespective of the source of contamination, all infested plants in the nurseries should be destroyed by burning them. Their continued presence in the nursery would only serve as a source of inoculum to plants that are not infested.

CONTROL IN THE FIELD

Once introduced into the field, it is not possible to eradicate pathogenic nematodes. The method adopted to control these species in tea fields is to ensure that there is minimal contamination and spread of nematodes in established field as well as by maintaining the plants in a vigorous state of growth by reducing the nematode population to levels below the damage threshold level, to help the plants to better tolerate nematode parasitism.

The damage threshold level varies with the respective species of nematode, the variety and age of tea, the soil type, the mean soil moisture levels and mean soil temperatures, etc. By using established damage threshold levels, along with an *in-situ* assessment of the existing environmental conditions, appropriate integrated management strategies are recommended to help mitigate nematode damage. These strategies are briefly described below :

Chemical Control

Various nematicides have been screened against the different species of pathogenic nematodes with special attention being paid to selecting only those that are not phytotoxic. Residue levels in both soils as well as plant tissues are also checked prior to recommending a chemical and such carefully screened nematicides are used only during specific vulnerable stages of growth of the tea plants.

In Sri Lanka, as an additional measure of insurance against possible contamination, in areas prone to nematode damage granular nematicides are routinely incorporated in the planting hole at planting. In view of the stimulatory effect, such treatments also help the plants to establish well. The recommended nematicides include fenamiphos (Nemacur 5%G) carbofuran (Furadan or Curaterr 3%G), at 7g per plant.

The efficacy of most recommended nematicide formulations lasts only for few weeks (3-24 weeks) depending on the nature of soil, organic matter content and the pH of the soil, etc. As such, chemical control of nematodes in a perennial crop such as mature tea is an uneconomic exercise. Nevertheless, in the case of tea, which is subject to periodical prunes, there is a critical stage of regrowth following prune, with newly formed feeder roots that can get readily infested with nematodes. As such it is recommended to treat fields with good yield potential with a single application of fenamiphos 5%G, at the rate of 10g/plant by mixing this with the first dose of fertilizer that is applied following pruning (Gnanapragasam, 1987a).

CULTURAL CONTROL

(i) Control in newly planted areas

Although mature tea can be maintained for several years (up to more than 50 years), uneconomic tea lands are routinely uprooted and replanted with cultivars suitable for the particular agro-climatic zones, as well as with special virtues such as high yield, good quality and resistance to pests and diseases, etc.

Infestation with plant-parasitic nematodes has been assessed to be as one of the main causes for failures of newly replanted tea areas in Sri Lanka (Sivapalan, 1967).

Such infestation in newly planted areas could originate from either the use of nematode contaminated nursery plants or from a residual population of nematodes left in the soil, following inadequate root removal during uprooting of the old infested tea.

The first step in the control of nematodes in replanted areas is by ensuring a very thorough uprooting of all old tea roots (up to pencil thickness). Nematodes can remain viable in root fragments for several months and can infest the newly planted tea. *Pratylenchus loosi* has been found to remain viable in host-free soil for up to 2-3 years (Hutchinson, 1962) and *Radopholus similis* for 14-15 months (Gnanapragasam, unpublished data). No records are yet available for other species of tea nematode.

In order to prevent the movement of infested soil into newly uprooted areas, the recommended cultural strategy is to commence uprooting and replanting from top of the hill downwards (Gnanapragasam, 1985a).

In Sri Lanka, such uprooted old tea lands are routinely reconditioned by planting deep-rooted grass species including Guatemala grass (*Tripsacum laxum*) and Mana grass (*Cymbopogon confertiflorus*) which help to improve the physical characteristics of the soil. Further, as a consequence of being non-hosts to certain parasitic nematodes, whatever residual populations that are left in the soil also gets depleted during the reconditioning period (Visser, 1959, Kerr, 1963, Kerr & Vythilingam, 1966, Sivapalan, 1972). Recent trials have, however, shown Guatemala grass to be a good host to *R. similis* (Gnanapragasam, 1986). As such, in areas prone to infestation with this species it has been recommended to plant Mana grass as this is non-host to *R. similis* (Anon. 1990).

As mentioned earlier, the next precautionary measure is to ensure that all plants used in the new clearing are healthy and uninfested. In order to avoid the risk of planting infested plants, the transport of nursery plants from plantation to plantation should be forbidden and every plantation should have its own nursery (Sivapalan, 1972).

(ii) Use of resistant/tolerant clone

One of the most economic long-term strategy for nematode management in perennial crops, is by planting of resistant or tolerant cultivars in those areas that are prone to nematode damage.

The cultivars selected against a specific species of parasitic nematode are seldom resistant/tolerant to other species of nematodes. The various cultivars have to be separately screened for resistance/tolerance to the different species of nematodes. Following years of study in this line of investigation in Sri Lanka, it has become possible to release several tea clones that are resistant/tolerant to the different species of nematodes that are pathogenic to tea in this country.

(iii) Shade trees and green manure crops

Shade trees and green manure crops help to provide shade to the tea plants, aid in nitrogen fixation, and provide regular supplies of much in the form of leaf litter and loppings. Some of these crops are good hosts for certain parasitic nematodes and hence their presence enhance the incidence of these pests.

The cultivation of such crops that are more susceptible than tea, to a given species of nematode, is discouraged in tea areas.

(iv) Planting of trap crops

Plants such as marigolds, *Tagetes erecta* and *T. patula*, (Visser & Vythilingam, 1959; Hutchinson, 1961; Hainsworth, 1970; Sivapalan, 1972). *Veti-veria zizanoides* and *Eragrostic curvula* (Gnanapragasam, 1981, 1985b), have been found to help reduce nematode populations. The vacant areas in fields heavily infested with nematodes are presently recommended to be planted with such crops, to filling the vacancies.

(v) Resting of tea fields

Tea bushes are pruned regularly, with pruning cycles ranging from 2-5 years. Pruning is a drastic operation and a tea bush in a vigorous state of growth will be able to recover normally, after pruning. However, when the bushes are subjected to various forms of stress including infestation with nematodes, they would not be in a position to accumulate sufficient amounts of root reserves to help recovery. Therefore, it is not recommended to rest the bushes in fields which are heavily infested with nematodes at least for a period of 4-8 weeks, prior to pruning.

(iv) Soil cultivation

Poorly drained soils, develop a compacted hard-pan resulting in poor soil aeration and poor growth of tea. Nematode damage is severe in such soils as a result of retardation in the rate of replenishment of damaged feeder roots. Forking to a depth of 30-40 cm helps to break such hard pan and improve soil aeration stimulating better feeder root growth. Infested areas have shown significant improvement in growth following such treatments (Sivapalan, 1972).

(vii) Fertilizer

The use of specific potassium-enriched fertilizers to both young and mature tea plants is known to retard population build-up of *P. loosi* and at the same time helps the plants to better tolerate nematode parasitism (Gnanapragasam, 1982).

(viii) Organic amendments

Soil amendments with organic substances, in addition to improving the soil tilth and organic matter content are also known to reduce nematode populations and/or help the host to better tolerate root invasion and damage caused by specific plant-parasitic nematodes. The practice of using soil amendments to control plant-parasitic nematodes has been in use since 1930's. Large inputs of organic matter including cow dung and well-decomposed plant residues have been reported to suppress populations of *P. loosi*. In Sri Lanka and India the growing concern for the environment has resulted in scaling up of the activities of screening various soil amendments in the form of farm-yard manures and industrial wastes, including oil cakes that could be readily procured in bulk, to assess their ability to suppress and/or help the plant to better tolerate nematode attack. In India, neem cake has been recommended to control root-knot nematodes in the tea field (Muraleedharan, 1991).

In studies carried out in Sri Lanka, oil cake of castor (*Ricinus communis*) has proved to be the most effective one for suppressing populations of *P. loosi*. This treatment also helps to boost the growth of the plant. The efficacy of castor oil cake is followed closely by the oil cakes of mahuva (*Madhuca indica*), neem (*Azdirachta indica*) and karanj (*Pongamia glabra*). Even though the other amendments (including poultry manure, coconut poonac, and tea waste) did not suppress nematode population, infested plants treated with these amendments too showed significant improvement in growth thereby enabling the plants to better tolerate nematode attack (Sivapalan, 1980; Gnanapragasam, 1987b, 1991). The above soil amendments, therefore, offer a cheap and environmentally acceptable method of treating small patches of infested areas in young and mature tea, in larger plantations, as well as in small holdings.

Other tested organic materials such as *Gliricidia* leaf much, gingerly/sesame oil cake and palm oil cake were found to be not effective in controlling nematode pathogenicity to tea. In recent studies addition of cow dung at economic rates was also not found to be effective.

CONCLUSION

As in the case of any other pest control programmes, any one of the above mentioned control strategy may not be itself be adequate to avert the economic damage in the field. Having ascertained the prevailing ecological conditions and various cultural practices adopted for the growth of tea in the nematode infested locations, the different control strategies need to be integrated and applied in a holistic manner, so as to maintain the population levels below the damage threshold level on the one hand, and at the same time maintain the plant in a vigorous state of growth thereby averting reduction in the productivity of tea.

REFERENCES

Anon. (1990). Replanting eelworm infested areas. *Advisory Circular* N4, Serial No. 4/90.

Barber, C.A. (1901). A tea eelworm disease in S. India. *Tropical Agriculturist* **21**: 239.

Basu, S.D. (1967). Eelworm in tea: Some important groups. *Two and a bud* **14:** 84-85.

Basu, S.D. (1968). Eelworms - A Progress Report on a few more groups found in N.E. Indian tea soils. *Two and a bud* **15**: 70-71.

Basu, S.D. (1978). Heat the soils and kill the nematodes. *Two and a bud* **25**: 42.

Basu, S.D. (1979). Control of *Meloidogyne incognita* on tea seedlings with some systemic nematicides. *Two and a bud* **23**: 58.

Basu, S.D. & Gope, B. (1985). Control of root-knot nematode, *Meloidogyne incognita* (Kofoid & White) Chitwood on stem of tea seedlings with some systemic nematicides. *Two and a bud* **32:** 32-35.

Gnanapragasam, N.C. (1981). The influence of cultivating *Eragrostis curvula* in nematode infested soils and on the subsequent build-up of populations in replated tea. *Tea Q.* **50**: 160-162.

Gnanapragasam, N.C. (1982). Effect of potassium fertilization and of soil temperature on the incidence and pathogenicity of the root-lesion nematode, *Pratylenchus loosi* Loof on tea. *Tea Q.* **51**: 169-174.

Gnanapragasam, N.C. (1985a). The strategy for replanting eelworm infested areas. *J. Tea Sci.* **54**: 46.

Gnanapragasam, N.C. (1985b). Report on the Tea Research Institute of Sri Lanka 1985. pp. 82-96.

Gnanapragasam, N.C. (1986). Report of the Tea Research Institute of Sri Lanka 1986. pp. 69-78.

Gnanapragasam, N.C. (1987a). Integrated strategies to manage nematodes in tea. *In : Proc. Int Conf. Tea Quality* and *Human Health, Hangzhou, China,* (November, 1987). pp. 117-183.

Gnanapragasam, N.C. (1987b). Report of the Tea Research Institute of Sri Lanka 1987. pp. 70-83.

Gnanapragasam, N.C. (1988). Report of Tea Research Institute of Sri Lanka 1988. pp. 64-67.

Gnanapragasam, N.C. (1989). Prevention of dissemination of nematodes pathogenic to tea into hitherto uninfested tea areas. *Tea Bull.* **9**: 20-22.

Gnanapragasam, N.C. (1990). Report of the Tea Research Institute of Sri Lanka 1990. pp. 72-83.

Gnanapragasam, N.C. (1991). Influence of soil amendments in reducing pathogenicity to tea by the root-lesion nematode, *Pratylenchus loosi. In: Proc. Int. Symp. Tea Sci.*, Shizuoka, Japan (August, 1991). pp. 684-688.

Gnanapragasam, N.C.; Dharmasena, W.A.M. & Liyanage, D.D. (1989). Development of age resistance in young tea to different species of *Meloidogyne. Int. Nematol. Network Newsl.* **6**: 4.

Gnanapragasam, N.C. & Herath, U.B. (1990). Pathogenicity of the burrowing nematode, *Radopholus similis* to young tea (susceptible clone, TRI 2025) at different density. *S.L.J.* **58**(2): 83-86.

Gnanapragasam, N.C. & Jebamalai, A.S. (1982). Nematode infestation of nursery plants through irrigation water. *Tea Q.* **51**: 31-33.

Gnanaprasam, N.C. & Manuelpillai, M.E.K. (1984). Influence of inoculum levels and temperature on the population build-up and pathogenicity of the root-lesion nematode, *Pratylenchus loosi* Loof of tea. *Tea Q.* **53**: 19-22.

Hainsworth, E. (1970). Tea. *J. Boards E. Africa* **11**: 15-17.

Hutchinson, M.T. (1960). Resistance and tolerance of tea nematodes. *Tea Q.* **34:** 68-84.

Hutchinson, M.T. (1961). Developments in the control of the meadow nematode. *Tea. Q.* **32:** 129-132.

Hutchinson, M.T. (1962). Rehabilitating tea soils susceptibility of plants now in use to the root-lesion nematode, *Pratylenchus loosi. Tea Q.* **34**: 68-84.

Kerr, A. (1963). Report of Tea Research Institute of Ceylon 1963. pp. 95-102.

Kerr, A. & Vythilingam, M.K. (1966). Replanting eelworm infested areas. *Tea. Q.* **37**: 67-72.

Muraleedharan, N. (1991). Non-arthropod pests - Nematodes and Rodents. *In: Pest Management in Tea* (Ed.-Director). United Planters' Association of Southern India, Tea Research Institute. pp. 77-81.

Rao, G.N. (1976). Control of nematodes in nursery soils. *Planters' Chronicle S. India* **7:** 257-259.

Rong, W.Z.; Juan, J. & Zheng, K.S. (1984). Study of tea seedlings disease caused by root-knot nematode, *Meloidogyne* spp. *Acta Phytopathologica Sinica* **14:** 225-232.

Sasser, J.N. (1989). *The Plant-parasitic Nematodes. The farmers Hidden Enemy*. Dept. of Plant Pathology, North Carolina State University & Consortium for International Crop Protection. pp. 115.

Sivapalan, P. (1967). Nematode problems in *Tea Q.* **38**: 260-268.

Sivapalan, P. (1972). Nematode Pests of Tea. *In: Economic Nematology* (Ed.-J.M. Webster). Academic Press, NY and London. pp. 285-310.

Sivapalan, P. (1980). Report of the Tea Research Institute of Sri Lanka 1980. pp. 83-96.

Sivapalan, P. & Gnanapragasam, N.C. (1991). Plant parasitic nematodes of tea (under chapter on Nematode Pests of Coffee, Cocoa and Tea). *In: Plant Parasitic Nematodes in Sub-Tropical Agriculture* (Ed.—M. Luc; R.A. Sikora & J. Bridge). Commonwealth Agricultural Bureau, London. pp. 404-430.

Sivapalan, P.; Gnanapragasam, N.C. & Jebamalai, A.S. (1980). Treatment of nursery soils with Dazomet 98G (Basamid) as an alternate method of control of rot-lesion nematode, *Pratylenchus loosi* in new plantings. *Tea Q.* **49**: 30-36.

Stuart Light, S. (1928). Pests of Nurseries : A warning. *Tea Q.* **1**: 19-22.

Visser, T. (1959). Observations on the prevalence and control of parasitic eelworms in tea. *Tea Q.* **30**: 96-107.

Visser, T. & Vythilingam, M.K. (1959). The effect of marigolds and some other crops on *Pratylenchus and Meloidogyne* populations in tea soils. *Tea. Q.* **30:** 30-38.

Zimmermann, A. (1899). Het voorkomen van nematoden in de wortels van sirihen thee. I. De aaltijesziekte der sirih (*Piper betel* L.). II. *Tylenchus acutocaudatus* Zn in de thee. *Teysmania* **10:** 523-531.

Chapter 10

Control of Coffee and Cocoa Nematodes

Vicente Paulo Campos

Secao de Namatologia, Departmento de Fitossanidade, Escola Superior de Agricultura de Lavras (ESAL), 37200 Lavras, Minas Gerais, Brazil

A. COFFEE

ROOT-KNOT nematode of the genus *Meloidogyne* are the major group of parasitic nematodes causing great economic damage or dizimating coffee plantations (Fig. 1) in some regions specially in Brazil (Luc *et al.*, 1990). Some species are widely distributed ehnancing the hazard to many coffee dependent economy countries (Table 1).

Table 1. Most common species of root-knot nematodes (*Meloidogyne* spp.) found on coffee and their distribution.

Meloidogyne species	Country
M.incognita	Brazil, Tanzania, Jamaica, Venezuela, Ivory Coast, India
M. exigua	Brazil, Guatemala, Dominican Republic, Nicaragua, Costa Rica, Puerto Rico, Colombia, Peru, El Salvador, Venezuela, Bolivia
M. coffeicola	Brazil
M. javanica	Brazil, Tanzania, Zaire, El Salvador, India
M. hapla	Brazil, Tanzania, Zaire, India
M. africana	Kenya, Zaire
M. decalineata	Tanzania, Sao Tome
M. kikuyensis	Tanzania
M. arenaria	Jamaica
M. megadora	Angola, Uganda
M. inornata	Guatemala
M. oteifae	Zaire
M. thamesi	India

Among the root-knot nematodes of coffee in Brazil, *M. incognita* causes the greatest losses, becoming a limiting factor to growing coffee in certain areas due to (i) its great capacity to destroy the root system, (ii) it being easily disseminated, (iii) it having a high persistence in soil, (iv) the inefficiency of chemical control measures, and (v) it having different races.

Fig. 1. Five-year-old coffee plantation killed by *Meloidogyne incognita* (Sao Paulo State, Brazil).

Lesion nematodes are the second nematode group of plant-parasitic nematodes causing economic damage to coffee. Many species are found attacking coffee (Kumar, 1982; Luc *et al.*, 1990). Other genera have been listed as pathogens but only eventually cause economic damage (Luc *et al.*, 1990).

CONTROL MEASURES

Since coffee is a perennial crop, the root-knot nematode keeps reproducing as long as the plant is growing in the field but at the end of its span which is about 17 years, the plantation in Brazil is renovated. During this range of time the nematode population fluctuates according to the climate and plant suitability. Within the same year the fluctuation is enormous (Fig. 2), which is important to consider on the timing of the chemical application to control the nematode.

Early diagnosis of the occurrence of important nematodes on coffee fields will avoid great dissemination by isolating the area by drenches and cleaning/cultivating machines after the work on infested spots. In fact diagnosis of the occurrence of *M. exigua* in field is not difficult because this nematode induces typical round galls on roots of infested coffee plants unlike the damage caused by *M. incognita* and *M. coffeicola*. With the latter species, laboratory diagnosis is required to search for these nematodes in non-galled sections of the root system. *M. coffeicola* is mostly found in older sections of the root specially the principal root. But, in all cases, extraction of second stage juveniles in soil helps to identify the disease in combination with the symptomatology on the plant.

The eradication of plant-parasitic nematodes from infested coffee field, specially *Meloidogyne* species, is impossible. The alternative will be the management of the population avoiding economic crop damage by using nematicides.

By the crop renovation time, the farmer may decide on turning the infested coffee area to pasture. Most graminicolous plants are non-hosts to *M. exigua* and *M. coffeicola* (Lima *et al.*, 1985). However, many weed species may prove to be good hosts for *M. exigua* and *M. incognita* (Lima *et al.*, 1985). The time of field disinfestation in other words, will depend upon the range of weed hosts of those nematodes growing in the field. The decision of growing forage crops and turning back again the area of coffee will certainly reduce drastically the nematode population specially *Meloidogyne* species but may not eradicate them. However, if the decision is to grow economic crop intensively the weed control should often be more efficient which may contribute to great decrease or eradication of nematodes in short period. Some research done in Brazil has established, 6 to 12 months as the time required to eliminate *M. exigua* by growing crop in previously infested area after a good

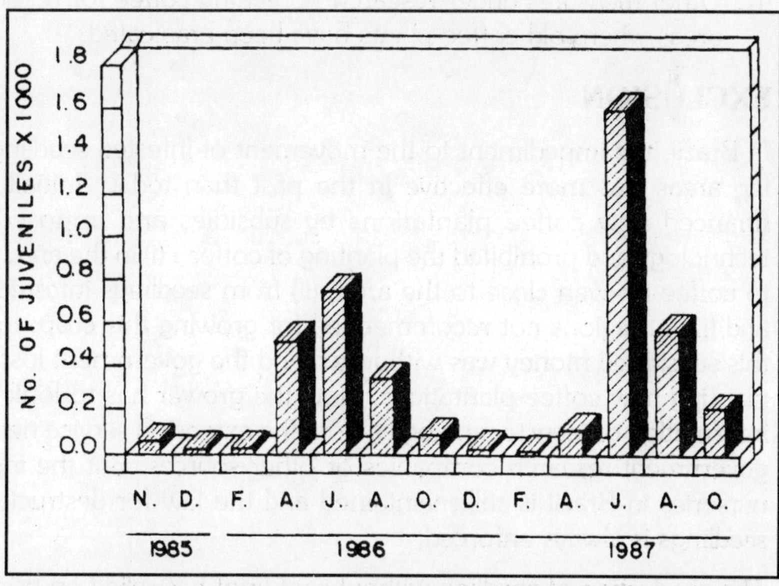

Fig.2. Fluctuation of second stage juveniles of *Meloidogyne exigua* in coffee cv. *Mundo Novo* during 1985, 1986 and 1987.

rootstock elimination (Almeida & Campos, 1991). In summary, old plants left in the fields, weed hosts, or surviving roots of excised plants provide a source for nematode multiplication and in part negate the effect of control practices.

Another situation regarding to coffee nematode control is the preoccupation of the farmer at the moment of planning the coffee activity. In traditional Brazilian coffee areas farmers try to establish the plantation where coffee has never been planted before. However, the land free of damaging nematodes is not the only way to avoid the problem, one must plant seedlings which are free of those pathogens. If the soil treatment is impossible, low income farmers use soil of known land never planted with coffee and discharge the top 30 cm layer. This procedure has avoided infestation only in many areas free of damaging nematodes in Brazil. Besides these cares, the irrigation water has to be collected from reservoir which sides hills not covered by coffee plantation.

The control measures of coffee root-knot nematodes that are used today by many farmers may be considered under four subgroups :

1) Exclusion, including the measures used to keep the parasite from entering the soil in which the host is growing.
2) Application of nematicides, for the elimination or reduction of the parasite level after it has become established in the soil where the host is growing.
3) Grafting on resistant or tolerant cultivars.
4) Other measures under research : breeding coffee for resistance, rotations in areas where old coffee plants have been eradicated.

EXCLUSION

In Brazil, the impediment to the movement of infested seedlings into new growing areas was more effective in the past than today. Initially the government financed new coffee plantations by subsidies and imposed the use of new technology and prohibited the planting of coffee : (i) in the area previously planted to coffee or even close to the area, (ii) from seedlings infested with nematodes, and (iii) in regions not recommended for growing this crop. In the last 13 years this subsidized money was withdrawn and the government lost their control over planting new coffee plantations. Now the grower has to look independently for information on new technologies from the extension service network, universities, government research companies or other sources. But the inspection of coffee nurseries in Brazil is still maintained and the law for destruction of the infested seedlings is always enforced.

The production of seedling without root-knot has relied on using soil in nurseries gathered from areas never grown previously with coffee, especially where pasture is currently grown. This soil can be sterilized with methyl bromide at the rate of

150 ml per m^2 of soil (Morais *et al.*, 1977b), placed under a plastic cover for 3-4 days and then aerated for 10 days before seeding. Other methods of sterilizing soil includes the uses of steam and exposure of nursery soil to sun for many weeks during the dry season period (Bridge, 1984). The water source has to be carefully selected avoiding dams in which run-off water comes from hillsides cultivated with infested coffee plants. Infected seedlings with root-knot nematodes should be burned and under no circumstances should they be planted into an area free of damaging nematodes.

The place to establish a new coffee crop has to be very carefully selected avoiding the recently eradicated old coffee plants, as well as the proximity of an infested field or on a site at a level below it, where the risk of contamination from run-off water is high. Sometimes a furrow has to be dug to prevent run-off water from the infested area. Care has to be taken to wash machines or farm implements used, or that have travelled through infested fields.

NEMATICIDES

Chemicals used today to control nematodes on coffee as on other crops have been mostly restricted to contact or systemic granular products. From the group of fumigants widely used for controlling nematodes in the past (Anon., 1968) only methyl bromide is still widely used to disinfest nursery soil, as mentioned in the exclusion section.

The systemic insecticides, the organophosphate and organocarbamate chemicals, that have potential for nematode control are rarely phytotoxic at concentrations used for field control. Nematicidal activity is usually confined to a shallow root zone of rhizosphere, and is often a result of naracotization and nematode behaviour modification rather than killing. Disruption of nematode infection, development and reproduction can temporarily slow or halt increases in nematode numbers. These chemicals give little or no control of fungal or bacterial diseases but provide insecticidal activity depending upon the chemical involved (Van Gundy & Mckenry, 1977).

In general the effective rates of aldicarb, carbofuran and phenamiphos range between 1.6 to 6.0 g of active ingredient per plant, in one or two applications during the year. The first application should be in the beginning of the rainy season followed by the second application 3 months later. In any case the soil should be wet for the application. A furrow is dug along the both sides of the plant row close to the tree where the product is applied and incorporated into the soil, by machine or by hand. Abrego (1974) showed the efficacy of oxamyl, phenamiphos and aldicarb for controlling *P. coffeae* in coffee nurseries in El Salvador. Increased yields of coffee were obtained in the second year in treated plots with carbofuran (Abrego, 1974). Good control of *P. coffeae* was also obtained with Nemacur and it remained effective under field conditions 90 days after application (Kumar, 1982). In India it was found that *Coffea robusta* is more tolerant to *P. coffeae* than *C. arabica* or *C. excelsa* (Anon., 1974), hence the use of *Coffea robusta* as rootstocks is the most promising means of nematode control (Palanichamy, 1973). In fact, Schieber & Grullon (1969) also suggested the use of *Coffea canephora* var. *robusta* as a source of resistance for root-stock in grafted plants.

Application of systemic or contact granular nematicides on severely damaged coffee plants infested by *M. incognita* has not been effective due to the rapid killing of greater parts of the root system by the nematode (Curi *et al.*, 1977). Poor control also occurs on seedlings infested by *M. incognita* (Jaehn *et al.*, 1984).

However, most of granular nematicides are effective in decreasing nematode populations a few months after application (Huang *et al.*, 1983; Campos *et al.*, 1988). After this time the populations may increase on treated plants, but the plants have good foliage which seems to be induced by some other action besides controlling the nematodes (Campos & Lima, 1986).

Granular nematicides when applied in coffee have to be incorporated into the soil under the edge of the foliage toward the stem. Different machines have been developed to do this work. Another aspect on the application is the timing; since the granular products require water to liberate the active ingredient, application at the beginning of the rainy (November, in Brazil) season has been indicated (Campos *et al.*, 1985).

GRAFTING

The widespread distribution and the aggressive parasitism of *M. incognita* in west of Sao Paulo forced the researcher in Brazil to search for an efficient control measure other than chemicals. An introduction of *Coffea canephora* C. 2258, from Turrialba, Costa Rica, showed high resistance to *M. exigua* and resistance and/or tolerance to several populations of *M. incognita* (Fazuoli, 1986). The rate of resistance was initially 70%, but through selections in the field highly infested with *M. incognita*, this rate had increased significantly. The line obtained by selections from C 2258 has been named LC 2258. On LC 2258 *C. arabica* var. Mundo Novo or Catuai Vermelho was grafted and then planted in infested fields. The growth and production were very good, but in many locations, the non-grafted plants of *C. arabica* did not survive. Even though the compatibility between scion and stocks still has to be studied (Fazuoli, 1986), the cooperatives of coffee growers have produced many thousand of grafted seedlings and given them to the farmers to show the possibility of this means of controlling *M. incognita* in coffee. It is a promising means of control in regions infested with *M. incognita*.

RESISTANCE

Many researchers are still seeking for resistance to *M. incognita* in germplasm of *Coffea*, notwithstanding the occurrence of three races of the nematode species on coffee plantations in Brazil (Moraes *et al.*, 1977 a, b; Medina Filho *et al.*, 1981; Fazuoli, 1986). Work also has been done on resistance to *M. exigua* (Curi *et al.*, 1970).

Several lines of *Coffea canephora, C. congensis* showed resistance to race 3 of *M. incognita*, and some progenies of *C. canephora*, e.g. Sarchimor (derived from crossing Vila Sarchi x Timor Hybrid) and Icatu (advanced line derived from crossing *C. arabica x C. canephora*) have shown moderate resistance. However, all germplasm of *C. arabica* tested were susceptible to race 3 of *M. incognita* (Goncalves & Ferraz, 1987).

Test of different *Coffea* species, crosses and selections against root-knot nematode in Tanzania has been done by Bridge (1984) which indicated that some resistance may occur. Reports from the Kenya Coffee Research Station cited by Whitehead (1968), suggests resistance to *Meloidogyne* sp. in *Coffea corrisoi, C. conuga* and some lines of *C. congensis* in Angola. Whitehead (1969) said that coffee is very resistant to *M. javanica* and *M. kikuyensis*.

CROP ROTATION

Moraes *et al.* (1977a,b), studied rotation with cotton, soybean and corn in *M. exigua* infested areas and concluded that, after one year's rotations with these crops, the grower can return to coffee cultivation. But Carneiro & Cameiro (1982), who screened 29 crops for rotation in *M. incognita* infested coffee fields, found that only *Arachis hypogaea* and *Ricinus communis* were immune. *Styzolobium deeringianum* and *Crotalaria spectabilis* showed resistance to this nematode. In other work, rotation was done with 9 crops on an eradicated coffee area previously infested by *Meloidogyne exigua*. The second stage larval population of *Meloidogyne exigua* was assessed in the soil for 23 months (Fig.3) when coffee seedlings were planted as a field bioassay but no second stage larva was found in the soil confirming the efficacy of all the crops planted in the area.

Some plants grown in the infested area were more efficient, in short period to decrease the population of *M. exigua* (Fig.3).

The survival of *M. exigua* during the period of growing those crops varied with the suitability of the plant tissues (Fig.4).

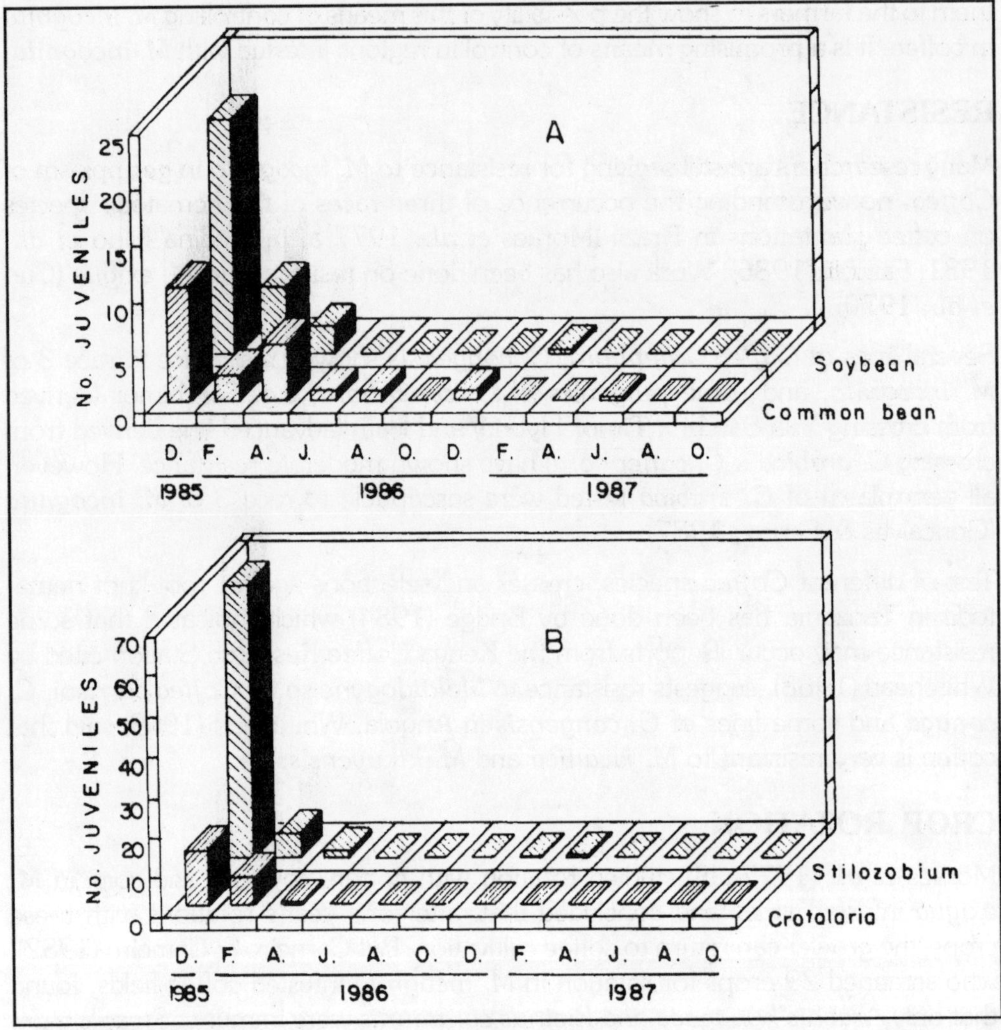

Fig.3. Number of second stage juveniles of *Meloidogyne exigua* per 200 cc of soil from the eradicated coffee area cultivated with soybean or common bean (A), and *Stilozobium* sp. or *Crotalaria* sp. (B).

B. COCOA

Meloidogyne spp. are the most important nematodes of cocoa due to their pathogenicity and wide distribution in cocoa producing regions. Root-knot nematodes have been found in cocoa since 1900 (c.f. Sosamma *et al.*, 1980a), and they have been reported from Zaire, Sao Tome, Java (Ghesquiere, 1921; Cotterel, 1930; Fluitter & Mulholland, 1941), Ghana, Malawi, Ivory Coast (Edwards, 1955; Luc & de Guiran, 1960; Martin, 1961), Nigeria (Caveness, 1967), Venezuela

(Torrealba, 1969), Brazil (Lordello, 1968) and India (Sosamma *et al.*, 1980b).

Meloidogyne incognita seems to be the most frequently found in cocoa (Luc & de Guiran, 1960; Sharma & Sher, 1974). It is common pest in West Africa (Whitehead, 1969), India (Sosamma *et al.*, 1980b) and is widespread in cocoa regions of Brazil (Sharma & Sher, 1974; Sharma, 1982). In the cocoa regions of Espirito Santo State, Brazil, it is most frequent nematode in sampled sites (Sharma & Sher, 1974).

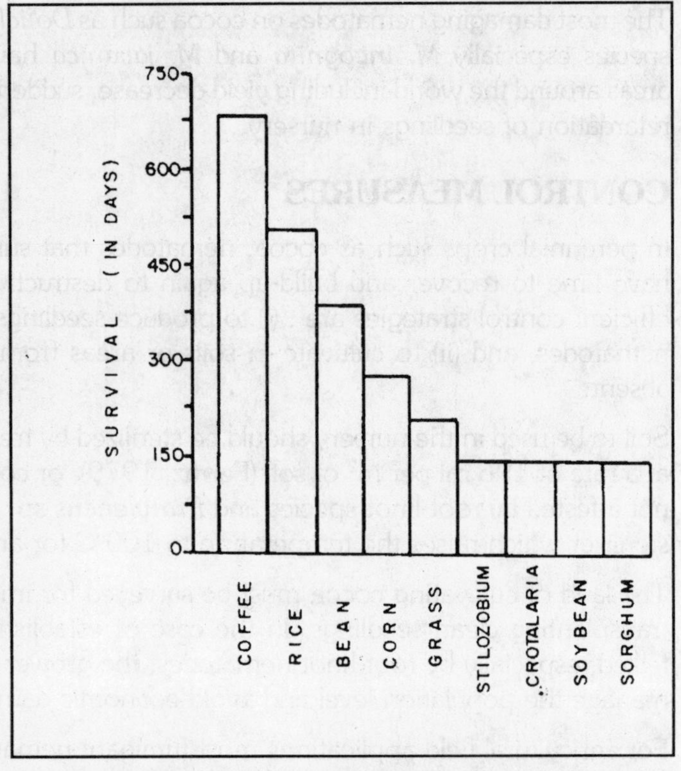

Fig.4. The range of survival of *Meloidogyne exigua* in field in rhizosphere of many plants in the field during 690 days of searching.

However, other species of *Meloidogyne* have also been found on cocoa: *M. exigua* in Bolivia (Bridge *et al.*, 1982), *M. javanica* in Malawi (Corbett, 1961) and in Central Africa (Martin, 1961), *M. arenaria* and *M. thamesi* in Brazil (Sharma, 1979).

In addition, many other genera and species of root feeding nematodes have been found in association with cocoa, although the pathogenic relationship, for most of them, has not been proved: *Criconemella* spp., *Dolichodorus minor, Helicotylenchus* spp., *Hemicriconemoides cocophillus, Hemicycliophora* spp., *Heterodera* spp., *Hoplolaimus* spp., *Longidorus* sp., *Meloidogyne arenaria, M. exigua, M. incognita, M. javanica, M. thamesi, Peltamigratus holdemani, Pratylenchus brachyurus, P. coffeae, Radopholus similis, Rotylenchulus reniformis, Rotylenchus microstriatus, Scutellonema brachyurus, S. clathricaudatum, Trichodorus monohystera, Trophurus imperialis, Tylenchorhynchus martini, Xiphinema* spp. (Bridge *et al.*, 1982; Sharma & Loof, 1974; Sharma & Sher, 1973, 1974; Sosamma *et al.*, 1980a,b; Thorold, 1975; Whitehead, 1969).

The most damaging nematodes on cocoa such as *Dolichodorus* and *Meloidogyne* species especially *M. incognita* and *M. javanica* have caused losses in cocoa areas around the world including yield decrease, sudden death of trees and growth retardation of seedlings in nursery.

CONTROL MEASURES

In perennial crops such as cocoa, nematodes that survive the control practices have time to recover and build-up again to destructive levels. Hence the most efficient control strategies are : (i) to produce seedlings free of major pathogenic nematodes, and (ii) to cultivate in soils or areas from which the nematode are absent.

Soil to be used in the nursery should be sterilized by treating with methyl bromide at a rate of 196 ml per m^2 of soil (Ferraz, 1979), or collected from areas that are not infested by root-knot species and *Pratylencus* sp. Another method is hot air sterilizer which raises the temperature to 100°C for an hour (Sharma, 1975).

The land of cultivating cocoa must be surveyed for important nematodes before transplanting clean seedlings. In the case of established plantations already infested, especially by root-knot nematodes, the grower should use namaticides to manage the population level and avoid economic damage.

For agricultural field applications most fumigant nematicides are no longer used. The emphasis has been on the production and uses of contact or systemic nematicides for controlling plant-parasitic nematodes.

Application of granular nematicides such as Nemacur 10G, Temik 10G at the rate of 50 mg of commercial product/plant and Terracur 5% at 100 mg/plant in plants infested by *M. incognita* in the glass-house reduced the nematode density and increased the numbers of leaves per plant (Sharma & Ferraz, 1977).

Tarjan *et al.* (1973) reported increased yield of 11 to 96% after field application of Mocap or Terracur at the rate of 34 kg of the commercial product/ha and Nemacur at 22 kg/ha. The products were applied within a cleared area of 1.0 or 1.5 m around each trunk and then incorporated into the upper 2.5 cm of soil. Sosamma *et al.* (1980a) have reported an increase in the number of pods by the application of Dasanit and Nemacur and an increase of yield by the application of Nemacur, Terracur and Mocap.

Care must be taken with the selection of shade plants, avoiding trees susceptible to root-knot or lesion nematodes, for example *Leucaena glauca* and banana (Corbett, 1961; Sosamma *et al.*, 1980a).

C. FINAL REMARKS

Growers must be aware of the nematode threat to the coffee and cocoa crops because of the perennial characteristic of those crops and the problems resulting from nematode infestation. Emphasis should be on the prevention of nematode infestation by disinfesting nursery soil and planting seedlings in non-infested field soil.

The control measures described for coffee against *Meloidogyne* species found in South American countries are likely to be effective for the control of African root-knot nematodes, but applicability of these measures on a practical basis in African countries is uncertain.

REFERENCES

Abergo, L. (1974). Ensayos de selectividade de nematicides en el combate del *Pratylenchus coffeae* en almaciqueras de cafe. *Nematropica* **4**: 17.

Almeida, V.F. de & Campos, V.P. (1991). Alternacia de culturas sobrevivencia de *Meloidogyne exigua* em areas de cafezal infestado e erradicado. *Namatologia Brasileira* **15**: (1): 30-42.

Anonymous (1968). *Principles of plant and animal pest control*. Vol. 4 — *Control of plant parasitic nematodes*. *National Academy of Sciences. Publ. No.* **1696**. Washington, D.C. p. 172.

Anonymous (1974). *Thirty third Annual Report*. 1972-1973 (Ist April, 1972 to 31st March, 1973). India Coffee Board, Bangalore, India. pp. XXIV + 269.

Bridge, J. (1984). Coffee nematode survey of Tanzania. Report on a visit to Examine Plant Parasitic Nematodes of Coffee in Tanzania, February-March 1984. Commonwealth Institute of Parasitology. p. 22.

Bridge, J.; Pages, S.L.J. & Waller, J.M. (1982). Plant-parasitic nematodes and diseases of crops in the Santa Cruz Department of Bolvia. U.K. Overseas Development Administration Report. p. 60.

Campos, V.P. & Lima, R.D. (1986). Nematoides parasitas do cafeeiro. *In*: *Cultura do Cafeeiro, fatores, que afetam a productividae* (Eds.-A.B. Rena, E. Malavolta, M. Rocha & T. Yamada). Associacao Brasileira para Pesquisa e do Fosfato. Piracicaba, SP. Brazil: 379-389.

Campos, V.P.; Lima, R.D. & Almelda, V.F. (1985). Nematoides parasitas do cafeeiro. *Informe Agropecuario*, Belo Horizonte **11**: 50-58.

Campos, V.P.; Resende, I.C. & de Faria, M.A. (1988). Flutuacao populacional de *Meloidogyne exigua* em cafezais tratados com nematicides. *Fitopatologia Brasileira* (In press).

Carneiro, R.G. & Carneiro, R.M.D.G. (1982). Selecao preliminar de plantas para rotacao de culturas em areas infestadas por *M. incognita* nos anos de 1979 e 1980. *Nematologia Brasileira* **6**: 141-148.

Caveness, F.E. (1967). Nematology studies 1960/1965. Ministry of Agriculture and Natural Resources, US AID Project, Lagos, Nigeria. p. 135.

Corbett, D.C.M. (1961). *Report of Department of Agriculture* 1959/1960. Nyasaland, Part II : 157-158.

Cotterel, G.S. (1930). Bulletin Department of Agriculture, Accra Gold Coast. p. 22.

Curi, S.M.; Carvalho, A.; Moraes, F.P.; Monaco, L.C. & Arruda, H.V. de (1970). Novas fontes de resistencia genetica de *Coffea* no controle de nematoide do cafeeiro, *Meloidogyne exigua. O. Biologico* **36**: 293-295.

Curi, S.M.; Silveira, S.G.P. da & Elias Jr., E.G. (1977). Resultados de producao e da protecao do sistema radicular de cafeeiros sob controle quimico do nematoide *Meloidogyne incognita* (Kofoid & White, 1919) Chitwood 1949, em condicoes de campo. *Nematologia Brasileira* **2**: 93-99.

Edwards, E.E. (1955). Further observations in the occurrence of nematodes of the genus *Meloidogyne* in Gold Coast. *J. Helminthol.* **29**: 153-170.

Fazuoli, L.C. (1986). Genetica melhoramento do cafeeiro. *In* : *Cultura do cafeeiro, factores* que afetam a produtividade (Eds.-A.B. Rena, E. Malavolta, M. Rocha & T. Yamada). Associacao Brasileira para Pesquisa da Potassa e do Fosfato. Piracicaba, SP. Brazil: 87-113.

Ferraz, E. (1979). Nematoides do cacaueiro. IV Curso Internacional de Cacau. *Centro de Cacau.* Centro de Pesquisa do Cacau, Itabuna, BA, Brazil. p. 24.

Fluitter, H.J. & Mulholland, J.J. (1941). Gegevens, Verkregen biji het onderzoek near de Waardplnten van *Tylenchus coffeae. Beragcultures* **15**: 1588-1593.

Ghesquiere, J. (1921). Nouveaux parasites du cacoyer. Maladie vermiculire du cacaoyer *Tylenchus (Heterodora radicicola)*. Greef. et relation aves la maladie du *Diplodia. Bulletin de Agriculture* **12**: 709-718.

Goncalves, W. & Ferraz, L.C.B. (1987). Resistencia do cafeeiro a nematoides II. Testes de progenies e hibridos para *Meloidogyne incognita* raca 3. *Nematologia Brasileria* **9**: 125-142.

Huang, S.P.; Resende, I.C.; Souza, P.E. De & Campos, V.P. (1983). Effect of aldicarb, ethoprop and carbofuran on control of coffee root-knot nematode, *Meloidogyne exigua. J. Nematol.* **15**: 510-514.

Jaehn, A.; Rebel, E.K. & Matiello, J.B. (1984) Viabilidade de recuperacao de mudas de cafeeiro infestadas por *Meloidogyne incognita* atraves de nematicidas. *Nematologia Brasileria* **8**: 295-300.

Kumar, A.C. (1982). Evaluation of Nemacur 5G against the coffee root-lesion nematode, *Pratylenchus coffeae. J. Coffee Res.* **12**: 1-7.

Lima, R.D. de; Campos, V.P.; Huang, S.P.; Melles, C.C.A. (1985). Reproductividade e parasitismo de *Meloidogyne exigua* em ervas daninhas que ocorrem em cafezais. *Nematologia Brasileira* **9**: 63-72.

Lordello, L.G.E. (1968). Nematoides associados a uma doenca do cacaueiro. *Revista de Agricultura, Piracicaba* (Brazil) **43**: 154.

Luc, M. & de Guiran, G. (1960). Les nematodes associes aux plantes de l'Ouest Africain. Liste preliminaire. *L'Agronomie Tropicale Noguet* **15**: 434-449.

Luc, M.; Sikora, R.A. & Bridge, J. (1990). *Plant-parasitic nematodes in subtropical and tropical agriculture*. C.A.B. International Institute of Parasitology. Wallingford, U.K. p. 629.

Martin, G.C. (1961). Plant species attacked by root-knot nematodes (*Meloidogyne* spp.) in the Federation of Rhodesia and Nyasaland. *Nematologica* **6**: 130-134.

Medina Filho, H.P.; Fazuoli, L.C. & Costa, W.M. da (1981). Identificacao das racas 2, 3 e 4 de *Meloidogyne incognita* parasitando cafeeiros. *In* : *Nono Congresso Brasileiro e Pesquisas* Cafeeiras, Sao Lourenco, MG: 161-168.

Moraes, M.V.; Lordello, L.G.E.; Reis, A.J.; Thomaziello, R.A. & Goncalves, W. (1977a). Ensaio de rotacao de culturas para reaproveitamento com cafeeiros de terras infestadas por *Meloidogyne exigua. Nematologia Brasileira* **2**: 257-265.

Moraes, M.V.; Thomaziello, R.A.; Lordello, L.G.E. & Krinski, J. (1977b). Ensaios de tratamento de solo por produtos quimicos para uso em viveiro de cafe. *Namatologia Brasileira* **2**: 231-244.

Palanichamy, L. (1973). Nematode problems of coffee in India. *Indian Coffee* **37**: 99-100.

Schieber, E. & Grullon, L. (1969). El problema de nematodos que atacam el cafe (*Coffea arabica* L.) en la Republica Dominicana. Turrialba **19**: 513-517.

Sharma R.D. (1975). Effect of partial soil sterilization on the growth of the cacao (*Theobroma cacao* L.) seedling in the nursery. *Arquivos do Instituto Biologico*, Sao Paulo, Brazil **42**: 23-30.

Sharma, R.D. (1979). Informe tecnico-Setor de nematologia 56-60. CEPEC-CEPLAC, Itabuna, Bahia, Brazil.

Sharma, R.D. (1982). Nematodes associated with cocoa hybrids and clones in Bahia, Brazil. *Nematologia Brasileira* **6**: 85-91.

Sharma, R.D. & Ferraz, E.C.A. (1977). Eficacia de nematicidas sistemicos e nao sitemicos no controle aos nematoides fitoparasitas associados a mudas de cacaueiro. *Nematologia Brasileira* **2**: 139-147.

Sharma, R.D. & Loof, P.A.A. (1974). Nematodes of the cocoa region of Bahia, Brazil. III-Plant Parasitic and free-living nematodes in the rhizospheres of six different plant species. Revista *Theobroma* **4**: 39-43.

Sharma, R.D. & Sher, S.A. (1973). Nematodes of the cocoa region of Bahia, Brazil. II. Occurrence and distribution of plant parasitic nematodes associated with cocoa (*Theobroma cacao* L.) *Revista Theobroma* **3**: 17-24.

Sharma, R.D. & Sher, S.A. (1974). Nematoides da regiao cacaueira do Espirito Santo, Brasil. I. Nematoides associados ao cacaueiro (*Theobroma cacao* L.) *Revista Theobroma* **4**: 26-31.

Sosamma, V.K.; Koshy, P.K. & Sundararaju, P. (1980a). Nematodes associated with cacao — A review. Plant Protection Committee for the South East Asia and Pacific Region, Technical Document no. 123. p. 14.

Sosamma, V.K.; Koshy. P.K. & Sundararaju, P. (1980b). Plant-parasitic nematodes associated with cocoa. *Cocoa Growers Bulletin* **29**: 27-30.

Tarjan, A.C.; Jimenez, M.G. & Soria, J. (1973). Increasing yields of cacao by application of nematocides. *Turialba* **23**: 138-142.

Thorold, C.A. (1975). *Disease of cocoa*. Oxford. Calrendon Press. p. 423.

Torrealba, P.A. (1969). Survey of plant parasitic and free-living nematode genera from Venezuela. *In: Nematodes of Tropical Crops* (Eds.-J.E. Peachey) St. Albans, U.K. Commonwealth Agricultural Bureaux, Technical Communication No. 40: 257-263.

Van Gundy, S.D. McKenry, M.V. (1977). Action of nematocides *In* : *Plant Disease — An advanced treatise* Vol. I. (Eds.-J.G. Horsfall & E.B. Cowling). New York, Academic Press : 263-283.

Whitehead, A.G. (1968). Nematodes. *In* : *Pests of Coffee* (Ed.-R.H. Le Pelley). Longmans, Green & Co. Ltd., London, Harlow: 407-422.

Whitehead, A.G. (1969). Nematodes Attacking Coffee, Tea and Cocoa and their control. *In* : *Nematodes of Tropical Crops* (Ed.-J.E. Peachey). Technical Communication no. 40. Commonwealth Bureaux of Helminthology, St. Albans, Herts, England: 238-250.

Chapter 11

Strategies for the Control of Citrus Nematodes

A. Mani.[1] & M.S. Rao[2]

1. Regional Agricultural Research Station, Andhra Pradesh Agricultural University, Palem-509 215, India.

2. Entomology & Nematology Division, Indian Institute of Horticultural Research, Hessaraghatta Lake Post, Bangalore-560 089, India.

CITRUS is a cosmopolitan crop grown in many soil types in various agro-climatic regions. Citrus is a perennial evergreen tree crop, harbouring and encouraging the build-up of nematode population throughout the year since there would be constant supply of new feeder roots required for the growth and multiplication of the nematodes. Hence nematodes attain a new dimension as one of the important factors limiting the production of citrus in the world.

Citrus decline, the complex disease syndrome has been observed in most of the citrus growing areas. Nematodes are actively involved in the etiology of the disease syndrome in addition to various other biotic and abiotic factors. In general, it would be difficult to identify nematode infected trees under field conditions, as they infect only the root system and do not produce specific above ground symptoms.

Neal (1889) for the first time recorded the association of plant-parasitic nematode, *Heterodera radicicola* (= *Meloidogyne* sp.) with citrus roots in Florida, U.S.A. Till today, about 250 nematode species have been reported from the rhizosphere of citrus, out of which seven species were found to be pathogenic to citrus, viz. *Tylenchulus semipenetrans* (Cobb, 1914), *Radophous similis* (Suit & DuCharme, 1953), *Belonolaimus gracilis* and *Trichodorus christiei* (Standifer & Perry, 1960), *Pratylenchus brachyurus* (Brooks & Perry, 1967), *Hemicycliophora arenaria* (Van Gundy & Rackham 1961) and *Meloidogyne javanica* (Mani, 1986). Out of 122 species of plant-parasitic nematodes reported to be in association with citrus in India, proof of pathogenicity is available for only six species, viz. *Tylenchulus semipenetrans* (Siddiqi, 1961), *Pratylenchus coffeae* (Siddiqi, 1964), *Basirolaimus indicus* (Gupta & Atwal, 1972), *Meloidogyne africana* (Chitwood & Toung, 1960), *M. indica* (Whitehead, 1968) and *M. javanica*

(Mani, 1986). However, in India *T. semipenetrans* and *Meloidogyne* spp. have been identified as the most potential nematode pests involved in the citrus decline syndrome in different citrus growing areas and hence gained importance in recent years.

CROP LOSS ASSESSMENT

The effect of nematode damage to citrus roots are often marked by certain soil and nutritional disorders, flooding, drought, pests, diseases, etc., and the damage would not be evident on citrus trees unless these disturbances are corrected (DuCharme, 1969). Cohn (1972) reported that the nematodes caused reduction in yield both in terms of quantity and quality. In addition, they also increase the cost of production by increasing the expenses incurred in nematode management practices. In general, it is very difficult to estimate the yield loss of citrus caused by nematodes because of the difficulty faced in assessing the exact crop loss, as several other factors are involved under field conditions besides the nematodes. However, an approximate estimate of loss can be obtained by measuring yield increase as a result of nematode control by comparing the performance of infested with that of healthy trees (Table 1).

Table 1: Effect of control of *Tylenchulus semipenetrans* on yield of citrus fruits.

Yield increase %	Region	Reference
10-30	Australia	Meagher (1969)
12-38	Arizona	Reynolds (1969)
10-50	California	Baines & Small (1969)
6-34	Florida	O'Bannon & Tarjan (1973)
100-200	India	Bindra & Chhabra (1972)
160-224	India	Baghel & Bhatti (1983)
40-44	India	Mukhopadhyaya & Dalal (1971)
5-40	Mediterranean countries	Scaramuzzi & Perrota (1969)
13-20	Texas	Timmer & Davis (1982)

Cohn (1972) reported that *T. semipenetrans* caused an average of 8.7-12.2 per cent annual loss in world citrus production. In India, nematode reportedly caused an annual crop loss of about 15 per cent (Reddy, 1988).

CITRUS NEMATODE (*Tylenchulus semipenetrans*)

Distribution and Dissemination

Since the first record of *T. semipenetrans* in California in 1912 on citrus roots (Thomas, 1913), the nematode has been reported from almost every citrus growing country in the world (Thorne, 1961). In India, the nematode was first recorded by Siddiqi (1961) and is now widely distributed in most of the citrus growing states. It has been recorded from Punjab, Delhi, Uttar Pradesh, Rajasthan, Maharashtra, West Bengal, Assam, Kerala, Orissa, Andhra Pradesh and Karnataka (Swarup *et al.*, 1964; Chona *et al.*, 1965; Nair *et al.*, 1969; Khuntia & Das, 1969; Reddy, 1979; Anon., 1983, 1984, 1985).

Dissemination of the nematode to different areas takes place through various means, but mainly takes place through planting material. Nursery stocks are usually placed with ball of earth along with the root system intact, harbouring nematodes (most of the times), reach new geographical areas. Commercial root stocks such as jambheri (*Citrus jambheri*), rangapur lime (*C. limonia*) and sathgudi (*C. sinensis*) were reported to be highly susceptible to the nematode (Mani & Murthy, 1989).

Symptoms

Symptoms expression may not be noticeable until several years after the peak of the nematode population is reached. In general, above-ground symptoms of nematode damage resemble that of malnutrition wherein leaves become chlorotic, often curling inward and some twigs become partially or totally defoliated giving rise to the upper most portion of the tree a naked appearance (O' Bannon & Esser, 1985). Infected trees shed more leaves than uninfected trees, or trees that have been treated with nematicides. In advanced stages fruit size and yield are reduced resulting in uneconomic returns.

Heavily infected roots are shortened, show irregular thickening (Swarup *et al.*, 1964) and exhibit extensive necrosis which give them rust brown colour. Soil particles cling to infected roots due to the gelatinous material in which eggs are embedded. Infected roots get slightly enlarged, owing to the surface being rough instead of smooth (Fig. 1.). Infected tissues get invaded by secondary microorganisms leading to dark necrotic lesions within the cortex. In advanced stages of infection, when the roots are gently rubbed between fingers the cortex of the root comes out in the form of a sleeve leaving the inner woody vascular stele intact, resulting the eventual death of that portion of the root (Schneider & Baines, 1964; Van Gundy & Kirkpatrick, 1964). Such feeding kills many roots over a period of years causing a gradual "Dieback" of the tree.

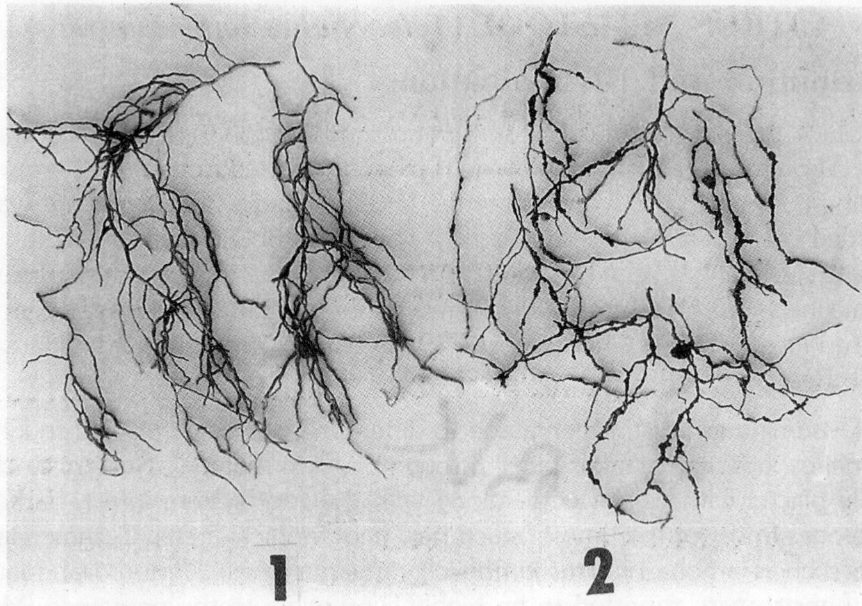

Fig.1. Citrus roots, 1) uninfected, and 2) infected with *Tylenchulus semipenetrans*.

Baines & Clarke (1952) suggested that the nematode affected the growth of citrus trees by : (1) injuring the bark of the roots, (ii) removing plant nutrients during feeding, (iii) impairing normal growth and functioning of the roots, and (iv) possibly injecting toxic material into the trees.

Damage to host tissue

Second-stage juveniles of *T. semipenetrans penetrate* epidermal and hypodermal cells and feed on the hypodermis and outer layers of cortical parenchyma. Reaction to nematode feeding reported to be localized. Larval feeding does not extend beyond the outer layer. Young females established a permanent feeding site which consist of 6-10 cortical cells or "nurse cells", resembling a cavity. The cavity is apparently a destroyed cortical cell that allows the nematode to move from cell to cell as it pierces the surrounding cells during feeding. Feeding enhances the cytoplasmic density and increases the volume of nucleus and nucleolus of each cell. Advanced stages of affected cells are characterised by thickening of the cell walls, very dense cytoplasm, and enlarged amoeboid nuclei (Van Gundy & Kirkpatrick, 1964).

Ecology

Information on ecological aspects not only help in understanding the citrus nematodes in totality but also important in conceiving a sound strategy of nematode management. Citrus crop being perennial in nature and citrus nematode occurring in all citrus-growing areas of the world, discussion on the relationship between citrus nematode and soil factors as well as other factors of citrus ecosystem is of special relevance.

Hung (1969) correlated the soil temperature with the nematode population and reported low larval and adult population when soil temperatures were low from December to March. High populations of larvae were observed in February, April and May during the period of high rainfall. In Florida, O'Bannon *et al.* (1972) observed maximum levels of population of citrus nematode during April-May and November-December coinciding with the periods of increased root growth. During April and August-September high densities of this nematode were reported from the rhizosphere of citrus in India (Chhabra, 1978). Several other researchers also recognised high and low population periods (Macaron, 1972; Toung, 1963; Prasad & Chawla, 1965). Certain soil factors influence infection and reproduction. Van Gundy *et al.* (1964) observed reproduction of citrus nematode and reduction in growth of citrus seedling in soils with the clay content ranging between 5-50%. Rate of reproduction was significantly lower in soils of 50% clay than in soils of 5, 15 or 30% clay. In general citrus nematode invasion and reproduction is lower in very sandy, coarse textured soils than in other soil types (Baines, 1974; O'Bannon, 1968; Van Gundy *et al.*, 1964).

Citrus nematode tolerate different soil salinity levels around the host plants. More citrus nematodes were recovered by Machmer (1958) from roots subjected to higher salinity levels than lower and positive correlation in the rate of reproduction of *T. semipenetrans* was found (Van Gundy *et al.*, 1964) to exist in the pH range of 6.0 to 7.5 Nematode populations were reported to exist at pH extremes and would influence tree growth (Van Gundy & Martin, 1961; Reynolds etal. 1974). In copper and zinc deficient soils the citrus crop was found to be more vulnerable for the attack by nematodes, than in nutritionally balanced soils (Van Gundy *et al.*, 1964).

Nematode parasites on their own do migrate through the soil, gradually expanding the area of infestation. The rate of spread varies with the kind of nematode, soil porosity, root density and soil moisture and it was found that *T. semipenetrans* could move 53.5 cm in two years (Tarjan, 1971). It was speculated that the interval between irrigations of citrus infected with *T. semipenetrans* is important in the life cycle of the nematode because free water is necessary to flush the juveniles from the egg masses (Van Gundy, 1958). It has been shown that citrus nematodes spread in the re-use of sub-soil drainage in irrigation supplies in Australia indicating that high number of nematodes may be transported along the tile drains beneath citrus crop (Meagher, 1967). It was found that nematode population levels varied with soil moisture. On the fine-textured soil they were favoured by soil suction fluctuating between 0-60 centibars and on the coarse-textured soil, they developed best when moisture suction did not exceed 10 centibars (Van Gundy *et al.*, 1964). It was indicated that wet soil enhances the damage done by nematode seven though it was not as favourable for their reproduction (Van Gundy *et al.*, 1964).

The most favourable temperatures for infectivity and development of the citrus nematode are between 25⁰C and 30⁰C. (Baines, 1950; O'Bannon *et al.*, 1966). The temperatures in unshaded moist soil around young trees at depths of 2.5.-5.0 cm were greater than 35⁰C and so the nematode populations remained low (Reynolds & O'Bannon, 1963). Populations increased when conditions became more favourable for the nematodes, as the trees became larger, providing more shade and thus changing the soil environment. Astonishingly citrus nematode was found to survive for at least 10 years in soils free of citrus (Baines *et al.*, 1962).

ROOT-KNOT NEMATODE (*Meloidogyne* spp.)

Distribution

Meloidogyne spp. have been recorded in association with citrus only in a few countries. Minz (1956) for the first time observed the damage to citrus trees caused by *M. javanica* (Treub) Chitwood in Israel. Subsequently, a few more species have been identified in association with citrus (Table 2), but further studies have not been made to confirm their pathogenic potential and economic significance.

Table 2: *Meloidogyne* species found associated with citrus.

Nematode species	Distribution	Reference
M. africana	Delhi (India)	Chitwood & Toung (1960)
M. incognita	Queensland (Australia)	Colbran (1958)
	South India	Kumar *et al.* (1971)
M. indica	India	Whitehead (1968)
	Andhra Pradesh (India)	Anon. (1988)
M. javanica	Andhra Pradesh (India)	Mani (1986)
	Israel	Minz (1956)
Meloidogyne sp.	Andhra Pradesh (India)	Perraju (1960)
	"	Reddy (1979)
	"	Reddy & Singh (1979)
	"	Singh et al. (1979)
	"	Tirumala Rao (1956)
	"	Inserra & Vovlas (1977)
	Taiwan, Delhi (India)	Chitwood & Toung (1960)
	Taiwan	Hung (1969)

In India, Tirumala Rao (1956) first reported the damage to lime trees caused by unidentified species of *Meloidogyne*. Mani (1986) reported that *M. javanica* caused severe crop loss of acidline in Nellore district of Andhra Pradesh, India and it was pathogenic to both acidline (*C. aurantifolia*) and sweet orange (*C. sinensis*). Recently, *M. indica* was recorded from the rhizosphere of sweet orange trees in Srikakulum district of Andhra Pradesh (Anon. 1988). It appeared that *M. indica* and *M. javanica* infecting citrus were confined to mostly coastal Andhra Pradesh and had not been recorded as a field problem elsewhere.

Symptoms

The root-knot nematode infected trees are poor in vigour, unthrifty in appearance and show severe stunted growth. They fail to flower and reduce fruits even after several years of planting. Infected root system shows the presence of conspicuous nematode galls on primary and fibrous roots (Fig.2). In advanced stages of infection, large cavities may be seen in the place of galls. Eggmasses can be seen as thin films spread over the root surface. Nematode damage is aggravated if vegetables such as okra, brinjal, cucurbits, tobacco, tomato, etc. are grown as intercrops in orchards or as a rotational crop in nurseries (Mani, 1986).

STRATEGIES FOR THE CONTROL OF NEMATODES

Citrus trees being perennial in nature, have deep-rooted and widespread root system which occupy considerably large volume of soil. As the nematodes are distributed around the root system in the voluminous rhizosphere soil, they easily escape detection. Besides, once the nematodes establish in orchard it would be practically impossible to exterminate them with the known methods of control. In ad-

Fig.2. Roots of acid lime infected with the root-knot nematode, *Meloidogyne javanica*.

dition, the various rootstock and scion combinations used in different regions react differently to nematode infections. Under these circumstances, no single method will be efficient and effective in controlling nematode pathogens of citrus, Hence, attempts should be made for the development and adoption of an integrated nematode management system to maintain the population within the economic threshold level.

CHEMICAL CONTROL

Nematode pathogens do occur on citrus at more than the economic threshold level (ETL) under field conditions and cause considerable loss in yield as observed in case of *Meloidogyne javanica* (Mani, 1986) and *Tylenchulus semipenetrans* (Anon., 1984). The primary advantage of chemical application is that the nematode population is reduced to less than 'ETL' within a short period of time. Hence, it will be economical to treat the orchards with chemicals if the area to be treated is fairly large. The cost : benefit ratio will also be high. However, chemicals should be applied at appropriate concentrations with extra care and caution.

Nursery bed treatment

The nematode population in nursery beds can be effectively checked by soil application of nematicides. In nurseries where nematode problems have been identified, aldicarb or carbofuran @ 4 kg a.i./ha (aldicarb @ 4 g/m^2 or carbofuran @ 13.3. g/m^2) may be applied for the control of *T. semipenetrans* (Reddy, 1988) and other nematode parasites.

Seedling bare root-dip treatment

Treating of growing plants in pesticidal solutions significantly reduces the load of nematode inoculum present in the root system. The method is economical as well as effective and can be employed for treating citrus nursery stock. The saplings can be treated as such in the same condition as they are packed. Usually, the saplings are removed along with a ball of earth and packed with gunny cloth. Experimental results revealed that the bare root-dip treatment of acidlime seedlings with monocrotophos or chlorpypriphos at 1000 ppm concentration for 45 minutes reduced the multiplication of *T. semipenetrans* by 90.0 per cent (Mani, 1990a).

Application of chemicals in orchards

(i) Pre-plant treatment

Seedlings/budlings planted in infested soils of old orchards are readily invaded by the surviving phytoparasitic nematodes. Therefore, pre-plant fumigation is important for the establishment of the young trees or for individual trees sites within an orchard. In U.S.A. pre-planting applications of nematicides have been found effective for nematode control. Applications by chisel injection of such chemicals as 1,3-dichloropropene give effective nematode control in most soils. Other compounds, such as methyl bromide, chloropicrin, ethylene dibromide, metham sodium and DD-MITC when properly applied, have also provided effective nematode control particularly in nursery or as tree site treatments. Unfortunately, several of these chemicals are not presently available or registered for use on citrus (O'Bannon & Essere, 1985).

(ii) Post-plant treatment

Treating living trees at their growing sites with chemical compounds do not toxic to trees but which provide effective nematode control has spurred the testing of many compounds. DBCP (1,2-dibromo-3-chloropropane) had been widely used by citrus industry and found very effective for the control of *T. semipenetrans* (Baines *et al*., 1977; Reyonlds & O'Bannon, 1958; Bindra & Chhabra, 1972; Mukhopadhyaya, 1970; Carter *et al*., 1971; Tarjan & O' Bannon 1974; Vilardebo *et al*., 1975; Timmer & French, 1979). But, since the chemical has been found detrimental to human beings, it is no longer registered for use. A search for alternate compounds has shown that several non-volatile nematicidal compounds which provide effective nematode control are non-phytotoxic and result in yield increases.

(iii) Non-fumigant nematicides

The non-fumigant nematicides most actively investigated include aldicarb, carbofuran, diazinon, fensulfothion, oxamyl, phnamiphos and prophos. These nematicides can be incorporated in soil, applied along with irrigation water or sprayed on foliage for the control of *T. semipenetrans* (Table 3).

Table 3: Effect of various non-fumigant nematicides on *Tylenchulus semipenetrans*.

Citrus cultivar	Chemical	Dosage	Type of treatment	Effectiveness
Acidlime	Aldicarb	4 kg a.i./ha	Soil application	Reduced the nematode population by 75-80%.
	Carbofuran	4 kg a.i./ha	Soil application	Reduced the nematode population by 65- 75%.

Citrus cultivar	Chemical	Dosage	Type of treatment	Effectiveness
Grape fruit	Aldicarb	5.7-11.4 kg a.i./ha	Soil application	Reduced the nematode population. Improved fruit size and quality. Increased fruit yield by 15.7-108.0%.
	Aldicarb, Accothion, Citrolane, Carbofuran, Ekalux, Fensulfothion, MIPC, Phorate	5 and 10 kg/ha	Soil application	Reduced the nematode population by 75-90% at lower concentration and by 90% at higher concentration.
Naval orange	Carbofuran Ethoprop Phenamiphos	22 kg a.i./ha	Soil application	Reduced the larval population in soil. Increased fruit yield by 108-137%.
Orange	Aldoxycarb, Carbofuran, Carbosulfan, Phenamiphos,	1.1 kg a.i./ha	Through drip irrigation	Reduced the number of female nematodes on roots and increased fruit yield.
	Phenamiphos	10 ml/m^2	With irrigation water	Reduced the nematode population. Increased the uptake of K, Ca and Cu. Gave extra yield of 6.1-7.7 tonnes.
Sweet lime	Prophos	40 kg a.i./ha	Soil application	Reduced the nematode population by 98%. Checked the population build-up for 16 months. Increased the yield @ 12 kg/tree.
Sweet orange	Aldicarb	4 kg a.i./ha	Soil application	Reduced the nematode population in soil.
		6 kg a.i./ha	Soil application	Reduced the nematode population by 84%. Increased fruit yield by 224% with a cost : benefit ratio of 1:11.
	Carbofuran	4 kg a.i./ha	Soil application	Reduced the nematode population in soil.
		6 kg a.i./ha	Soil application	Reduced the nematode population by 76%. Increased fruit yield by 152% with a cost : benefit ratio of 1:8.

Citrus cultivar	Chemical	Dosage	Type of treatment	Effectiveness
	Phorate	6 kg a.i./ha	Soil application	Reduced the nematode population by 66%.
	Oxamyl	4 μg/ml	Foliar spray	Decreased nematode population in pots. Ineffective in orchards.
Unidentified	Fensulfothion Phenamiphos	-	Soil application	Significantly reduced the nematode population.

There is no information available on the control of root-knot nematodes in citrus orchards. However, an observational trial carried out for the control of *M. javanica* on lime indicated that aldicarb and carbofuran at 4 kg a.i./ha reduced the larval population in soil by 77.8 and 62.8 per cent respectively (Mani, unpublished).

Keeping in view, the available experimental evidences it is suggested that the granular nematicides, viz. aldicarb and carbofuran, which are commonly available in market, may be applied @ 4-6 kg a.i./ha depending on the severity of the nematode problem, age of the trees and soil type so as to effectively check the multiplication of nematodes and obtain increased yield. It is preferable to apply the chemical in two split doses at 4-6 months interval to achieve maximum reduction in nematode populations. Besides, these chemicals should be applied at the time of formation of new roots to protect them against nematode infection.

NON-CHEMICAL CONTROL

Though the nematicides bring about quick reduction of the target nematode population, they are not commonly recommended due to high cost involved and hazards to the environment. The most important adverse effect of the chemicals is that they cause imbalances among the biotic factors in the soil ecosystem which may even aggravate the nematode problems in due course of time. It had been observed that citrus nematode was reduced initially following nematicidal applications and the nematode population reached its original level about six months after the application of the chemicals. Hence, it is necessary that non-chemical and low cost technologies should be evolved which are practically feasible, environmentally safe, economical having long term effects, and familiar to the farmer. Integration of two or more of these practices is quite feasible and necessary for maintaining nematode populations at a level lower than the economic damage threshold. The manageable components of the ecosystem and certain crop husbandry practices offer great potential in evolving low-monetary input strategies for nematode management.

Ploughing and fallowing

These methods are recommended only under selected situations where these can be easily followed to bring down the nematode populations at minimum cost. Farmers as well as government agencies have taken-up citrus nursery cultivation in different parts of the country. In general, cultivators prefer to grow a short duration crop or keep the land fallow for a short period after the disposal of nursery stock. Crop rotation can be followed as recommended elsewhere in the chapter. If the land is to be kept fallow, the following measures may be taken up for reducing the populations of *T. semipenetrans, M. indica,* and *M. javanica* in citrus nurseries.

All the discarded and week seedlings/budlings left in nursery and the sprouts growing from broken stumps or root pieces may be removed along with the root system and destroyed. At least two or three deep ploughings should be given during bright sunny days at an interval of 10-15 days.

Crop rotation

Crop rotation attempts to keep the nematode populations at a low level at which the crop suffers only a minimum damage. In general, non-host crops are included in rotation to reduce population build-up of the major pathogenic nematodes. Crops recommended for rotation should be economical to grow, useful, suitable to the region and have the acceptability among farmers. As such, there is limited scope for following crop rotation strictly when perennial fruit crops are grown. However, it can be followed under certain selected situations.

Root-knot nematodes (*Meloidogyne* spp.) are commonly present in commercial citrus nurseries in Andhra Pradesh, India. It is a common practice to grow vegetables in nursery areas for a season after the disposal of the seedlings. This rotation (citrus nursery-vegetables-citrus nursery) not only introduces the nematode inoculum into the nursery areas but also helps in quick build-up of the nematode populations. Hence, it should be discouraged. Instead, cereals and millets may be grown according to cultivator's choice. Green manure crops like *Crotalaria juncea* and *C. spectabilis* may also be grown and used for either composting or manuring the main crops.

Once the life of the citrus orchard comes to an end, it is highly preferable to grow cereals and millets for at least 2-3 seasons before initiating new orchard in the same field. This will help in reducing the populations of *Meloidogyne* spp. and *T. semipenetrans* which are common in citrus orchards.

Nursery bed treatment

It has been often observed that nematode-infected nursery stocks were the primary sources of nematode inocula for orchards (Anon. 1984). It is also well known that nematode management in nurseries is much more easy, effective and economical than in orchards. Hence, it is important to take necessary steps for treating nursery beds.

Solarization is one of the effective ways of suppressing nematode populations and can be employed mostly during hot weather days. The nursery areas should be ploughed well and levelled after breaking soil clods. The prepared field may be covered with white or black polyethylene sheet of 100 gauge thickness for about 7-10 days before sowing in primary nursery or transplanting in the secondary nursery.

Incorporation of organic amendments into soil has been found to reduce the population of certain major parasitic nematodes like, *Tylenchulus semipenetrans, Meloidogyne* spp. and *R. reniformis*. In addition, the organic amendments will enhance the growth and vigour of the seedlings by improving the soil fertility.

Farm yard manure (FYM) or compost at 20-25 tonnes/ha or oil cakes, particularly neem or karanj cakes at 2000-3000 kg/ha (8-12 kg/cent) may be incorporated in soil before sowing or transplanting.

Selection of healthy planting material

In fruit crops, propagating materials like seedlings and budlings are of special importance as they harbour potential nematode pathogens which often go unnoticed. All nursery stock should be obtained from reliable sources, fruit nurseries maintained by Government agencies, Agricultural Universities and registered fruit nurseries (private) which practice strict nematode control measures. The damage of root-knot nematodes (*Meloidogyne* spp.) on citrus can be easily identified based on nematode galls and eliminated by selecting nematode-free planting stocks.

Hotwater treatment of planting materials

It is recommended by killing nematodes within plant tissues before planting and has proved to be highly effective in treating nematode-infected planting stocks of citrus (Table 4). Hot water treatment of planting materials is found to be an excellent method for controlling *Meloidogyne* spp., *Tylenchulus semipenetrans* and *Radopholus similis*. However, this method is not practised by farmers due to the difficulty involved in preparing large quantity of hot water and maintaining the temperature of water precisely.

Table 4 : Schedule of hot water treatment (HWT) for the control of *Tylenchulus semipenetrans*.

Planting stock	Temperature (°C)	Duration of treatment (min)	Type of treatment
Citrus seedlings/ budlings	45	25	Root-dip
	46.7	10	Root-dip
	50	10	Immersion followed by cooling for 10 minutes.

Intercropping

It is common practice that farmers grow vegetables as intercrops in citrus orchards and at times in citrus nurseries during the initial stages of crop growth to get additional income. Growing of vegetables, particularly transplanted crops like brinjal, chillies, tomato, tobacco, etc., not only introduces the inoculum into main field but also helps in fast multiplication of root-knot nematodes (*Meloidogyne* spp.). In due course, the nematode populations may infect citrus crops and cause considerable economic loss (Mani & Dakshinamurti, 1986).

It was also observed, on the other hand, that intercropping with marigold and mustard markedly reduced the populations of *T. semipenetrans* (Mani, 1988b).

Keeping in view of the above findings, the practice of growing vegetables as intercrops in nurseries as well as orchards should be strongly discouraged. Instead, cereals and millets can be safely grown. Green manure crops can be grown and incorporated in soil before flowering which will increase the soil fertility.

Application of organic amendment

Application of certain indigenous plant products, non-edible oilcakes and other organic wastes for the control of nematodes has received attention in recent years. The oilcake and other plant products increased the level of phenolic contents in the roots and induced certain degree of resistance in plants against nematode infestation (Alam *et al.*, 1977). Further, Mankau & Minteer (1962) suggested that organic amendments reduced the soil population of *T. semipenetrans* apparently by increasing the microbial activity which was unfavourable for nematode survival. The higher contents of formaldehyde in oilcakes (Alam *et al.*, 1978) and release of ammonia from decomposing products were found to be responsible for the nematode mortality (Khan *et al.*, 1974). Cohn *et al.* (1965) observed that in an orchard irrigated with sewage water, nematode population was very low while in neighbouring fresh water-irrigated orchards, nematode infestation was very heavy. It was also observed that addition of refuse compost to seedlings infected with *T. semipenetrans* increased the growth and vigour of seedlings (Tarjan, 1977). It was further noticed that application of castor pomace and non-edible oilcakes like neem, sesame, olive, soybean and cotton seed cakes improved the growth of *Citrus reticulate* and significantly reduced the multiplication of *T. semipenetrans* (Mankau & Minteer, 1962).

Experiments carried out under controlled conditions revealed that the application of neem cake at 4% concentration not only increased the shoot weight of acidlime by 91.7% but also reduced the population of *T. semipenetrans* by 51.2 per cent. Field experiments carried out in nematode infested lime orchards with castor, gingelly and neem cakes indicated that the application of neem cake @ 20 kg per tree basin reduced the population of *T. semipenetrans* as well as *M. javanica* by 25.0 per cent after 3 months of application. Repeated applications of the cake once in 4 months significantly reduced the population of both the nematodes compared to untreated control (Anon., 1983, 1985). Hence, application of oilcakes at regular intervals would be one of the most feasible ameliorative measures which could be taken up for the management of nematodes infesting citrus orchards.

Nutrition

It was observed that the effect of nematode damage on the yield of citrus crops could be mitigated to some extent by improving irrigation and nutrition (Anon., 1988). Hence, it is important to manage the orchard by giving timely irrigations and applying manures and fertilizers. Excess application of organic manures will improve the growth and vigour of the trees by increasing the soil fertility status, moisture retention capacity, etc. and reduce the nematode populations by direct effects or by increasing the antagonistic microbial activity in soil.

It is also preferable to apply a major portion of the nutritional requirement of citrus through organic manures in the form of FYM compost, green manures or oilcakes as they supply nutrients as well as check the multiplication of pathogenic nematodes.

HOST RESISTANCE

The use of the host resistance is a cost-effective, economically viable and environmentally safe method of reducing crop losses, caused by nematode pathogens. In India, a number of attempts have been made and sources of resistance identified in citrus against *T. semipenetrans.* Trifoliate orange (*Poncirus trifoliata*) and its hybrids and certain rough lemon strains were found to be either resistant or moderately resistant to citrus nematode (Mani, 1990b; Mani & Reddy, 1986; Reddy & Singh, 1979). However, the resistant varieties identified so far will have a direct bearing in citrus production only when data on the horticultural traits like yield potential, availability of the resistant root stocks for commercial cultivation, acceptability of such root-stocks by nursery growers, susceptibility of the root-stocks to other pests and diseases, etc. are also generated. Meanwhile, it is suggested that horticulturists and nematologists should classify the available resistant sources and recommend for various purposes, viz. field use, breeding programme, further evaluation, etc. after thorough understanding.

BIOLOGICAL CONTROL

The use of microbial pesticides in nematode management has gained momentum in recent years. This is mainly due to the awareness of the danger of using toxic pesticides, the time required for the development of resistant cultivars and other cultural methods. Since many of the commonly used nematicides are expensive and/or being withdrawn from market because of their harmful effect on human beings as well as their persistence in soil or contamination of ground water, efforts are being concentrated on the integration of biocontrol agents into the overall nematode management system (Jatala, 1986).

The fungal parasite, *Paecilomyces lilacinus* has been found to be effective in controlling nematode populations of *Meloidogyne* spp. and *Globodera* spp. Studies on the efficacy of *P. lilacinus* on the biocontrol of *T. semipenetrans* revealed that the nematode population in roots and soil around citrus trees inoculated with *P. lilacinus* was significantly lower than that in the untreated or the nematicide treated trees (Jatala, 1986). In addition, the diameter of orange fruits from fungus treated trees was greater than that from trees treated with nematicides.

Recent studies carried out under local conditions in Andhra Pradesh, India indicated that *P. lilacinus* was very effective against *T. semipenetrans* and checked the multiplication of the nematode to the maximum extent of 71.7 to 97.8 per cent depending on the type of inoculum. Application of *P. lilacinus* in the form of spore suspension was found to be more effective than fungal infected wheat grains in controlling citrus nematode (Mani & Murthy, 1989).

Attempts were made to evaluate the natural substrates of plant origin for mass multiplication of *P. lilacinus* and found that wheat, bajra, jowar, ragi and rice grains and potato tubers were suitable for multiplying *P. lilacinus* as they supported good growth and spore load of the fungus (Mani & Murthy, 1989). Certain commonly available plant products such as leaves of neem, pongamia, subabool and tanner's cassia, peels of tapioca tubers and 'thippi (a waste product from sago industry) soaked in 0.75 per cent of potassium dihydrogen phosphate and sodium nitrate were also found to be equally good for mass multiplication of the fungus. Further, it was observed that oilcakes such as neem, castor, groundnut and gingelly did not affect the fungal growth, instead supported good growth of the fungus. Reddy *et al.* (1990) indicated that integration of castor and neem cakes with *P. lilacinus* was most effective in increasing plant growth characters and in reducing the citrus nematode population both in soil and roots. Further, in an attempt to manage *T. semipenetrans* infecting acid lime under glasshouse conditions, it was found that *P. lilacinus* at 4 g in combination with carbofuran at 1 kg a.i./ha gave maximum shoot length and shoot weight (Reddy *et al.*, 1989b). These studies suggest that the fungus can be applied in citrus nurseries as well as orchards even along with oilcakes or chemicals for the management of *T. semipenetrans*. As *P. lilacinus* has all the attributes of a successful biocontrol agent, there is good scope for developing it as a potential biocontrol agent of citrus nematode.

Recently, a bacterial spore parasite, *Pasteuria penetrans* has been identified as an effective biocontrol agent of *Meloidogyne javanica* (Stirling, 1984; Jayaraj & Mani, 1988). It is an obligatory spore parasite resistant to desiccation, viable for more than a year and compatible with oilcakes and nematicides. Preliminary tests revealed that *P. penetrans* infected and multiplied well on *T. semipenetrans* indicating that the bacterial spore parasite can be employed for controlling citrus nematode (Mani, 1988a). Hence, investigations should be carried out in detail to determine the efficacy of *P. penetrans* in controlling *T. semipenetrans* as well as *Meloidogyne* spp. in citrus orchards.

REGULATORY MEASURES

One of the most important means of nematode management in crops is prevention and it is cheaper and more effective than cure. It will be more so in case of citrus orchards. Nursery stocks are generally carried to orchards with a ball of earth packed along with root system.

Information on the occurrence and distribution of pathogenic nematodes in citrus nurseries and enforcement of domestic quarantine measures at inter- or intra-state level to prevent the movement of nursery stocks from such infested nurseries are essential prerequisites for effectively checking the spread of nematodes.

The results of the survey of commercial citrus nurseries carried out in Andhra Pradesh revealed that most of the private nurseries in different regions were severely infested with either *T. semipenetrans* or *M. javanica* (Anon., 1984;

Mani & Murthy, 1989). Observations made in citrus growing areas of Andhra Pradesh, India clearly revealed that *T. semipenetrans* as well as *M. javanica* were frequently transported to new geographical areas through infested seedlings and soil (Anon., 1984; Mani, 1986). It is suggested that infested nurseries may be inspected and nematicides may be applied at regular intervals for the control of nematodes.

(i) Certification of planting materials

The movement of nematodes along with nursery stocks can be effectively checked by establishing an appropriate Government institution which can certify the planting materials produced by Government as well as private nursery growers. The nursery growers should be brought under the control of Government institution through a registration act. The Government authority may periodically inspect the nursery farms for identifying the nematode populations and giving suitable recommendations for controlling them. The growers should be permitted to sell only certified saplings, though slightly at a higher cost than the present, which will be free from potential nematode pathogens. Such a system exists in Andhra Pradesh, India for preventing the spread of viral diseases of citrus from nurseries and works satisfactorily.

(ii) Establishment of Government fruit nurseries

As there is growing demand for healthy nursery stocks of fruit crops, suitable Government institution may establish fruit nurseries on commercial scale at selected centres for multiplying and distributing pest- and disease-free nursery stocks. Necessary steps may also be taken to certify the planting materials produced by them. This will not only help in meeting the needs of fruit growers but also preventing the indiscriminate movement of planting materials.

CONCLUSION

Tylenchulus semipenetrans and *Meloidogyne* spp. (*M. indica* and *M. javanica*) are the major pathogenic nematodes found associated with citrus and cause significant crop loss. *T. semipenetrans* is wide-spread in almost all the citrus growing areas while *Meloidogyne* spp. have been recorded mostly in coastal Andhra Pradesh, India. It is evident from the observations that nematode-infected nursery stocks are the main sources of nematode inocula for the orchards. Besides, pathogenic nematodes are introduced in new geographical areas through infected planting materials. If the various strategies discussed are suitably incorporated in the integrated nematode management system, it will help in a big way to contain the pathogenic nematodes of citrus and avoid the crop loss caused by them. However, work should be intensified so as to include nematode resistant rootstocks and potential biocontrol agents in the overall nematode management system. Keeping in view the foregoing discussion, the following research and implementation strategies are suggested to maximise citrus production:

1. Research should be intensified to identify the major nematode parasites associated with various citrus species in different parts of the country and determine their role on the etiology of citrus decline.
2. As there is only limited information available on root-knot nematodes infecting citrus, studies should be initiated to determine their distribution in other parts of the country, to study their pathogenic threshold levels and evolve suitable methods for their management.
3. Efforts should be made to estimate the crop loss caused by the major pathogenic nematodes under local conditions to understand and demonstrate their economic importance.
4. The sources of nematode resistance identified may be analysed and suitable recommendations may be given for field testing, field use and incorporation in crop improvement programmes.
5. Attempts should be made to test the potential biocontrol agents under field conditions for their efficacy in controlling nematodes.
6. All the nematode management techniques may be evaluated under field conditions for identifying the most simple, economic and effective methods of controlling nematodes of citrus.
7. Certain selected nematode management techniques may be demonstrated in farmers' field for convincing farmers, scientists, extension workers and above all the scientific administrators of the importance of managing nematode pathogens.
8. The existing extension network in various organisations should be revitalised for effective transfer of the available plant protection technologies for the management of insect pests, nematodes and diseases.
9. Government bodies in the states, may establish commercial fruit nurseries at selected centres for multiplying pest- and disease-free propagating materials of fruit crops.
10. Government may evolve a working system in collaboration with Agricultural University and Department of Horticulture in different states for certifying planting materials produced by private nursery growers.
11. Government organisations like Agricultural Universities and Departments of Agriculture/Horticulture existing in various states, may make efforts for publishing extension leaflets, booklets, bulletins, technical bulletins, books, etc. in national and regional languages on various aspects of plant protection to popularise the pest and disease management techniques among farmers, students, scientists and extension workers.

REFERENCES

Alam, M.M.; Khan, A.M. & Saxena, S.K. (1978). Mechanism of control of plant parasitic nematodes as a result of the application of organic amendments to the soil. IV. Role of formaldehyde and acetone. *Indian J. Nematol.* **8**: 172-174.

Alam, M.M.; Siddiqui, S.A. & Khan, A.M. (1977). Mechanism of control of plant-parasitic nematodes as a result of the application of organic amendments to the soil. III. Role of phenols and amino acids in host roots. *Indian J. Nematol.*; **7**:27-31.

Anonymous (1983). *Annual Report.* Citrus Improvement Project, Andhra Pradesh Agricultural University, Tirupati.

Anonymous (1984). *Annual Report.* Citrus Improvement Project, Andhra Pradesh Agricultural University, Tirupati. p. 55.

Anonymous (1985). *Annual Report.* Citrus Improvement Project, Andhra Pradesh Agricultural University, Tirupati.

Anonymous (1988). *Annual Report.* Citrus Improvement Project, Andhra Pradesh Agricultural University, Tirupati, p. 31.

Baghel, P.P.S. & Bhatti, D.S. (1983). Evaluation of pesticides for the control of phytonematodes on citrus. Proc. *Third Nematol. Symp.* Solan, India. pp. 38-39.

Baines, R.C. (1950). Nematodes on citrus. Calif Agric. **4**:7.

Baines, R.C. (1974). The effect of soil type on movement and infection rate of larvae of *Tylenchulus semipenetrans. J. Nematol.* **6**: 60-62.

Baines, R.C. & Clarke, O.F. (1952). Citrus root nematode. Effects on young lemon and orange tree studies in inoculation tests under controlled conditions. *Calif. Agric.* **6**: 9.

Baines, R.C.; Martin, J.P.; Dewolf, T.A.; Boswell, S.B. & Gorber, M.J. (1962). Effect of high doses of D-D on soil organisms and the growth and yield of lemon trees. *Phytopathology* **52**: 723.

Baines, R.C. & Small, R.H. (1969). Efficacy of some new compounds for control of the citrus nematode, *Tylenchulus semipenetrans* on established trees. *Proc. First Int. Citrus Symp.* Riverside, California **2**: 973-977.

Baines, R.C.; Van Gundy, S.D. & Small, R.H. (1977). Efficacy of non-fumigant and low volatile nematicides for control of *Tylenchulus semipenetrans* on Naval and Valentia oranges. *J. Nematol.* **9**: 262.

Bindra, C.S. & Chhabra, H.K. (1972). Control of citrus nematode, *Tylenchulus semipenetrans* Cobb by DBCP application irrigation water. *Indian J. Hort.* **29**: 302-306.

Brooks, T.C. & Perry, V.G. (1967). Pathogenicity of *Pratylenchus brachyurus* to citrus. *Plant Dis. Reptr.* **51**: 569-573.

Carter, W.W.; Reynolds, H.W. & Rodney, D.R. (1971). Response of debilitated grape fruit trees to manure, phosphate fertilizer and DBCP fumigation against *Tylenchulus semipenetrans. J. Nematol.* **3**: 306.

Chhabra, H.K. (1978). Distribution pattern of *Tylenchulus semipenetrans* population and determination of suitable distance and depth for soil sampling. *Indian J. Nematol.* **8**: 66-68.

Chitwood, B.G. & Toung, N.C. (1960). Host-parasite interactions of the Asiatic pyroid citrus nema. *Plant Dis. Reptr.* **44**: 848-854.

Chona, B.L.; Sethi, C.L. & Swarup, G. (1965). Citrus nematode (*Tylenchulus semipenetrans* Cobb, 1913) associated with die-back disease of citrus trees in India. *Indian J. Hort.* **22**: 370-373.

Cobb, N.A. (1914). Citrus-root nematode. *J. Agric. Res.* **2**: 217-230.

Cohn, E. (1972). Nematode diseases of citrus. *Economic Nematology* (Ed.—JM. Webster). Academic Press, New York. pp. 215-244.

Cohn, E.; Minz, G. & Monselise, S.P. (1965). The distribution, ecology and pathogenicity of the citrus nematode in Israel. *Israel J. Agric. Res.* **15**: 187-200.

Colbran, R.C. (1958). Studies of plant and soil nematodes. 2. Queensland host records of root-knot nematodes (*Meloidogyne* spp.). *Qd. J. Agric. Sci.* **15**: 101-135.

DuCharme, E.P. (1969). Temperature in relation to *Rodopholus similis* (Nematoda) spreading decline of citrus. *Proc. First Int. Citrus Symp.*, Riverside, California **2**: 979-983.

Gupta, J.C. & Atwal, A.S. (1972). Biology and ecology of *Hoplolaimus indicus* (Hoplolaiminae: Nematoda). 3. Influence of population density of tomato and citrus plants. *Proc. XI Int. Nematol. Symp*. p. 25.

Hung. I.P. (1969). Occurrence of plant parasitic nematodes in Taiwan citrus plantations. *Proc. First Int. Citrus Symp*. Riverside, California, **2**: 1007-1011

Inserra, R.N. & Volvas, N. (1977). Nematodes other than *Tylenchulus semipenetrans* pathogenic to citrus. *Proc. Int. Soc. Citriculture* **3**: 826-831.

Jatala, P. (1986). Biological control of plant parasitic nematodes. *Ann. Review Phytopathol.* **24**: 453-489.

Jayaraj, M.A. & Mani, A. (1988). Biocontrol of *Meloidogyne javanica* with the bacterial spore parasite, *Pasteuria penetrans. Int. Nematol. Network Newsi.* **5**(1): 3-4.

Khan, A.M.; Alam, M.M. & Ahmad, R. (1974). Mechanism of the control of plant-parasitic nematodes as a result of the application of oilcakes to the soil. *Indian J. Nematol*. **4**: 93-96.

Khuntia, N. & Das. S.N. (1969). Plant-parasitic nematodes associated with fruit trees in Orissa. *Proc. All India Nematol. Symp. Indian Agricultural Research. Institute*, New Delhi. pp. 22-23.

Kumar, A.C.; Viswanathan, P.R.K. & D'Souza, G.I. (1971). A study on plant-parasitic nematodes of certain commercial crops in coffee tracts of South Indian. *Indian Coffee* **35**: 222-224.

Macaron, J. (1972). Contribution to the study of plant-parasitic nematodes *Tylenchulus semipenetrans* Cobb, 1913 (Nematoda-Tylenchida). Ph. D. Thesis, Univ. Sci. Technol. Languedoc, Montpellier.

Machmer, J.H. (1958). Effect of soil salinity on nematodes in citrus and papaya plantings. *J Rio Grande Valley Hort. Soc.* **12**: 57-60.

Mani, A. (1986). Occurrence of *Meloidogyne javanica* in citrus on Andhra Pradesh (India). *Int. Nematol. Network. Newsl.* **3**(2): 9-10.

Mani, A. (1988a). Studies on the bacterial parasite *Pasteuria penetrans*. I. Spore viability after storage. II. culture on citrus nematode, *Tylenchulus semipenetrans. Int. Nematol. Network Newsl.* **5**(2): 24-25.

Mani, A. (1988b). Effect of interculture of marigold and mustard with acid lime on citrus nematode, *Tylenchulus semipenetrans. Int. Nematol. Network Newsl.* **5**(4): 14-15.

Mani, A. (1990a). Control of citrus nematode, *Tylenchulus semipenetrans*, by bare-root dip treatment of acid lime seedlings in pesticides. *Int. Nematol. Network Newsl* **6** (3): 20-21.

Mani, A. (1990b). Reaction of rangapurlime and rough lemon strains to citrus nematode, *Tylenchulus semipenetrans* Cobb. *Int. Nematol. Network Newsl.* **6**(3): 22-23.

Mani, A. & Dakshinamurthy, V : (1986). Citrus nematode and root-knot nematodes. Distribution, crop loss, symptoms and management. *Plant Parasitic Nematodes of India : Problems and Progress* (Eds—G. Swarup & D.R. Dasgupta). Allied Publishers, New Delhi. pp. 249-260.

Mani, A. & Murthy, I.R. (1989). National symposium on insect pests and disease management on citrus (Sept. 18-20). J.N.K.V.V., Chindwara.

Mani, A. & Reddy, G.S. (1986). Studies on the varietal reaction of some species of *Citrus* and *Poncirus* to *Tylenchulus semipenetrans* Cobb. *Indian J. Nematol.* **16**: 267-268.

Mankau, R & Minteer R.J. (1962). Reduction of soil population of the citrus nematode by the addition of organic materials. *Plant Dis. Reptr.* **46**: 375-378.

Meagher, J.W. (1967). Observations on the transport of nematodes in sub-soil drainage and irrigation water. *Australian J. Exp. Agric. Animal Husb.* **7**: 577-579.

Meagher, J.W. (1969). Nematodes as a factor in citrus production in Australia. *Proc. First Int. Citrus Symp*. Riverside, California **2**: 999-1006.

Minz, G. (1956). Cyst forming nematodes in Israel. *Plant Dis. Reptr.* **40**: 971-973.

Mukhopadhyaya, M.C. (1970). Efficacy of D-D, EDB. DBCP and V.C.-13 in controlling plant parasitic nematodes. *J. Res. Punjab Agric. Univ.* **7**: 656-658.

Mukhopadhyaya, M.C. & Dalal, M.R. (1971). Effects of two nematicides on *Tylenchulus semipenetrans* and on sweet lime yield. *Indian J. Nematol.* **1**: 95-97.

Nair, M.R.G.K.; Nair, K.R. & Visalakshy, A. (1969). Studies on the plant parasitic and soil nematodes of Kerala. *Proc. All India Nematol. Symp*. Indian Agricultural Research Institute, New Delhi. pp. 4-5.

Neal, J.C. (1889). The root-knot disease of peach, orange and other plants in Florida due to the work of *Anguillula. Bull. U.S. Bureau Entomol.* **20**: 31.

O'Bannon, J.H. (1968). The influence of an organic soil amendments on infectivity and reproduction of *Tylenchulus semipenetrans* on two citrus root stocks. *Phytopathology* **58**: 597-601.

O'Bannon, J.H. & Esser, R.P. (1985). Division of Plant Industry, Florida Department of Agricultural and Consumer Service. *Nematol. Cir.* **114**: 4.

O'Bannon, J.H.; Radewald, J.D. & Tomerlin, A.T. (1972). Population fluctuation of these parasitic nematodes in Florida citrus. *J. Nematol.* **4**: 194-199.

O'Bannon, J.H.; Reynolds, H.W. & Leathers, C.R. (1966). Effects of temperature on penetration, development and reproduction of *Tylenchulus semipenetrans. Nematologica* **12**: 483-487.

O'Bannon, J.H. & Tarjan, A.C. (1973). Preplant fumigation for citrus nematode control in Florida *J. Nematol.* **5**: 88-95.

Perraju, A. (1961). Preliminary observations in the trials for chemical control of tobacco and citrus nematode for root galls. *Andhra Agric. J.* **7**: 43-48.

Prasad, S.K. & Chawla, M.L. (1965). Observations on the population fluctuations of citrus nematode, *Tylenchulus semipenetrans* Cobb, 1913. *Indian J. Entomol* **27**: 450-454.

Reddy, P. P. (1979). Plant parasitic nematodes associated with citrus in Karnataka state. *Indian Phytopath.* **32**: 534-537.

Reddy, P.P. (1988). Technical Bulletin No. 4. Indian Institute of Horticultural Research, Bangalore. p. 52.

Reddy, P.P.; Khan, R.M. & Rao, M.S. (1989). Biological control of the citrus nematode *Tylenchulus semipenetrans* using the parasitic fungus *Paecilomyces lilacinus, Symp. Plant Dis, Problems & Their Management in South India* (Dec. 7-9). p. 26.

Reddy, P.P.; Khan, R.M. & Rao, M.S. (1990). Integrated Management of the citrus nematode *Tylenchulus semipenetrans* using oilcakes and *Paecilomyces lilacinus. Proc. Int. Seminar on New Frontiers in Hort*. Bangalore (Dec. 25-27).

Reddy, P.P. & Singh, D.B. (1979). Nematode pests of citrus in India. *Pest Articles and News Summaries* **25**: 409-419.

Reynolds, H.W. (1969). Control of the citrus nematode under southwest irrigation conditions. *Proceeding of First Int. Citrus Symp.* Riverside, California. **2**: 969-971.

Reynolds, H.W. & O'Bannon. J.H. (1958). The citrus nematode and its control on living citrus in Arizona. *Plant Dis. Reptr.* **42**: 1288-1292.

Reynolds, H.W.; O'Bannon, J.H. (1963). Factors influencing citrus replants in Arizona. *Nematlogica* **9**: 337-340.

Reynolds, H.W.; O'Bannon, J.H. & Nigh, E.L. (1974). The citrus nematode and its control in the southwest. *USDA Tech. Bull. No.* 1478.

Scaramuzzi, G. & Perrotta, G. (1969). Research on the control of citrus nematode in the Mediterranean area. *Proc. First Int. Citrus Symp.* Riverside, California **2**: 957-960.

Schneider, H & Baines, R.C. (1964). *Tylenchulus semipenetrans* parasitism and injury to orange tree roots. *Phytopathology* **54**: 1202-1206.

Siddiqi, M.R. (1961). Occurrence of the citrus nematode, *Tylenchulus semipenetrans* Cobb, 1913 and the reniform nematode, *Rotylenchulus reniformis* in India. *Proc. 40th Indian Sci. Cong.* p. 504.

Siddiqi, M.R. (1964). Studies on nematode root-rot of citrus in Uttar Pradesh, India. *Proc. Zool. Soc.* (Calcutta) **17**: 67-75.

Singh, D.B.; Rao, V.R. & Reddy, P.P. (1979). Plant parasitic nematodes associated with horticultural crops in South India. *Indian J. Nematol.* **9**: 193-196.

Standifer, M.S. & Perry, V.G. (1960). Some effects of sting and subby root nematodes on grapevine roots. *Phytopathology* **50**: 152-158.

Striling, G.R. (1984). Biological control of *Meloidogyne javanica* with *Bacillus penetrans. Phytopathology* **74**: 55-60.

Suit, R.F. & DuCharme, E.P. (1953). The burrowing nematode and other parasitic nematodes in relation to spreading decline of citrus. *Plant Dis. Reptr.* **37**: 379-383.

Swarup, G.; Sethi, C.L. & Gill, J.S. (1964). Some records of plant parasitic nematodes in India. *Curr. Sci.* **33**: 593.

Tarjan, A.C. (1971). Migration of three pathogenic citrus nematodes through two Florida Soils. *Soil Crop Sci. Soc. Proc.* **31**: 253-255.

Tarjan, A.C. (1977). Application of municipal solid waste compost to nematode-infected citrus *Nematropica* **7**: 53-56.

Tarjan, A.C. & O'Bannon, J.H. (1974). Post-plant fumigation with DBCP for citrus nematode control in Florida. *J. Nematol.* **6**: 41-48.

Thomas, E.E. (1913). A preliminary report of a nematode observed on citrus roots and its possible relation with the mottled appearance of citrus trees. *Calif. Agric. Exp. Sta. Cir.* 85.

Thorne, G. (1961) I *Principles of Nematology*. McGraw Hill, New York, p. 553.

Timmer, L.M. & Davis, R.M. (1982). Estimate of yield loss from citrus nematode in Texas grape fruit. *J. Nematol.* **14**: 582-584.

Timber, L.W. & French, J.V. (1979). Control of *Tylenchulus semipenetrans* on citrus with aldicarb, oxamyl and DBCP. *J. Nematol.* **11**: 387-394.

Tirumala Rao, V. (1956). Stray notes on some crop pest outbreaks of South India. *Indian J. Entomol.* **18**: 123-126.

Toung, M.C. (1963). A study of seasonal influence in quantitative variation of the citrus nema. *Tylenchulus semipenetrans* Cobb. *J. Soc. Plant Protec.* (Taiwan) **5**: 323-327.

Van Gundy, S.D. (1958). The life history of citrus nematode *Tylenchulus semipenetrans* Cobb. *Nematologica* **3**: 283-294.

Van Gundy, S.D. & Kirkpatrick, J.D. (1964). Nature of resistance in certain root stocks nematode. *Phytopathology* **54**: 419-427.

Van Gundy, S.D. 7 Martin, J.P. (1961). Influence of *Tylenchulus semipenetrans* on the growth and chemical composition of sweet orange seedlings in soils of various exchangeable cation ratios. *Phytopathology* **51**: 146-151.

Van Gundy, S.D.; Martin, J.P. & Tsao, P.H. (1964). Some soil factors influencing reproduction of the citrus nematode and growth reduction of sweet orange seedlings. *Phytopathology* **54**: 294-299.

Van Gundy, S.D. & Rackam, R.L. (1961). Studies on the biology and pathogenicity of *Hemicycliophora arenaria. Phytopathology* **51**: 393-397.

Vilardebo, A.; Squalli, A. & Devaux, R. (1975). Possible use of DBCP, phenamiphos and prophos against *Tylenchulus semipenetrans* in orchards in Marocco. *Fruits (Paris)* **30**: 313-327.

Whitehead, A.G. (1968). Taxonomy of *Meloidogyne* with description of four new species. *Trans. Zool. Soc.* London **31**: 263-401.

Chapter 12

Nematode Pests of Species and Condiments and their Control

P.K. Koshy
Division of Nematology, Central Plantation Crops Research Institute, Regional Station, Kayangulam, Krishnapuram-690 533, India.

BLACK PEPPER (*Piper nigrum* L.)

BLACK pepper has its origin in the hills of South Western India. The major pepper producing countries are Indonesia, Malaysia and Brazil. In India, the total area under pepper cultivation during 1987-88 was 1,58,500 ha with an annual production of 49,200 tonnes. The export earnings from pepper during the above period was Rs. 2,400 million (Source : Statistics Unit, CPCRI, Kasaragood, 7/89). Of the 36 species of nematodes reported on pepper, the two most important ones are *Rodopholus similis* and *Meloidogyne* spp.

RADOPHOLUS SIMILIS

Association of the burrowing nematode with the yellows disease of pepper was first reported by Van der Vecht (1950). It caused 'yellows disease' in black pepper which wiped out 22 million pepper vines from the Bangka Island in Indonesia within 20 years. In Kerala and Karnataka states of India it causes the 'slow-wilt' disease in black pepper which is almost identical with the pepper yellows of Bangka Island. In Kerala, surveys showed that high population of *R. similis* occurred in roots of slow-wilt disease-affected vines (Ramana *et al.*, 1987a).

Coconut and arecanut on which black pepper vines are trailed are good hosts of the nematode. Intercrops like banana, ginger and turmeric are also susceptible to *R. similis*.

The life cycle, biotypes and influence of environmental factors is the same as given in the chapter 'Nematode Pests of Palms and their Control.

NATURE OF DAMAGE

Symptoms

Primary symptom of slow-wilt disease is the appearance of a few pale, yellow, dropping leaves whose number gradually increases and within a year or two the entire foliage may become yellow. This is followed by shedding of leaves, cessation of growth and die-back symptoms. In the very early stage, the symptoms may disappear with the onset of south-west monsoon. However, within three to five years of initiation of yellowing all the leaves are shed and death of the vine occurs. In bearing vines spike shedding takes place. In large plantations affected patches are conspicuous with many barren standards. Young and old plants are affected.

The tender, thin, white, feeder roots show typical orange to purple coloured lesions. The root system exhibits extensive rotting with the main roots devoid of feeder roots. Extensive necrosis of large roots develops subsequently.

Histopathology

R. similis is found to penetrate roots within 24 hours of inoculation and the cells around the site of penetration become brown. It does not enter the stelar portion. Nematodes are found in inter and intracellular positions within the cortex. The xylem vessels are seen plugged with a "gum like substance".

Losses

A population level of 250 nematodes per g of root was constantly recorded with slow-wilt affected pepper vines in Kerala, India (Ramana, 1986).

In Sumatra, Indonesia, a nematode population of 2/100g of soil and 25/10g of roots were recorded from vines with yellows disease (Mustika, 1978).

CONTROL STRATEGIES

Cultural

Symptoms of slow-wilt/pepper yellows is known to ameliorate with mulching.

Plant resistance

A total of 106 cultivated germplasm, 36 wild related *Piper* spp., 20 intercultivar hybrids, 90 selections of cv. 'Karimunda' and 12,200 open pollinated seedlings of popular pepper cultivars were screened against *R. similis* and found no resistance/tolerance in any of them (Ramana *et al.*, 1987b).

Chemical

Aldicarb sulphone at 8 kg a.i./ha have best control of *R. similis* followed by fensulfothion and DBCP (Venkitesan, 1976). Under Indian conditions, Aldicarb/Carbofuran/Phorate @ 3 g a.i./vine applied in May/June and again in September/October result in the remission of foliar yellowing and reduction in nematode population. Among these phorate was found to be superior to others (Ramana, 1986; Mohandas & Ramana, 1987). But studies on the cost-benefit ratio and residue analysis of these chemicals have not been carried out.

Integrated

For heavily infested areas, an integrated method of management is suggested :

1. Planting nematode-free rooted cuttings.
2. Uprooting of affected vines and replanting after 9-12 months.
3. Use of non-living standards and exclusion of *R. similis* — susceptible trees as standards for trailing black pepper vines and avoiding susceptible intercrops such as banana, ginger, turmeric, etc.

MELOIDOGYNE sp.

Meloidogyne sp. was the first nematode to be recorded on black pepper (Delacriox, 1902) in Cochin-China. In 1906, Butler reported root-knot nematodes from black pepper in Wynad, Kerala, India.

Diagnostic features (Jepson, 1987)

Female : Body white, globular to pearshaped, 295-4250 μm long, sometimes elongate, usually with short neck, cuticle thin, annulated with characteristic terminal striations forming perineal pattern around vulva and anus. Head has cuticular framework with a 10-20 μm long robust stylet. Gonads are paired, elongate and convoluted. Eggs are laid in gelatinous matrix secreted by rectal glands.

Male : Vermiform, migratory with a length about 700-1900 μm. Strong cephalic framework. Robust, 13-30 μm stylet. Metacorpus weakly developed. Tail bluntly rounded with paired spicules. One or two testes.

Second-stage juvenile : Vermiform, infective, migratory, of about 250-600 μm long. Head with labial cap, robust stylet. Tail tapering narrow with rounded tip and hyaline terminal portion.

Third and fourth stage juveniles swollen, sedentary within roots, without stylet and within cuticle of second stage which retains the tail spike.

Systematic position

Nematoda, Secernentea, Tylenchida, Tylenchina, Heteroderoidea, Meloidogynidae, Meloidogyninae, *Meloidogyne*.

DISTRIBUTION AND HOSTS

Meloidogyne javanica and *M. incognita* have been reported from India, Brazil, Sarawak, Borneo, Cochin-China, Malaysia, Brunei, Kampuchea, Indonesia, Thailand and Vietnam, and *M. arenaria* from Sri Lanka.

Among the commercially used standards, *Oroxylon indicum* Vent, *Erythrina lithosperma* Blume, *Ceiba pentandra* (Linn.) Gaerth and *Bombax malabaricum* DC. are highly susceptible to root-knot nematode. Large numbers of weeds found in pepper gardens are also highly susceptible to this nematode.

BIOLOGY AND ECOLOGY

Life cycle

The eggs are laid in a gelatinous matrix. Its contents undergo a series of determinate cleavage giving rise to the first juvenile stage which molts within the egg and gives rise to the second stage juvenile which hatches out. The J2 juveniles penetrate host plants above the meristematic zone and orient parallel to the stele. It starts feeding on the pericycle cells for several weeks, gradually swelling. A series of three moults occur in quick succession. Reproduction may be parthenogenetic or bisexual with a female laying 400-500 eggs. Egg laying is completed within a week. The life cycle under optimum conditions takes 3-4 weeks.

Interaction with other micro-organisms

Meloidogyne incognita and *Fusarium solani were* found associated with black pepper vines in Brazil. Both the organisms together were found to do more harm than either of them alone (Lopes & Lordello, 1979). Increased susceptibility of *M. incognita* and *M. javanica* infested pepper cultivar 'Kuching' to *Phytophthora* infection was reported in Malaysia (Winto, 1972).

NATURE OF DAMAGE

Symptoms

The black pepper vines on infestation by root-knot nematodes prominently show unthrifty growth and yellowing of leaves. The interveinal areas on leaves of such vines show dense yellowish discolouration making the leaf veins appear quite distinct with deep green colour. The roots show prominent galls.

Biochemical

The pepper variety 'Cingapura' recorded high concentrations of total phenols on inoculation with 6,000 *M. incognita* juveniles/pot, 95 days after planting although no resistance was shown (Ferraz *et al*., 1984). Changes in levels of amino acids, organic acids and sugars in *M. incognita* infected plants, compared to uninfected plants were reported by Freire & Bridge (1985a).

Losses

An initial inoculum level of ten second-stage juveniles per rooted cutting was found to reduce growth by 16% while at a level of 1,00,000 juveniles 50% reduction in growth was observed over one year period (Koshy *et al.*, 1979a). Root-knot nematode infestation is a major problem in black pepper in several departmental nurseries in Kerala.

CONTROL STRATEGIES

Cultural

Growing of non-host cover plant siratro (*Macroptilium atropurpureus*) in the interspaces and mulching with Guatemala grass are recommended to reduce population of *M. incognita* on black pepper in the Amazonian region (Ichinohe, 1980).

Plant resistance

Among the several popular cultivars screened, the cultivar 'Panniyur-1' was the most susceptible and 'Valiakaniakadan' the least (Koshy & Sundararaju, 1979). The intensity of damage on infestation with *M. incognita* was less in cultivar 'Kerimunda' compared to the variety 'Panniyur-1' (Mohandas & Ramana, 1983). The cultivars 'Kalluvalli', 'Balankotta', 'Narayakodi' and 'Padapan' had fewer galls when compared with 'Panniyur-1' (Jacob & Kurian, 1979). Of the 101 cultivars, 74 accessions of wild *Piper* sp. and 140 intercultivar hybrids screened, one cultivar, 'CLT-P-812' was found resistant to *M. incognita* (Ramana & Mohandas, 1986, 1987; Koshy, 1987).

Biological

The only attempt made in controlling the root-knot nematode using bio-control agents is by Freirie & Bridge (1985b). The rates of infection of *Paecilomyces lilacinus* and *Verticillium chlamydosporum* on *M. incognita* egg masses on black pepper seedlings were 15 and 12% respectively when the fungi were introduced to sterile soil before inoculation of nematodes to potted plants.

Chemical

Most of nematicides have been found effective in reducing root-knot nematode population on black pepper. Under Indian conditions when a live standard is used, the dosage differs depending on the susceptible/resistant reaction of the standard to the nematode. Application of aldicarb @ 1 g a.i./vine twice a year (May/June and October/November) integrated with fertilizers (N = 100g, P = 40g, K = 140g/vine) in two equal split doses, earthing up to 50 cm radius at the base of vines and mulching the base of the vines with leaves reduces foliar yellowing by 83% and *M. incognita* larval population by 33-88% (Venkitesan & Jacob, 1985).

CARDAMOM (*Elettaria cardamomum* (L.) Maton)

Cardamom has its origin in the ever green rain forests of Western hills of South India. Other cardamom growing countries include Guatemala, Brazil, Tanzania, Sri Lanka, El Salvador, Vietnam, Laos, Kampuchea and Papau New Guinea. The area under cultivation in India during 1987-88 was 1,05,00 ha with an annual production of 3,200 metric tonnes. The export earning during 1987-88 was Rs. 34 million. The most important nematode problem is the root-knot nematode, *Meloidogyne* spp.

MELOIDOGYNE sp.

The two *Meloidogyne* spp. widely occurring in cardamom nurseries and plantation in Kerala, Karnataka and Tamil Nadu are *Meloidogyne incognita* and *M. javanica*.

Many annual weeds and common shade trees like *Erythrina indica* and *E. lithosperma* present in cardamon plantations are susceptible to root-knot nematodes.

NATURE OF DAMAGE

Symptoms

On infestation, mature plants in a plantation show stunting, yellowing, reduced tillering, delay in flowering, immature fruit-drop and reduction in yield. Roots do not show galling. Instead, they show excessive branching.

In primary nurseries germination is reduced by more than 50%. Infested seedlings fail to establish in secondary nurseries. The roots show heavy galling. In secondary nurseries, infested plants exhibit stunting, yellowing, poor tillering and heavy galling of roots (Ali & Koshy, 1982).

In nurseries, presence of *M. incognita* increases incidence of rhizome rot and damping off disease caused by *Rhizoctonia solani* (Ali, 1986; Eapen, 1987).

Losses

Thirty two to forty seven per cent yield loss occurs due to root-knot nematode infestation (Ali, 1984, 1986). An initial population level of 100 nematodes per plant causes considerable damage to cardamon (Eapen, 1987).

CONTROL STRATEGIES

Plant resistance

The popular cardamom cultivars, 'Malabar', 'Mysore and Vezhuka' are susceptible to root-knot nematodes.

Chemical

Disinfesting nursery beds with methyl bromide @ 500g/10m^2 is effective in controlling root-knot nematodes in cardamom nurseries. Seed beds can also be drenched with two per cent formalin to a depth of 20-30 cm and covered with polythene sheets for three to seven days. Seeds can be sown two weeks after formalin application when the soil is free from formalin fumes. Application of aldicarb @ 5 kg a.i./ha three times after every three months results in increase in growth and vigour of seedlings both in primary and secondary nurseries (Koshy *et al.,* 1979b; Ali, 1986). Aldicarb at 15 kg/ha reduced nematode population by 90 per cent (Ali, 1987). An increase of 88% in yield of cardamom plants infested with *M. incognita* was obtained on application of aldicarb/carbofuran/phorate @ 5 g and 10 g a.i./plant and neem oil cake @ 500 g and 1000 g/plant twice a year (Ali, 1984).

Integrated

The following measures are suggested for an integrated management schedule for *Meloidogyne* spp. infestation on cardamom :

1. Changing nursery sites frequently.
2. Disinfesting nursery beds.
3. Introduction of biocontrol organisms at nursery level.
4. Control of susceptible weeds.
5. Exclusion of susceptible shade trees.
6. Destruction of infested crop residues.
7. Application of nematicide and neem oil cake.
8. Mulching with dead leaves.

GINGER (*Zingiber officinale* Rosc.)

The country of origin of ginger is not known with certainty. It is presumed to be in the region of India or China. Apart from India and China other ginger producing countries are Jamaica, Siera Leone, Nigeria, Japan, Taiwan and Australia. India is the largest producer (1,27,000 metric tonnes) and exporter of dry ginger in the world (export earning Rs. 191 million). The total area under cultivation during 1986-87 was 5,24,600 ha.

A large number of nematode species have been recorded from ginger of which the most important are the root-knot and burrowing nematodes.

MELOIDOGYNE sp.

The first report of *Meloidogyne* sp. on ginger was by Nagakuar (1930). Later, it was recorded from other areas also. Several weeds found in ginger growing areas are known hosts of root-knot nematodes.

In fleshy and fibrous roots the nematode develops to maturity in 24 days, while in rhizomes it takes 43 days at a soil temperature of 30°C (Huang, 1966). Infested rhizomes help in long distance dissemination of nematodes.

NATURE OF DAMAGE

Symptoms

Heavily infested plants show stunting and chlorosis and marginal necrosis of leaves. Roots and underground rhizomes exhibit galling and rotting. The rhizome is attacked by second stage juveniles through axils of leaf sheaths in the shoot apex. Fleshy roots are invaded along the entire length, while in fibrous roots it is in the area of differentiation. Infested rhizomes have brown, water-soaked areas in the outer tissues especially in angles between shoots.

Histopathology

Abnormal xylem and hyperplastic parenchyma are observed in all infested tissues except in rhizome meristems. Giant cells are found in all the infected tissues. At infection sites in differentiated rhizomes and fleshy roots, wound cork and lignified wall thickening of endodermis and pericycle are formed.

Interaction between *Pythium myriotylum* and *M. incognita* on ginger results in fungal antagonism against the nematode, but has no influence on the soft-rot disease syndrome (Lenjewar & Shukla, 1985).

Losses

Under potted conditions an initial inoculum level of 10,000 nematodes/plant over a period of six months resulted in 74% reduction in rhizome weight. A population level of one infective juvenile/30g of soil can cause significant reduction in yield. Under potted conditions with an initial level of 50 larvae/100 ml soil with *M. incognita* and *M. hapla* cause significant reduction in growth characters. At higher initial inoculum levels both the species cause withering of aerial shoots. *M. arenaria* on the other hand caused typical symptoms of drying and twisting of leaves. At an inoculum level of two nematodes per g of soil the fibrous roots are very much reduced.

CONTROL STRATEGIES

Cultural

Application of well decomposed cattle manure or compost @25-30 tonnes/ha, neem cake @ two tonnes/ha, and mulching with green leaves @ 10-12 tonnes/ha at planting and mulching again during the growth period help in reducing nematode population (Kaur, 1987).

Chemical

In Himachal Pradesh, application of phenamiphos @ 3 kg a.i./ha resulted in 70-144% increase in yield of ginger in fields infested with *M. incognita* (Kaur, 1987).

RADOPHOLUS SIMILIS

Rodopholus similis infestation in ginger was first reported by Hart (1956) in Florida. In India, it was reported from Kerala by Charles & Kurian (1979). The coconut isolate of *R. similis* in Kerala reproduced well on ginger (Koshy & Sosamma, 1975, 1977). The perpetuation and dissemination of the nematode take place through infested rhizomes used for planting.

NATURE OF DAMAGE

Symptoms

Affected plants exhibit stunting, reduced vigour and tillering and mature and dry out faster than healthy plants. The topmost leaves become chlorotic with scorched tips. The infected rhizomes exhibit small, shallow, sunken, water-soaked lesions.

Histopathology

The nematodes penetrate through cell walls and are found coiled within a single cell. Due to infection large channels or galleries are found within the rhizomes.

Losses

An initial inoculum level of 10,000 nematodes per plant causes 74% reduction in rhizome weight (Sundararaju *et al.*, 1979).

CONTROL STRATEGIES

No study appears to have been carried out on control aspects of burrowing nematode on ginger. But the measures adopted for control of root-knot nematode could be adopted to reduce loss till definite measures become available.

TURMERIC (*Curcuma domestica* Val. (*C. longa* L.)

Turmeric is best known as an important spice crop. In India its area of cultivation is 1,07,700 ha and total production is 2,94,000 tonnes during 1987-88. India's export earning from turmeric was Rs. 92 million annually (Source : Statistic Unit, Kasaragod 7/89). Of the many plant-parasitic nematodes reported in association with turmeric, the *Meloidogyne* sp., *Radopholus similis* and *Pratylenchus coffeae* are important.

MELOIDOGYNE sp.

Affected plants show stunted growth, yellowing, marginal and tip drying of leaves and reduced tillering. Galling and rotting are exhibited on the root system. Infested rhizomes tend to lose their bright yellow colour.

CONTROL STRATEGIES

Plant resistance

Among 60 germplasm collections screened against *M. incognita* the varieties '5379-1-2' '5363-6-3', 'Kodur', 'Chayapasupu', 5335-1-7' and '5335-2-7' were found to be resistant (Gunasekharan *et al.*, 1987).

Chemical

Aldicarb and carbofuran applied @ 1 kg a.i./ha increased yield 71 and 68 per cent, resistively. The cost benefit ratio in aldicarb treatment was 1:6 and in carbofuran, 1:2.

RADOPHOLUS SIMILIS

Turmeric was reported as a host of *Radopholus similis* by Koshy & Sosamma (1975). Nematodes get disseminated through infested planting material. Infested rhizomes lose the golden yellow colour seen in healthy rhizomes. They exhibit shallow water soaked brownish area. Scale leaves also harboured *R. similis*. Roots show rotting and most of the decayed roots are devoid of cortex and stellar portion. The infested plants dry faster than healthy plants.

CONTROL STRATEGIES

Avoid turmeric as an intercrop in *R. similis* infested coconut and arecanut based farming systems. There appears to have been no studies on control of *R. similis* on turmeric.

PRATYLENCHUS COFFEAE

Pratylenchus coffeae was reported to cause discolouration and rotting of mature rhizomes of turmeric. As infection advances, the rhizomes become deep and red to dark brown in colour, less turgid, wrinkled and exhibit dry-rot symptoms. The fingers are more severally affected. Infected rhizomes show dark brown necrotic lesions.

OTHER SPICES

A number of spice crops including tree spices and seed spices are cultivated over large areas in India (Table 1), but there is very little information available on the damage and yield loss caused by plant-parasitic nematodes on some of these crops. This is not because nematode problems do not exist on these crops, but it

Table 1 : Area, production, export and import details of important tree and seed species.

Crop	Year	Total area (hectares)	Production (metric tonnes)	Export value (Rs.)	Import quality (metric tonnes)	Import value (Rs.)
Coriander	1987-88	3,29,899	1,34,433	1,39,47,000	-	-
Cumin	1987-88	1,51,469	78,500	2,48,22,000	-	-
Funnel	1987-88	17,433	20,042	1,53,82,000	-	-
Fenugreek	1987-88	30,079	30,696	1,99,84,000	-	-
Clove	1985-86	1,600	1,300	-	1,133 (1987-88)	5,01,00,000 (1987-88)
Nutmeg	1985-86	-	-	-	331	5,05,700
Nutmeg Mace	1985-86	-	-	-	128	2,09,600
Cinnamon	1984-85	-	-	2,43,000	254 (1985-86)	8,58,30 (1985-86)
Cassia	1984-85	-	-	83,94,000	733 (1987-88)	2,28,94,000 (1987-88)

is only due to lack of nematological investigations. The plant-parasitic nematodes reported in association with these crops are given in Table 2.

Table 2 : Economically important plant-parasitic nematodas reported on important tree and species.

Nematode	Nutmeg	Clove	innamon	Coriander	Cumin	Fenugreek
Caloosia sp.	-	+	-	-	-	-
Criconemoides sp.	-	-	+	-	-	-
Dolichodorus sp.	-	+	-	-	-	-
Helicotylenchus dihystera	-	+	-	-	-	-
Helicotylenchus sp.	+	+	+	-	-	-
Hoplolaimus indicus	-	-	-	-	-	-
Hoplolaimus sp.	+	+. -	-	-	-	+
Meloidogyne incognita	+	+	-	+	+	+
M. incognita acrita	-	-	-	+	+	-
M.javanica	-	-	-	+	+	+
Meloidogyne n. sp.	-	-	-	-	-	+
Meloidogyne sp.	+	+	+	+	+	+
Paratrichodorus mirzai	-	-	-	-	-	+
Paratylenchus sp.	+	-	-	-	-	-
Pratylenchus exilis	-	-	-	+ -	-	
Pratylenchus sp.	+	-	-	-	-	-
Radopholus similis	-	-	-	+	+	+
Rotylenchulus reniformis	+	+	+	-	+	+
Rotylenchus sp.	+	-	+	-	+	-
Trichodorus sp.	-	+	-	-	+ -	
Tylenchorhynchus brevidens	-	-	-	-	+	-
T. brevilineatus	-	-	-	-	+	-
T. swarupi	-	-	-	-	-	+
T. vulgaris	-	-	-	+	-	-
Tylenchorhynchus sp.	-	+	-	-	-	-
Xiphinema sp.	+	+	+	-	-	-

REFERENCES

Ali, S.S. (1984). Preliminary observations on the effect of some systemic nematicides and neem oil cakes in a cardamom field infested with root-knot nematodes (Abstr.) *In : Proceedings of the Sixth Symposium on Plantation Crops,* Kottayam, 16-20 December, pp. 215-223.

Ali, S.S. (1986). Root-knot nematode problem in cardamom and its management. *In : Proceedings of the Second Group Discussion on the Nematological Problems of Plantation Crops,* April 24-25, 1987, Central Coffee Research Station, Balehonnur, pp. 10-12.

Ali. S.S. (1987). Effect of three systemic nematicides against root-knot nematodes in a cardamom nursery. *Nematol. medit.* **15**: 155-158.

Ali, S.S. & Koshy, P.K. (1982). Occurrence and distribution of root-knot nematodes in Cardamom Plantations of Kerala. *Nematol. medit.* **10**: 107-110.

Buttler, E.J. (1906). The wilt disease of pigeon-pea and pepper. *Agric. J. India* **1**: 25-36.

Charles, J.S. & Kurian, K.J. (1979). Studies on nematode incidence in ginger. *In: Proceedings of the Second Symposium on Plantation Crops. Proc.* PLACROSYM II. pp. 50-57.

Delacriox, G. (1902). A malady affecting pepper (*Piper nigrum*) in Cochin China. *Agr. Prat. Des. Pays Chands* **1**: 672-680.

Eapen, S.J. (1987). Plant-parasitic nematode problem in small Cardamom. *In: Proceedings of the Third Group Discussions on the Nematological Problems on Plantation Crops,* October 29-30, 1987, Sugarcane Breeding Institute, Coimbatore. pp. 43-44.

Ferraz, E.D.; De, A.; Orchard, J.E. & Lopez, A.S. (1984). Respota da pimenteira-do-reino a infestacao per *Meloidogyne incognita* un relacao ao teor de jenois totais. *Revista Theobroma* **14**: 217-227.

Freire, F.C.O. & Bridge, J. (1985a). Biochemical changes induced in roots and xylem sap of black pepper by *Meloidogyne incognita. Fitopathologia Brasileira* **10**: 483-497.

Freire, F.C.O. & Bridge, J. (1985b). Parasitism of eggs, females and juveniles of *Meloidogyne incognita* by *Paecilomyces lilacinus* and *Verticillium chlamydosporium. Fitopatologia Brasileira*. 10: 577-596.

Gunasekharan, C.R.; Vadivelu, S. & Jayaraj, S. (1987). Experiments on Nematodes of Turmeric—A Review. *In: Proceedings of the Third Group Discussions on the Nematological Problems of Plantation Crops,* October 29-30, 1988, Sugarcane Breeding Institute. pp. 45-46.

Hart, W.H. (1956). Spreading decline of citrus caused by the burrowing nematode. *Bull. Calif. Dept. of Agric.* **45**: 263-267.

Huang, C.S. (1966). Host-parasite relationships of the root-knot nematode in edible ginger. *Phytopathology.* **56**: 755-759.

Inchinohe, M. (1980). Studies on the root-knot nematode of black pepper plantation in Amazon. *Annual Report of the Society of Plant Protection, North Japan.* **31**: 1-8.

Jacob, J.A. & Kurian, K.J. (1979). Screening of pepper varieties for resistance against root-knot nematode *Meloidogyne incognita. Agric. Res. J. Ker.* **17**: 90.

Jepson, B.S. (1987). *Identification of Root-knot Nematodes* (*Meloidogyne* species). CAB Inter. Wallingford Oxon OX 108 DE U.K. p. 265.

Kaur, D.J. (1987). Studies on nematodes associated with ginger (*Zingiber officinale* Rosc.). in Himachal Pradesh. Thesis submitted to Himachal Pradesh University, Shimla. p. 198.

Koshy, P.K. (1987). Integrated Nematode Management in Plantation Crops. *In: Proceedings of the Third Group Discussion on the Nematological Problems of Plantation Crops,* October, 29-30, 1987, Sugarcane Breeding Institute, Coimbatore. pp. 49-51.

Koshy, P.K.; Premachandran, D.; Sosamma, V.K. & Premkumar, T. (1979a). Effect of *Meloidogyne incognita* population on black pepper. *Indian Phytopath.* **32**: 221-225.

Koshy, P.K. & Sosamma, V.K. (1975). Host-range of the burrowing nematode *Radopholus similis* (Cobb, 1893) Thorne, 1949. *Indian J. Nematol.* **5**: 255-257.

Koshy , P.K. & Sosamma, V.K. (1977). Identification of the physiological race of *Radopholus similis* populations infesting coconut, arecanut and banana in Kerala, South India. *Indian Phytopath.* **30**: 415-416.

Koshy, P.K. & Sundararaju, P. (1979). Response of seven black pepper cultivars to *Meloidogyne incognita. Nematol. medit.* **7**: 123-125.

Koshy, P.K.; Sundararaju, P. & Wilson, K.I. (1979b). Efficacy of certain systemic nematicides against nematicides against *Meloidogyne incognita* in a cardamom nursery. *Nematol. medkt.* **7**: 195-198.

Lanjewar, R.D. & Shukla, V.N. (1985). Parasitism and interaction between *Pythium myriotylum* and *Meloidogyne incognita* in soft rot complex of ginger. *Indian J. Nematol.* **15**: 170-173.

Logdes, E.B. & Lordello, L.G.E. (1979). *Meloidogyne incognita Fusarium solani f. Piperi*, Associados A murcha da Pimenteria-do-reino na paraiba. *Revista de Agricultura* **54**: 266-268.

Mohandas, C. & Ramana, K.V. (1983). Effect of different levels of *Meloidogyne incognita* on plant growth of two cultivars of black pepper (*Piper nigrum* L.). *In*: Abstracts of papers, *Third Nematology Symposium Organised by Nematological Society of India,* Himachal Pradesh, Krishi Viswa Vidyalaya, Solan, Himachal Pradesh, India. pp. 9-10.

Mohandas, C. & Ramana, K.V. (1987). Slow wilt disease of black pepper and its control. *Indian Cocoa, Arecanut* and *Spices Journal* **11**: 10-11.

Mustika, I. (1978). Observations on the relationship between nematode population and yellow disease or black pepper in Bangka. *Pamberitaan Lembaga Penelitian Tenamaan Industry,* No. **30**: 11-12.

Nagakura, K. (1930). "Ubar den Ban und die Lebensgeschichte der *Heterodera radicicola* (Greef) Muller". *Jap. J. Zool.* **3**: 95-160.

Ramana, K.V. (1986). Slow wilt disease of black pepper and the role of plant-parasitic nematodes in its etiology (Abstr.) *In: Proceedings of the Second Group Discussion on Nematological Problems of Plantation Crops*, April 25-26, 1986 Central Coffee Research Station, Balehonuur. pp. 17-21,

Ramana, K.V. & Mohandas, C. (1986). Reaction of black pepper germplasm to root-knot nematode, *Meloidogyne incognita. Indian J. Nematol.* **16**: 138-139.

Ramana, K.V. & Mohandas, C. (1987). Nematological Problems in Black pepper (*Peiper nigrum* L.). *In: Proceedings of the Third Group Discussion on the Nematological Problems of Plantation Crops,* October, 29-30, 1987, Sugarcane Breeding Institute, Coimbatore. pp. 43-44.

Ramana, K.V.; Mohandas, C. & Balakrishnan, R. (1987a). Role of plant parasitic nematodes in the slow wilt disease complex of black pepper (*Piper nigrum* L.) in Kerala. *Indian J. Nematol.* **17**: 225-230.

Ramana, K.V.; Mohandas, C. & Ravindran, P.N. (1987b). Reaction of black pepper germplasm to the burrowing nematode (*Radopholus similis*).. *J. Plant Crops.* **15**: 65-66.

Sundararaju, P.; Sosamma, V.K. & Koshy, P.K. (1979). Pathogenicity of *Rodopholus similis* on ginger. *Indian J. Nematol.* 9: 91-94.

Van der Vecht, J. (1950). Plant Parasitic Nematodes. *In : Disease of cultivated Plants in Indonesia*. Col. I (Eds.-L.G.E. Karshovan & J.V.d. Vecht). W. van Woeve, S' gravenhage. pp. 16-45.

Venkitesan, T.S. (1976). Studies on the burrowing nematode, *Radopholus similis* (Cobb, 1893) Thorne, 1949 on pepper (*Piper nigrum* L.). and its role in slow-wilt disease. Ph. D. thesis. Department of Plant Pathology, University of Agricultural Sciences, Bangalore. p. 122.

Venkitesan, T.S. & Jacob, A. (1985). Integrated control of root-knot nematode, infecting pepper vines in Kerala (Abstr.). *Indian J. Nematol.* **15**: 261-262.

Winto, S.R. (1972). Effect of *Meloidogyne* species on the growth of *Piper nigrum* L. *Mal. Agric. Res.* **1**: 86-90.

Chapter 13

Nematode Pests of Palms and their Control

P.K. Koshy

Division of Nematology, Central Plantation Crops Research Institute, Regional Station, Kayangulam, Krishnapuram-690 533, India

COCONUT (*Cocos nucifera* L.)

THE coconut palm has its origin in South East Asia and is cultivated in the warm damp regions of the tropics. It is particularly distributed on all tropical shores. The largest producers of coconut are the Philippines, Indonesia, India and Sri Lanka. The total area under coconut cultivation in India is 14,28,700 ha with an annual production of 7,562.3 million nuts during 1987-1988. Coir and coir products exported from India earned Rs. 322 million during the above period (Source: Statistics Unit, CPCRI, Kasaragod 7/89).

Of the several nematode species reported on coconut (Govindankutty & Koshy, 1979) the important ones are *Radopholus similis* and *Rhadinaphelenchus cocophilus.*

RADOPHOLUS SIMILIS

The burrowing nematode, *Radopholus similis* was first described by Cobb (1893) as *Tylenchus similis* from necrotic root lesions of banana *Musa sapientum* received from Fiji in 1891. It was renamed as *Radopholus similis* by Thorne (1949) and Sher (1968), who confirmed it as the type species of the amended genus. The nematode occurs in most tropical and sub-tropical areas of the world.

Diagnostic features (Thorne, 1961)

Female : Vermiform, migratory, endoparasitic. Lip region rounded, strong well-developed stylet and oesophagus. Two outstretched ovaries, tail conoid to blunt rounded terminus.

Male : Vermiform, migratory, not parasitic. Lip region sub-spheriod, offset, slender stylet, degenerated oesophagus. Single testis spicules paired with bursa extending two thirds length of the tail.

Juvenile : Vermiform, migratory, parasitic. Lip region rounded, well-developed stylet and oesophagus.

Systematic position

Nematode, Secernentea, Tylenchida, Tylenchoidea, Pratylenchidae, Radopholinae, *Radopholus*.

DISTRIBUTION AND HOSTS

R. similis has been reported from coconut plants in Florida (U.S.A.), Jamaica, Sri Lanka and India (van Weerdt *et al.*, 1959a, b; Latta, 1966; Weischer, 1967; Koshy *et al.*, 1978). Surveys conducted in coconut plantation in South India recorded 24 per cent incidence of *R. similis* in coconut (Sosamma, 1984; Koshy, 1986a).

The coconut isolate of *R. similis* has a wide host range which includes many weeds, crops, fruit trees and many palms (Table 1). Forty eight species of plants belonging to forty five genera of seventeen families were recorded as host in India (Koshy & Sosamma, 1975; Sosamma & Koshy, 1977; Holdeman, 1986).

Table 1: List of palms reported as hosts of the burrowing nematode *Radopholus similis*.

Scientific Name	Common Name
Archontophoenix cunninghamiana	Seaforthia palm, Piccabeen bungalow palm
Areca (Actinorthytis) calapparia	
Areca catechu	Betel-nut palm
Areca macrocalyx	
Areca normanbyii	
Areca triandra	
Arecastrum romanzoffianum	Queen palm
Chamaedorea cataractarum	
Chamaedorea elegans (Collina elegans)	Parlor palm, neanthebella palm
Cocos nucifera	Coconut palm
Elaeis guineensis	Oil palm
Phoenix canariensis	Canary Island date palm
Phoenix dactylifera	Date palm
Raphis excelsa	Large lady palm
Roystonea regia	Royal palm

BIOLOGY AND ECOLOGY

Life Cycle

R. similis is a migratory endoparasite and is capable of spending its entire life within roots. All juvenile stages and adult females including graved females except the fourth stage and adult males are found to be infective. One life cycle was found to be completed within 25 days at a temperature range of 25-28°C (Koshy, 1986b).

Survival and Dispersal

Burrowing nematode populations survive under field conditions for six months in moist soil (27 to 36°C) and one month in dry soil (29 to 39°C), whereas it survives for 15 months in moist soil (25.5 to 28.5°C) and 3 months in dry soil 27 to 31°C) under greenhouse conditions. The nematode survives in roots of stumps of felled coconut palms up to six months (Sosamma & Koshy, 1986). Studies conducted at the Central Plantation Crops Research Institute (Regional Station), Kayangulam, Kerala, India, show that adult females are the most persistent form in coconut roots and soil during summer months and caused annual recurrence of infection.

In coconut plantations in Kerala, coconut seedlings are raised by sowing seednuts in the interspaces of coconut palms. The burrowing nematode is widely prevalent in most of the nurseries in Kerala and Tamil Nadu. One-year-old coconut seedlings raised in such infested nurseries harbour large populations of the burrowing nematode within roots both internal and external to the husk. These infested seedlings when distributed to distant places through Government as well as private agencies for planting help in the dissemination of the nematode. Moreover, in the present day fast trade between countries for foliage and other ornamental plants such as *Anthurium* spp., *Calathea* spp., etc. chances of dissemination of *R. similis* through these non-conventional hosts are greater (Koshy, 1986b).

Biotypes/Pathotypes

Two morphologically indistinguishable races of *R. similis* are known. One is the 'banana race' which parasitises banana and not citrus and 'citrus race' which parasitises both banana and citrus (Du Charme & Birchfield, 1956). But, the citrus race has been elevated to species rank and named *Radopholus citrophilus* because it has a haploid number of five chromosomes (n=5) against four (n=4) in banana race (*R. similis*) and it differed in protein patterns and pheromone mediated behaviour (Huettel *et al.*, 1984). In addition there is a report of an *R. similis* population in Puerto Rico which has a haploid number of five (n-5) chromosome not infesting five species of citrus (Rivas & Roman, 1985). The coconut isolate of

R. similis from Kayangulam and Kasaragod in Kerala, India is the banana race (Koshy & Sosamma, 1977) with a haploid number of four (n=4) chromosomes (Koshy, 1986b).

Influence of Environmental Factors

Studies on population fluctuations of the burrowing nematode in coconut plantations in Kerala, India, showed that infested coconut roots yielded maximum number of *R. similis* during October to November and minimum during March to July. A mean soil temperature below 25°C and light rainfall coupled with availability of tender fleshy roots are favourable for nematode multiplication (Koshy & Sosamma, 1978). The burrowing nematode multiplies well on coconut in loamy sand followed by riverine alluvium and least in 'Kari' soil and causes maximum plant damage in riverine alluvium and least in laterite soil (Sosamma & Koshy, 1985).

Interaction with other Microorganisms

The fungi *Cylindrocarpon effusum, C. lucidum and Cylindrocladium clavatum* have been reported in association with lesions produced by *R. similis* in coconut roots. In pathogenicity studies. *C. effusum* did not cause any appreciable damage to inoculated seedlings. But when it was inoculated simultaneously with the nematode, the rate of multiplication of the nematode and damage to coconut seedlings was reduced (Sosamma & Koshy, 1978, 1983; Koshy & Sosamma, 1987).

NATURE OF DAMAGE

Symptoms

Burrowing nematode infested coconut palms exhibit general decline symptoms like yellowing, button shadings, reduction in leaf size, yield, etc. which are non-specific. Symptoms on roots are more specific. *R. similis* on infestation produces isolated elongate orange-coloured lesions on tender and semi-hard roots. Consequent to nematode parasitisation and multiplication these lesions enlarge and coalesce to cause extensive rotting of roots. Tender roots on heavy infestation become spongy in texture. On semi-hard orange coloured roots surface cracks are commonly seen. Lesions are not usually seen on the old, hard, dark brown roots. As high as 4,000 nematodes were recovered from one gram (one inch length) of main roots. The nematode also attacks the plumule, leaf bases and haustoria of seedlings. The drastic reduction in the number and mass of tertiary feeder roots on parasitisation by the nematode limits plant growth (Koshy & Sosamma, 1987).

Histopathology

Burrowing nematode penetrates the absorbing region covered by very delicate epidermis behind the root cap by lysis of cells. Such entry points are of 1-2 cells in diameter and surrounded by sclerenchymatous cells to a depth of 10-15 cells. The cavities formed in the outer cortex are always surrounded by deeply stained and heavily suberised cells of irregular shape unlike those formed in the inner cortex. Maximum number of nematodes and cavities are seen in the outer cortex. Nematodes are not observed in the stellar region. In the early stage of infection, roots show cavities of independent origin separated by several cells. Consequent to nematode multiplication and lysis of cytoplasm and cell walls, adjacent cavities merge with each other thus destroying the cortex to a great extent. All stages of the nematode are seen in cavities in longitudinal sections (Koshy & Somamma, 1987).

Losses

R. similis causes heavy crop loss to many fruit, spice, and agricultural plantation crops. It is notorious for having wiped out 22 million black pepper vines in the Banka Island of Indonesia within two decades (Christie, 1957). Similarly spreading decline in citrus caused by *R. similis*, first observed in Florida in 1928, spread to 6000 ha within 35 years (Kaplan, 1985). It causes root rot, blackhead toppling diseases and decline in banana (Blake, 1972). It causes 30% yield loss in coconut. The threshold inoculum density required to cause significant reduction in various growth parameters of coconut is 100 nematodes/625 cc sandy loam soil over a period of five years under field conditions (Koshy, 1986a).

CONTROL STRATEGIES

Cultural

The cultural practices which exist in Kerala and Karnataka, India such as application of oilcakes, farmyard manure and growing of sannhemp in the basins and interspaces, and their incorporation as green manure may help in the inhibition of nematode multiplication.

Plant Resistance

Of the 29 exotic and 15 indigenous coconut cultivars ad 15 hybrids screened, the dwarf cultivars 'Kenthali' and 'Klappawangi' and the hybrids such as 'Java Giant x Kulasekharam Dwarf Yellow', 'Java x Malayan Dwarf Yellow' and 'San ramon x Gangabondam' recorded the least nematode multiplication and lesion indices (Sosamma *et al.*, 1980, 1986; Sosamma, 1984).

Chemical

Increased incidence of *R. similis* was noticed in coconut nurseries when banana was used as shade crop. More than 50 per cent of the seedlings raised in such nurseries failed to establish on transplantation to the main field. Treatment of nurseries with nematicides is the only way to release nematode-free seedlings.

A dip in 1000 ppm DBCP for 15 minutes was found effective in controlling nematode population and to ensure release of nematode-free coconut seedlings from *R. similis*-infested coconut nurseries (Koshy & Sosamma, 1979). Soil application of phenamiphos or phorate @ 25 kg a.i./ha during September, December and May in infested coconut nurseries completely eliminated *R. similis* (Koshy & Nair, 1979; Koshy *et al.*, 1983; Sosamma *et al.*, 1986).

Application of phorate @ 10g a.i./palm in June-July and again in October-November increased 30 per cent yield of *R. similis* L infested palms.

Integrated

The following measures are suggested for an integrated management schedule for *R. similis* infestation on coconut palms :

1. Application of cowdung, farmyard manure, oilcakes and green manure to the basins. *Crotolaria juncea* may be cultivated in the basin and interspaces and used as a green manure.
2. Application of phorate @ 10 g a.i./palm in June-July and October-November.
3. Avoid banana as a shade crop in coconut nurseries.
4. Use of nematode-free planting material of coconut and other intercrops.
5. Use of less susceptible/tolerant cultivars or their hybrids in infested areas.

RHADINAPHELENCHUS COCOPHILUS

The "Red ring disease" of coconut caused by *Rhadinaphelenchus cocophilus* was first reported as occurring in Trinidad by Hart (1905). Later it was reported in Grenada (Nowell, 1918). It is a major problem faced by the coconut industry in Caribbean, Central and South America. The disease does not occur in Africa, South East Asia, Malaysia, the Pacific Islands, Florida or Cuba (Dean, 1979). *R. cocophilus* was first described by Cobb (1919) as *Aphelenchus cocophilus*. Goodey (1960) designated it as *R. cocophilus*. Nowell (1920) demonstrated its pathogenicity.

Diagnostic features (Thorne, 1961)

Female : Body very slender. Offset head, prominent cephalic framework composed of strong sclerotized arches. Spear with strong basal knobs. Median bulb elongated. Vulva covered by a flap of cuticle leading into a curved vagina.

Male : Body slender. Offset head, prominent cephalic framework. Spear with strong basal knobs. Elongated median bulb. Most characteristic feature is the tail, which on death curves to about four-fifths of a circle. Spicules are slightly arcuate and have prominent rostrum. Tail bears a terminal bursal flap. Four pairs of ventrosubmedian papillae are present.

Juveniles : Pre-adult juvenile has conical rounded head, not set off from the body. Tail shorter than that of female with a short sharp terminal bearing a mucron.

Systematic position

Nematoda, Secernentea, Tylenchida, Aphelenchina, Aphelenchoidea, Aphelenchoidie, Rhadinaphelenchinae, *Rhadinaphelenchus*.

DISTRIBUTION AND HOSTS

At present *R. cocophilus* has been reported from the West Indian Islands of Tobago, Grenada and St. Vincent and from the Dominican Republic, Venezuela, Guyana, Surinam, French Guyana, Colombia, Ecuador, Mexico, Brazil, Panama, Costa Rica, Hunduras and El Salvador. It has not been reported from India (Dean, 1979).

Though *R. cocophilus* is primarily found on coconut and oil palm, it has also been found to occur naturally in the date palm and cabbage palm (Hagley, 1962). Successful inoculation have been reported in a number of palms (Brathwaite & Siddiqi, 1975; Dean, 1979).

BIOLOGY

Life Cycle

Life history studies conducted by inoculating immature nuts of coconut palm showed that the nematodes completed one life cycle in 9 to 10 days which is probably one of the shortest life cycle reported for plant-parasitic nematodes (Blair, 1966, 1970)

Survival

R. cocophilus can survive for sufficient time and move over sufficient distances in soil (Fenwick, 1969). It survived for a maximum of 15 days in soil (Blair, 1964). In fresh water films it survived for 7-8 days and in sea water for 3 days (Dean, 1979). The third stage juvenile is the most persistent form.

Dispersal

Fragments of diseased tissue from contaminated bodies of the palm weevil *Rhynchophorus palmarum* are deposited into the bases of leaf axils of healthy plants. The nematode is carried by weevil larvae, pupae and adults both internally and externally. The nematode persists without change or multiplication through the metamorphosis of the insect. On emergence of the adult insect, large numbers of nematodes located in the region of the ovipositor of the female are injected into the soft tissue of the coconut palm when the insect deposits eggs (Ashby, 1921; Hagley, 1962; Fenwick & Mohammad, 1962; Griffith 1967, 1968).

Natural cannibalism in the larvae also affects the number of emerging weevils. The bacterium *Micrococcus roseus* in cedros wilt diseased palms produces disease in affected palm weevils. Moreover, some ground lizards also feed on the adult insects. The nematode dies in decomposing tissue of diseased palms and an unidentified fungal parasite exists for the nematode (Griffith & Koshy, 1989).

NATURE OF DAMAGE

Symptoms

Palms of bearing age (5-7 years) are more susceptible to infestation. Primary symptom is the appearance of yellowing of the lower leaves starting at the tips of distal pinnae which progresses towards the base of the pinnae and leaves. The yellowing is succeeded by browning and later by death of the leaf. The inner leaves are also affected and the affected trees die within 3-4 months of the appearance of the primary symptoms.

Internal symptom of a redring nematode infested palm is characteristic. At first, about 2.5 cm beneath the stem surface scattered reddish dots of about 1 mm diameter are seen which eventually coalesce to form an orange-red ring about 3 cm in width. Hence the name 'red ring disease' and 'red ring nematode'. The ring extends the whole length of the stem and roots and in petioles it assumes a crescent-like shape. Large number of juveniles are seen in the centre of the discoloured areas and adults in the periphery (Fenwick, 1969; Brathwaite & Siddiqi, 1975; Dean, 1979). Shedding of green nuts of all sizes takes place. In advanced stages of infection putrefaction of the soft tissue takes place.

R. cocophilus causes 'little leaf disease' of coconut and oil palm in Surinam and Guyana (Van Hoof & Seinhorst, 1962).

Histopathology

A thermostable phytotoxin was produced due to breakdown of coconut tissue (Goberdhan, 1963). Water uptake of infested coconut palms was less due to occlusion of xylem vessels (Goberdhan, 1964; Blair, 1964; Blair & Darling, 1968). In the upper parts of the stem, the nematodes are found intercellularly (Goberdhan, 1964; Blair, 1964). In coconut roots, the nematode attacks cortical tissues.

Losses

At present, most Latin American and Caribbean countries show losses ranging from less than 1% to more than 20% of 2 to 10-year-old trees. Thirty-five per cent mortality of young coconut palms has been reported in Trinidad and 80% loss in a single plantation in Tobago (Esser, 1969). Within a span of 10 years, 35% of oil palms died in Venezuela (Oostenbrink, 1963).

CONTROL STRATEGIES

Biological

The vector weevil is found to be parasitised by several species of Rhabditidae or Heterorhabditidae throughout Latin America. Since the vector insects can be highly parasitised with the above nematodes, selective pressure can be introduced against the vectors. Such measures are being employed in Trinidad with a species of Rhabditidae (Griffith & Koshy, 1989).

Chemical

The leaf axils of diseased palms should be sprayed with 0.1 per cent Lannate (Methomyl) insecticide suspension when early symptoms appear. Then the palm should be poisoned by adding one tablespoon of weedicide "100" in three holes (2 cm diameter and 10 cm deep) bored around the trunk of the tree at a level of 15 cm above the soil. In advanced stages the palms should be cut down and the pieces and stumps sprayed with at least 4.5l of 0.1 per cent Lannate suspension.

Guard baskets made of 2 cm mesh wire are used to protect frequent outbreaks of the disease. These baskets are filled with fresh infected tissue and sprayed with 0.1 per cent Lannate suspension. The palm weevils are attracted to the tissues in the basket. After two weeks the tissue in the basket is burnt. One guard basket is used per 0.4 ha of palms.

Integrated

Even though no effective method of control is known at the moment, the following methods are being adopted :

1. Phytosanitation by using arsenic preparations in diseased trees.
2. Leaf axil filling with a granular nematicide.
3. Control of the palm weevil by applying insecticide.
4. Use of resistant varieties with short internodes.
5. Plant quarantine measures to check its spread to newer areas.

ARECANUT (*Areca catechu* L.)

Arecanut has its origin in the hot damp regions of Asia and the Malay Islands, India is the largest producer of arecanut in the world covering an area of 2,00,000 with an annual production of 2,28,600 ha tonnes. It earned Rs. 16 million from export during 1987-1988 (Source: Statistics Unit, CPCRI, Kasaragod 7/89).

RADOPHOLUS SIMILIS

DISTRIBUTION

Radopholus similis was first reported from soil around arecanut roots in Mysore, India by Kumar *et al.* (1971). Later Koshy *et al.* (1975, 1976) recorded 22 genera of plant-parasitic nematodes from the root zone of arecanut during their survey. Among them, *R. similis* was the only endoparasite encountered in more than 50 per cent of the root samples collected.

NATURE OF DAMAGE

Symptoms

Burrowing nematode infested areca palms exhibit non-specific above ground symptoms like general yellowing and visible reduction in growth, vigour and yield. The most conspicuous symptom is the appearance of lesion and rotting of roots. The nematode produces small, elongate, orange-coloured lesions on young, succulent, creamy-white to light-orange coloured portion of the main and lateral roots. Later the adjoining lesions coalesce and cause extensive root rotting. The thick primary roots produced from the bole region of the palm exhibit large, oval, sunken, dark lesions. Unlike in coconut the tips of lateral and tertiary roots on infestation become black.

Histopathology

The burrowing nematodes are found in inter- and intra-cellular positions in the cortex. They do not enter the stelar region (Sundararaju & Koshy, 1988a).

Losses

R. similis on infesting arecanut roots caused reduction in growth parameters. The threshold inoculum level causing significant damage to growth of arecanut was found to be 100 nematodes per seedling or one nematode in 800 g laterite soil (Koshy, 1986b). Ten-fold increase in yield was recorded by treatment with aldicarb @ 10 g a.i./palm, DBCP @ 10 ml a.i./palm or fensulphothion 50 g a.i./ palm (Sundararaju & Koshy, 1986).

ECOLOGY

Interaction with other microorganisms

The fungus *Cylindrocarpon obtusisporum* was found in consistent association with lesions caused by *R. similis* in arecanut roots. Though the fungus in combination with the nematode caused more damage, it inhibited the rate of multiplication of the nematode (Sundararaju & Koshy, 1984, 1987).

CONTROL STRATEGIES

Plant resistance

None of the 46 accessions of arecanut germplasm screened, is immune or highly resistant to *R. similis*. The cultivars 'Indonesia-6' ('VTL - 11'), Mahuva B' and 'Andaman-5' ('VTL - 29e') are tolerant to *R. similis* (Koshy *et al.*, 1979); Sundararaju & Koshy, 1982). The cultivar ('Indoneisa - 6' ('VTL - 11') and 'Singapore' ('VTL - 17') are known to yield 50% more nuts over local South Kanara variety (Anonymous, 1974). The hybrid 'VTL - 11 x VTL 17' is highly resistant to *R. similis* (Sundararaju & Koshy, 1988b). Hence, these cultivars could be recommended for *R. similis*-infested areas.

Chemical

Application of aldicarb or fensulfothion @ 1 g a.i./seedling thrice a year for three years controlled *R. similis* population at the seedling stage. In adult palms reduction in nematode population and 10 fold increase in yield was obtained by aldicarb @ 10 g a.i/palm, DBCP @ 10 ml a.i./palm or fensulfothion 50 g a.i./palm (Sundararaju & Koshy, 1986).

Integrated

Integrated management of the burrowing nematode infesting arecanut roots can be done as suggested below :

1. Use of nematode-free planting materials of arecanut and other intermixed crops.
2. Avoid *R. similis* susceptible intermixed crops like black pepper and banana in infested areas.
3. Apply phorate @3 g a.i. per plant to the root zones of banana, black pepper and arecanut in arecanut based farming system in June-July and in Sept-Oct.

OIL PALM (*Elaeis guineensis* Jacq.)

The oil palm has its origin in Central Africa and has been introduced throughout the entire tropics in large groves. It is being cultivated in Kerala, India on a plantation scale since 1970. In India, it is being cultivated over an area of 4,300 ha and produced 200 tonnes of oil during 1987-88 (Source: Statistic Unit, CPCRI, Kasaragod 7/89). Among the different plant-parasitic nematodes found in the rhizosphere of the oil palm, the most important is *Rhadinaphelenchus cocophilus* (Cobb, 1919; Goodey, 1957, 1960; Shier, 1966; Pizarro, 1968).

RHADINAPHELENCHUS COCOPHILUS

DISTRIBUTION

The red ring disease on oil palm caused by *Rhadinaphelenchus cocophilus* was reported from Venezuela (Webster & Gonzales, 1959). Later, it was reported from Latin America (Malaguti, 1953) and Surinam (Van Hoof & Seinhorst, 1962). It has not yet been reported from India.

NATURE OF DAMAGE

Diseased palms in young groves are initially found clustered which gradually expand. In older groves, such palms have a random distribution.

In Surinam 'little leaf' was found to be a prominent symptom in diseased palms (Van Hoof & Seinhorts, 1962). Diseased palms had erect, short and deformed leaves. The top part of the main vein bore suberized patches. The pinnae were shorter, wavy and necrotic at the tips and yellow patches were present on the petiole and leaf bases. The disease palm had a brownish discolouration. *R. cocophilus* was found in discoloured tissue and thrived more in the petioles of young folded leaves than in stem and roots. They were found outside the necrotic zones. A notable feature was that the band of necrotic tissue is always very small.

In diseased oil palms an average of less than 500 nematodes was found in one gram of infested tissue. The disease could be produced experimentally (Mass, 1970).

Losses

R. cocophilus was found to cause economic loss in oil palms in South Africa. Malaguti (1953) had observed that in a seven-year-old oil palm grove suffering from little leaf, many trees had died.

RADOPHOLUS SIMILIS

Radopholus similis was found to infest roots of oil palm on artificial inoculation under greenhouse conditions (Koshy & Sosamma, 1975).

REFERENCES

Anonymous (1974). Central Plantation Crops Research Institute. Annual Report for 1973, Kasaragod, Kerala, India. p. 262.

Ashby, S.F. (1921). Some recent observations on red ring disease. *Agri. News* (Bridgetown) **20**: 350-351.

Blair, G.P. (1964). Red ring disease of the coconut palm. *J. Agric. Soc. Trinidad & Tobago* **64**: 31-49.

Blari, G.P. (1966). The use of immature nuts *Cocos nucifera* for studies on *Rhadinaphelenchus cocophilus. Nematologica* **11**: 590-592.

Blair, G.P. (1970). Studies on red ring disease of the coconut palm. Oleagineux. **25**: 19-22, 79-83.

Blair, G.P. & Darling, H.M. (1968). Red ring disease of the coconut palm, inoculation studies and histopathology. *Nematologica* **14**: 395-403.

Blake, C.D. (1972). Nematode diseases of banana plantation. *In* : *Economic Nematology* (Ed.—J.M. Webster). Academic Press, London, U.K. pp. 245-267.

Brathwaite, C.W.D. & Siddiqi, M.R. (1975). *Rhadinaphelenchus cocophilus*, C.I.H. descriptions of plant parasitic nematodes, Set. 5, No. 72. p. 4.

Christie, J.R. (1957). The yellows disease of pepper and spreading decline of citrus. *Plant Dis. Reptr.* **41**: 267-268.

Cobb, N.A. (1983). Nematodes, mostly Australian and Fijian *Macleay Mem. Vol. Linn. Soc. N.S.W.* pp. 252-308.

Cobb, N.A. (1919). A newly discovered nematode, *Aphelenchus cocophilus* n. sp. connected with a serious disease of the coconut palm. *W.I. Bull.* **17**: 203-210.

Dean, G.G. (1979). Red Ring Disease of *Cocos nucifera* L. caused by *Rhadinaphelenchus cocophilus* (Cobb, 1919) Goodey, 1960. *Tech. Commun. Commonw, Bur, Helminth*. St. Albans No. 47. p. 70.

Du Charme, E.P. & Birchfield, W. (1956). Physiologic races of the burrowing nematode. *Phytopathology* **46**: 615-616.

Esser, R.P. (1969). *Rhadinaphelenchus cocophilus* a potential threat to Florida palms. *Nematology Circular, Division of Plant Industry, Fla. Dept. Agric.* No. 9.

Fenwick, D.W. (1969). Red ring disease of the coconut palm. *In : Nematodes of Tropical Crops* (Ed.-J.E. Peachey). *Tech. Commun. Commonw. Bur. Helminth.* No. 40. pp. 89-98.

Fenwick, D.W. & Mohammed, M.(1962). The presence of *Rhadinaphelenchus cocophilus* (Cobb, 1919) Goodey, 1960 on the body of the palm weevil, *Rhynchophorus palmarum* (L.) *A. Rep. Res. Stu. Trinidad & Tobago Coconut Research Ltd.* 1961. pp. 1-15.

Goberdhan, L.C. (1963). Possible presence of a phytotoxin in the red ring infected trees. *Nature* **197**: 619-620.

Goberdhan, L.C. (1964). The comparative distribution of dye in healthy and red ring infected coconut palms. *Tropl. Agric.* **41**: 169-174.

Goodey, J.B. (1957). *Hoplolaimus proporicus n.* sp. (Hoploaiminae : Tylenchida). *Nematologica* **2**: 108-113.

Goodey, J.B. (1960). *Rhadinaphelenchus cocophilus* (Cobb, 1991) n. Comb. the nematode associated with 'red ring' disease of coconuts. *Nematologica* **5**: 98-102.

Govindankutty, M.P. & Koshy, P.K. (1979). Nematodes associated with the coconut palm (*Cocos nucifera* L.). Bull. 2 Cent. Plant. Crops Res. Inst., Kasaragod, India. pp. 1-50.

Griffith, R. (1967). Progress on the entomological aspects of red ring disease. *J. Agric. Soc. Trin.* **67**: 231-244.

Griffith, R. (1968). The relationship between the red ring nematode and the palm weevil. *J. Agric. Soc. Trinidad & Tobago* **68**: 342-356.

Griffith, R. & Koshy, P.K. (1989). Nematode parasites of coconut and other palms. *In: Plant Parasitic nematodes in Sub-tropical and Tropical Agriculture* (Eds.-M. Luc, R. Sikora & J. Bridge). CAB Internatl., U.K. pp. 363-386.

Hagley, E.A.C. (1962). The palm weevil *Rhyncophorus palmarum* L. a probable vector of red ring disease of coconuts. *Nature* **193**: 499.

Hart, J.H. (1905). Coconut disease, *Bull. Misc. Inform. Roy. Bot. Gdns. Trinidad* **6**: 241-243.

Holdeman, Q.L. (1986). The Burrowing Nematode, *Radopholus similis*, Sensu Lato. Division of Plant Industry, California Department of Food and Agriculture and the Agricultural Commissioners of California. p. 52.

Huettel, R.N.; Dickson, D.W. & Kaplan, D.T. (1984). *Radopholus citrophilus* sp. *n.* (Nematoda), a sibling species of *Radopholus similis. Proc. Hel. Soc. Wash.* **51**: 32-35.

Kaplan, D.T. (1985). Plant parasitic nematode affecting citrus. *Fruit Varieties Journal* **39**: 26-39.

Koshy, P.K. (1986a). Research on plant parasitic nematodes at the Central Plantation Crops Research Institute, October 1972 to June, 1986. p. 45.

Koshy, P.K. (1986b). The burrowing nematode *Radopholus similis* (Cobb 1893). Thorne, 1949. *In : Plant Parasitic Nematode of India—Problems and Progress.* (Eds.-G. Swarup & D.R. Dasgupta). Allied Publishers, New Delhi. pp. 223-248.

Koshy, P.K. & Nair, C.P.R. (1979). Control of *Radopholus similis* (Cobb, 1893) Thorne, 1949 in coconut nursery. *Indian J. Nematol.* **9**: 15-19.

Koshy, P.K. & Sosamma, V.K. (1975). Host-range of *Radopholus similis* (Cobb, 1893) Thorne, 1949. *Indian J. Nematol.* **5**: 255-257.

Koshy, P.K. & Sosamma, V.K. (1977). Identification of the physiological race of *Radopholus similis* populations infesting coconut, arecanut and banana in Kerala, South India. *Indian Phytopath.* **30**: 415-416.

Koshy, P.K.& Sosamma, V.K. (1978). Studies on the population fluctuations of *Radopholus similus* in coconut and arecanut roots. *Indian Phytopath.* **31**: 180-185.

Koshy, P.K. & Sosamma, V.K. (1979). Control of the burrowing nematode, *Radopholus similis* (Cobb, 1893) Thorne, 1949 on coconut seedlings with DBCP. *Indian J. Nematol.* **9**: 32-33.

Koshy, P.K. & Sosamma, V.K. (1987). Pathogenicity of *Radopholus similis* on coconut (*Cocos nucifear* L.) seedlings under greenhouse and field conditions. *Indian J. Nematol.* **17** : 108-118.

Koshy, P.K.; Sosamma, V.K. & Nair, C.P.R. (1975). Preliminary studies on *Radopholus similis* (Cobb, 1893) Thorne, 1949 infesting coconut and arecanut palms in South India. *Indian J. Nematol.* **5** : 26-35.

Koshy, P.K.; Sosamma, V.K. & Sundararaju, P. (1976). Yellow leaf disease: Nematological studies. *Arecanut & Spices Bulletin.* **8**: 37-41.

Koshy, P.K.; Sosamma, V.K. & Sundararaju, P. (1979). Reactions of thirty one areca germplasm collections to the burrowing nematode, *Radopholus similis. Plant Dis. Reptr.* **63**: 433-435.

Koshy, P.K.; Sundararaju, P. & Sosamma, V.K. (1978). Occurrence and distribution of *Radopholus similis* (Cobb, 1893) Thorne, 1949 in South India. *Indian J. Nematol.* **8**: 49-58.

Koshy, P.K.; Sundararaju, P.; Sosamma, V.K. & Ravikumar, K. (1983). Efficacy of four systemic nematicides against *Radopholus similis* in coconut nursery. *Third Nematol. Symp.* HPKVV, Solan. pp. 40-41.

Kumar, A.C.; Viswanathan, P.R.K. & D'Souza, C.I. (1971). A study on plant parasitic nematodes of certain commercial crops in coffee tracts of South India. *Indian Coffee* **35**: 222-224.

Latta, R. (1966). Attempts to relate nematodes to lethal yellowing disease of coconut palms in Jamaica. *Tropical Agriculture* (Trinidad) **43**: 59-58.

Maas, P.W.T. (1970). Tentative list of plant parasitic nematodes in Surinam, with description of two new species of Hemicycliophorinae. *Bull. Landbouwproefstation Surinam No.* 87. p. 9.

Malaguti, G. (1953). Pudricon del cogello de la palmera de accite africana (Elaeis guinensis Jacq.). *Agr. Trop. Venezuela* **3**: 13-21.

Nowell, W. (1918). Root disease of coconut palms in Grenada. *Agric. News.* (Bridgetown) **17**: 398-399.

Nowell, W. (1920). The red ring disease of coconut palms. Infection experiments. *W. Indian Bull.* **18**: 74-76.

Oostenbrink, M. (1963). Symptomen in aaltyespopulaties in 'red ring' — zieke cocos-an olie-palmen. *Mededelingen van de Landbouwhogeschool an de Opzoekingsstations vas de staatte Gent* **28**: 663-671.

Pizarro, A.C. (1968). Observation of *Helicotylenchus* sp. parasitic on coconuts and African oil palms. *Philipp. J. Pl. Ind.* **32**: 77-81.

Rivas, X. Roman, J. (1985). Investigations on the host-range of a population of *Radopholus similis* from Puerto Rico. *Nematropica* **15**: 165-170.

Sher, S.A. (1966). "Revision of the Hoplolaiminae (Nematoda). VI *Helicotylenchus* (Steiner, 1945)". *Nematologica* **12**: 1-56.

Sher, S.A. (1968). "Revision of the genus *Radopholus similis*, Thorne, 1949 (Nematoda: Tylenchoidae)" *Proc. Helminth. Soc. Wash.* **35**: 219-237.

Sosamma, V.K. (1984). Studies on the burrowing nematode of coconut. Ph.D. Thesis, Kerala University, Trivandrum, Kerala, India. p. 166.

Sosamma, V.K. & Koshy, P.K. (1977). Additional hosts of the burrowing nematode : *Radopholus similis* infesting coconut palm in South India. *Plant Dis. Reptr.* **61**: 760-761.

Sosamma, V.K. & Koshy, P.K. (1978). A note on the association of *Cylindrocarpon* spp. with

Radopholus similis in coconut. *Indian Phytopath.* **31**: 381-382.

Sosamma, V.K. & Koshy, P.K. (1983). A note on the occurrence of Cylindrocladium clavatum Hodges & May in lesions caused by *Radopholus similis* on coconut roots. *Curr. Sci.* **52**: 438.

Sosamma, V.K. & Koshy, P.K. (1985). Effect of different types of soils on multiplication of *Radopholus similis* and growth of coconut seedlings. *Indian J. Nematol.* **15**: 217-222.

Sosamma, V.K. & Koshy, P.K. (1986). Survival of *Radopholus similis* in host free soil. *Indian J. Nematol.* **16**: 74-76.

Sosamma, V.K.; Koshy, P.K. & Bhaskara Rao, E.V.V. (1980) Susceptibility of coconut cultivars and hybrids to *Radopholus similis* in the field. *Indian J. Nematol.* **10**: 250-252.

Sosamma, V.K.; Koshy, P.K. & Bhaskara Rao, E.V.V. (1986). Response of coconut cultivars to the Burrowing Nematode *Radopholus similis*. *In:* Abstracts of papers: National Conference — Plant Parasitic Nematodes of India — Problems and Progress, 17-20 December, 1986, IARI, New Delhi, India. p. 107.

Sundararaju, P. & Koshy, P.K. (1982). Response of *Areca* collections against *Radopholus similis*. *Indian J. Nematol.* **12**: 402-404.

Sundararaju, P. & Koshy, P.K. (1984). A note on the association of *Cylindrocarpon obtusisporum* in lesions caused by Radopholus similis on arecanut (*Areca catechu* L.) roots. *Indian Phytopath.* **37**: 578.

Sundararaju, P. & Koshy, P.K. (1986). Control of *Radopholus similis* on arecanut seedlings with Aldicarb, Aldicarb sulfone, Carbofuran and Fensulfothion. *Indian J. Nematol.* **16**: 4-7.

Sundararaju, P. & Koshy, P.K. (1987). Separate and combined effects of *Radopholus similis* and *Cylindrocarpon obtusisporum* on arecanut seedlings. *Indian J. Nematol.* **17**: 301-305.

Sundararaju, P. & Koshy, P.K. (1988a). Histopathology of arecanut roots infected with *Radopholus similis. Indian J. Nematol.* **18**:378-380.

Sundararaju, P. & Koshy, P.K. (1988b). Screening of arecanut cultivars and hybrids for resistance to *Radopholus similis. Indian J. Nematol.* **18**: 376-377.

Thorne, G. (1949). On the classification of the Tylenchida, new order (Nematoda-Phasmidia). *Proc. Helminth. Soc. Wash.* **16**: 37-73.

Thorne, G. (1961). *Principles of Nematology*, New York McGraw Hill Book Co., Inc. pp. XIV + 553.

Van Hoof, H.A. & Seinhorst, J.W. (1962). *Rhadinaphelenchus cocophilus* associated with little leaf of coconut and oil palm. *Tijdschr Plziekt.* **68**: 251-256.

Van Weerdt, L.G.; Birchfield, W. & Esser, R.P. (1959a). Observations on some subtropical plant parasitic nematodes in Florida. *Proc. Soil Crop Sci. Soc. Fla.* **19**: 443-451.

Van Weerdt, L.G.; Martinez, A.P. & Esser, R.P. (1959b). Results of a survey designed to determine the etiology of a 'Lethal yellowing of *Cocos nucifera* L. *Proc. Fla. St. Hort. Soc.* **72**: 421-430.

Webster, B.N. & Gonzales, J. (1959). Investigaciones sobre la enfermedad 'amillo-rojo en cocos oy palma des aceite. IX Conv. de la assoc. Venezolana. el avance de la ciencia, service shell paa el agr. Establa Arajua.

Weischer, B. (1967). Plant Parasitic Nematodes. Report to Govt. of India, UNDP, F.A.O. No. T.A. of the United Nations, Rome.

Chapter 14

Control Strategies for Nematodes Damaging Mushroom

Naval Kumar Sharma
Regional Horticultural Research Station, Dr. Y.S. Parmar University of Horticulture & Forestry, Jacch (Nurpur), Kangra - 176 201, India

NEMATODE problems in mushroom are unique in a way that this is probably the only example in nematology where ectoparasitic nematode forms inflict as heavy losses as total crop failures. This is mainly because of the fact that two basic requirements of mushroom growing, viz. temperature and moisture, suit well to nematode multiplication. The other factors which favour higher nematode build-up are : capability of female to lay good number of eggs, short life cycle of the nematode and easy availability of food in mushroom growing trays or bags.

Many countries are producing mushroom, making "mushroom growing" a popular industry particularly in U.S.A., France, China, Holland, Great Britain and Taiwan. In India though mushroom growing is becoming popular day by day, growers with lesser facilities are still not sure of a successful crop. This is mainly due to the lack of purposely built mushroom houses and unhygienic conditions prevailing in and around the mushroom farms. Such conditions provide congenial atmosphere for many pests and diseases. Of these, the most dangerous are nematodes which once entered into the crop cannot be eradicated without completely destroying the crop. Their presence in the beds simply means very poor yields or total crop failures. In this regard an intensive survey of a mushroom farm in India revealed 84.4 per cent nematode infestation, which led to extremely poor flushes in initial stages followed by total crop failure (Seth, 1984).

Since, nematodes cause serious crop losses in mushroom and are difficult to control once enter the mushroom growing trays, it is important to understand them in detail before the control strategies are discussed.

HISTORY

The first report of a nematode damage to mushroom dates back to 1906, when Newstead reported the destruction of mushroom crop by a species of *Ditylenchus*.

But no attention was paid to identify this nematode species until Lambert *et al.* in 1949 again observed a *Ditylenchus* species in mushroom farm, which they reported was causing injury to mushroom mycelium resulting losses in yield. The species was later identified and described by Goodey (1958b) as *Ditylenchus myceliophagus*. Several workers later confirmed this species to be one of the most destructive nematodes associated with mushroom.

During 1952-56, another important nematode species belonging to *Aphelenchoides* group causing considerable losses to mushroom yields was reported by various workers from U.S.A. and England. This species was later described and named by Franklin in 1957 as *Aphelenchoides composticola*. After the importance of nematodes as a real menace to mushroom cultivation was recognised, various scientists started working on nematodes of mushroom and identified many new species under this group. Hooper (1958a) found *Aphelenchoides saprophilus* and *Aphelenchus avenae* to be parasitic on mushroom mycelium. Later in the same year, he described a new species *Aphlenchoides sacchari* from the rhizosphere of *Saccharum officinalis* (Hooper, 1958b), which he also reported to be parasitic on mushroom mucelium. Recent survey reports have shown that this species is widely prevalent on mushroom in Poland and India. Paesler (1957) reported seven species of genus *Aphlenchoides*, viz. *A. cyrtus, A dactylocercus, A helophilus, A. parietinus, A. spinosus, A. sacchari* and *A. winchesi* (now *Seinura winchesi*) from mushroom beds. The other species reported as parasites of mushrooms are *Paraphelenchus myceliophthorus* (Goodey, 1958a), *Aphelenchoides subtenius* (Hooper, 1962), *Aphelenchoides bicaudatus* (McLeod, 1967) and *Aphelenchoides coffeae* (McLeod, 1968) from various parts of the world so far.

Though heavy populations of various saprozoic nematode species belonging to *Rhabditis, Diplogaster* and *Cephalobus* have been reported from mushroom trays, their harmful or pathogenic effects of mushroom growing are still doubtful.

DISTRIBUTION

The species of *Aphelenchoides* and *Ditylenchus* are most commonly associated with mushroom growing in various parts of the world. Among these, *D. myceliphagus* is the only and most commonly occurring nematode species under genus *Ditylenchus*, whereas there are about 16 nematode species under *Aphelenchoides* which have been reported to be associated with mushroom cultivation and about half among these are highly pathogenic. It has been observed that their occurrence vary greatly, not only from place to place but even in two crops at same place. There are also instances where more than one *Aphelenchoides* species have been recorded from the same sample drawn from a single tray, therefore these should be identified properly before the control measures are undertaken. In general the distribution of mushroom nematodes depends upon the source and type of the mushroom growing ingredients used, since unlike other crops it is an indoor crop. However, the observations recorded so far reveal that the occurrence of *Ditylenchus myceliophagus* in mushroom is more common in northern than in southern India.

MUSHROOM NEMATODES

Based on their feeding habits, the nematodes associated with mushroom can be grouped into five categories (Table 1). Of these, mycophagous and saprozoic nematodes are of main concern being harmfully associated with mushroom growing.

Table 1: Feeding groups of nematodes associated with mushroom.

Feeding group	Source of contamination	Remarks
Mycophagous (fungal feeders)	Unpasteurized compost, casing soil,, irrigation water, etc.	These suck the cell sap from the fungal hyphae and devitalize the mycelium, thus responsible for reduction in sporophore yields.
Saprozoic	-do-	These are the most abundant nematode forms in the mushroom beds that feed upon microbes (mainly bacteria) and exert indirect harmful effects on mycelial growth and sporophore yields.
Predatory	Compost and casing soil	These feed upon other nematodes including mycophagous and are thus beneficial for the crop.
Animal parasites	Horse dung and chicken manure	These are mainly the parasites of higher animals and do not affect the mycelium in any way.
Plant parasites	Casing soil	These are more contaminants which do not feed upon fungal hyphae.

Fungal feeding nematodes

In all, 21 nematode species representing two orders (*Aphelenchida* and *Tylenchida*) have been reported to be harmfully associated with mushroom cultivation from various parts of the world. Among these, 20 species belonging to four genera (*Aphelenchoides, Aphelenchus, Paraphelenchus* and *Seinura*) fall in order Aphelenchida, whereas the only species in Tylenchida is *Ditylenchus myceliophagus*.

The different fungal feeding nematode species associated with mushroom are : *Ditylenchus myceliophagus*, Aphelenchoides agarici*, A. asterocaudatus*, A. bicaudatus, A. coffeae, A. composticola*, A. cyrtus, A. dactylocercus, A. helophilus, A. limberi, A. minor*, A. myceliophagus*, A. parietinus, A. sacchari*, A. spinosus, A. swarupi*, A. saprophilus, Aphelenchus avenae, Paraphelenchus myceliophthorus*, and *Seinura winchesi*. The asterisked species have also been reported from India.

In India, so far eight species of *Aphelenchoides* and *Ditylenchus myceliophagus* have been found to damage mushroom mycelium. *Aphelenchus avenae* though often found in cropping beds occur in low number, hence its pathogenicity is of doubtful validity. *Paraphelenchus myceliophthorus* and *Seinura winchesi* are two other genera which though important as pathogen have not so far been reported from India.

Following nematode species on mushroom are of economic importance in India:

***Aphelenchoides agarici* :** This species is recorded from mushroom farms in Himachal Pradesh (Seth & Sharma, 1986). It is one of the most destructive nematode species and multiply freely on mushroom mycelium. An initial inoculum of 10 nematodes has been shown to cause 100% mycelial depletion within 25 days at 28°C resulting over 800 times multiplication rate under laboratory conditions (Khanna & Sharma, 1988b). It completes its life cycle within 8 days at 28±1°C. The ratio between female to male was found to be 1.5:1 (Seth, 1984).

***Aphelenchoides composticola* :** It is one of the most widely distributed species present throughout the mushroom growing countries of the world. In India, it is known to occur in Himachal Pradesh and Punjab. This nematode also has a short life cycle of 8 days (at 23°C) and multiplies very fast on mushrooms (Cayrol, 1967). An initial inoculum of nine nematodes per 300 g compost could destroy the mycelium completely causing great yield losses (McLeod, 1967). Its pathogenicity when tested under Indian conditions revealed that an initial number of 10 nematodes completely depleted the mycelium with 40 days resulting into about 1000 times multiplication rate (Khanna & Sharma, 1988b).

***Aphelenchoides myceliophagus* :** The species has been recorded from district Solan in Himachal Pradesh. It is again one of the highly destructive nematode species infesting mushroom, showing the pathogenicity status of the level of *A. composticola* (Khanna & Sharma, 1988b).

***Aphelenchoides neocomposticola* :** The nematode was reported from Shimla district of Himachal Pradesh. The nematode though pathogenic in nature was found to be less damaging than *A. agarici, A. composticola* and *A. myceliophaus* (Khanna & Sharma, 1988b).

Aphelenchoides sacchari : In India the occurrence of this species from mushroom farm was reported by Sharma *et al.* (1981). The nematode is highly pathogenic on white button mushroom, *Agaricus bisporus*. Depending upon the source and time of inoculation an average of 50-100 per cent yield losses in sporophores of white button mushroom have been reported due to this nematode species (Sharma *et al.*, 1984). The species is quite prevalent on mushroom in Poland and India.

Ditylenchus myceliophagus : It is widely distributed nematode species present in almost all mushroom growing belts of the world. In India, this species has been found to be distributed in mushroom farms of Himachal Pradesh, Jammu & Kashmir, and Punjab (Thapa *et al.*, 1981). It is one of the highly destructive species as an initial inoculum of 10-1000 individuals could cause 15-70 per cent mycelial depletion within 60 days (Sharma *et al.*, 1985). Goodey (1960) reported 30 per cent yield losses in mushroom even when 3 nematodes were initially present at the time of spawning. However, it is considered to be less pathogenic than *A. composticola* and *A. sacchari*.

Saprozoic nematodes

Among saprobes, Rhabditids are most commonly found in mushroom compost, sometimes in very high population. The most commonly present genera are *Rhabditis, Caenorhabditis, Caphalobus, Panagrolaimus, Diplogaster* and *Acrobeloides*. Generally mixed population of some of these genera occur in the beds. As per their feeding apparatus these nematodes are unable to feed upon the mycelium directly and thus are considered to be non-pathogenic. However, many workers have observed that when present in huge numbers, these can severely damage the mycelium and thus affect sporophore yields (Cairns & Thomas, 1950). Though Chandel (1982) demonstrated the mycelial damage by *Rhabditis* sp. under sterilized conditions, so far no direct correlation has been established between these nematodes and mycelial destruction. Possibly these affect the crop indirectly by producing unsuitable conditions for proper mycelial growth. This is merely a speculation and needs further experimentation.

Origin of nematode contaminants

Nematode contamination in a mushroom farm may occur at any time from compositing to cropping. Various components of compost (e.g., wheat straw, chicken manure, horse dung, etc.), casing soil (F Y M, forest soil, spent compost used as casing soil), platform soil, water, used infested trays, handling implements, arthropod vectors (mainly sphaerocerid flies) are the possible sources of nematode contamination in cropping beds (Table 2).

Table 2: Nematodes as contaminant in different components of mushroom cultivation.

Source	No. of samples analysed	No. of samples possessing nematodes			
		A. sacchari	*D. myceliophagus*	*Rhabditis* sp.	Others*
Casing soil					
FYM + Forest soil	15	4	2	10	5
Spent compost	5	2	1	6	2
Compost					
Wheat straw	10	3	-	6	3
Chicken manure	10	3	-	4	1
Horse dung	5	2	-	3	-
Brewer's grain	5	-	-	-	-
Gypsum	5	-	-	-	-
Cotton seed meal	3	-	-	-	-
Used empty trays	10	10	-	10	-
Platform soil	3	1	-	3	1
Water	10	6	-	8	1

* Other nematodes include species of *Tylenchus, Paratylenchus, Meloidog yne, Sienura, Psilenchus, Helicotylenchus* and *Monochids* etc.

(Source : Thapa *et al.*, 1981)

HOST-PARASITE RELATIONSHIP

The mycophagous nematodes thrust their hollow stylet into the hyphal cell. The mycelium bleeds and its sap is sucked upon by to and fro movement of the stylet. After feeding upon one cell the nematode shifts to another cell with the help of moisture film present in the compost. In the process the entire mycelium is fed upon by nematodes.

SYMPTOMS

Due to somewhat specific method of cultivation, mushroom beds are generally not disturbed after casing and this hides the early symptoms of nematode attack. First symptom appears only in the form of highly decreased sporophore yields.

However, if observed in time, the following symptoms appear in the affected trays in succession. Initially whiteness of spawn slowly starts turning brown followed by sparse and patchy appearance of the mycelium. This mycelium is stingy in nature. The compost surface sinks. Alternately medium and low quantity of sporophores (fruiting bodies) are obtained in succession of flushes. There are extremely poor yields in the later flushes finally leading to total failure of sporophore production thus reducing the total yield as well as duration of the crop.

LOSSES

Economic losses due to decline in yields although generally very high are not estimated for most of the myceliophagous nematodes infesting mushroom in India. Extent of losses mainly depend upon the time of nematode infestation, nematode species involved and population of the nematodes. Losses caused by *Aphelenchoides sacchari* are only so far worked out and have been given in Table 3. The mycelial damaging potentials of some other nematode species have been given in Tables 4-7.

Table 3: Estimation of losses due to *Aphelenchoides sacchari* in mushroom beds.

Time of inoculation	Initial inoculum	Yield loss (%)
Spawning	1×10^1	98.20
	1×10^2	100.00
	1×10^3	100.00
	1×10^4	100.00
Casing	1×10^1	90.60
	1×10^2	92.10
	1×10^3	87.30
	1×10^4	100.00

(Source : Sharma *et al.*, 1984)

NEMATODE MANAGEMENT

Nematode management in mushroom is to be dealt in a different way because of its unique growing pattern, its short cropping duration and fresh sporophore consumption immediately after harvest. Some of the work done for nematode control in mushroom and control strategies to be adopted for managing/eliminating nematode in mushrooms have been discussed hereunder.

Table 4: Effect of different population levels of *Aphelenchoides agarici* on mycelial growth of *Agaricus bisporus* and nematode multiplication (inoculation at the time of spawning).

No. of nematodes	% mycelial damage after			No. of nematodes ($x10^4$) after	
	20 days	30 days	45 days	30 days	45 days
10	0.7	50.0	45.0	7.7	26.8
	(1.3)	(2.5)	(8.7)		
100	13.3	40.0	95.0	12.2	56.7
	(3.8)	(6.4)	(9.8)		
500	36.7	78.3	100.0	24.8	82.4
	(6.1)	(8.9)	(10.1)		
1000	46.7	91.7	100.0	26.9	91.9
	(6.9)	(9.6)	(10.1)		
Uninoculated	0	0	0		
	(1.0)	(1.0)	(1.0)		
C.D. (P=0.01)					
For population levels		0.33			$2.81x10^4$
for days		0.23			$0.71x10^4$
for interaction		0.57			$1.62x10^4$

(Source : Khanna & Sharma, 1988a)

Table 5: Effect of various population levels of *Aphelenchoides agarici* on mycelial growth of *Agaricus bisporus* and nematode reproduction (inoculation at time of casing).

No. of nematodes	% mycelial depletion after			No. of nematodes ($x10^3$) after		
	20 days	40 days	60 days	20 days	40 days	60 days
10	6.7	26.7	58.3	6.8	72.0	9.9
	(2.8)	(5.3)	(7.7)			
100	13.3	31.7	81.9	9.9	88.6	12.8
	(3.8)	(5.7)	(9.1)			
500	16.7	40.0	88.3	14.9	102.0	15.0
	(4.2)	(6.4)	(9.4)			

(Continued)

No. of nematodes	% mycelial depletion after			No. of nematodes ($x10^3$) after		
	20 days	40 days	60 days	20 days	40 days	60 days
1000	18.3	50.0	91.7	21.2	131.1	21.7
	(4.4)	(7.1)	(9.6)			
Uninoculated	0	0	0			
C.D. (P=0.01)						
for population levels		0.36			$3.92x10^3$	
for days		0.28			$3.38x10^3$	
for interaction		1.29			$6.83x10^3$	
(Source : Khanna & Sharma 1988a)						

Table 6: Multiplication rate and damaging potential of various *Aphelenchoides* species on *Agaricus bisporus* mycelium.

Nematode species (Initial inoculum = 10 individuals)	Temperature range (°C)	Days after inoculation	Per cent mycelial damage	Nematode population	Multiplication rate
A.agarici	12-25	15	15.00	114.33	11.43
		30	56.67	482.00	48.20
		40	98.00	6349.67	634.96
A.agarici	28±1	5	6.67	61.00	6.10
		15	48.33	272.33	27.23
		25	100.00	8175.00	817.50
A.myceliophagus	12-25	15	25.00	179.00	17.90
		30	86.67	8016.67	801.66
		40	100.00	10450.00	1045.00
A.neocompost-icola	12-25	15	23.33	123.33	12.33
		30	56.67	6773.33	677.33
		40	80.00	9420.00	942.00
A.composticola	12-25	15	33.33	167.67	16.76
		30	80.00	7740.00	774.00
		40	100.00	10953.33	1095.33
(Source : Khanna & Sharma, 1988b)					

Table 7: Mycelial damage and multiplication of *Ditylenchus myceliophagus* ("Indian population") on *Agaricus bisporus* at different inoculum levels.

Initial nematode inoculum	Av. nematode population after		% mycelial damage after 60 days	Rate of nematode multiplication after	
	60 days	90 days		60 days	90 days
10 nematodes	12533.33	52666.66	15	1253	5246
100 nematodes	63233.33	57333.33	40	632	573
1000 nematodes	167833.33	82333.33	70	167	87
C.D. at 5%	42568.79	20603.74			

(Source : Sharma *et al.* 1985)

CHEMICAL CONTROL

The nematicides to be used for nematode control should be non-toxic and/or non-persistent, easy in application and economically cheap. All these factors make the use of chemicals practically near impossible, at least under Indian conditions.

Though many chemicals have been tested and found to be effective against nematodes by many workers in India and abroad (Lengwieler-Rey, 1966; Cayrol & Ritter, 1971; Farkas & Balazs, 1975; McLeod & Khair, 1978; Chabbra & Kaul, 1982), none except Thionazin could be recommended for one or the other reason. Thionazin if applied at 80 ppm concentration is quite effective in controlling the myceliophagous forms without producing any adverse effect on mycelial growth and also does not leave any residue in sporophores. Unfortunately this chemical is not available in India. The other available nematicides tested for nematode control in mushroom trays resulted into residues in sporophores (Kaur *et al.*, 1987).

BIOLOGICAL CONTROL

Bio-control, if exploited, has a great impetus in this crop. The possible biological control agents are predatory nematodes, fungi, bacteria and certain plant products.

Among the biological control agents utility of parasitic fungi has only been exploited to some extent. Nematode-trapping fungus *Arthrobotrys robusta* commercially available as 'Royal-300' strain *Antiopolis* in France has been recommended (Cayrol, 1987). Besides, *Arthrobotrys oligospora* and *A. superba* have also been known to reduce nematode populations and increase the yields (Kermarrec, 1973). In India, recently two fungi *Arthrobotrys irregularis* and *Candelalretta musiformis* isolated from spent compost have shown promising results against myceliophagous nematodes in mushroom.

The bacterium *Bacillus thuringiensis* is found to suppress the population of *Ditylenchus myceliphagus* (Cayrol, 1974). The population of predatory nematodes that feed upon mycophagous forms may prove to be an important control measure, if exploited.

RESISTANCE

Cayrol & Ritter (1971) reported *Agaricus edulis* as resistant to nematode (*Ditylenchus myceliophaus*) attack. He pointed out that the resistance was probably because of hyphal size of the fungus. Cayrol & Combettes (1973) while comparing the susceptibility of *Agaricus sylvicola* with *A. bisporus*, however reported that although *A. sylvicola* was comparatively poor host for *D. myceliophagus*, it had virtually similar hyphal diameter to that of *A. bisporus*. In India, Thapa *et al*. (1987) tested eight edible fungi against *Aphelenchoides sacchari* and *Ditylenchus myceliophagus* and reported that *Pleurotus sajor caju* and *Stropharia rugoso annulata* were resistant to both these nematode species. In case of *Flammulina velutipes*, *Agrocybe aegrita* and *Lentinus edodes*, though the mycelium was not destroyed, it changed its colouration to reddish brown. Among the various strains of *A. bisporus* and *A. bitorquis* tested, K 30 strain of *A. bitorquis* showed resistance to *A. sacchari* only but not to *D. myceliophagus* (Table 8). Resistance to *P. sajor caju* against *A. sacchari, A. agarici* and *A. composticola* has been reported by several workers.

PHYSICAL CONTROL

Use of heat has been the most successful method of nematode control in mushroom cultivation. Its use has been particularly advised during composting and casing process as well as at the end of cropping (Sharma & Seth, 1986). Heat treatment though very effective is costly for small growers and require purposely built mushroom houses for its effective use. Dips of used infested trays and handling implements in boiling water for 1-2 minutes kill the nematodes completely and this practice must be adopted before re-using these things.

Looking into the limitations of control measures in this crop, it is better to prevent the nematode entry into the beds rather than controlling them afterwards in the beds. The following precautions and control measures should therefore be practiced to manage nematodes in mushroom cultivation :

1. Cropping should be done in purposely built mushroom houses with proper ventilation.

Table 8 : Mycelial damage in case of different edible fungi and the population of two myceliophagous nematodes inoculated in spawn run compost bottles.

Edible fungi	Per cent mycelial damage after								Population after 80 days	
	20 days		40 days		60 days		80 days			
	A	D	A	D	A	D	A	D	A	D
Agaricus brunnescens										
S11	40	60	100	80	-	100	-	-	81, 500	1,00,0000
TM7	40	60	100	100	-	-	-	-	24, 500	17,775
649	40	60	100	100	-	-	-	-	1, 54, 600	95, 590
Pleurotus sajorcaju	0	0	0	0	0	0	0	0	0	2,650
Pleurotus eryngii	20	30	40	40	50	60	90	90	13,650	13,000
Agaricus bitorquis										
K_1	10	20	20	60	40	80	100	100	13, 875	42, 000
K_3	10	20	20	60	40	80	100	100	7,425	60,000
K_{30}	6	15	8	30	10	60	30	90	13, 375	2,00,000
Flammulina velutipes	x	x	x	x	x	x	x	x	22,110	21,950
Agrocybe aegerita	x	x	x	x	x	x	x	x	20,200	42,500
Stropbaria rugosoannulata	0	0	0	0	0	0	0	0	14,500	13,320
Lentinus edodes	x	x	x	x	x	x	x	x	40,670	50,310

O = Nil, x = Reddish brown colouration.
A = *Aphelenchoides sacchari* D = *Ditylenchus myceliophagus*

(Source : Thapa *et al.*, 1987)

2. Cleanliness should be maintained inside the surroundings of mushroom farms so that flies, mites and other contaminants are avoided.
3. Composting platform must be cemented to prevent the direct contact of compost with soil.
4. No person or worker should be allowed to enter into the farm without proper disinfestation of his/her hands and feet.

5. Wooden trays and handling tools should be disinfected before use by dipping them in boiling water for 2 minutes.
6. Steaming of compost at 60°C for minimum of two hours should be done before spawning.
7. Casing material should be well sterilized before use.
8. Irrigation water should be nematode-free.
9. Proper cook out of mushroom houses with steam at 70°C for 5-6 hours or at 80°C for 30-60 minutes is essential after the crop is over.
10. In case the nematode menace is not completely checked with above given steps, the dry leaf powder of neem may be mixed with compost at 2-5% conc. This has been found useful not only to check nematode multiplication in mushroom growing trays but it also helps in proper mycelial growth and improved sporophore yields.

REFERENCES

Cairns, J. & Thomas, C.A. (1950). Nematodes causing mushroom crop losses. *Mushr. Sci.* **1**: 89-91.

Cayrol, J.C. (1967). Etude Du Cycle Evolutif D' *Aphelenchoides composticola. Nematologica* **13**: 23-32.

Cayrol, J.C. (1974). Action of *Bacillus thuringiensis* toxins on the mycophagous nematode *Ditylenchus myceliophagus. In* : Simposo. XII Int. De. Nematologia, Socieded Europea de Nematologos, 1-7 Septimbro, Granada, Spain. pp. 21-22.

Cayrol, J.C. (1978). Mise an point d' use methods de lutte biologigue centre les nematodes des champignonniers a laide dun champignon predateur *Arthrobotrys robusta* Souche "Antipolis" (Royal-300). *Butt. Fed. Nat. des. Syndicates Agricoles des cultivaleurs de Champignons* No. **15**: 1445-1451.

Cayrol, J.C. & Combettes, S. (1973). Differences in tolerance of 2 species of mushroom to *Ditylenchus myceliophagus* Goodey, 1958. *Comptes Rendus Hebdomadaires des Seances del Academic Agriculture de Franca* **59** (13): 1022-1036.

Cayrol, J.C. & Ritter, M. (1971). Nematode et nematicides en myciculture. *Mush. Sci*. **8**: 867-878.

Chandel, Y.S. (1982). Studies on the nematodes associated with mushroom and their control. M.Sc. Thesis, Deptt. of EAP, HPKVV, Solan, India.

Chhabra, H.K. & Kaul, V.K. (1982). Occurrence and control of *Aphelenchoides composticola* on mushrooms. *Indian J. Nematol.* **12**(1): 191-192.

Farkas, K. & Balazs, S. (1975). The nematode species of horse manure used for cultivation of the field mushroom and possibility of control. *Phytopathologica Academic Scientitarum Hungaria* **10** (1-2): 153-163.

Franklin, M.T. (1957). *Aphelenchoides composticola n.* sp. and *A. saprophilus n.* sp. from mushroom compost and rotting plant tissues. *Nematologica* **1**: 306-313.

Goodey, J.B. (1958a). *Paraphelenchus myceliophthorus n.* sp. (Nematoda: Aphelenchidae). *Nematologica* **3**(1): 1-5.

Goodey, J.B. (1958b). *Ditylenchus myceliophagus n.* sp. (Nematoda: Tylenchidae). *Nematologica* **3**(2): 91-96.

Goodey, J.B. (1960). Observations on the effects of the parasitic nematodes *Ditylenchus myceliophagus, Aphelenchoides composticola* and *Paraphelenchus myceliophthorus* on the growth and cropping of mushrooms. *Ann. Appl. Biol.* **48**: 655-663.

Hooper, D.J. (1958a). Mushroom as host of *Aphelenchoides saprophilus* and *Aphelenchus avenae. In: Report Rothamsted Exp. Stn.* for 1957. p. 125.

Hooper, D.J. (1958b). *Aphelenchoides dactylocercus* n. sp. and *A. sacchari* (Nematoda : Aphelenchoides). *Nematologica* **3**: 229-235.

Hooper, D.J. (1962). Effects of a nematode on the growth of mushroom mycelium. *Nature* No. **193**: 496-497.

Kaur, D.J.; Thapa, C.D.; Nath, A. & Sharma, N.K. (1987). Evaluation of some nematicides against *Aphelenchoides sacchari. Indian J. Agric. Sci.* **57**(11): 867-868.

Kermarrec, A. (1973). Investigations on the enemies of cultivated mushroom. II. Preliminary results on the utilisation of predatory fungi in biological control of nematodes of cultivated composts. *Annals De Zoologic. Ecologic. Animale* **5**(30): 645-480.

Khanna, A.S. & Sharma, N.K. (1988a). Effect of population levels and time of infestation of *Aphelenchoides agarici* on the mycelial growth of *Agaricus bisporus. Nematol. Medit.* **16**: 125-127.

Khanna, A.S. & Sharma, N.K. (1988b). Pathogenicity of various *Aphelencoides* species on *Agaricus bisporus. Indian Phytopath.* **41**(3): 472-473.

Lambert, E.B.; Steiner, G. & Drechsler, C. (1949). The 'Cephalothecium disease' of cultivated mushrooms caused by a nematode (*Ditylenchus* sp.) evidenced by surface development of predaceous fungi. *Pl. Dis. Reptr.* **33**(6): 252-253.

Lenaweiler-Rey, V. (1966). The control of nematodes in mushroom production using tray system. *Z. Pflkrankh.* **73** (1-2): 46-56.

McLeod, R.W. (1967). *Aphelenchoides bicaudatus* a parasite of cultivated mushroom. *Nature* No. **214**: 1164-1165.

McLeod, R.W. (1968). The effects of *Aphelenchoides saprophilus, A. coffeae, Aphelenchus avanae* and *Panagrolaimus* sp. on the cropping of cultivated mushroom. *Nematologica* **14**: 573-576.

McLeod, R.W. & Khair, G.T. (1978). Control of *Aphelenchoides composticola* in mushroom compost with nematicides. *Ann. Apl. Biol.* **88**: (1): 81-88.

Newstead, R. (1906). A large crop of mushrooms destroyed. *Quart. J. Inst. Comml. Res. Trop.* **1**(1): 22-23.

Paesler, F. (1957). Beschreibung Einiger nematoden Aus Champignon beeten. *Nematologica* **2**: 314-328.

Seth, A. (1984). Studies on the incidence, pathogenicity, biology and control of *Aphelenchoides* species associated with mushroom. Ph. D. Thesis, Deptt. of Entomology and Apiculture, H.P.K.V.V., Solan (H.P.), India.

Seth, A. & Sharma, N.K. (1986). Five new species of genus *Aphelenchoides* (Nematoda : Aphelenchida) infesting mushroom, in northern India. *Indian J. Nematol.* **16** (2): 205-215.

Sharma, N.K. & Seth, A. (1986). Nematode problems in mushroom, losses, symptomatology and management, *In: Plant parasitic nematodes of India* (Eds.—Gopal Swarup & D.R. Dasgupta). pp. 384-399.

Sharma, N.K.; Thapa, C.D. & Kaur, D.J. (1984). Estimation of loss in mushroom yield due to myceliophagous nematode *Aphelenchoides sacchari*. *Indian J. Nematol.* **14** (2): 121-124.

Sharma, N.K.; Thapa, C.D. & Kaur, D.J. (1985). Pathogenicity and identity of *Ditylenchus myceliophagus* (Indian population) infesting *Agaricus bisporus* (Lange) Singh in India. *Indian J. Nematol.* **15**(2): 233-234.

Sharma, N.K.; Thapa, C.D. & Nath, A. (1981). Pathogenicity and identity of myceliophagous nematode infesting *Agaricus bisporus* (Lange) Singh in Himachal Pradesh (India). *Indian J. Nematol.* **11**(2): 230-231.

Thapa, C.D.; Jandaik, C.L. & Sharma, N.K. (1987). Comparative resistance in some cultivated edible fungi against two mushroom nematodes occurring in India. *Proceeding of the International Conference on Science and Cultivation Technology of Edible Fungi*. pp. 169-172.

Thapa, C.D.; Sharma, N.K. & Nath, A. (1981). Mushroom nematodes and their source of infestation in mushroom beds in India. *Mushr. Sci.* **11**(1): 443-448.

Chapter 15

Nematode Diseases of Cotton and their Control

P.K. Sakhuja, Inderjit Singh & M. Sarwat Sultan
Department of Plant Pathology, Punjab Agricultural University, Ludhiana-141 004, India.

COTTON lint is world's most important plant fibre, which plays key role in economic, social and political affairs. Cotton crop is grown on about 35 million hectares in nearly 80 countries. India ranks first in terms of acreage and fourth in production in the world. Cotton fibre is obtained mainly from four species of *Gossypium*, though more than 30 of them are known. Two of these are allotetraploid (*G. hirsutum, G. barbadense*) and two diploid (*G. arboreum, G. herbaceum*). American upland cotton, *G. hirsutum*, which was restricted till the last century now accounts for over 80% of the global cotton production.

A large number of insects and diseases attack cotton throughout the world. Out of several pathogens responsible for diseases on cotton, plant-parasitic nematodes are unique. They not only cause diseases by themselves, but also aggravate the diseases caused by fungi. Nematodes attack plant roots, survive in soil and cause recurring losses year after year. Cotton Disease Council estimated yield reduction of 1.63, 2.04 and 2.04% during 1966, 1971 and 1975 in U.S.A. due to nematodes, while the total yield reduction due to all the diseases was 15.28, 15.2 and 13.6% respectively in these years (Anon., 1967, 1972b, 1976). Similarly a loss of 17% in cotton has been attributed to root-knot nematode alone in Brazil (Sasser, 1979).

Several plant-parasitic nematode species belonging to different genera have been encountered in the rhizosphere of cotton (Goodey *et al.*, 1965; Alam *et al.*, 1978, 1979; Sitaramaiah, 1984), but the species of root-knot, reniform, lance and sting nematodes which cause significant losses in cotton production are discussed in detail here. Little information is available on the role and control of other species associated with cotton, which are chiefly ectoparasitic in nature.

ROOT-KNOT NEMATODE

It is the major nematode pest of cotton prevalent in most of the cotton producing areas either alone or as part of wilt complex. Atkinson (1889) was first to observe the occurrence of this nematode on cotton and referred it as *Heterodera radicicola*. Out of more than 60 species of root-knot nematodes, *Meloidogyne* spp. described, only two species, viz. *M. incognita* (race 3 and 4) and *M. acronea* attack cotton. The former species is prevalent on cotton in Brazil, China, Egypt, India, U.S.A., USSR, etc. In India, it is a problem on cotton in Punjab, Haryana, and Gujarat (Bajaj & Bhatti, 1982; Sakhuja *et al.*, 1986; Thakar *et al.* 1986). *M. acronea* is restricted to southern Africa (Bridge *et al.*, 1976).

The root-knot nematode has been found to cause a quarter million bale loss annually in U.S.A. (Crawford, 1976) and 17% in Brazil (Sasser, 1979). Orr & Robinson (1984) on the basis of 80 field trials conducted over 16 years found an increase of 25.7% in cotton yield by the control of *M. incognita*. Regression analysis has revealed that this nematode reduces the lint yield by 35.7 kg/ha for every unit increase in galling index (Minton & Meredith, 1987).

Root-knot nematode forms galls on tap as well as lateral roots (Fig. 1). However, galls on cotton are not as big and numerous as on other susceptible crops. Infestation results in uneven, pale, stunted and sick crop giving the look of deficiency symptoms. High population density of the nematode at sowing can kill the plants at seedling stage. Mature plants exhibit temporary wilting in afternoons. It thrives well in sandy soils but is rarely a problem in clay soils.

Root infested with *M. macronea* produce extensive rootlets at the point of infestation and females, when mature, almost bulge out of the root. *M. acronea* prefers alluvial soils but is not found in vertisol in the same cotton field (Page, 1984).

LIFE HISTORY

Second stage larvae of root-knot nematode penetrate the roots, usually in the first 2 cm of root tip. They migrate to stele injuring cortex inter- and intra-cellularly. The larvae become sedentary, feed and damage the vascular bundles by forming giant cells, which impair the translocation process. The nematode completes one life cycle in 33-38 days and produces 5-6 overlapping generations in crop season. Veech & Starr (1986) found no difference between race 3 and 4 of *M. incognita* in terms of development, reproduction or aggressiveness on cotton. Inoculation of 100 or more larvae of *M. incognita* per plant in *G. arboreum* (cv. G-27) and 500 or more in *G. hirsutum* (cv. H-777) reduces the plant yield significantly (Rai & Jain, 1987).

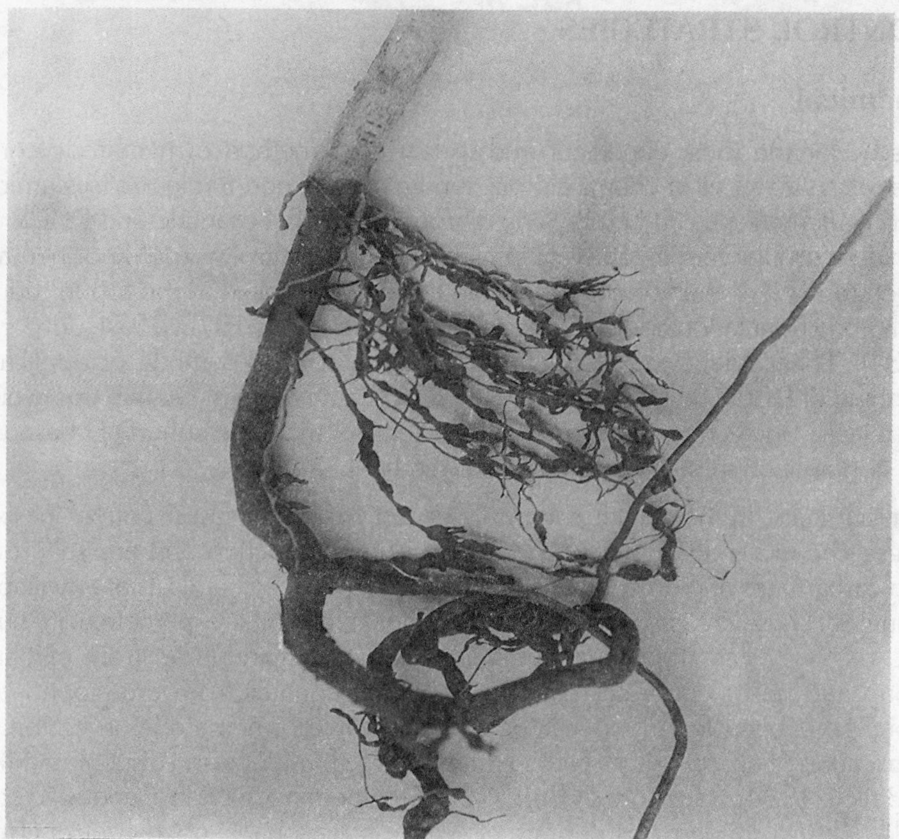

Fig. 1. Cotton root infected with the root-knot nematode, *Meloidogyne incognita.*

DISEASE COMPLEXES

Root-knot nematode interacts synergistically with microorganisms like *Fusarium oxysporum f. vasinfectum, Rhizoctonia solani, Verticillium albo-atrum, Pythium debaryanum*, etc. inhabiting the same ecological niche. The involvement of *M. incognita* with *Fusarium*-wilt of cotton was noticed 100 years ago and has been complicating the breeding programme. Atkinson (1896). concluded that *Fusarium* could not enter cotton roots unaided and Garber *et al.* (1979) found that as many as 50 larvae of *M. incognita* are sufficient to induce wilt at even very low inoculum level of *Fusarium* sp.

M. incognita advances the wilt symptoms, aggravates them and breaks the resistance in varieties to *Fusarium*-wilt, but has limited role in inoculating *Fusarium* sp. (Perry, 1961; Karaev *et al.*, 1980). The role of *M. incognita* and *R. solani* in seedling disease has been well elucidated (Reynolds & Hanson, 1957; Brodie & Cooper, 1964; Carter, 1975 a,b).

CONTROL STRATEGIES

Chemical

It is by far the most expedient and widely used method of nematode control. Basically, two types of chemicals, viz. fumigants and non-fumigants (systemic) are used as nematicides. D-D, ethylene dibromide, methyl bromide and DBCP from fumigant group were extensively explored for the control of root-knot nematode on cotton. All of them proved effective but DBCP ranked at the top in reducing nematode population and in increasing cotton yield (Orr, 1973, 1974; Jorgenson, 1984). Their chief demerits were high cost and tedious mode of application. Banning of DBCP in 1981 and EDB in 1983 came as a last blow to this group of chemicals. Shell chemical company has stopped manufacturing D-D because of the contamination of ground water caused by it.

Non-fumigant (non-volatile) chemicals began to be tested on cotton in sixties. These chemicals are systemic, act through contact action and are subgrouped into carbamates and organophosphates. Most of these chemicals are available as granules. These are easy to apply, water soluble, exhibit low phytotoxicity but are not, of course, as effective as fumigants. Aldicarb, carbofuran from carbamate group and nellite, phorate, phenamiphos and counter from organophosphate group have been tested extensively against root-knot nematode on cotton. Soil application of aldicarb gives better results over fenamiphos and disulfoton (Minton & Bailey, 1984; Ratchford & Burris, 1985) in nematode control and seed cotton yield. Nellite, counter and phorate also gave a reasonable good control (Turner, 1963; Rodriguez-Kabana & King, 1976; Farlow *et al.*, 1988). In China, carbofuran @ 5 kg a.i./ha has been found effective for the control of race 4 of *M. incognita*. Sakhuja *et al.* (1987) observed excellent control of root-knot nematode and 41% increase in yield by applying carbofuran at 1 kg a.i./ha at sowing and followed by additional 2 kg a.i./ha 50 days thereafter. Systemic nematicides give better results when applied in seed furrow as compared to band application in cotton.

Seed treatment of cotton with systemic nematicides has been tried by some workers and has been found to confer protection from root-knot nematode for 3-4 weeks, which gives a healthy start to the plant (Tayar, 1979). Besides protecting plants from nematodes, systemic nematicides improve the quality of fibre also. Aldicarb improves fibre strength and maturity and carbofuran improves fibre length and maturity when applied @ 2 kg a.i./ha (Lordello & Sabino, 1985). Multi-advantages of systemic compounds have led to their widescale use these days. Aldicarb is being used on 78% cotton fields in areas surveyed in U.S.A. in contrast to only 2% fields exposed to 1,3-D as soil fumigant (Hadden & Crawford, 1991).

Several other compounds of diverse origin have been tested for the control of nematodes. Avermectins, the macrocyclic lactones derived from the mycelia of *Streptomyces avermitilis* appear to have definite potential. Avermectin B_2a is 10-30 times more potent than several organophosphate and carbamate nematicides against *M. incognita* on tobacco and *Tylenchulus semipenetrans* on citrus (Stretton *et al.*, 1987). However, they have not been tried in case of cotton so far.

Cultural

All the cultural practices adopted on the farm, exert some influence on the biotic community of soil. Careful manipulation of these practices can be utilized to reduce the population density of root-knot nematode. The foremost practice, which is quite effective and is being followed by the farmers is crop rotation. Inclusion of a non-host crop or resistant variety of even susceptible crop in rotation with cotton, reduce the nematode population for the next cotton crop and can also make it more amenable to chemical control in future. Bengal bean, wheat, barley and alfalfa reduce the root-knot nematode population considerably in cotton fields. Cultivation of peanut for one or two years before raising cotton, reduces the population of *M. incognita* to negligible levels (Kirkpatrick & Sasser, 1984). Similarly alternating susceptible with resistant cotton (var. Auburn-623) also helps in checking the nematode and gives good returns. Variety Auburn-623 is, however, susceptible to *M. acronea* and hence should not be sown in fields having infestation of this species.

Timely removal of weeds, serving as collateral hosts, from the infested fields also helps substantially in reducing nematode population. Exposing the soil to sun by deep ploughing during summer before sowing of cotton also plays a definite role in population reduction specially at places which experience temperatures above 40°C. The efficacy of this method can be greatly enhanced by soil solarization with the help of LDPE and HMHDPE films and the results are comparable to nematicidal treatment as has been observed in tobacco nursery (Patel *et al.*, 1991).

Resistance

Moderate resistance in varieties Auburn-56 and Clevewilt 6-3-5 of *G. hirsutum* and Darwinii of *G. barbedense* was spotted nearly 30 years ago. Subsequently resistant stocks were identified at many places and sustained breeding programmes resulted in the release of resistant varieties. Multiple disease resistant line HYC-76-59 had resistance to root-knot nematode (Sappenfield, 1981). Resistance to race 2 of *M. incognita* was observed in cultivars Giza-68, Giga-70, Giza-75 of *G. barbadense* and Acala-4-42 and Acala-67A of *G. hirsutum* in Egypt (Ibrahim *et al.*, 1982). Highly resistant varieties Auburn-623 RNR, N-6072, N-8577 and

634 RNR were identified out of which the last one was expected to be much stable over time and space (Hyer *et al.*, 1979, Shepherd, 1982; 1983). However, the yield of these was not at par with susceptible varieties. Lines like Aub-82, RNR Ne, Aub.-244 RNR and Aub.-299 RNR have been identified which allow 45-50 times lesser egg production of *M. incognita* and root galling than Auburn-56 and have lint yield and quality traits comparable to Stoneville-213 (Shepherd, 1987). At present resistant varieties Auburn-634 and N-6072 are in use in U.S.A. (Klump & Thomas, 1987). Some resistant stock have been identified in other countries also, e.g. GDH-22, G-2637, G-Cot in India and Liaoyang-4, Chuan-7327, Jihe-328, Zhekang-110 in China (Thakar *et al.*, 1986; Rong *et al.*, 1987).

Some studies have been carried out to understand the mechanism of resistance in cotton. Veech (1978) suspected the involvement of infection induced synthesis of methoxy-substituted terpenoid aldehyde in resistance mechanisms. Terpenoid aldehyde accumulation in the endodermis and outer stele was found to be more in resistant varieties. Resistance has been found to be characterized by root necrosis, retarded gall development and failure of majority of nematodes to reach maturity. Larvae enter resistant varieties as many as in susceptible ones, but their development is sharply retarded. Breeding varieties having resistance to root-knot nematode and *Fusarium*-wilt complex is very important aspect and has been well explained by Shepherd (1986) and Shepherd & Kappelman (1986). Varieties Bayon, N-6072 and breeding lines CABCR-1-81, CAHUS-2-81 and LEBOCAS-3-81 have been reported to have field resistance against this complex, though they are not truly resistant to root-knot nematode (Hyer *et al.*, 1979; Starr & Veech, 1986).

Biocontrol

Root-knot nematode has been successfully controlled by certain specific isolates of *Paecilomyces lilacinus* and *Verticillium chlamydosporium* in laboratory and on some crops in the field and is being explored on cotton also (Jatala, 1985; Morgan-Jones & Rodriguez-Kababa, 1987). Davide & Batino (1985) reported control of *M. incognita* by treating or soaking cotton seeds in spore and mycelial suspension of *P. lilacinus* or *Gliocladium roseum*, the former fungus was effective as granules of phenamiphos @ 0.5 kg a.i./ha

Some reports suggest the usefulness of endomycorrhizal fungi in reducing the damage caused by root-knot nematode. Roncadori & Hussey (1978) observed that *Gigaspora margarita* nullified the stunting effect of nematode. *Glomus interradices* has also been found to reduce nematode population and the yield loss caused by it on cotton (Smith *et al.*, 1986a, b). Substantial work has to be done on allied factors, before these results are utilized in the field for the control of root-knot nematode.

RENIFORM NEMATODE

Reniform nematode, *Rotylenchulus reniformis* was first observed on the roots of cotton in Baton Rouge, Louisiana (Smith & Taylor, 1941). It is primarily an inhabitant of tropical and subtropical areas and has been found to attack different plants in more than 38 countries. More than 200 plant species have been recorded as hosts of *R. reniformis*. This nematode has a wide distribution in India also. Another species of Reniform nematode *R. parvus* has also been found attacking cotton in Africa (Starr & Page, 1990).

R. reniformis has been found to be capable of reducing cotton yields by 40-60% (Birchfield & Jones, 1961). The relationship between logarithmic densities of *R. reniformis* and damage has been found to be governed by J.W. Seinhorst's equation $Y = m+(1=m)Z^{P-T}$ (Varaprasad *et al.*, 1987).

Injury to cotton caused by *R. reniformis* becomes evident with reduction in emergence of seedlings. At 3rd or 4th leaf stage, seedlings cease rapid growth and become light green to pale in colour. Severely infested plants may die while others remain stunted giving uneven crop outlook. Roots of such plants are smaller, fewer and sometimes show browning. Affected plants bear lesser and smaller bolls. Roots of affected plants when examined under stereoscopic microscope show several semiendoparasitic female nematodes, which get darkly stained.

LIFE HISTORY

Two races (A & B) of *R. reniformis* have been reported from India (Dasgupta & Seshadri, 1971), out of which only race A parasitizes cotton. The first stage larva moults in the egg and comes out as second stage larva. The second, third and fourth stages do not feed. After 4th moult, the immature females penetrate the roots and become reniform (kidney like) in shape in 5 days. Males also penetrate but do not feed. Birchfield (1962) studied host-parasite relationship between *R. reniformis* and cotton. Entry of females resulted in destruction of epidermal cells and browning as well as necrosis of other surrounding cells. Cortical parenchymatous cells were destroyed as the nematode extended its head through the endodermis and pericycle, and began to feed in phloem. Necrosis of phloem and collapse of parenchyma ultimately resulted in severe root pruning. It took 17-23 days for the completion of one life cycle on cotton.

High population density (6000 to 8000 larvae per 250 cc soil) occur in cotton rhizosphere (Nath *et al.*, 1969). The per cent increase in the population of *R. reniformis* on cotton has been found to be 49 in comparison with 136, 84, 73, 40 and 6.7 on tomato, castor, papaya, okra and *Amaranthus* respectively (Sivakumar & Seshadri, 1971). The juvenile females can survive for 6 months in desiccated soil and 7 months in moist soil; and a high percentage of them remain infective even after 4 months of starvation (Sivakumar & Seshadri, 1976).

DISEASE COMPLEXES

Reniform nematode interacts synergistically with important fungal species prevalent in cotton fields. The involvement of *R. reniformis* in *Fusarium*-wilt of cotton was proved by the systematic studies of Neal (1954). He conclusively proved that high incidence of *Fusarium*-wilt was dependent upon the presence of heavy reniform nematode population. Tchatchoua & Sikora (1983) demonstrated the synergistic interaction between *R. reniformis* and *Verticillium dahliae* in growth reduction as well as per cent wilt incidence. They noticed that in some cases *V. dahliae* favoured nematode population development also. In India, Krishna Prasad & Padaganur (1980) observed the association of this nematode with *Verticillium*-wilt on four varieties of cotton. Varieties resistant to *V. dahliae* lose their capacity to resist in the presence of *R. reniformis*.

CONTROL STRATEGIES

Chemical

For mitigating the damage caused by *R. reniformis* to cotton, pesticides have been tried by several workers. As with other nematode problems on crops, fumigant nematicides were the first to be explored in combating this nematode pest. Dichloropropene type fumigants gave significant increase in yield of cotton in nematode infested soils. In untreated control plots the population peak reached in 2.5 months whereas in treated plots it took 5.1 months (Thomas *et al.*, 1970).

D-D and EDB at 280 l and 320 kg/ha respectively, DBCP at 38 l/ha before sowing followed by 18 l/ha two months thereafter reduced population of *R. reniformis* and increased seed cotton yield (Vadivelu *et al.*, 1974). Combination of DBCP with PCNB performed better than DBCP alone (Birchfield & Pinckard, 1964).

The many fold advantages of systemic nematicides over fumigants promoted their widespread testing in field trials. Nellite, an organophosphatic compound @ 1 lb/acre was found to reduce nematode population by 42% and increase the yield by 30% (Pinckard, 1964). In Egypt, Temik was found to be most effective, followed by Thimet+Nemaphos and Nemaphos alone in reducing peak nematode density in July, average reproduction rate and cumulative reduction in numbers. Carbofuran at 2 or 4 kg and aldoxycarb at 4 kg a.i./ha also proved quite effective in controlling *R. reniformis* in Brazil (Lordello *et al.*, 1984). A combination of 1-3D + Aldicarb reduced nematode population and increased plant height and yield over other nematicides and control (Lawrence *et al.*, 1990). In spite of effectiveness and benefits derived, the chemical control of *R. reniformis* has not become so popular.

Cultural

Several cultural means of control have also been tried by workers at different places. Castillo (1985) found the application of *Gliricidia sepium* (Kakawate) leaves and coir dust @ 2.5 and 5 tonnes/ha effective against this nematode and the effect was found to be equivalent to that obtained with 7.5 kg a.i./ha of Phenamiphos. *Chloris gayana* (rhodes grass), *Tagetes patula* (marigold) and *Crotolaria juncea* (sunhemp) are not only non-hosts of *R. reniformis* but also reduce its population more in comparison to fallow. Root exudates of marigold and margosa have been reported to be toxic to *R. reniformis*, *M. incognita* and many other nematodes (Alam *et al.*, 1975). Toxic thiophenes have been recovered from marigold root extract and from undisturbed rhizosphere of marigold, which are directly toxic to this nematode (Caswell *et al.*, 1991). Studies have revealed that cultivation of resistant soybean (cv. Pickett-71) for two years in infested fields gave good control and was rather better than some nematicides (Williams *et al.*, 1983).

Resistance

Studies on the resistance in wild species of *Gossypium* have yielded useful information. *G. longicalyx* has been found to be immune, while *G. somalense, G. stocksii* approached immunity and *G. anomalum, G. raimondi, G. thurberi* and *G. davidsonii* also had sufficient resistance to *R. reniformis* (Muralidharan & Sivakumar, 1975; Yik & Birchfield, 1984). Amongst the cultivated species, cv. K-7, K-8, Pl-417891, Pl-417895 and CB-3839 of *G. arboreum*, T-10 of *G. barbadense*, Pl-408775 of *G. herbaceum* and T-874, T-893 and T-903 of *G. hirsutum* possess resistance (Muralidharan & Shivkumar, 1975; Yik & Birchfield, 1984). *G. hirsutum* lines La RN 910, Auburn-612 NR and Mo 19 RNR have high degree of resistance. Their resistance has been found to be controlled by ≥ 2 pairs of genes (Mohammad & Jones, 1990) and holds promise for the future.

LANCE NEMATODES

Ten species belonging to *Hoplolaimus* group occur in the rhizosphere of cotton. *Basirolaimus seinhorsti, B. indicus, B. columbus* and *B. aegypti* are most widespread and pathogenic. Columbia lance nematode, *B. columbus* attacks cotton in southern U.S.A. and Egypt, while *B. seinhorsti* in Africa, Asia and South America. In Tanzania, *Scutellonema bradys* is a predominant nematode on cotton. In India, reports suggest the infestation by *B. indicus* and *B. seinhorsti*.

Losses caused by lance nematode have not been estimated. However, studies have been conducted on the damage caused by different inocula to cotton plants in pots. Three hundred females of *B. columbus* per plant can cause an economic loss to seed cotton yield (Koura & Osman, 1984).

Occurrence of high population of lance nematode in soil causes stunting and irregular growth of cotton. Nematode infested fields, when experience low moisture conditions results in yellowing and severe defoliation of the crop. Nematodes feed both ecto- and endo-parasitically. Feeding of lance nematode has been observed on cortex and vascular tissues, and phloem appears to be most preferred (Krusberg & Sasser, 1956).

Nematodes belonging to this group also interact with phyto-pathogenic fungi occurring in cotton fields *B. seinhorsti* interacts synergistically with *F. oxysporum* f. *vasinfectum* and *V. dahliae* in advancing wilt (Shanmugam *et al.*, 1977).

CONTROL STRATEGIES

Chemical

Basirolaimus columbus can be efficiently controlled by applying methyl bromide or 1,3-D before sowing of cotton (Noe, 1990; Schmitt & Bailey 1990). Damage due to this species can also be reduced by applying aldicarb or phenamiphos, which increase the yield by 10-25% (Mueller & Sullivan, 1988). Similar results have also been obtained for the control of *B. aegypti* by the application of aldicarb (Koura & Naser, 1984). For the control of *B. seinhorsti*, fensulfothion is better even over D-D and DBCP, but increase in yield in nematode infested plots is higher with DBCP (Shanmugam *et al.*, 1976).

Other methods

Little information is available on the occurrence of resistance in cotton germplasm against *Hoplolaimus/Basirolaimus* spp. Tolerance in varieties Suvin and Giza-75 to *B. seinhorsti* and *B. aegypti* respectively has been observed (Rajaram, 1983; Koura, 1984). In spite of the efforts, neither resistance nor tolerance has been identified against *B. columbus*. Control of lance nematodes can be best achieved by exploitation of cultural methods. Subsoiling to a depth of 35 cm under the planting rows for 3 consecutive years increased the seed cotton yield by 50-200% (Hussey, 1977).

STING NEMATODE

The sting nematode, *Belonolaimus longicaudatus* is a highly devastating pest of cotton. It occurs in whole of the coastal plain of southeastern U.S.A. High population of this nematode can kill quite a large percentage of plants. Concomitant occurrence of *B. longicaudatus* and *F. oxysporum* f. *vasinfectum* can result in even total failure of cotton crop. This nematode is restricted to soils with more than 85% sand. Extreme low temperature of winter reduce the population drastically (Starr & Page, 1990). Occurrence of 3 races in *B. longicaudatus* has been demonstrated (Abu-Garbieh & Perry, 1970).

Sting nematode is one of the longest nematode and feeds ectoparasitically on epidermal and cortical cells and sucks their contents. It causes stunting, chlorosis, wilting and premature death of cotton plants. The nematode prefers feeding on root tips. New roots are poorly developed and exhibit dark, sunken lesions, which may girdle the whole root and break it off.

CONTROL STRATEGIES

Chemical

Fumigant nematicides exhibit a high toxicity towards *B. longicaudatus*. Nemagon (DBCP) and Dowfume 40 are better than other fumigants (Good & Parkam, 1957). Nemagon is so effective that plots treated with it hardly have any nematode even at the end of crop. Integrated control of sting nematode, wilt and weeds is obtained by applying 8:2:1 mixture of aldicarb, PCNB and trifluralin. However, aldicarb when replaced by fensulfothion renders the mixture ineffective (Brodie & Hauser, 1970).

Cultural

Sting nematode has a wide host range, which includes cereals, vegetables, peanut and soybean. Tobacco is not a host of this nematode and has been successfully used in rotation with cotton for reducing the nematode population (Holdeman & Graham, 1952). Likewise *Crotolaria spectabilis* and *Tagetes minuta* have given good results when used as cover crops.

LESION NEMATODE

Pratylenchus steineri, *P. brachyurus, P. coffeae* and *P. sudanenesis* are associated with cotton, but their distribution is restricted to certain areas only. Reproduction of *P. zeae* was investigated on three species of cotton and was found to be poor on all of them (Sudheer & Sivakumar, 1983). Similarly, involvement of *P. brachyurus* in cotton stunt has also not been substantiated further (Bird *et al.*, 1971).

Use of DBCP against *P. brachyurus* resulted in 15% increase in yield (Anon. 1972a). 2, 3, 4, 5- tetrachloronitrobenzene and Dowfume-40 which are effective against root-knot and sting nematodes respectively, proved quite less effective against *P. brachyurus* (Graham & Holdeman, 1953). Control of lesion nematodes on cotton appears to be difficult in comparison to other nematodes.

REFERENCES

Abu-Garbieh, W.I. & Perry, V.G. (1970). Host differences among Florida populations of *Belonolaimus longicaudatus*. *J. Nematol.* **2**: 209-216.

Alam, M.M.; Masood, A. & Husain, S.I. (1975). Effect of margosa and marigold root-exudates on mortality and larval hatch of certain nematodes. *Indian J. Exp. Biol.* **13**: 412-414.

Alam, M.M.; Saxena, S.K. & Khan, A.M. (1978). Suitability of crops to certain ectoparasitic nematodes. *Acta Bot. Indica* **6** (Suppl.): 205-208.

Alam, M.M.; Saxena, S.K. & Khan, A.M. (1979). Suitability of crops to the reniform, stunt and root-knot nematodes. *In* : *Recent Researches in Plant Sciences* (Ed.-S.S. Bir). Kalyani Publishers, New Delhi. pp. 630-634.

Anon. (1967). Reduction in yield of cotton caused by parasitic diseases in 1966. Cotton Disease Council. *Plant Dis. Reptr.* **51**: 370.

Anon. (1972a). Rhodesia; Secretary for Agriculture. Report for Agriculture for the period Ist October, 1970 to 30th September, 1971, Salisburg, Rhodesia.

Anon. (1972b). Reduction in yield of cotton caused by parasitic diseases in 1971. Cotton Disease Council. *Plant Dis. Reptr.* **56**: 328.

Anon. (1976). Reduction in yield of cotton caused by parasitic diseases in 1975. Cotton Diseases Council. *Plant Dis. Reptr.* **60**: 862.

Atkinson, G.F. (1889). Albama Polytech. Inst. *Agricultural Experimental Station Bulletin* 9.

Atkinson, G.F. (1896). Diseases of cotton, USDA Office. *Experimental Station Bulletin* 33.

Bajaj, H.K. & Bhatti, D.S. (1982). Nematodes associated with cotton in Haryana and Punjab with description of two new leptonchid nematodes. *Indian J. Nematol.* **12**: 6-13.

Birchfield, W. (1962). Host-parasite relationship of *Rotylenchulus reniformis* on *Gossypium hirsutum*. *Phytopathology* **52**: 862-865.

Birchfield, W. & Jones, J.E. (1961). Distribution of reniform nematode in relation to crop failure of cotton in Louisiana. *Plant Dis. Reptr.* **45**: 671-673.

Birchfield, W. & Pinckard, J.A. (1964). New combination of nematicides for control of reniform nematode of cotton. *Phytopathology* **54**: 393-394.

Bird, G.W.; McCarter, S.M. & Roncadori. R.W. (1971). Role of nematodes and soil borne fungi in cotton stunt. *Indian J. Nematol.* **3**: 17-22.

Bridge, J.; Jones, E. & Page, S.L.J. (1976). *Meloidogyne acronea* associated with reduced growth of cotton in Malawi. *Plant Dis. Reptr.* **60**: 5-7.

Brodie, B.B. & Cooper, W.E. (1964). Relation of parasitic nematodes to post emergence damping off of cotton. *Phytopathology* **54**: 1023-1027.

Brodie, B.B. & Hauser, E.W. (1970). Multiple pest control in cotton with mixtures of selective pesticides. *Phytopathology* **60**: 1609-1612.

Carter, W.W. (1975a). Effects of soil temperature and inoculum levels of *Meloidogyne incognita* and *Rhizoctonia solani* on seedling disease of cotton. *J. Nematol.* **7**: 229-233.

Carter, W.W. (1975b). Effects of soil texture on the interaction between *Rhizoctonia solani* and *Meloidogyne incognita* on cotton seedlings. *J. Nematol.* **7**: 234-236.

Castillo, M.B. (1985). Some studies on the use of organic soil amendments for nematode control. Philippine Agriculturist **68**: 76-93.

Caswell, E.P.; De Frank, J.; Apt, W.J. & Tang, C.S. (1991). Influence of non-host plants on population decline of *Rotylenchulus reniformis. J. Nematol.* **23**: 91-98.

Crawford, J.L. (1976). Cotton disease loss estimate committee report. *Plant Dis. Reptr.* **60**: 862.

Dasgupta, D.R. & Seshadri, A.R. (1971). Races of the reniform nematode, *Rotylenchulus reniformis* Linford & Oliveira 1940. *Indian J. Nematol.* **1**: 21-24.

Davide, R.G. & Batino, E. (1985). Biological control of root-knot nematodes on cotton through the use of fungi *Paecilomyces lilacinus* (Thom.) Samson and *Gliocladium roseum* Bainier as seed treatment. *Philippine Agriculturist* **68**: 159-167.

Farlow, R.A.; Zummo, L.A. & Zummo, G.R. (1988). Counter systemic insecticide-nematicide use in cotton. *Proc. Beltwide cotton Prod. Conf. New Orleans* (3-8 Jan.)

Garber, R.H.; Jorgenson, E.C.; Smith, S. & Hyer, A.H. (1979). Interaction of population levels of *Fusarium oxysporum* f. sp. *vasinfectum* and *Meloidogyne incognita* on cotton. *J. Nematol.* **11**: 133-137.

Good, J.M. & Parkam, S.A. (1957). Control of sting nematode on upland cotton by soil fumigation. *Phytopathology* **47**: 312.

Goodey, J.B.; Franklin, M.T. & Hooper, D.J. (1965). *The Nematode Parasites of Plants Catalogued Under Their Hosts.* Commonwealth Agriculture Bureaux, Farnham Royal, Bucks, England, pp. iv+214.

Hadden, J.F. & Crawford, J.L. (1991). Initial survey of plant parasitic nematode distribution in Georgia cotton fields. *J. Nematol.* **23**: 530-531.

Holdeman, Q.L. & Graham, T.W. (1952). The association of the sting nematode with some persistent cotton wilt spots in north eastern South Carolina. *Phytopathology* **42**: 283-284.

Hussey, R.S. (1977). Effect of sub soiling and nematicides on *Hoplolaimus columbus* and cotton yield. *J. Nematol.* **9**: 83-86.

Hyer, A.H.; Jorgenson, E.C.; Garber, R.H. & Smith, S. (1979). Resistance to root-knot nematode in control of root-knot nematode—*Fusarium* wilt disease complex in cotton. *Crop Science* **19**: 898-901.

Ibrahim, I.K.A.; Rezk, M.A. & Khalil, H.A.A. (1982). Reaction of fifteen malvaceous plant cultivars to root-knot nematode, *Meloidogyne* spp. *Nematol. medit.* **10**: 135-139.

Jatala, P. (1985). Biological control of nematodes. *In: An Advanced Treatise on Meloidogyne Vol. I. Biology and Control* (Eds.—C.C. Carter & J.N. Sasser) North Carolina State Univ. Graphics.

Jorgenson, E.C. (1984). Nematicides and non-conventional soil amendments in the management of root-knot nematode on cotton. *J. Nematol.* **16**: 154-158.

Karaev, K.K.; Mavlanov, O.M. & Nigmanova, S.N. (1980). Study of the association of plant nematodes and *Fusarium*-wilt on cotton. *Trudy Sredneaziaticheskogo Naucho — Issledovatels' kogo Inst. Zashchity Rastenii* No. **14**: 67-70.

Kirkpatrick, T.L. & Sasser, J.N. (1984). Crop rotation and races of *Meloidogyne incognita* in cotton root-knot management. *J. Nematol.* **16**: 323-328.

Klump, R.S. & Thomas, S.H. (1987). Comparative resistance of selected Acala 1517 cotton cultivars of *Meloidogyne incognita* race 3. *Ann. Appl. Nematol.* **1**: 113-115.

Koura, F.H. (1984). Reaction of Egyptian cotton cultivars to *Hoplolaimus aegypti* infection. *Ann. Agri. Sci. Ain. Shams University* **29**: 513-519.

Koura, F.H. & Naser, S.A. (1984). Nematicidal efficacy of aldicarb in the control of *Hoplolaimus aegypti* and relative susceptibility of cotton cultivars to nematode infections. *Ann. Agri. Sci. Ain. Shams University* **29**: 531-538.

Koura, F.H. & Osman, H.A. (1984). The relation between population density of *Hoplolaimus columbus* and damage to cotton and soybean. *Ann. Agri. Sci. Ain. Shams University* **29**: 521-529.

Krishna Prasad, K.S. & Padaganur, G.M. (1880). Observations on the association of *Rotylenchulus reniformis* with *Verticillium* wilt of cotton. *Indian J. Nemtol.* **10**: 91-92.

Krusberg, L.R. & Sasser, J.N. (1956). Host-parasite relationship of the lance nematode in cotton roots. *Phytopathology* **46**: 505-510.

Lawrence, G.W.; McLean, K.S.; Batson, W.E.; Miller, D. & Borbon, J.C. (1990). Response of *Rotylenchulus reniformis* to nematicide applications on cotton. *Suppl. J. Nematol.* **22**: 707-711.

Lordello, A.I.L.; Lordello, R.R.A. & Sabino, N.P. (1984). Effects of nematicides on the economic characteristics of cotton grown in soil infested by *Helicotylenchus* sp. and *Rotylenchulus reniformis. Nematologica Brasileira* **8**: 47-55.

Lordello, R.R.A. & Sabino, N.P. (1985). Effect of nematode control on fiber properties of upland cotton. *Bragantia* **44**: 695-699.

Minton, E.B. & Bailey, J.C. (1984). Effects of aldicarb and Telone II on cotton production. *Phytopathology* **74**: 830.

Mohammed, N. & Jones, J.E. (1990). Genetic of resistance to reniform nematode in Upland cotton. *Crop Science* **30**: 13-16.

Monton, E.B. & Meredith, W.R. Jr. (1987). Root-knot nematode effect on nine cotton cultivars in Mississippi. *Crop Science* **27**: 1001-1004.

Morgan-Jones, G. & Rodriguez-Kababa, R. (1987). Fungal biocontrol for the management of nematodes. *In: Vistas on Nematology* (Eds.—J.A. Veech & D.W. Dickson). Society of Nematologists, Maryland.

Mueller, J.D. & Sullivan, M.J. (1988). Response of cotton to infection by *Hoplolaimus columbus. Ann. Appl. Nematol.* **2**: 86-89.

Muralidharan, R. & Shivakumar, C.V. (1975). Susceptibility of certain Indian varieties of cotton and wild species of *Gossypium* to the reniform nematode, *Rotylenchulus reniformis. Indian J. Nematol.* **5**: 116-118.

Nath, R.P.; Swarup, G. & Ramarao, G.V.S.V. (1969). Studies on the reniform nematode, *R. reniformis* Linford & Oliveira, 1940. *Indian Phytopath.* **22**: 99-104.

Neal, D.C. (1954). The reniform nematode and its relationship to the incidence of *Fusarium*-wilt of cotton at Baton, Ronge, La. *Phytopathology.* **44**: 447-450.

Noe, J.P. (1990). Efficacy of fumigant nematicides to control *Hoplolaimus columbus* on cotton. *Suppl. J. Nematol.* **22**: 718-723.

Orr, C.C. (1973). Economics of root-knot nematode control on cotton by DBCP fumigant on the Texas High Plains. *J. Nematol.* **5**: 69-71.

Orr, C.C. (1974). Interaction of nematicides and herbicides for pest control of cotton. *Plant Dis. Reptr.* **58**: 120-121.

Orr, C.C. & Robinson, A.F. (1984). Assessment of cotton losses in western Texas caused by *Meloidogyne incognita. Plant Dis.* **68**: 284-285.

Page, S.L.J. (1984). Effects of the physical properties of two tropical cotton soils on their permanent wilting point and relative humidity in relation to survival and distribution of *Meloidogyne acronea. Revue Nematol.* **7**: 227-232.

Patel, D.J.; Patel, H.R.; Patel, S.K. & Makwana, M.G. (1991). Soil solarization through clear polythylene tarping for management of root-knot disease in a tobacco nursery. *Tobacco Sci.* **35**: 63-64.

Perry, D.A. (1961). The interaction of *Fusarium oxysporum* f. *vasinfectum* and *Meloidogyne* spp. on cotton. *Sixth Commonwealth Mycological Conference*, London, 1960.

Pinckard, J.A. (1964). N, N-dimethyl-phosphorodiamidata (nellite) a new nematicide, systemic insecticide and growth-promoting substance on cotton seedlings. *Phytopathology* **54**: 626.

Rai, B.S. & Jain, R.K. (1987). Pathogenicity of root-knot nematode (*Meloidogyne incognita*) on cotton (*Gossypium* spp.). *Indian J. Nematol.* **17**: 296-300.

Rajaram, B. (1983). Studies on the reaction of cotton varieties to the infection of the lance nematode, *Hoplolaimus seinhorsti* Lnc, 1958; *3rd Nematology Symposium. D.P.K.V.V.* Solan (May 24-26): 29.

Ratchford, K. & Burris, E. (1985). Effects of early season pesticides in cotton. *Louisiana Agric.* **28**: 14-15.

Reynolds, H.W. & Hanson, R.G. (1957). *Rhizoctonia* disease of cotton in presence or absence of the cotton root-knot nematode in Arizona. *Phytopathology* **47**: 256-261.

Rodriguez-Kabana, R. & King. P.S. (1976). Activity of insecticide phorate against root-knot nematodes. *J. Nematol.* **8**: 300.

Roncadori, C.W. & Hussey, R.S. (1978). A joint action between the endomycorrhizal fungus *Gigaspora margortia* and root-knot nematode on cotton *Proc. American Phytopathol. Soc.* **4**: 124.

Rong, W.Z.; Huan, J. & Shentu, G.R. (1987). A study of cotton disease caused by root-knot nematode (*Meloidogyne incognita*) in Zhejiang province. *Scientia Agricultura Sinica* **20**: 13-18.

Sakhuja, P.K.; Chopra, B.L.; Sharma, J.R. & Singh, I. (1987). Evaluation of carbofuran and subufos against root-knot nematode, *Meloidogyne incognita* on American cotton (*Gossypium Hirsutum* L.) *J. Cotton. Res. Development* **1**: 213-215.

Sakhuja, P.K.; Jbooty, J.S. & Kang, M.S. (1986). Widespread incidence of root-knot nematode, *Meloidogyne incognita* on cotton in Punjab (India). *Curr. Sci.* **55**: 740-741.

Sappenfield, W.P. (1981). Registration of HYC 76-59 cotton germplasm (Reg. No. GP-163). *Crop Sci.* **21**: 991-992.

Sasser, J.N. (1979). Economic importance of *Meloidogyne* in tropical countries. *In:* Lamberti, F. & Taylor, C.E. *Root-knot nematodes* (*Meloidogyne* spp.), *Systematics, biology and control* (Eds.-F. Lamberti & C.E. Taylor). Academic Press.

Schmitt, D.P. & Bailey, J.E. (1990). Chemical control of *Hoplolaimus columbus* on cotton and soybean. *Suppl. J. Nematol.* **22**: 689-694.

Shanmugam, N; Rajendran, G.; Vinayagamurthi, A.; Muthukrishnan, T.S. & Kandaswamy, T.K. (1976). Studies on chemical control of cotton wilts. *Madras Agric. J.* **63**: 386-388.

Shanmugam, N.; Vinayagamurthi, A.; Rajendran, G.; Muthukrishnan, T.S. & Kandaswamy, T.K. (1977). Occurrence of nematode fungus interactions on cotton around Coimbatore. *Sci. Cult.* **43**: 346.348.

Shepherd, R.L. (1982). Genetic resistance and its residual effects for control of root-knot nematode—*Fusarium*-wilt complex in cotton. Crop. Sci. 22: 1151-1155.

Shepherd, R.L. (1983). New sources of resistance to root-knot nematode among primitive cottons. *Crop Sci.* **23**: 999-1002.

Shepherd, R.L.(1986). Cotton resistant to the root-knot — *Fusarium*-wilt complex. II. Relation to root-knot resistance and its implications on breeding for resistance. *Crop Sci.* **26**: 233-237.

Shepherd, R.L. (1987). Registration of 3 root-knot resistant cotton germplasm lines. *Crop Sci.* **27**: 153.

Shepherd, R.L. & Kappelman, A.J. (1986). Cotton resistance to root-knot — *Fusarium*-wilt complex. I. Relation to *Fusarium* wilt resistance and its implications on breeding for resistance. *Crop Sci.* **26**: 228-232.

Sitaramaiah, K. (1984). *Plant Parasitic Nematodes of India*. Today & Tomorrow's Printers & Publishers, New Delhi. pp. xxiii+292.

Sivakumar, C.V. & Seshadri, A.R. (1971). Effect of host on the population development and fecundity of the reniform nematode, *Rotylenchulus reniformis Indian J. Nematol.* **1**: 91-93.

Sivakumar, C.V. & Seshadri, A.R. (1976). Longevity of the reniform nematode, *Rotylenchulus reniformis* in host free soil. *Indian J. Nematol.* **6**: 138-144.

Smith, A.L. & Taylor, A.L. (1941). Nematode distribution in the 1940 regional cotton wilt plots. *Phytopathology* **31**: 771.

Smith, G.S.; Hussey, R.S. & Roncadori, R.W. (1986a). Penetration and postinfection development of *Meloidogyne incognita* on cotton as affected by *Glomus intraradices* and phosphorus. *J. Nematol.* **18**: 429-435.

Smith, G.S.; Roncadori, R.W. & Hussey, R.S. (1986b). Interaction of endomycorrhizal fungi, superphosphate, and *Meloidogyne incognita* on cotton in microplot and field studies. *J. Nematol.* **18**: 208-216.

Starr, J.L. & Page, S.L.J. (1990). Nematode parasites of cotton and other tropical fibre crops. *In: Plant parasitic nematodes in subtropical and tropical agriculture* (Eds.—M. Luc, R.A. Sikora & J. Bridge). CAB International Institute of Parasitology.

Starr, J.L. & Veech, J.A. (1986). Susceptibility to root-knot nematode in cotton lines resistant to the *Fusarium*-wilt/Root-knot complex. *Crop Sci.* **26**: 543-546.

Stretton, A.O.W.; Campbell, W.C. & Babu, J.R. (1987). Biological activity and mode of action of Avermectins. *In: Vistas on Nematology* (Eds.—J.A. Veech & D.W. Dickson). Society of Nematologists, Maryland, USA.

Sudheer, J.M. & Sivakumar, C.V. (1983). Host-range of *Pratylenchus zeae* Graham, 1951. 3rd Nematology Symposium, HPKVV, Solan (May, 24-26) 3.

Tayar, A. (1979). Seed treatment for control of *Meloidogyne incognita* on cotton, Proc. 2nd Res. *Plan. Conf. on Root-knot Nematode Meloidogyne* spp. Region VII. Athens, Greece (Nov. 26-30).

Tchatchoua, J. & Sikora, R.A. (1983). Alterations in susceptibility of wilt resistant cotton varieties to *Verticillium dahliae, Rotylenchulus reniformis. Zeitschrift fur Pflanzenkrankheiton* und Pflenzenschutz **90**: 232-237.

Thakar, N.A.; Patel, H.R. & Patel, C.C. (1986). Screening of cotton varieties/hybrids against root-knot nematode *Meloidogyne* spp. *Indian J. Nematol.* **16**: 265-266.

Thomas, W.H.; Heald, C.M. & Amador, J. (1970). Population of Rotylenchus reniformis and yield of cotton following soil fumigation in Texas. *Phytopathology* **60**: 1542.

Turner, G.U. (1963). Nellite nematicide used in irrigation waters increases cotton yields. *Biokemia Midland Michigan* **1**: 14-17.

Vadivelu, S.; Rajendran, G.; Buckner, A.H. & Muthukrishnan, T.S. (1974). Plant parasitic nematodes associated with cotton. *Madras Agric. J.* **61**: 962-965.

Varaprasad, K.S.; Sud, U.C. & Swarup, G. (1987). Relationship between *Rotylenchus reniformis* densities and host damage with reference to tolerance levels. *Indian J. Nematol.* **17**: 11-16.

Veech, J.A. (1978). An apparent relationship between methoxy substituted terpenoid aldehyde and resistance of cotton to *Meloidogyne incognita*. *Nematologica* **24**: 81-87.

Veech, J.A. & Starr, J.L. (1986). Comparison of development, reproduction and aggressiveness of *Meloidogyne incognita* races 3 and 4 on cotton. *J. Nematol.* **18**: 413-415.

Williams, C.; Gilman, O.F.; Jones, J.E. & Birchfield, W. (1983). Cotton-soybean rotation for reniform nematode control. *Bull. Louisiana Agric. Exptl. Sta.* **741**: 1-22.

Yik, C.P. & Birchfield, W. (1984). Resistant germplasm in *Gossypium* spp. and related plants to *Rotylenchulus reniformis*. *J. Nematol.* **16**: 146-153.

Chapter 16

Nematode Control in Forage Crops - An Overall Appraisal

N. Hasan
Indian Grassland and Fodder Research Institute, Jhansi-284 003, India.

FORAGE crop production in all of its ramification is of major importance to agriculture. This is another way of saying that forage crops are of vital concern to man. It is not only the cheapest source of nitrogen and palatable nutrient for livestock, but is also valuable in restoring and maintaining soil fertility and structure. Moreover, in the milkshed areas where the major enterprise of the farmers centres around the milk production, continuous supply of green fodders round the year has become the basis for the success of such industry. Therefore, cultivation of nutritive fodder crops as such or as a part of the normal crop rotation practices has gained importance in India and elsewhere in the world.

Forage crops virtually in every part of the world are also impaired to a greater or lesser extent by nematode pests in the same way as in other crops. However, the most evident damage may occur in tropical and subtropical regions where, besides other favourable conditions, crops are grown all the year round without a vegetative rest period. In such a case the nematodes, particularly the polyphagous ones, get the opportunity of uninterrupted food supply and thus multiply to high population resulting in severe crop losses. The plant damage is sustained in several ways which together lead to poor growth, low yield and impaired quality.

Few reviews are available on nematode diseases of forage crops which have already been recognised as major agricultural problems especially in the temperate countries of the world (Bingefors, 1969; Ericksson, 1972; Hasan, 1990). But a comprehensive information on nematode problems in forage crops from India as well as most of the tropical countries is perhaps lacking. The objective of this review is, therefore, to focus the importance of nematodes in relation to forage crop production alongwith an overall appraisal of the problems and progress made in various facets of nematode control with special reference to India and other environmentally similar countries.

FORAGE CROP PRODUCTION IN INDIA

India is one of the tropical countries where a large population of farm animals do not commensurate with limited supply of feed and fodders. This is because of the fact that the country has only about 4% of the cultivated area under fodder production in addition to about 25% of the total land area under natural grassland. Therefore, fodder production has been given consideration recently and about 16.5 million hectare irrigated area has been earmarked for fodder production during 2000 AD which is about 6 times more than the area under fodder during 1971. The estimates of feed and fodder requirements are shown in Table 1 (Pathak & Jakhmola, 1986).

Table 1 : Estimates of feed and fodder availability and requirement in India (in million tonnes).

Feed stuffs	Requirement	Availability	Deficit (%)
Green fodder	347.57	214.50	37.57
Dry fodder	347.41	207.00	40.42
Concentrates	19.57	11.05	43.54
Expected by 2000 A.D.			
Green fodder	594.80	575.00	3.33
Dry fodder	373.00	356.00	4.34
Concentrates	82.80	77.05	6.94

(Source : Pathak & Jakhmola 1986).

The principal forages are largely in the two botanical families, the grasses (Graminae) and the legumes (Leguminosae). Some of the forage crops, viz. berseem, lucerne, oat, stylo are exclusively cultivated for fodder production, whereas others such as cowpea, clusterbean, moth bean, maize, sorghum and bajra are grown both for grain as well as forage in various states of India (Table 2).

Table 2 : Some of the important tropical forage crops suited to different agroclimatic zones of India.

Species	Common Name	Habit & Habitat	Area
Legumes:			
Trifolium alexandrinum	Berseem	annual, erect	Central, Western, North-Western, Subtemperate
Medicago sativa	Lucerne or alfalfa	annual to biennial, erect	Western
Vigna unguiculata	Cowpea	annual, semi-erect	North-West, South-West
Cyamopsis tetragonoloba	Cluster bean	annual, erect	Central, Western

Species	Common Name	Habit & Habitat	Area
Vigna aconitifolius	Moth bean	annual, erect	Western
Lablab purpureus	Field bean	annual to biennial semi-erect, trailing	Central, South-Central
Stylosanthes hamata	Stylo	perennial, erect	Central, South-Central
S. humilis	Townsville stylo	annual to biennial prostrate to ascending	Central, South-Central
Macroptilium atropurpureum	Siratro	perennial, creeping	Central, South-Central
Clitoria ternatea	Clitoria	biennial, semi-erect	Throughout the country
Grasses :			
Sorghum bicolor	Sorghum	annual, erect	Southern, South-West
Pennisetum typhoides	Pear millet	annual, erect	North-Western
Zea mays	Maize	annual, erect	North-Western, South-West
Avena sativa	Oat	annual, erect	North and North-Central
Pennisetum pedicellatum	Dinanath grass	annual, erect	Eastern
Cenchrus ciliaris	Buffel or Anjan grass	perennial, erect	Central, South-Central
Dicanthium annulatum	Marvel grass	perennial, erect	Central, South-Central
Pennisetum purpureum	Napier or elephant grass	perennial, erect	Throughout the country

First two legumes are adapted to subtropics and tend to behave as winter annuals.

NEMATODE PROBLEMS

Cultivated Fodders

A number of plant-parasitic nematode genera belonging to various species have been found associated with a variety of fodder crops grown in different parts of the country. It is not possible to include every nematode pest which might injure a crop. Therefore, only major nematode species are covered which have been selected on the basis of their actual or potential economic importance or on their frequency of occurrence (Table 3).

Table 3 : Major nematode parasites infesting forage crops and their distribution in India.

Nematode	Host Crops	Distribution in India
Meloidogyne incognita	Cowpea, fieldbean, guar, moth bean, lucerne, clitoria	Throughout
Meloidogyne javanica	Cowpea, fieldbean, guar, mothbean, lucerne, clitoria	Throughout
Rotylenchulus reniformis	Cowpea, fieldbean, moth bean, songhum, bajra	Throughout
Heterodera cajani	Cowpea, fieldbean, moth bean, bajra	U.P., Rajasthan, Delhi, Haryana, Bihar, A.P., Tamil Nadu
Heterodera zeae	Maize	U.P., M.P., Bihar, Rajasthan, Maharashtra, Delhi, H.P., J&K
Heterodera sorghi	Sorghum, maize, millets	U.P., Delhi, Haryana, Punjab, H.P., J&K
Heterodera avenae	Oat	Rajasthan, Haryana, Punjab, Part of U.P.
Pratylenchus zeae	Maize, sorghum, bajra, lucerne & some range grasses	Throughout
Tylenchorhynchus vulgaris	Berseem, maize, sorghum, oat & some range grasses	Throughout
Helicotylenchus dihystera	Sorghum & many range grasses	Throughout

Nematodes can be a real problem in forage production when they cause a reduction in the growth and vitality of the crop plant leading to an economic loss. Moreover, certain nematode have the additional ability to transmit viruses. Others alter the physiology of their hosts so that it becomes more susceptible to fungal diseases, or they provide avenues of entry for pathogenic bacteria and these are often of great economic importance than the direct feeding activity of the nematode. Examples of certain disease complexes involving nematodes and other pathogens in forage crops are given in Table 4. In addition many nematode species were also reported to cause strong decline of *Rhizobium* nodules resulting into reduced nitrogen fixing capacity of forage legumes. It is difficult to assess the exact loss caused by nematodes in forage crops. However, in some temperate countries losses of 3-11% equivalent to several million dollars per year were reported to occur in some important forage legumes (Sesser & Freckman, 1987). Presumably the losses in most of the tropical and subtropical countries with less advanced agricultural techniques may even be higher.

Table 4 : Examples of disease complexes involving nematodes and other pathogens in forage crops.

Crop	Nematode	Interacting organism (Fungus/Bacteria)	Disease Syndrome
Cowpea	*Meloidogyne incognita*	*Fusarium oxysporum*	Wilt
		Rhizoctonia solani	Root-rot
	M. javanica	*Fusarium sp.*	Wilt Root-rot
		R. solani	Root-rot
	Rotylenchulus reniformis	*R. solani*	Root-rot
Lucerne	*Ditylenchus dipsaci*	*Corynebacterum insidiosum*	Wilt
	M. incognita, M. hapla	*C. insidiosum*	Wilt
		Phytophthora megasperma var. medicagnis	Root-rot
	M. incognita acrita	*F. roseum, F. oxysporum var. batatas*	Wilt
	Pratylenchus penetrans	*F. oxysporum*	Wilt
		Trichoderma viride	Root-rot
		F solani	Wilt
White clover	*D. dipsaci* *Meloidogyne spp.*	*Pythium middletonii*	Death of stolon
Berseem	*Tylenchorhynchus vulgaris*	*F. semitectum, R. solani*	Root-rot
Soybean	*M. arenaria*	*Sclerotium rolfsii*	Southern blight
	Tylenchorhynchus claytoni, Tylenchus agricola	*F. roseum, Pythium ultimum*	Root-rot
	T. vulgaris, P. zeae	*F. moniliforme*	Root-rot
	P. brachyurus,	*F. moniliforme*	Seedling rot
	Hoplolaimus indicus	*F. moniliforme*	Root-rot
Pearl millet	*M. incognita*	*Sclerospora graminicola*	Downy mildew
Kentucky blue grass	*Tylenchorhynchus dubius*	*F. roseum*	Blight

Grasslands

Plant-parasitic nematodes are widely distributed in both tropical and temperate grasslands of the world. These plant-parasitic groups in grassland include ectoparasitic migratory, *Longidorus*, *Xiphinema*, *Tylenchorhynchus* and *Helicotylenchus*; endo-parasitic migratory, *Pratylenchus* and *Ditylenchus* spp. and non-migratory cyst forming *Heterodera*, root-knot forming *Meloidogyne* and seed gall forming *Anguina*. Although grassland normally supports and can tolerate large number of nematodes, recent evidences suggest that there are some situations where nematode damage is more likely to occur, e.g. on light sandy soil, especially during the establishing period of young seedlings. Similarly, high populations of certain important nematode species of the grassland soil have also been implicated in the resowing problem of old pastures.

Precise losses in grasslands due to nematodes are poorly documented. However, ample data showing per cent yield reduction has appeared in the literature and though mainly derived from pot experiments, illustrate the detrimental effect on plant growth caused by nematodes (Ericksson, 1972; Jain & Hasan, 1986b). In additional herbage losses of 28-59% have been attributed to root feeding nematodes in prairie grassland (Smolik, 1974), as well as 5-10% in hay in U.S.A. (Hodged, 1973) and significant depression in production in new *Lolium* plots in England (Spaull *et al.*, 1983).

CONTROL STRATEGIES - PRINCIPLES, PROBLEMS & PROGRESS

Over the years a number of methods have been experimented to suppress nematode population in forage crops with varying degree of success (Bingefors, 1969; Ericksson, 1972; Hasan, 1990). In this article it is proposed to deal in detail the principles, problems and progress of controlling nematodes in forage crops as well as control strategies adopted with special reference to India and other tropical countries.

CULTURAL PRACTICES

Crop Rotation

Too frequent cropping of land with the same crop can lead to serious build-up of nematodes in the soil. Among various cultural practices, crop rotation as nematode management technique particularly in annual crops continues to be one of the important recommendation all over the world. Sequential alternation of crops with varying degree of susceptibility to the namatode pests can produce striking results in the reduction of crop losses especially in the tropical developing countries. The basic principles underlying crop rotation for nematode control are: (a) the differential host preference to nematodes, or host status of the crops classified as good, poor, resistant, tolerant, non-host (immune) or antagonistic, and (b) the rate of population decline in the absence of the host. In addition to these the non-host crops (alternative crops) to be grown between the host crops must satisfy certain agronomical and economical requirements of the grower. However, incorporation of resistant cultivars of the susceptible crops normally grown in the cropping system is considered as the first alternative.

A number of reviews have dealt with the question whether cropping systems or crop rotations are sufficient as a means to manage nematodes. With very few exceptions, the importance of crop rotation in regulating nematode population has been universally admitted (Norton, 1978; Nusbaum & Ferris, 1973; Sethi & Gaur, 1990). In consequence, research efforts in evolving suitable cropping sequences consisting forage as main or alternative crop for nematode population regulation to different parts of the world have been studied and reviewed (Ericksson, 1972; Hasan, 1990; Sethi & Gaur, 1990).

Red clover and lucerne were found to be highly susceptible to *Ditylenchus dipsaci* and *Pratylenchus penetrans* in most of the temperate countries and advised to be grown only after 8-10 years or after beet which was found to suppress the population of these nematodes (Decker, 1969; Oostenbrink, 1961). Similarly, *Heterodera trifolii* infecting red clover may be eliminated if rotated with corn or soybean (Norton, 1967). In U.S.A. lecerne is usually rotated with cotton for two years because of the damage caused by *Meloidogyne arenaria* (Reynold & O'Banon' 1960). Moreover, this crop was found to consistently reduce the population of pin nematode that has been increased by presowing with birds foot trefoil (Hawn & Kuzub, 1978). Shagalina (1973) reported that two mediterranean lucerne (*Medicago polia*) varieties grown after cotton together with elimination of weeds considerably reduced *M. incognita* and *M. javanica* population. Populations of *Tylenchorhynchus* and *Pratylenchus* species were reported to be the lowest following berseem in the rotation (Shafshak *et al.*, 1974). Kobe lespedeza has been advised not to be grown in rotation with soybean because of the susceptibility to *H. glycines* (Ericksson, 1972). Population of *M. naasi* was found to be 30 times greater after Italian rye grass but reduced after potato and wheat (Gooris & Herde, 1967).

In most of the tropical and subtropical countries including India, work on this approach of nematode control particularly in forage crops are limited. This is perhaps due to: (a) the multiple growing seasons in which 3-4 or even more crops are grown in succession on the same piece of land during a year; in such a case not only the crop replacing the main crop but also the intervening crops have to meet the requirements of a rotation crop, (b) the nematodes, particularly the polyphagous and continuously reproducing ones get almost uninterrupted food supply and thus multiply to high population making selection of suitable non-host crop a difficult task, and (c) the poor recognition of the nematode problem.

In spite of the above facts, recently in Nigeria intercropping of maize with leuceana or growing *Stylosanthes gracilis* prior to tomato, reduced root-knot nematode (Egunjobi, 1985). However, in India all the three species of stylo (*S. gracilis*, *S. similis*, *S. hamata*) grown after banana were heavily infested with *Radopholus similis* (Subramaniyan, 1985). Chawla & Prasad (1973) observed a little or no difference in the nematode population following either sorghum-wheat-cowpea or bajra-wheat-cowpea cropping sequences. As a consequence, Prasad & Gaur (1974) suggested that often the combined damage caused by polyspecific communities present is more important than that due to any one species. Moreover, Oostenbrink (1964) observed that the composition of the population is markedly influenced by the last grown crop or crops and thus it is this principle which probably contributes more towards the utility of crop rotation in nematode control. Population of *Tylenchorhynchus* following berseem was observed to be relatively lower than after maize, wheat or sugarcane in a cropping sequence (Mukhopadhya & Prasad, 1968). Similarly, Srivastava & Sethi (1986) reported that stunt and reniform nematodes maintained their population without major fluctuation while *Heterodera zeae* showed host specifity to maize in maize-cowpea-wheat rotations. Sethi & Gaur (1986) suggested that employing crop rotation especially against the above nematode species with limited or delineated host range would prove to be more effective in population reduction. In India some of the major nematode species like *H. avenae, H. zea, H. sorghi* are found to be mainly parasitic on graminaceous forages while *H. cajani*, *M. incognita* and *M. javanica* generally prefer leguminous crops (Table 5). Therefore, it would not be impossible to find a suitable and acceptable alternative crops to replace main host crop in most parts of the country for short or even long rotations.

Research efforts during past few years on the role of certain important tropical pasture grasses in effective nematode population suppression when, introduced as alternative crop in the crop rotation system or inter-cropped with susceptible crops has also been recently reviewed (Hasan, 1990; Sethi & Gaur, 1990). Among these grass species or varieties, Pangola grass *Digitaria decombans*, Love grass *Kragrostis curvula* cv. Ermalo, Guinea grass *Pannicum maximum* cv. Umtali and Sabi, Rhode grass *Chloris gayana* cv. Katombora and coastal Bermuda grass have been successfully utilized in tobacco or cotton rotations for effective control of dominant root-knot nematode species, in different countries of the world (Ayala *et al.*, 1967; Burton & Johnson, 1987; Daulton, 1962; Koen & Grobbelaar, 1968; Shephard, 1978; Shephard & Coombs, 1979; Winchester, 1960). Similarly, many legume cover crops or green manure crops which also give valuable

Table 5 : Host suitability of forage crops to major nematode species.

Crops	Nematode species*/Host suitability**								
	Mi	Mj	Rr	Hc	Hs	Ha	Pz	Tv	Hd
Legumes :									
Berseem	+	+	0	-	0	0	++	+++	0
Lucerne	+++	+++	0	-	0	0	+++	0	0
Cowpea	+++	+++	+++	+++	0	0	0	0	++
Cluster bean	+++	+++	++	+++	0	0	0	0	0
Moth bean	+++	+++	+++	+++	0	0	0	0	0
Field bean	+++	+++	+++	+++	0	0	0	0	++
Stylo	+	+	0	+	0	0	0	0	0
Siratro	+	+	0	-	0	0	0	0	0
Clitora	+++	+++	+	-	0	0	0	0	0
Grasses :									
Sorghum	+	+	++	-	+++	-	+++	+++	+++
Bajra	++	++	+++	-	++	-	+++	+++	++
Maize	++	++	+	-	+	+	+++	+++	++
Oat	+	++	++	-	+	+++	+++	+++	++
Buffel grass	-	-	0	0	0	-	++	+++	+++
Marvel grass	-	-	0	0	0	-	++	+++	+++
Setaria grass	-	-	0	0	0	-	++	+++	+++

(Source : Alam *et al.*, 1979; Ericksson, 1972; Hasan, 1985b, 1990; Jain & Hasan, 1986b; Seth & Gaur 1990).

*Mi = *Meloidogyne incognita*; Mj = *M. javanica*; Rr = *Rotylenchulus reniformis*, Hc = *Heterodera cajani*, Hs = *H. sorghii*, Ha = *H. avenae;* Pz = *Pratylenchus zeae*: Tv= *Tylenchorhynchus vulgaris*; Hy = *Helicotylenchus dihystera.*

** 0 = Not known; - = Non-host; + = Poor; ++ = Fair; +++ = Good hosts.

forage for livestock were found to reduce a wide range of nematode species. Of these *Crotalaria spectabilis, C. brevidens, Mucuna deeringiana, Indigofera hirusta* were reported to be the best alternative crops for efficient control of the *Meloidogyne* spp., *Belonolaimus longicaudatus*, *Pratylenchus brachyurus*, *Criconemoides* spp. and *Trichodorus* spp. (Brodie *et al.*, 1970; Ferraz, 1985; Reddy, 1983; Rhoades, 1976; Soffes *et al.*, 1980). These crops can reduce or eliminate the use of chemical nematicides and nitrogen fertilizers in cropping systems (Soffes *et al.*, 1980). In addition, certain tropical pasture legumes have also been identified as non-hosts or resistant to some important nematode species which may be utilized in crop rotation systems (Hasan, 1990; Kretschmer *et al.*, 1980). Further it needs to be emphasised that incorporation of the above legume and grass species or varieties would be economically attractive in localities where livestock form a part of farming system. Noe (1986) also suggested that the most profitable management alternative for particular plant-parasitic nematode infestation would be to use the land as pasture for a number of years, thus rotating to a profitable non-host.

Weeds as foci of infestation within crops and as carry over hosts usually negate the effect of crop rotation for nematode control (Hasan, 1985a; Hooper & Stone, 1981). Morerover, nematode control via weed control presents greater problem in tropical and subtropical agriculture because of the greater abundance of the weeds together with the wide host range of some important nematode species. Therefore, thorough weed control either with weedicides or by mechanical means must be essentially combined with crop rotation.

Host Resistance

Potentially the most economic and effective method of nematode control in forage crops is to use nematode resistant plants or varieties. The primary objective, therefore, is to develop crop varieties resistant to one or more nematode species alongwith other desirable agronomical characters. Considerable progress in resistant breeding in forage crops has already been made resulting into a number of resistant varieties (Table 6) throughout the world particularly in temperate countries (Bingefors, 1969, 1971; Hasan, 1990).

Genetic variability for susceptibility to root-knot in cowpea was demonstrated early as in 1902 by Webber & Orton and genetic resistance could be transferred to a susceptible cultivar. As a result, the first root-knot resistant variety was developed in cowpea. Similar genetic variation for nematode susceptibility/resistance of various other forage legumes like alfalfa, white clover, red clover and lespedeza to major nematodes, e.g. *Meloidogyne*, *Heterodea* spp. and D. *dipsaci* were also demonstrated from different parts of the world particularly from America and most European countries (Bingefors, 1969; Ericksson, 1972). Genetic variation for susceptibility to certain nematode species were also reported in some grass species and varieties, but resistance breeding in this important forage family is an almost virgin area of research (Bingefors, 1969). Availability of genetic resistance in plants is, therefore, more common in forage legumes. During the last few years considerable research has been made in understanding the genetic basis of plant-nematode interaction (Sidhu & Webster, 1981a) and also current methods of developing nematode resistant cultivars (Bingefors, 1971; Kaplan, 1982; Sidhu & Webster, 1981b). As a consequence genetic improvement in

Table 6 : Examples of nematode resistant sources/varieties in forage legumes.

Crop and nematode species	Resistant sources/varieties	Reference
Alfalfa :		
M. incognita	Moapa, Sonora	Bingefors, 1969, 1971
M. javanica	Moapa	
M. Hapla	Vernal and cv. Hilmar	
D. dipsaci	Alfa II, Vertus, Nematol (Europe); Frontier, Pergamino, San Martin, Nemasint (S. America); Nemastan, Lahontan, Talent, Washoe (U.S.A.)	
Red clover :		
Meloidogyne spp.	Plants in different varieties	
D. dipsaci	Merkur, Resistanta, Hermes, Hermes II, Britta, Disa, Reko (Sweden), Corona, Resident, Otofte, Hera (Denmark); Dorset, Marl, Cotswold, Essex, S123, Montogomery (Great Britain)	
White clover :		
M. incognita	Plants in Swedish and Dutch varieties.	
H. trifolii	Plants in cultivated varieties	
D. dipsaci	Wild English and other local varieties, Milka, Milkanova, Lodi, Mira, Trifo, Zero, Mega, Adefa (Denmark)	
Lespedeza :		
M. incognita	Rowan	
M. hapla	Lines in *L. cuneata*	
Soybean :		
Meloidogyne spp.	Jackson, Hill, Delmar, Palmetto, Laredo Dyer, Forest, JS-2, Selec. Ds 74-20-1	Bingefors, 1971; Dasgupta & Ganguly, 1986
H. glycines	Dyer, Picket, Custer, Nemasherazu	
R. reniformis	Dyer, Pickett, Peking, Centennial, Cluster	
Cowpea :		
M. incognita	Iron, Clay, Chinese, Red, Groit, New era, Mississippi purple, Mississippi silver, Viktor K 798, Broweya 7, Brabham, Monetta, Catjang, Columbia	Dastupta & Ganguly, 1986; Hartman, 1976; Singh *et al.*, 1974

Crop and nematode species	Resistant sources/varieties	Reference
R. reniformis	V-16	
H. cajani	Sel. A, Sel. 1-2, Sel. 4-1, 24-8-2, P33-2-1, P85-2	
Cluster bean :		
M. incognita	ILO, 1539-1, IL-79-1028-1, ILO 77-4-1 (germplasm lines)	Hasan & Jain, 1988
Mothbean :		
M. incognita	Il 1134, IL 1202 (germplasm lines from indigenous collection)	Hasan & Jain, 1984c
Clitoria :		
M. incognita	ILO-160-1, ILO 1442 (germplasm lines from indigenous collection)	Hasan & Jain, 1985a
Sesbania :		
M. incognita	*S. tetraptera*, *S. macrocarpa* 1535	Hasan & Jain, 1985b

nematode resistance through various breeding techniques has been successfully applied resulting in a number of nematode resistant varieties.

Although some success has been achieved in the identification of combined resistance to different nematode species and other organisms such as insects, fungi, bacteria and viruses (Table 7), but further progress on this important aspect of resistant breeding is meagre. Availability of genetic resistance is more common in case of specialized endo and/or semi-endoparasites but for many other ectoparasitic nematodes very little is known.

In India most of the efforts have been on screening of available germplasm for resistance against a number of important nematodes. As a result large number of lines exihibiting moderate to high degree of resistance have been identified. However, efforts for resistant breeding especially in forage crops have been rather scanty.

Where specific nematode species is a limiting factor to the realization of the economic potential of the crop, resistant varieties have acquired greater importance than any other method of nematode control. However, in view of the possibility of breakdown of resistance or evolution of resistant breaking biotypes, caution is necessary against repeated growing of the same resistant variety in the same field year after year.

Organic Amendments

The agroecosystem can be modified in many ways to improve the benefits obtained from biological control organisms. Populations of natural enemies can be

Table 7 : Examples of multiple pests and disease resistant sources/ varieties in forage legumes.

Crop and nematode species	Other pests and disease (insects, fungi, bacteria viruses)	Resistant source/variety	Reference
Alfalfa :			
D. dipsaci, *M. hapla*	*Achyrthosiphon pisum*, *Therioaphis maculata*, *Colletotricum falcatum*, *Glomerella tucumensis*, *Phytophthora megasperma*, *C. insidiosum*	Arc x Agate x Apalachee W 10	Elgin, 1982
M. hapla	*P. megasperma*, *F. oxysporum* and two insect spp.	Plants in *M. sativa* lines U.C. Cibola	Lehman *et al.*, 1983
D. dipsaci	*Verticillium albo-atrum*	Plants in *M. sativa* lines, Vernema	Peaden *et al.*, 1983
D. dipsaci	*C. insidiosum*	Plants in cv. Beaver Trek	Hanna & Hawn, 1975
Red clover :			
D. dipsaci	*Sclerotinia trifoliorum*	Plants in *M. sativa* lines Britta	Jonsson, 1976
Cowpea :			
M. incognita	*Fusarium* sp.	Iron	Orton & Webber, 1902
Lespedeza :			
M. incognita acrita	*Rhizoctonia* spp.	Beltsville 23-864 x Al 2193, Al-Lotan	Donnelly, 1981
Soybean :			
M. incognita	*P. sojae*	Plants in *M. sativa* lines Mack, Centennial	Caviness *et al.*, 1977
Race 3 H. glycines M. incognita Race 1, 3 *R. reniformis* *H. Glycines* Race 1, 2, 3	*P. megasperma*, *X. phaseoli* *P. tabaci*, *Corynespora cassicola*	D64-4636 (Hill x [D49 -2491(4) x Jackson]) x Picket 71, Centennial	Hartwig & Epps 1977
Snap bean :			
M. incognita	*Uromyces phaseoli*, NY 15 Strain of mosaic	PI 165426, Nema snap	Wyatt *et al.*, 1983

enhanced by increasing the sphere of necessary habitat and/or by altering habitats to be more suitable for the colonization or subsequent population. The addition of organic matter in the soil appears to be the only factors that can rapidly alter the physical, chemical and biological properties of the soil and may be the reason as to why the additional organic matter succeeds in retarding nematode populations. Therefore, until recently the principal approach to manage nematodes by organic soil amendments has been to study the possible mechanism involved in the ensuing control. The two most prevalent hypotheses are: (a) the decomposition products released from amendments into the soil are directly toxic (Alam, 1976; Hasan, 1977; Sitaramaiah, 1990), and (b) the increase of the soil microbial and faunal population by addition of amendments stimulates parasitic and predacious activity (Godoy *et al.*, 1983; Van Gundy, 1985). Both are the important and functional hypotheses, largely attributed to the biological control of plant-parasitic nematodes.

Unfortunately most organic additives must be used in such large quantities to make a significant change in the ecosystem that they are neither practical nor economical unless there is an abundant supply available locally. Nevertheless, Mullar & Gooch (1982) and Alam (1993) have emphasised the advantages of using organic amendments for nematode control particularly in tropical countries and may not be ignored altogether.

Recently in India of the various oilcakes used, application of neem cake @ 20 g/ha once in a year, considerably reduced the nematode population in berseem-bajra and oat-cowpea in two successive cropping seasons. Moreover, this treatment also resulted in the increase of the photosynthetic pigments of the crops (Hasan & Jain, 1984a; Jain & Hasan, 1986a). Similarly in another field trial incorporation of chopped leaves of *Parthenium* @ 50 g/ha, sawdust @ 30 g/ha, neem cake @ 20 g/ha, 7-10 days before sowing not only reduced the nematode population but also increased both the forage yield and nutritional quality in two successive cropping seasons (Hasan & Jain 1986, 1987a). Further in a pot experiment incorporation of leuceana leaves into the root-knot infested soil resulted into significant reduction of the gall index in cowpea (Jain & Hasan, 1984b). Neem cake @500 kg/ha was also found to be effective against *R. reniformis* and *H. dihystera* infecting field bean (Sundaram & Velayutham, 1988). In a recent study Akhtar & Alam (1992) found that increase in the populations of plant-parasitic nematodes was suppressed when *berseem*/Egyptian clover (*Trifolium alexandrinum*) and *rizka*/lucerne, alfalfa (*Medicago sativa*) was treated with oilcakes.

BIOLOGICAL CONTROL

Since long scientists have adopted various definitions for biological control. However, the most preferred (49%) one is Baker & Cook's (1974) definition of which a shortened version is found in their recently published book (Cook & Baker, 1983): "Biological control is the reduction of the amount of inoculum or disease producing activity of a pathogen accomplished by or through one or more organisms other than man."

Soil Antagonists

The significance and current state of development in biological control of nematodes through number of pathogens, parasites and predators like viruses, fungi, bacteria, turbellarians, enchytrieds, collembola and mites, etc. has been extensively studied and reviewed in recent years (Jatala, 1986; Kerry, 1984; Alam & Jairajpuri, 1990; Swarup & Gokte, 1986; Van Gundy 1985). However, of these 200 soil antagonists identified as potential biocontrol agents, at least 75% are soil fungi, that parasitise nematode juveniles, eggs and females or cysts, and on which main emphasis appears to be concentrated.

There has been very little effort and incentive to search for practical biocontrol alternatives in nematology. The recent information of CSRS, Science and Education, USDA, on biological control research suggest that approximately only 16% of the effort in 1982 was in plant pathology and nematology of which less than 2% was in nematology (Van Gundy, 1985). Another deterrent to biological control of nematodes has its foundation in nematode control philosophy based on the use of synthetic chemicals designed to reduce nematode population by 90% or more. Also biological control of nematodes was considered only a curiosity instead of a serious practical approach to nematode control. Fortunately, the expansion of IPM strategy in the 1970's has encouraged management approach with integrated multiple control practices, one of which may be the biological control.

In addition there are a number of scientific barriers to the commercial development of biological control agents of plant-parasitic nematodes. These include: (a) stability of the natural biological balance of the ecosystem, (b) antagonism of soil organism to new introduction through the soil profile, (c) establishment of the introduction in agro-ecosystems, (d) the zone of activity of the biological control agents, and (e) the assessment of biological activity and synchronization with nematode biology.

Despite of the above facts, however, recently few products were marketed commercially, like Royal 300, Royal 350 which contain nematode trapping fungi of the genus *Arthrobotrys* in France (Mankau, 1981) and Biocon of *Paecilomyces lilacinus* an egg parasite in Philippines (Jatala, 1986). Of the several antagonists of nematode pests, most of the success has been attributed to *Arthrobotrys, Nematophthora gynophila, Dactylella oviparasitica, Paecilomyces lilacinus* and *Bacillus penetrans*. Moreover, since these organisms have not only shown promising results in initial studies, but fortunately also parasitise the most destructive groups of nematodes, e.g. *Meloidogyne* spp. and *Heterodera* spp. and, therefore, deserve to be called as promising biocontrol agents.

Reports on the efficiency of the potential biocontrol agents on nematodes infecting forage crops either in the green house or field experiments are very few as compared to other economically important crops. However, recently in India and elsewhere *P. lilacinus* has been successfully cultured at mass scale on maize, bajra, sorghum, stylo seeds or on chopped leuceana and water hyacinth leaves (Davide, 1985; Hasan, 1988; Sharma & Trivedi, 1986). Cowpea seed treatment with *P. lilacinus* spores were reported to be more effective in reducing the root-knot infestation level when either egg masses or the eggs were used as source of inoculum (Midha, 1985). Further in a green house experiment Hasan (1988) observed that application of certain organic matters, viz. leuceana leaves, neem cake or sawdust prior to the application of *P. lilacinus* not only enhanced the growth and multiplication of the fungus resulting into high percentage of egg mass infection but also effectively reduced the soil incidence in cowpea. Subsequently, in a similar field experiment application of this fungus culture @ 15 kg/ha alongwith certain organic matters effectively reduced the root-knot nematode incidence in cowpea followed by lucerne (two cropping seasons), reflecting the persistency and efficiency of the fungus in the subsequent crop also (Hasan & Jain, 1992).

Vesicular-Arbuscular Mycorrhizal (VAM) Fungi

Plant-parasitic nematodes and vesicular-arbuscular mycorrhizal (VAM) fungi commonly occur together in the roots or rhizosphere of the same plant, each having a characteristic but opposite effect on plant vigour. VAM fungi increase the supply of nutrients particularly phosphorus to crops by better exploration of the soil beyond the depletion zone and receiving the fixed carbon from the plants, thus act as symbionts. Whereas the nematodes as pathogens or parasites take out vital nutrients from the roots resulting in reduced plant growth. The single most common effect of VAM fungi on nematode-susceptible plants is promoting tolerance to nematodes. Thus, plants heavily colonised by mycorrhizal fungi are able to grow well in spite of the presence of damaging levels of plant-parasitic nematodes. The interaction between VAM fungi and soil borne pathogens particularly plant-parasitic nematodes has been extensively studied and reviewed by several workers (Husey & Roncardori, 1982; Schenck *et al.*, 1975; Schonbeck, 1979; Smith, 1987), but their effect on nematodes varied. Many workers have demonstrated a reduction in nematode population densities (Cooper & Grandison, 1986; Grandison & Cooper, 1986; Jain & Hasan, 1988; Sikora & Schonbeck, 1975), but cases where nematode population remain unaffected (Hasan & Jain, 1987b; Jain & Hasan 1988) or even increase (Atilano *et al.*, 1981) under the influence of mycorrhiza are not uncommon. However, Saleh & Sikora (1984) suggested that these differences in the interaction studies may therefore, be the result of variable root colonization by VAM fungi and not necessarily due to differences in nematode-VAM-host combination. He has further emphasised that each nematode-VAM-host combination is unique and therefore, not subject to generalization and thus requires study of each combination.

On the basis of the findings of a number of workers it has been observed that regardless of the species, less nematodes were found in most of the mycorrhizal plants (Table 8). However, the mechanism involved in the reduction of nematode numbers in the mycorrhizal plant may vary (Schonbeck, 1979) in the following ways: (a) penetration of the larvae in the root was not inhibited but their development within the mycorrihizal roots was restricted, (b) number of females within egg sacs of mycorrhizal plants was half than on non-mycorrhizal plants, (c) condition for nematode development in the roots appear to be determined due to mycorrhizae. In addition to this, recently Smith (1987) also presented hypotheses in detail on the mechanism of mycorrhizal induced host tolerance or resistance to nematodes.

Though, largely based on green house tests, considerable success has been achieved towards the possibility of utilizing VAM fungi to mitigate or reduce the adverse effect of nematodes throughout the world. However, the future use of mycorrhizal involvement as biological means of plant protection at field levels lies mainly: (a) in the mass production of fungal inocula whose properties are known and predictable, and (b) they must be genetically stable and competitive as symbionts so that they persist in the crop. At the moment, the greatest drawback is perhaps the inability to culture it on synthetic media. Also we know too little of the genetic variation and variability of mycorrhizal fungi and almost nothing of their breeding system and genetics. Nor do we know clearly what properties are required for efficient symbiotic relationship between any host and its fungus (Shonbeck, 1984).

Several methods for commercial production of mycorrhizal inoculum has been attempted and described in different parts of the world. However, their use is currently restricted to situations where the natural populations of VAM fungi have been destroyed, as in fumigated, chemically treated areas, green houses or disturbed areas (Reeves *et al.*, 1979). Daniels & Menge (1981) presented results suggesting that VAM fungus *Glomus epigaeus* shows great potential for commercial application. It is capable of producing extremely large numbers of spores in pot cultures. The spores can be stored for at least three months with little reduction in germinating capacity. Recently in India considerable efforts have been made to identify the efficient indigenous strains of VAM fungi in relation to both host and nematode species, alongwith the methods of producing inoculum.

From the foregoing it is concluded that such interactions may some day lead to the manipulation of fungi forming VA mycorrhizae by altering specific crop management system for the purpose of minimising crop losses. Learning to deal with VAM might permit some form of nematode control requiring less reliance on pesticides and might even allow reduction in the high rates of phosphorus fertilizers currently used for certain crops (Hussey & Roncadori, 1982).

NEMATICIDES

The use of synthetic chemicals to control nematodes in forage crops holds little promise of ever being an economic possibility. Moreover, it has also other problems like toxic residues, development of resistance in pests, killing of beneficial

Table 8 : Effect of VAM fungi on nematodes infecting forage crops.

Host and nematode species	Effect in mycorrhizal plant	Reference
Alfalfa :		
M. hapla	Improved plant growth, reduced nematode numbers and adult development	Grandison & Cooper, 1986
White clover :		
M. hapla	Improved plant growth, fewer nematode numbers	Cooper & Grandison, 1986
Barseem :		
T. vulgaris	Improved plant growth, nematode population unaffected	Hasan & Jain, 1987b
Soybean :		
M. incognita	Increased yield, fewer galls	Schenck *et al.*, 1975
Cowpea :		
M. incognita	Increased plant growth, fewer nematodes and galls	Jain & Hasan, 1988; Jain & Sethi, 1988a
H. cajani	Improved plant growth, reduced larval penetration and cyst production	Jain & Sethi, 1987, 1988b
Oat :		
M. incognita *M. hapla*	Increased plant growth, fewer nematode.	Sikora & Schonbeck, 1975
Sorghum :		
P. zeae *T. vulgaris* *H. dihystera*	Increased plant growth, nematode population unaffected	Jain & Hasan, 1987, 1988

organism and many environmental and health hazards. However, since all the methods of nematode control offer far less certainty of killing or stalling the action of nematodes, the nematicides are not likely to be dispensed with altogether. Nematicides by themselves, have been most important demonstration tools giving spectacular yield increases. Therefore, at present chemical methods can not altogether replace other conventional methods and vice-versa. These will have to form part of an overall integrated programme (Gaur & Prasad, 1986; Sethi & Gaur, 1986). It is normally more cost effective to use a nematicide in high risk situations, for example, (a) susceptible varieties (b) high infestation level in or nearby crops, (c) suitable environment conditions for multiplication and spread, (d) predisposition factor operative. Moreover, to achieve benefit/cost ratio, efforts are now being directed towards the development of techniques aimed at improving the effectiveness of nematicide usage so that smaller quantities may be used in a more selective manner, for example: (a) restricting their use for spot and row application, nursery bed, seed or bare-root dip treatments coupled with additional fertilizers, non-host or resistant varieties, etc., (b) establishing economic thresholds alongwith forecasting the abundance at which control is justified.

During the past many years several chemicals at different doses have been evaluated for their efficiency in managing nematodes infesting forage crops throughout the world. A brief account of present state of knowledge on chemical control in forage crops with special reference to India is described below :

The seed treatment technique is a useful way of giving seeds on inbuilt protection against nematodes, besides the advantages of lower dose rate and placing the chemicals, precisely where they are needed. Therefore, it has been extensively used by crop protection specialists. Further, this technique was found to be useful during the first half of the plant's life especially in case of cultivated annual forages or during establishing phase of pastures, when the plants are young, tender and even small number of nematodes in association with other organism can cause considerable damage. The chemicals used were found to be persistent enough to give effective protection against nematodes in later critical stage of plant growth (Ahmad *et al.*, 1988; Hasan *et al.*, 1986; Hoveland *et al.*, 1977). In early sixties complete control of seed borne *Ditylenchus dipsaci* was observed using phorate @ 1-4% as seed treatment (Gostik, 1963). Later, Hoveland *et al.* (1977) reported that seed treatment with nematicides holds considerable promise both in reducing the nematode population and higher forage yield of pasture grasses. Recently in India also, certain systemic nematicides like, carbofuran, aldicarb, phorate, fensulphothion @ 1-2% were successfully employed to control certain key nematode species infecting cowpea, lucerne, berseem, oat and maize (Ahmad *et al.*, 1988; Hasan, 1988; Khan & Husain, 1988; Sethi & Srivastava, 1986).

Application of vydate (oxamyl) as foliar spray @ 20 kg a.i./ha, 4 times at 2-week intervals increased the herbage by 28-31% and reduced nematodes by 87-96% (Smolick, 1978). Similarly several other chemicals either in the form of fumigant or systemic nematicide have also been extensively applied into the soil resulting into spectacular reduction in nematode population around a variety of forage crops throughout the world (Ericksson, 1972; Hasan, 1990). In India a number of nematicides as main field treatment have been evaluated against some major nematode species infecting cowpea, lucerne, berseem, mothbean, maize, sorghum, and oat. Of these carbofuran, aldicarb, phorate and phenamiphos @ 2-2.5 kg a.i./ha applied in furrows at the time of sowing proved to be most promising in nematode population reduction and higher forage yield (Hasan, 1988; Jain, 1980a, b, 1982; Jain & Hasan, 1984a; Verma *et al.*, 1981).

Chemical control involving more than one chemical may also some time leads to additive effects. For example Labruyere (1979) reported large responses in increased herbage yield with the combined application of a nematicide and fungicide treatment as compared to alone or either of them. Similarly in an another investigation a three way combination of herbicide (ethofumesate) + insecticide/nematicide (phorate) + fungicide (benomyl) has been reported as giving improved yield of sown grass compared with each of the component chemicals used on its own (Williams, 1984). Moreover, because of the recent growing trend towards finding multiple pest and disease control methods, identification and application of pesticides capable of killing nematodes, insects, fungi, etc. together is also required (Webster, 1972). For example carbofuran an insecticide/nematicidies was found to significantly reduce the damping off caused by *Pythium* sp. in alfalfa alongwith nematodes and insects (Grybauskas *et al.*, 1986).

NATURAL PLANT PRODUCTS/BOTANICAL PESTICIDES

In view of various problems explicit to chemical control of plant-parasitic nematodes, interest on the possibility of evolving nematicides of plant origin had recently been revived both in India and abroad. As a consequence nematotoxic potential of different kinds of natural plant products (NPP) especially in the form of crude extracts and certain chemical constituents of many plant species have been extensively investigated and reviewed (Gommers, 1981; Suryanarayan, 1990; Alam, 1993). Besides, a number of bioassay methods were also developed for these NPP, though largely based on laboratory or green house tests. However, application in the main field at large scale is still seems to be the biggest limiting factor for their success in nematode control. In the following paragraph an attempt has been made to mention briefly some of the representative references related to this aspect of nematode control.

The water extract of neem (*Azdirachta indica*), leucaena (*Leucaena leucocephala*), parthenium (*Parthenium hysterophorus*) alongwith several other plants were found to be toxic to some important nematode species especially *Meloidogyne in vitro* tests. Moreover, incorporation of either the extracts or leaves of these plants into the nematode infested soil has resulted into reduced nematode incidence of the host plants (Hasan & Jain, 1984b; Jain & Hasan, 1984b; Kali Ram & Gupta, 1980). The water soluble fractions or the bitter principles, i.e. nimbidin, thionimone, azadirachtin isolated from neem and parthinin from parthenium were reported to be effective both as inhibitors of egg hatching and larval mortality of root-knot nematode (Hasan, & Jain, 1984b; Khan *et al.*, 1984; Siddiqui & Alam, 1988, 1989a, b; Warthen, 1989).

Certain plants are known to have allelochemicals such as terthienyls, acetylenes, terpenoids, aldehydes, phenoxypropionic acid derivatives, etc. which have been recently reported to exhibit high nematicidal activity (Hayashi *et al.*, 1984; Kimura *et al.*, 1981; Mori *et al.*, 1982; Veech, 1979). Further, terpenoids in general and terpenoid lactones especially their synthetic derivatives and certain sesquiterpenoid phytoalexins for example rishitin, coumestro, glyceollin, cucurbitacin in particular are now emerging as a new group of nematotoxic compounds. These compounds have been shown to either repel or kill the nematode at a very low concentration; moreover, their presence in the plants have been attributed to both pre-infectional and post-infectional resistance to nematodes (Alphey & Robertson, 1988; Mahajan *et al.*, 1986; Malik *et al.*, 1987; Veech, 1981). Unfortunately the biochemical mode of action of most of the compounds are not yet known, however, Alphey & Robertson (1988) have emphasised that the potential nematotoxic properties shown by them may be sufficient to protect a crop system from nematode attack.

Sangwan *et al.*, (1985) and Hasan (unpublished) have reported nematicidal properties of essential oils isolated from certain plant species. Of these *Trachyspermum ammi* oil exhibited maximum toxicity to root-knot nematode. In addition, the infectivity of egg masses was found to be greatly reduced when exposed in the above oil resulting in a very low gall intensity of cowpea. Sesquiterpenes are the major chemical component of all essential oils, therefore, the potential nematotoxic properties shown by *T. ammi* oil has been attributed to the presence of some specific biologically active sesquiterpenes in it (Hasan, unpublished).

Some of the major problems in the successful use of NPP are: (a) their large scale extraction, synthesis and availability, (b) their storage and field stability because of rapid degradation and destruction of the extracts or chemical constituents (Larson, 1987). Perhaps due to these reasons none of the NPP have yet attained commercial status. Nonetheless, such biodegradable and environmentally acceptable natural plant products particularly from neem tree may be of potential value in nematode control in several ways: (a) they may be a source of structures for screening, (b) they may possess activity that is applicable to nematode control directly or after structural modification of the original structure and finally, recognition of their function in nature may suggest new approaches to nematode control.

INTEGRATED NEMATODE MANAGEMENT (INM)

As obvious from the proceeding discussions, in many, if not most cases, various control measures by themselves are not free from one or the other limitations to provide safe, economically acceptable and efficient nematode population suppression in agricultural cropping systems. Further it needs to be emphasised that nematode diseases rarely reach the epidemic level. However, severe nematode infestations of the crops are likely to result if: (a) susceptible host is repeatedly sown, (b) a high population of polyspecific/a particular species or a race of nematode occur, (c) environment is favourable for nematode multiplication, development, attack and spread. If any of these three factors is limiting in any way, crop damage is not likely to be severe. Therefore, there is a great scope of developing and implementing integrated pest management system (IPM), a subject of considerable current interest (Fig.1).

The definition of IPM that has been most used was coined by Smith & Reynolds (1966) in an FAO symposium on integrated pest control as "Integrated pest control is a pest population management system that utilizes all suitable techniques in a compatible manner to reduce pest population and maintain them at levels below those causing economic injury". Integrated control achieves this idea by harmonising techniques in an organised way, by making control practices compatible, and by blending them into a multifaceted flexible evolving system. Subsequently Bird (1981) emphasised that integrated nematode management (INM) must also be implemented in conjunction with the broader objectives of IPM. Later Sethi & Gaur (1986) suggested that the aims of INM approach should be: (a) to utilise several compatible control techniques in combination, (b) to maximise natural environmental resistance to plant parasitic nematodes, (c) to apply specific and drastic control measures only as and when necessary, (d) to maximise profit to the particular grower with location and resource specific recommendation.

Of these, economic considerations, in INM system has gained more importance in recent years perhaps because of the involvement of farmer's economic interest. For example, Reis *et al.* (1983) considered 24 different methods of controlling soybean cyst nematode with combination of varieties, fertilizers, nematicide and crop rotation to select the control method that maximise the income over variable costs from the farm at Illinois, U.S.A. A sample of the solution indicated that part of the farm would be in continuous soybean with a resistant cultivar and the balance was to be in corn-soybean rotation using a susceptible cultivar and nematicide, thus maximising the net return. Bird (1981) has described in more

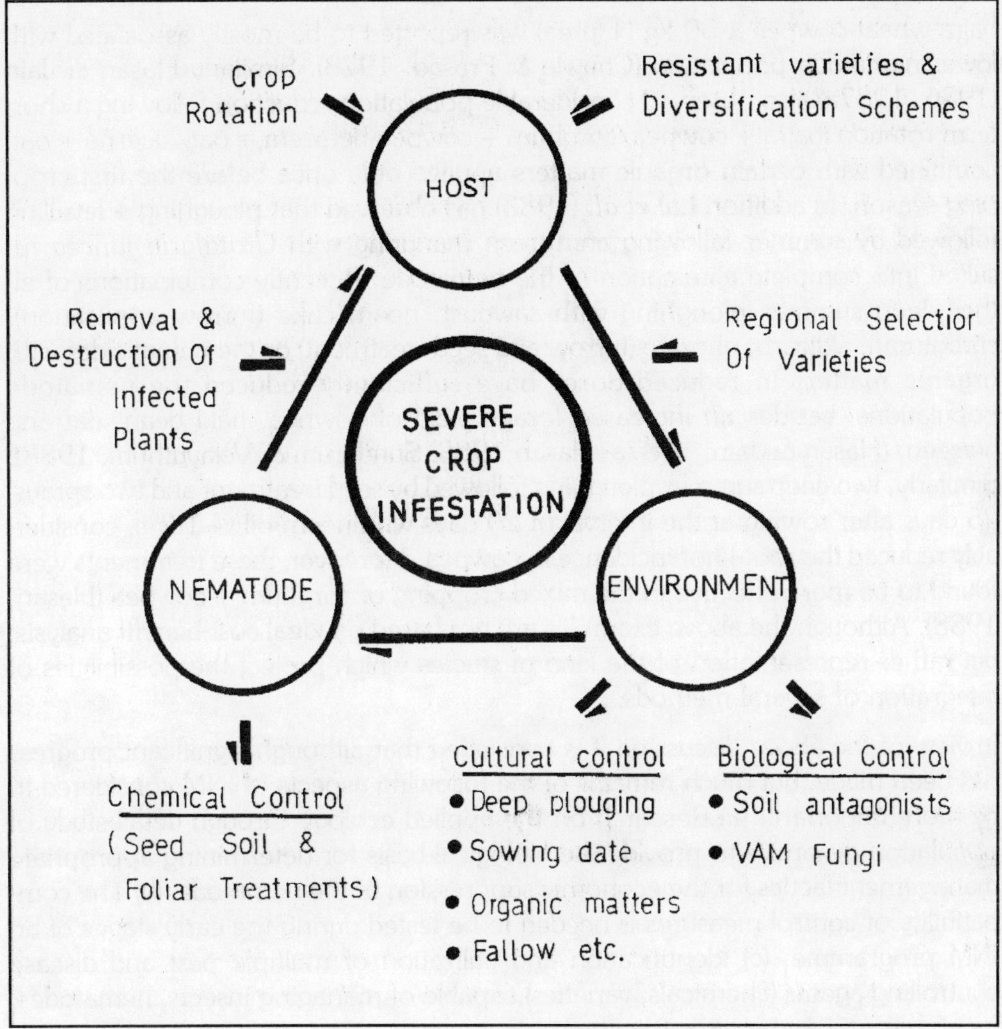

Fig.1. A hypothetical nematode infestation and its integrated control.

detail the components of INM for plant protection as: (a) Biological and environmental monitoring, (b) Agricultural production system (APS), (c) Pest crop ecosystem models and system design and implementation. Further he has also emphasised that the INM system consists of the principle of exclusion, population reduction and tolerance in relation to threshold level and not eradication of nematodes which does not appear to be a practical proposition.

In India substantial efforts have been made to combine several methods to manage nematodes in forage crops, though on limited experimental stage. For instance certain intensive rotation combined with fertility management (sorghum/

bajra-wheat-cowpea x 50 kg N (urea) was reported to be mostly associated with lower nematode population (Chawla & Prasad, 1973). Similarly Hasan & Jain (1986, 1987a) also observed considerable population reduction following a short term rotation (bajra + cowpea/sorghum + cowpea-berseem + oat/lucerne + oat) combined with certain organic matters applied only once before the first cropping season. In addition Lal *et al.* (1983) has observed that ploughing + levelling followed by summer fallowing and green manuring with *Crotalaria juncea* resulted into complete elimination of the nematode. Recently combinations of either deep summer ploughing with sawdust, neem cake (furrow application), carbofuran, aldicarb, phorate (furrow and seed treatment) or the nematicides with organic matters in reduced doses have sufficiently reduced the nematode populations, besides an increased forage yield of cowpea, field bean, oat and berseem (Hasan & Jain, 1987a; Hasan 1988; Sundarum & Velayuthum, 1988). Similarly, two deep summer ploughing followed by seed treatment and two sprays, 25 days after sowing at the interval of 20 days with neem oil @ 1-2%, considerably reduced the root-knot incidence in cowpea. Moreover, these treatments were found to be more effective in the mixed cropping of sorghum + cowpea (Hasan, 1988). Although the above examples are not based on ideal cost-benefit analysis, but rather representation of the kind of studies which project the possibilities of integration of several methods.

In view of the above discussion it is concluded that although significant progress has been made, but much remains of the following aspects of INM considered to be more important: (a) Research on the applied ecology through detail study of population dynamics to provide the biological basis for determining appropriate management tactics for the economic suppression of the nematode, (b) The compatibility of control measures is needed to be tested during the early stages of an INM programme, (c) Identification and utilization of multiple pest and disease controlling agents (chemicals, varieties) capable of managing insects, nematodes, fungi, bacteria and viruses together.

REFERENCES

Ahmed, S.T.; Faruqui, S.A. & Hasan, N. (1988). Disease and pest management in forages. *Indian Farming* **37**: 26-27.

Akhtar, M. & Alam, M.M. (1992). Integrated nematode control with organic amendments/nematicide and intercropping of some fodder crops. *First Afro-Asian Nematol. Symp. Aligarh (Nov. 29-Dec.3)* : 2.

Alam, M.M. (1976). Organic amendments in relation to nematodes. Ph. D. thesis, *Aligarh Muslim University, Aligarh, India.*

Alam, M.M. (1993). Bioactivity against phytonematodes. *In: Neem Research & Development.* (Eds.—N.S. Randhawa & B.S. Parmar). Soc. of Pesticide Sci., India. Pub. No. 3. pp. 123-143.

Alam, M.M. & Jairajpuri, M.S. (1990). Natural enemies of nematodes. *In* : *Nematode Biocontrol: Aspects & Prospects* (Eds.—M.S. Jairajpuri, M.M. Alam & I. Ahmad). CBS Publishers & Distributors, Delhi, pp. 17-40.

Alam, M.M. Saxena, S.K. & Khan, A.M. (1979). Suitability of crops to the reniform, stunt and root-knot nematodes. *In* : *Recent Researches in Plant Sciences* (Eds.-S.S. Bir). Kalyani Publisher, New Delhi, India. pp. 630-634.

Alphey, T.J.W. & Robertson, W.M. (1988). Assessment of nematicidal potential of natural plant products. *Nematologica* **34:** 253.

Atilano, R.A.; Menge, J.A. & Van Gundy, S.D. (1981). Interaction between *Meloidogyne arenaria* and *Glomus fasciculatum* in grape. *J. Nematol.* **13:** 52-57.

Ayala, A.; Roman, J. & Gonzalex, J. (1967). Pangola grass as a rotation crop for pineapple nematode control. *J. Agric. Univ. Puerto Rico* **51:** 94-96.

Baker, K.F. & Cook, R.J. (1974). *Biological Control of Plant Pathogens*. W.H. Freeman, San Francisco.

Bingefors, S. (1969). The use of nematode resistant varieties of grasses and legumes. *Herbage Abstract* **39:** 107-111.

Bingefors, S. (1971). Mutation breeding for disease resistance. *In* : *Proceeding of the International Atomic Energy Authority, Vienna*. pp. 209-235.

Bird, G.W. (1981). Integrated nematode management for plant protection. *In*: *Plant Parasitic Nematodes*, Volume III (Ed.—B.M. Zuckerman & R.A. Rohde). Academic Press, London, New York, pp. 355-357.

Brodie, B.B.; Good, J.M. & Jaworski, C.A. (1970). Population dynamics of plant parasitic nematodes in cultivated soil: Effect of summer cover crops in old agricultural land. *J. Nematol.* **2:** 147-151.

Burton, G.W. & Johnson, A.W. (1987). Coastal bermuda grass rotation for control of root-knot nematodes. *J. Nematol.* **19:** 138-140.

Caviness, C.E.; Walter, H.J.; Riggs, R.D. & Beatty, K. (1977). Soybean varietal performance. *Arkansas Farm Research* **26:** 2.

Chawla, M.L. & Prasad, S.K. (1973). Multiple cropping and nematodes. I: Effect of fertility management and intensive rotations on nematode population and crop yield. *Indian J. Nematol.* **3:** 34-49.

Cook, R.J. & Baker, K.F. (1983). *The Nature and Practice of Biological Control of Plant Pathogens*. American Phytopathological Society, Saint Paul, Minnesota. pp. 539.

Cooper, K.M. & Grandison, G.S. (1986). Interaction of vesicular-arbuscular mycorrhizal fungi and root-knot nematode on cultivars of tomato and white clover susceptible to *Meloidogyne hapla*. *Ann. Appl. Biol.* **108:** 355-365.

Daniels, B.A. & Menge, J.A. (1981). Evaluation of the commercial potential of the vesicular-arbuscular mycorrhizal fungus *Glomus epigaeus*. *New Phytologists* **87:** 345-454.

Dasgupta, D.R. & Ganguly, A.K. (1986). Plant resistance to nematodes. *In* : *Plant Parasitic Nematodes of India, Problems & Progresss* (Eds.-G. Swarup, & D.R. Dasgupta). Indian Agricultural Research Institute, New Delhi, India. pp. 458-475.

Daulton, R.A.C. (1962). Survey of land use and plant parasitic nematode control practices in the southern flue cured tobacco areas of Rhodesia. *Rhodesian J. Agric. Res.* **59:** 216-217.

Davide, R.G. (1985). International *Meloidogyne* project in region VI. *In*: *An Advance Treatise on Meloidogyne. Vol. I. Biology & Control* (Eds.—J.N. Sasser & C.C. Carter). North Carolina State University, U.S.A. pp. 369-372.

Decker, H. (1969). *Phytonematology*. Veb Deutscher Landwirtschafts verlag, Berlin.

Donnelly, E.D. (1981). Registration of All-Lotan *Sericea lespedeza* (Rg. No. 11). *Crop Science* **21:** 474.

Egunjobi, O.A. (1985). The international *Meloidogyne* project in region IV. *In* : *An Advance Treatise on Meloidogyne. Vol. I. Biology & Control* (Eds.—J.N. Sasser & C.C. Carter). North Carolina State University, U.S.A., pp. 353-360.

Elgin.Jr., J.H. (1982). Registration of W10 multiple pest resistant alfalfa germplasm (Reg. no. GP 116). *Crop Sci.* **22:** 163.

Ericksson, K.B. (1972). Nematode diseases of pasture legumes and turf grasses. *In* : *Economic Nematology* (Eds.—J.M. Webster). Academic Press, London, New York, pp. 66-69.

Ferraz, S. (1985). International *Meloidogyne* project in Brazil (Region III). *In*: *An Advance Treatise on Meloidogyne. Vol. I. Biology & Control* (Eds.—J.N. Sasser & C.C. Carter). North Carolina State University, U.S.A., pp. 351-352.

Gaur, H.S. & Prasad, S.K. (1986). *In*: *Proceeding of National Conference on short term and long term hazards of pesticides and strategies for their safe use*. National Academy of Science, India: pp. 66-77.

Godoy, G.; Rodriguez-Kabana, R.; Shelby, R.A. & Morgan-Jones, G. (1983). Chitin amendments for control of *Meloidogyne arenaria* in infested soil. II. Effect on microbial population. *Nematropica* **13:** 63-74.

Gommers, F.J. (1981). Biochemical interactions between nematodes and plant and their relevance to control. *Helminthol. Abstr. Series B, Plant Nematol.* **50:** 9-24.

Gooris, J. & Herde, J.D. (1967). Distribution, Host plants and Biology of the grass root-knot nematode, *Meloidogyne naasi*. *Revue Agric. Brux*. **20:** 1055-1065.

Gostik, K.G. (1963). Control of seed borne lucerne stem eelworm with phorate. *Ann. Appl. Biol.* **51:** 503-507.

Grandison, G.S. & Cooper, K.M. (1986). Interaction of vesicular-arbuscular mycorrhizae and cultivars of alfalfa susceptible and resistant to *Meloidogyne hapla*. *J. Nematol.* **18:** 141-149.

Grybauskas, A.P.; Decker, A.M.; Vaugh, L.R. & Abdul-Gawad, K.I. (1986). A non-target effect of carbofuran on damping off in alfalfa. *Phytopathology* **76:** 563.

Hanna, M.R. & Hawn, E.J. (1975). Treck, alfalfa. *Canadian J. Plant Sci.* **55:** 847-849.

Hartman, R.W. (1976). Breeding for nematode resistance in vegetables. *Sabrao J.* **8:** 1-10.

Hartwig, E.E. & Epps, J.M. (1977). Registration of centennial Soybeans (Reg. No. 114). *Crop Sci.* **17:** 979.

Hasan, N. (1977). Effect of oil-cakes, sawdust and certain chemical compounds on the development of root-knot on eggplant and tomato. *Ph. D. thesis, Aligarh Muslim University, Aligarh*, India.

Hasan, N. (1985a). New host records of the root-knot nematode (*Meloidogyne incognita*). *Indian J. Plant Science* **3:** 80-82.

Hasan, J. (1985b). Host status of some important cultivated legumes in relation to root-knot nematode *Meloidogyne incognita. Proc. 5th Nematol. Symp. Udaipur, India*: p. 79.

Hasan, N. (1988). Plant parasitic nematodes associated with herbaceous forages and their management. *Annual Report*, Indian Grassland and Fodder Research Institute, Jhansi, India. pp. 88-90.

Hasan, N. (1990). Nematode problems of forage grasses and legumes. *In* : *Progress in Plant Nematology* (Eds.-S.K. Saxena *et al.*) CBS Publishers & Distributors, Delhi, India, pp. 433-459.

Hasan, N.; Ahmed, S.T. & Pandey, K.C. (1986). Pest and diseases of rangelands and their importance in forage production. *In*: *Proc. 3rd Int. Rangeland Cong.* Delhi, India, p. 393.

Hasan, N. & Jain, R.K. (1984a). Effect of soil amendments on fodder production, photosynthetic pigments and nematodes associated with berseem (*Trifolium alexandrinum* L.) followed by bajra (*Pennisetum typhoides* L.). *Agric. Sci. Digest* **4:** 12-14.

Hasan, N. & Jain R.K. (1984b). Biotoxicity of *Parthenium hysteroporus* L. extracts against *Meloidogyne isncognita* and *Helicotylenchus dihystera*. *Nematologia Medit.* **12:** 239-242.

Hasan, N. & Jain, R.K. (1984c). Screening of mothbean (*Vigna aconitifolius* Jacq.) germplasm for resistance to root-knot nematode (*Meloidogyne incognita*). *Agric. Sci. Digest* **4:** 198-199.

Hasan, N. & Jain, R.K. (1985a). Preliminary assessment of the response of *Clitoria ternatea* lines to the root-knot nematode, *Meloidogyne incognita*. *Nematologica* **31:** 236-238.

Hasan, N. & Jain, R.K. (1985b). Response of some selected *Sesbania* species to root-knot nematode *Meloidogyne incognita*. *Nematol. Medit.* **13:** 15-19.

Hasan, N. & Jain, R.K. (1986). Plant parasitic nematodes associated with herbaceous forage legumes and their management. *Annual Report*, Indian Grassland and Fodder Research Institute, Jhansi, India, pp. 178-179.

Hasan, N. & Jain, R.K. (1987a). Plant parasitic nematodes associated with herbaceous forage legumes and their management. Annual Report, Indian Grassland and Fodder Research Institute, Jhansi, India, pp. 124-125.

Hasan, N. & Jain R.K. (1987b). Parasitic nematodes and vesicular-arbuscular mycorrhizal (VAM) fungi associated with berseem (*Trifolium alexandrinum L*). in Bundelkhand region. *Indian J. Nematol*. **17:** 184-188.

Hasan, N. & Jain, R.K. (1988). Screening clusterbean *Cyamopsis tetragonoloba* L. (Taub.) germplasm for resistance to root-knot nematode *Meloidogyne incognita* (Kofoid & White) Chitwood and studies on host nematode relationship. *Indian J. Botany* **11:** 33-39.

Hasan, N. & Jain, R.K. (1992). Field application of *Paecilomyces lilacinus* in combination with certain organic matters for controlling *Meloidogyne incognita* infecting cowpea followed by lucerne. *First Afro-Asian Nematol. Symp. Aligarh* (Nov. 29-Dec. 3): 10-11.

Hawn, E.J. & Kuzub, G.C. (1978). Influence of *Pratylenchus projectus* on alfalfa sickness in alberta. *Canadian Plant Dis. Survey* **58:** 1-4.

Hayashi, M.: Wada, K. & Munakata, K. (1983). Synthesis and nematicidal activity of phenoxypropionic acid derivatives. *Agricultural Biological Chemistry* **47:** 2653-2655.

Hodges, L.R. (1973). Nematodes and their control. *Union Carbide Corporation*, Salinas, California.

Hooper, D.J. & Stone, A.R. (1981). Role of wild plants and weeds in the ecology of plant parasitic nematodes. *In* : *Pest, Pathogen and Vegetation* (Ed.-M. Thresh). Pitman Publisher, London, Boston. pp. 199-215.

Hoveland, C.S.; Haaland, R.L. & Rodriguez-Kababa, R. (1977). Soil nematode and forage production of some cool season grasses. *Proceeding XIII International Grassland Congress*. Leipzig, G.D.R. pp. 1257-1259.

Hussey, R.S. & Roncadori, R.W. (1982). Vesicular-arbuscular mycorrhizae may limit nematode activity and improve plant growth. *Plant Dis.* **66:** 9-14.

Jain, R.K. (1980a). Control of *Tylenchorhynchus mashoodi* on berseem. *Indian J. Nematol*. **10:** 98-100.

Jain, R.K. (1980b). Effect of granular nematicides on green fooder production of M.P. chari sorghum and nematode population. *Indian J. Mycol. Plant Pathol.* **10:** 2.

Jain, R.K. (1982). Control of *Tylenchorhynchus vulgaris* on maize. *Agric. Sci. Digest* **2:** 143-144.

Jain, R.K. & Hasan, N. (1984a). Increased fodder production by chemical control of lesion nematode, *Pratylenchus zeae* on maize. *Pesticides* **18:** 48.

Jain, R.K. & Hasan, N. (1984b). Toxicity of Koo-babool extracts to *Meloidogyne incognita* and *Helicotylenchus dihystera*. *Indian J. Nematol.* **14:** 179-181.

Jain, R.K. & Hasan, N. (1986a). Efficacy of neem cake on fodder production, photosynthetic pigments and nematodes associated with oats and its residual effect on cowpea. *Indian J. Nematol.* **16:** 98-100.

Jain, R.K. & Hasan, N. (1986b). Indian Grassland and Fodder Research Institute, Jhansi. *In* : *Nematology in India* (Institutes and Universities). Indian Agricultural Research Institute, New Delhi, India. pp. 28-32.

Jain, R.K. & Hasan, N. (1987). Plant parasitic nematodes associated with herbaceous forage grasses and their management. *Annual Report*, Indian Grassland and Fodder Research Institute, Jhansi, India. pp. 125-127.

Jain, R.K. & Hasan, N. (1988). Role of vesicular-arbuscular mycorrhizal (VAM) fungi and nematode activities in forage production. *Acta Botanica Indica* **16:** 84-88.

Jain, R.K. & Sethi, C.L. (1987). Pathogenicity of *Heterodera cajani* on cowpea as influenced by the presence of VAM fungi *Glomus fasciculatum* or *G. epidaeus*. *Indian J. Nematol* **17:** 165-170.

Jain, R.K. & Sethi, C.L. (1988a). Interaction between vesicular-arbuscular mycorrhiza and *Meloidogyne incognita*, *Heterodera cajani* on cowpea as influenced by time of inoculation. *Indian J. Nematol.* **18:** 263-268.

Jain, R.K. & Sethi, C.L. (1988b). Influence of endomycorrhizal fungi *Glomus fasciculatum* and *G. epigaeus* on penetration and development of *Heterodera cajani* on cowpea. *Indian J. Nematol.* **18:** 89-93.

Jatala, P. (1986). Biological control of plant parasite nematodes. *Ann. Review Phytopathol.* **24:** 453-489.

Jonsson, H.A. (1976). Britta — a red clover variety with considerable disease resistance. *Plant Breeding Abstact* **46:** 461.

Kali Ram & Gupta, D.C. (1980). A note on the efficacy of fresh neem leaf extract in the control of *Meloidogyne javanica* infecting chickpea (*Cicer arietinum*). *Indian J. Nematol.* **10:** 96-98.

Kaplan, D.T. (1982). Plant resistance to nematodes: Symposium Introduction. *J. Nematology* **14:** 1-2.

Kerry, B.R. (1984). Nematophagus fungi and the regulation of nematode population in soil. *Helminthol. Abstr. Series B* **53:** 1-14.

Khan, M.W.; Alam, M.M.; Khan, A.M. & Saxena, S.K. (1974). Effect of water soluble fractions of oil-cakes and bitter principles of neem or some fungi and nematodes. *Acta Botanica Indica* **2:** 120-128.

Khan, T.A. & Husain, S.I. (1988). Efficacy of the seed treatment with pesticides, oil-cakes, neem leaf and cultural filtrate of *Paecilomyces lilacinus* for the control of disease caused by the presence of *Rotylenchulus reniformis*, *Meloidogyne incognita* and *Rhizotonia solani* either individually or concomitantly on cowpea. *Indian J. Nematol.* **18:** 192-198.

Kimura, Y.; Mori, M.; Hyeons, S.; Suzuki, A. & Mitsui, Y. (1981). A rapid and simple method for the assay of nematicidal activity and its application to measuring the activities of dicarboxylic esters. *Agricultural Biological Chemistry* **45:** 249-251.

Koen, H. & Grobbelaar, N. (1968). The detrimental effect of *Eragrotris curvula* on the *Meloidogyne javanica* population of soil. *Nematologica* **11** : 573-580.

Kretschmer, A.E.; Sonoda, R.M. & Synder, G.H. (1980). Resistance of *Desmodium heterocarpon* and other tropical legumes to root-knot nematodes. *Tropical Grassland* **14:** 115-120.

Labruyere, R.E. (1979). Resowing problems of old pastures. *In: Soil borne plant pathogens* (Eds.—B. Schippers & W. Gams.). Academic Press, London, New York. pp. 313-326.

Lal, A.; Sanwal, K.C. & Mathur, V.K. (1983). Changes in the nematode population of undisturbed land with the introduction of land development practices and cropping sequences. *Indian. J. Nematol.* **13:** 133-140.

Larson, R.O. (1987). Development of Margosan-O, a pesticide from need seed. *In : Proceeding 3rd International Neem Conference* (Eds.—H. Schmutterer & K.R.S. Ascher) Eschborn, Germany. pp. 243-250.

Lehman, W.F.; Ede, L.; Marble, V.L.; Nelson, M.W. & Radewald, J.D. (1983). Registration of U.C. Cibola alfalfa. *Crop Sci.* **23:** 1216.

Mahajan, R.; Singh, P.; Bajaj, K.L. & Kalsi, P.S. (1986). Nematicidal activity of some sesquiterpenoids against root-knot nematode *Meloidogyne incognita. Nematologica* **32:** 201-209.

Malik, M.S.; Sangwan, N.K.; Dhindsa, K.S. & Bhatti, D.S. (1987). Nematicidal efficacy of some monoterpenes and related derivatives. *Pesticides* **21:** 30-32.

Mankau, R. (1981). Microbial control of nematodes. *In: Plant Parasitic Nematodes* (Eds.—B.M. Zuckerman, W.F. Mai & R.A. Rhode). Academic Press, London, New York. pp. 475-494.

Midha, S.K. (1985). Efficacy of *Paecilomyces lilacinus* in controlling root-knot infestation on cowpea and mungbean. *Indian J. Nematol.* **15** : 267.

Mori, M.; Hyeons, S.; Kimura, Y. & Suzuki, A. (1982). The nematicidal activity of acetylence compounds. *Agricultural Biological Chemistry* **46:** 309-311.

Mukhopadhyaya, M.C. & Prasad, S.K. (1968). Population dynamics of *Tylenchorhynchus. Nematologica* **14:** 404-418.

Muller, R. & Gooch, P.S. (1982). Organic amendments in nematode control : An examination of literature. *Nematropica* **12:** 319-326.

Noe, *J.P. (1986). Cropping system analysis for limiting losses due to plant parasitic nematodes. In: Guide to Research Methodology*. Releigh, North Carolina, U.S.A. pp. 1-17.

Norton, D.C. (1967). Relationship of *Heterodera trifolii* to some forage legumes. *Phytopathology* **57:** 1305-1308.

Norton, D.C. (1978). *Ecology of Plant Parasitic Nematodes*. John Wiley & Sons, New York. p. 268.

Nusbaum, C.J. & Ferris, H. (1973). The role of cropping system in nematode population management. *Ann. Review Phytopathol.* **11:** 423-448.

Oostenbrink, M. (1961). Nematode in relation to plant growth. III. *Pratylenchus penetrans* (Cobb) in tree crops, potato and red clover. *Netherland J. Agric. Sci.* **9:** 188-209.

Oostenbrink, M. (1964). Hormonious control of nematode infestation. *Nematologica* **10:** 49-56.

Orton, W.A. & Webber, H.J. (1902). The wilt disease of cowpea and its control. *United States Department of Agriculture, Bureau Plant Industry Bulleting* **17:** 9-20.

Pathak, N.N. & Jakhmola, R.C. (1986). *Forages and livestock production*. Vikas Publishing House Private Limited, Delhi, India.

Peaden, R.N.; Evans, D.W.; Elgin Jr., J.H. & Rincker, C.M. (1983). Vernema alfalfa. *Crop Sci.* **23:** 1009.

Prasad, S.K. & Gaur, H.S. (1974). Relation between nematode population at harvest and yield of wheat. *Indian J. Nematol.* **4:** 152-159.

Reddy, K.C. (1983). Nitrogen fixation and nematode resistance of 13 tropical legumes. *Dissertation Abstr. Int. B.* **44:** 380.

Reeves, F.B.; Wagner, D.; Moorman, T. & Kiel, J. (1979). The role of endomycorrhiza in revegetation practices in the semi-arid west. I. A comparison of incidence of mycorrhiza in severely disturbed natural environment. *American J. Botany* **66:** 6-13.

Reis, R.P.; Noel, G.R. & Swanson, E.R. (1983). Economic analysis of an alternative control method for soybean cyst nematode in Southern Illinois. *Plant Dis.* **67:** 480-483.

Reynolds, H.W. & O' Banon, J.H. (1960). Reaction of sixteen varieties of alfalfa to two species of root-knot nematodes. *Plant Dis. Reptr.* **44:** 441-443.

Rhoades, H.L. (1976). Effect of *Indigofera hirsuta* on *Belonolaimus longicaudatus*, *Meloidogyne incognita* and *M. javanica* and subsequent crop yield. *Plant Dis. Reptr.* **60:** 384-386.

Saleh, H. & Sikora, R.A. (1984). Relationship between *Glomus fasciculatum* root colonization of cotton and its effect on *Meloidogyne incognita. Nematologica* **30:** 230-237.

Sangwan, M.K.; Verma, K.K.; Verma, B.S.; Malik, M.S. & Dhindsa, K.S. (1985). Nematicidal activity of essential oils of *Cymbopogon* grasses. *Nematologica* **31:** 93-99.

Sesser, J.N. & Freckman, D.W. (1987). *Vistas on Nematology* (Eds.—J.A. Veech & D.W. Dickson). Hayattsville, Maryland, U.S.A. pp. 7-14.

Schenck. N.C.; Kinloch, D.W. & Dickson, D.W. (1975). *In*: *Endomycorrhizas* (Eds.- Sanders *et al.*). Academic Press, London, New York. pp. 607-617.

Schonbeck, F. (1979). Endomycorrhiza in relation to plant disease. *In* : *Soil borne plant pathogens* (Eds.-B. Schippers & W. Gams). Academic Press, London. pp. 271-280.

Schonbeck, F. (1984). Endomycorrhizae. *In*: *Progress in Botany Vol.* 46. (Eds.—Bochum *et al.*) Springer-verlag, Berlin, Heidelberg, New York, Tokyo. pp. 382-390.

Sethi, C.L. & Gaur, H.S. (1986). Nematode management: An overview. *In* : *Plant parasitic nematodes of India — Problem and Progress* (Eds.—G. Swarup & D.R. Dasgupta). Indian Agricultural Research Institute, New Delhi, India. pp. 424-445.

Sethi, C.L. & Gaur, H.S. (1990). Population dynamics of plant parasitic nematodes in relation to cropping systems. *In*: *Progress in Plant Nematology*. (Eds.-S.K. Saxena *et al.*). CBS Publishers & Distributors, Delhi, India. pp. 223-262.

Sethi, C.L. & Srivastava, A.N. (1986). Management of *Heterodera* zeae on maize through pesticidal seed treatment. *Indian J. Nematol.* **16:** 82-86.

Shafshak, S.E.; El-Debaby, A.S.; Sary; G.A. & Salem, M.S. (1974). Effect of preceding winter crops on the growth and yield of two succeeding crops, maize and wheat. *Zeitschrift Fuer Ackerund Pflanzenban* **140:** 44-53.

Shagalina, L.M. (1973). On lucerne resistance to some pathogenic nematode. *Izvestiya Akademii* Nauk **1:** 48-51.

Sharma, J.L. & Trivedi, P.C. (1986). Biological control of *Meloidogyne incognita* using fungus *Paecilomyces lilacinus In*: *Proceeding of National Conference on Plant Parasitic Nematodes of India, Problems and Progress*. Indian Agricultural Institute, New Delhi, India. p. 86.

Shephard, J.A. (1978). T.R.B. Tests of new love grass strain and recommends it. *Rhodesian Tobacco Today* **1:** 15.

Shephard, J.A. & Coombs. R.F. (1979). The effect of four *Meloidogyne* species (Nematoda: Meloidogynidae) on *Panicum maximum* cv. Umtali, *Rhodesian J. Agric. Res.* **17:** 155-156.

Siddiqui, M.A. & Alam, M.M. (1988). Effect of seed treatment with azadirachtin on root-knot development on, and growth of, some vegetables. *Tests Agrochem. Cult.* **9** (*Ann. appl. Boil.* 112 Suppl.): 20-21.

Siddiqui, M.A. & Alam, M.M. (1989a). Seed treatment with azadirachtin for the control of the stunt nematode attacking cabbage and culiflower. *Tests Agrochem. Cult.* **10:** (*Ann. appl. Biol. 113 Suppl.*) : 4-5.

Siddiqui, M.A. & Alam, M.M. (1989b). Efficacy of azadirachtin as seed treatment for the control of reniform nematode on some vegetables. *Tests Agrochem. Cult.* **10** (*Ann. appl. Biol. 133 Suppl.*): 20-21.

Sidhu, G.S. & Webster, J.M. (1981a). The genetics of plant nematode parasitic systems. *Botanical Review* **47:** 387-419.

Sidhu, G.S. & Webster, J.M. (1981b). Genetics of plant nematode interaction. *In* : *Plant parasitic nematodes Vol. III* (Eds.-B.M. Zuckerman & R.A. Rhode). Academic Press, London, New York. pp. 61-83.

Sikora, R.A. & Schonbeck, F. (1975). Mycorrhiza (*Endogone mosseae*) on the population dynamics of the root-knot nematode (*Meloidogyne incognita* and *M. hapla*). *8th Int. Cong. Plant Protect.* **5:** 158-166.

Singh, B.; Bhati, D.S. & Singh, K. (1974). Resistance to root-knot nematodes (*Meloidogyne* spp.) in vegetable crops. *PANS* **20:** 58-67.

Sitaramaiah, K. (1990). Mechanism of reduction of plant parasitic nematodes in soil amended with organic materials. *In*: *Progress in plant nematology* (Ed.—S.K. Saxena *et al.*). CBS Publishers Distributors, Delhi. pp. 263-295.

Smith, G.S. (1987). Interaction of nematodes with mycorrhizal fungi. *In* : *Vistas on Nematology* (Eds.—J.A. Veech & D.W. Dickson). Hayattsville, Maryland, U.S.A. pp. 292-300.

Smith, R.F. & Reynolds, H.T. (1966). Principles, definition and scope of integrated pest control. *In*: *Proc. FAO Symp. Integrated Pest Control,* Rome, Italy. **1:** 11-17.

Smolik, J.D. (1974). Nematode studies at the cotton-wood site, *US/IBP Grassland Biome Technical Report No. 251*, Fort Collins, Colorado; Colorado State University, U.S.A.

Smolik, J.D. (1978). Effect of nematicides treatment on growth of range grasses in field and glasshouse studies. *Science Series, Range Science Department*, Colorado State University, U.S.A., No. **26:** 257-260.

Soffes, A.R.; Prine, G.M. & Dunn, R.A. (1980). Nitrogen fixation and effect on plant parasitic nematodes by eight legume cover crops. *72nd Annual meeting, American Soc. Agronomy*, Madison, Wisconsin, U.S.A. pp. 105.

Spaull, A.M.; Clement, R.O. & Newton, P.G. (1983). The effect of root ectoparasitic nematodes on grass establishment and productivity. *In: Proceeding of the 9th General meeting of the European Grassland Federation,* Reading, U.K. (Ed.-J. Corrall). pp. 310-311.

Srivastava, A.N. & Sethi, C.L. (1986). Seasonal population fluctuation of *Heterodera zeae, Rotylenchulus reniformis* and *Tylenchorhynchus vulgaris* in maize, cowpea and wheat rotations. *Indian J. Nematol.* **16:** 16-18.

Subramaniyan, S. (1985). Stylo (*Stylosanthes* spp.) a natural host of *Radopholus similis* (Cobb, 1893) Thorne, 1949. *Proc. 4th Nematol. Symp.* Udaipur, India. p. 78.

Sundaram, R. & Vellayutham, B. (1988). Relative efficacy of some insecticides and neem cake in the control of *Rotylenchulus reniformis* and *Helicotylencus dihystera. Indian J. Nematol.* **18:** 329-331.

Suryanarayana, Y.V. (1990). Botanical pesticides: a glance. *In*: *Souvenir National Symposium on*

Problems and Prospects of Botanical Pesticides in Integrated Pest Management, Indian Society of Tobacco Science, Rajahmundry, India. pp. 19-73.

Swarup, G. & Gokte, N. (1986). Biological control. *In* : *Plant Parasitic Nematodes of India — Problems and Progress* (Eds.—G. Swarup & D.R. Dasgupta). Indian Agricultural Research Institute, New Delhi, India. pp. 476-479.

Van Gundy, S.D. (1985). Biological control of nematodes: Status and prospects in agricultural IPM System. *In* : *Biological Control in IPM System* (Eds.- M.A. Hoy & D.C. Herzog). Academic Press, New York, London. pp. 467-478.

Veech, J.A. (1979). Histopathological localization and nematotoxicity of terpenoid aldehydes in cotton. *J. Nematol.* **11:** 240-246.

Veech, J.A. (1981). Plant resistance to nematodes. *In*: *Plant Parasitic Nematodes Vol. III* (Eds.— B.M. Zuckerman & R.A. Rohde). Academic Press, London, New York. pp. 377-400.

Verma, M.K.; Sharma, R.C. & Mathur, J.R. (1981). Efficacy of granular nematicides against *Meloidogyne incognita* in moth bean *Vigna aconitifolius. Nematol. Medit.* **9:** 109-110.

Warthen Jr., J.D. (1989). Neem (*Azadirachta indica H. Juss.*). Organisms affected and references list uptodate. *Proc. Entomol. Soc. Wash.* **91:** 367-388.

Webber, J.J. & Orton, W.A. (1902). A cowpea resistant to root-knot *Heterodera radicicola*. *US Department of Agriculture, Bureau of Plant Industry Bulletin* **17:** 23-26.

Webster, J.M. (1972). Nematodes and biological control. *In* : *Economic Nematology* (Ed.— J.M. Webster). Academic Press, London. pp. 469-496.

Winchester, J.A. (1960). Preliminary investigation on the mode of action of Pangola grassroot in reducing cotton root-knot nematode *Meloidogyne incognita acrita* populations. *Proc. Soil & Crop. Sci. Florida* **20** : 178-182.

Wyatt, J.E.; Fassuliotis, G.; Hoffman, J.C. & Deaking, J.R. (1983). 'Nema snap' Snap bean. *Horticultural Sci.* **18** : 776.

Chapter 17

Nematode Problems of Medicinal and Aromatic Plants and their Control

Akhtar Haseeb* & Farzana Butool

Central Institute of Medicinal and Aromatic Plants (CIMAP), Post Bag No. 1, P.O.-Indra Nagar, Lucknow 226 016, India.

IN recent years, there has been an increased interest in the cultivation of medicinal and aromatic plants to meet the requirements of cosmetic, flavouring, perfumery and pharmaceutical industries and to earn foreign exchange by way of export (Husain, 1983). While developing improved agro-technologies for higher yields, protection against biotic stresses (diseases/pests) is of utmost interest. Attention has also been made on the study of plant-parasitic nematodes which have been recognized as serious constraints to the productivity of medicinal and aromatic plants (Alam *et al.*, 1978; 1979; Haseeb *et al.*, 1984, 1985, 1988a, b, 1990; Haseeb & Pandey, 1987, 1989a, b, c; Haseeb & Butool, 1989, 1991; Butool & Haseeb, 1993).

MINT *(Mentha* spp.)

Five *Mentha* species, viz. *M. arvensis* L. sub sp. *haplocalyx* Briquet var. *piperascens* Holmes, *M. cardiaca* (S.F. Gray) Baker, *M. citrata* Ehrh., *M. piperita* L. and *M. spicata* Huds. are important oil bearing plants which are cultivated on a large scale in tropical and subtropical countries of the world. The oils and their principal constituents of these mint species have great demand as they are valued in flavouring, perfumery, cosmetics and pharmaceutical applications. Oil of *M. arvensis* is being produced and traded in larger quantities than any other mint oil. About two and a half decades ago, the entire requirements of Japanese mint oil and menthol in India were met through imports. Owing to the efforts made by CIMAP, Lucknow at present more than 800 tonnes oil and 400 tonnes menthol are produced annually with the result that the country has not only become self sufficient but foreign exchange to the extent of about Rs. 120 million per annum is also saved. Second in order is *M. spicata* oil, the current production being 80-

*Present Address : Department of Plant Protection, RAK Institute of Agriculture, Aligarh Muslim University, Aligarh-202 022, India.

100 tonnes valued at about Rs. 30 million per annum. Annual production of *M. piperita* and *M. citrata* oils are 20 tonnes and 10 tonnes, respectively.

ROOT-KNOT NEMATODES

Previous accounts of plant-parasitic nematodes associated with the *Mentha* species are few and widely scattered in the literature. Buhrer (1938) for the first time listed *M. arvensis* and *M. piperita* as a host for *Meloidogyne*. During 1952 and 1953, high populations of *Meloidogyne* hapla were found closely associated with or causing disease in mints in Western Oregan (Horner & Jensen, 1954). The pathogenic effect of *M. hapla* on Scotch spearmint, *M. cardiaca* was also reported for the first time. Skotland & Menzies (1957) reported that *M. hapla* was the most prevalent nematode species associated with *M. piperita* var. Mitcham, *M. cardiaca* var. Scotch and *M. spicata* var. Native in the Yakima Valley and in the Columbia basin.

With the introduction of Japanese mint and Bergamot mint in India and subsequent monoculture for several years, crop production decreased. Haseeb & Pandey (1989b) for the first time observed the root-knot disease of Japanese mint and Bergamot mint caused by *Meloidogyne incognita* and *Meloidogyne javanica* in the experimental plots of CIMAP headquarters, Lucknow and Tarai region of Uttar Pradesh. Generally mixed infection of both these species was found on the same root system, however, *M. incognita* dominated over *M. javanica*.

Initial symptom of the disease is occasional yellowing of leaves. Within a month's time, a large portion of the foliage become completely yellow. Growth seems to cease soon after yellowing. In general, leaves turn yellow and thin, scorch easily and eventually turn brown whereas the veins remain green. Below ground symptoms include large numbers of galls on the roots (Haseeb & Pandey, 1989b). Variation in the severity of infection differed among locations and cultivars to cultivars. The most severe infection occurred at Lucknow and Moradabad whereas the least at Pantnagar.

For the first time Haseeb & Pandey (unpublished) and Anon. (1986, 1987) established the pathogenicity of *M. incognita* on different cultivars of *M. arvensis* (cvs. MA-1, MA-2, MA-3, Ma-4, MAS, MAS-1, MAS-77). An increase in initial inoculum level of the nematode resulted in corresponding decrease in fresh and dry plant weight, chlorophyll content, rate of photosynthesis and oil yield. The maximum decrease in fresh and dry plant weight was observed at the highest inoculum level.

Seasonal changes in population of root-knot nematodes associated with different cultivars and species of mints have indicated variation in population from season to season and cultivar to cultivar (Anonymous, 1984, 1985, 1986). Highest number of larvae of root-knot nematodes was found in soil from the rhizosphere of *M. arvensis* and lowest from that of *M. citrata*.

So far no published report is available on the control of root-knot disease of mints. Only recently Haseeb & Pandey (unpublished) evaluated different pesticides and oil-cakes for the control of *M. incognita* infesting Japanese mint in field conditions. The treatments improved the growth of the plant coupled with oil yield and reduced root-knot development and nematode population. Best results were achieved when the soil was amended with neem cake or treated with furadan.

ROOT-LESION NEMATODE

Skotland & Menzies (1957) for the first time reported the association of *P. minyus* Sher & Allen, *P. penetrans* Filipjev & Schuurmons Stekhoven and *P. thornei* Sher & Allen with *M. cardiaca* var. Scotch, *M. piperita* var. Mitcham and *M. spicata* var. Native from the Yakima Valley and the Columbia basin. Bergeson (1963) for the first time established the pathogenicity of *P. penetrans* on peppermint. He reported that the weight of root and foliage decreased up to 34% and 64% respectively, in inoculated plants and the population of *P. penetrans* increased ranging from 20x to 80x after 8 month of inoculation. Later, Bergeson & Green (1979) found that in Indiana the growth of *M. piperita* and *M. spicata* was considerably reduced due to *P. penetrans* and they subsequently proved that *P. penetrans* was equally pathogenic to reduce the growth of all the cultivars of peppermint grown in Indiana. Rhoades (1983) reported that in Sherbakaff and Stanelly in Central Florida the growth of *M. spicata* was reduced and the plants were stunted due to the high population of *P. scribneri*. He established the pathogenicity of the nematode on *M. spicata*.

High population of *P. thornei* was found to be associated with different mints and caused wilting and chlorosis of the plants and lesions on the root system (Anonymous, 1984, 1985, 1986; Haseeb,. 1992, 1994). Studies on seasonal changes in the population of *P. thoreni* around different mints indicated high population during February through April and the lowest during June through August *M. arvensis* favours the build-up of high population of nematode followed by *M. cardiaca, M. piperita, M. spicata* and *M. citrata*. For the first time studies were also made to establish the pathogenicity of *P. thornei* (Anonymous, 1985, 1986; Haseeb & Shukla, 1994, 1995).

Pratylenchus sp. can cause severe damage but it also increased the severity of *Verticillium* wilt on a number of crops. Bergeson (1963) for the first time studied the effect of *P. penetrans* and *V. alboatrum* alone and in different combinations on peppermint. Foliage weight was considerably reduced when plants were inoculated with *P. penetrans* alone. However, the diagnostic symptoms of

Verticillium wilt appeared 2-3 weeks earlier in nematode-infected plants inoculated with *V. alboatrum* than in plants inoculated with fungus alone. This did not occur with simultaneous inoculations. Faulkner & Scotland (1963) reported that in the Yakima Valley of Washington, *P. minyus* and *V. dahliae* (Kleb) f. *menthae* Nelson were associated with *M. piperita*. Later, Faulkner & Skotland (1965) reported that the *P. minyus* increased both incidence and severity of *Verticillium* wilt. The period for the development of disease symptom was reduced 2-3 weeks by the nematode. The rate of multiplication of nematode increased in plants inoculated with nematode and fungus than that with nematode alone. Faulkner & Bolander (1969) have determined the effect of *V. dahliae*. f. sp. *menthae* and *P. minyus* in disease development on peppermint at 18, 21, 24, 27 and 30°C soil temperatures. Characteristic symptoms were produced at 27°C, when the plants were inoculated with both the pathogens. Whereas, at soil temperature of 24°C, the most severe symptoms were produced in plants inoculated with the fungus alone. The highest population of nematode was found in plants inoculated with both the pathogens at 24°C. Whereas, in plants inoculated with the nematode alone, the highest population was achieved at 30°C.

Several reports are available on the control of root-lesion nematode infesting various crops. However, information with regard to control of *Pratylenchus* is lacking. Rhoades (1984) reported the control of *P. scribneri* infesting *M. spicata* by using carbofuran, fenamiphos, and oxamyl @ 5.6 and 11.2 kg/ha and terbufos @ 11.2 kg/ha. All the treatments were effective in reducing nematode population. The best results were achieved in plants treated with fenamiphos followed by terbufos. Recently, Pinkerton *et al.* (1988) reported that the treatment of oxamyl, aldicarb and carbofuran when applied as broadcast sprays during spring reduced the *Pratylenchus penetrans* densities and increased yields. Single treatment of 5.5. kg a.i./ha in early April was as effective as multiple treatments in April and May, or April, May and June. Treatment effects were still evident after two seasons. Full applications of oxamyl and carbofuran neither reduced over wintering nematode populations nor enhanced spring growth.

Studies conducted at CIMAP, Lucknow, revealed that neem oil-cake together with nematicides not only improved the growth of *M. citrata*, *M. piperita* and *M. spicata* but also reduced the *P. thornei*.

OTHER NEMATODES

Unlike root-knot and root-lesion nematodes other plant-parasitic nematodes, associated with different crops, did not produce very remarkable or specific single symptom on the hosts.

Goodey (1940) reported *Aphelenchoides ritzemabosi* and *A. olesistus* on *M.*

piperita and *M. spicata* respectively. Horner & Jenson (1954) observed *A. parientinus* and *Aphelenchoides* sp. in the meristematic cells of the underground portion of *M. piperita* shoots. They further reported that single shoot tips possessed about 400 specimens of the nematode.

Horner & Jensen (1954) reported that in Western Oregan *Pratylenchus macrophallus* parasitized commercial peppermint with plants becoming stunted and chlorotic. About 8,000 nematodes were isolated per quart soil. Skotland & Menzies (1957) observed that in addition to other plant-parasitic nematodes, *P. hamatus* Thorne & Allen was also associated with *M. cardiaca* var. Scotch, *M. piperita* var. Mitcham and *M. spicata* var. Native in the Yakima Valley and in the Columbia basin. Scotch and Native spearmint in Washington was infested with *P. hamatus* (Faulkner, 1962). Generally 3000-10,000 nematodes per pint soil were isolated from the rhizosphere of infected plants. Severely infected plants did not show any specific symptoms other than occasional dwarfing where the root system was much reduced with varying stages of decay. Faulkner (1964) further reported 10-20% and 20-36% reduction in fresh and dry weight of plants respectively and delay in flowering in inoculated plants.

Horner & Jensen (1954) reported that the *Longidorus elongatus* (de Man) Thorne & Swanger was most prevalent in flood plain along the Santiam and Willamette rivers of Western Oregan affecting peppermint. Infected plants were stunted and usually exhibited a reddish colour. The root system was much reduced and the feeder roots looked like short stubby remnants. Further, Jensen & Horner (1956) established the pathogenicity and successfully controlled the disease by soil fumigation. Jatala & Jensen (1974) found that foliar application of oxamyl was effective in reducing the population of *L. elongat* affecting peppermint. Pinkerton & Jenson (1983) reported that foliar sprays of oxamyl and subsequent application of aldicarb and oxamyl as soil drench, significantly increased hay and oil yield one year after the treatment. Two broadcast sprays or one granular incorporation of oxamyl was the most effective treatment. Nematode population was only slightly and temporarily reduced, however, it did not correlate with yield response.

Tylenchus sp., *Tylenchorhynchus* sp., *Hoplolaimus* sp., *Helicotylenchus* sp., *Rotylenchulus reniformis, Xiphinema* and *Longidorus* sp. were consistently isolated from the rhizosphere of different cultivars of *M. arvensis*, *M. citrata*, *M. piperita* and *M. spicata*. All the cultivars/species of *Mentha* in general supported high populations of tylenchids during October-November and February through April.

BASIL (*Ocimum* spp.)

About 60 species of *Ocimum* are found in tropical Asia, Africa, America and subtropical regions of the world. Out of these, sweet basil, *O. basilicum* L. is considered to be the most important crop for its high quality of essential oil. Major components of its oil are linalool (43.50%), menthyl chavicol (18-33%), eugenol and isoeugenol (5-6%) and the minor constituents are alpha and beta pinene, camphor, geraniol, etc. The oil of sweet basil owes its importance to its extensive use in condemnatory products, cosmetics, toiletry, perfumery and confectionery industries, particularly in European countries.

Earlier the *Ocimum* was considered as wild and it received very little attention from plant pathological point of view. Only scattered reports are available on the association of *Meloidogyne* species and *Aphelenchoides ritzemabosi* with *O. basilicum* and *O. Canum* (Breda de Haan, 1899; Goffart, 1931; Buhrer, 1938; Ustinov, 1939).

For the first time Rangaswami *et al.* (1961) reported that *O. sanctum* is a host of *M. javanica*. In 1967, Krishnamurthy & Elias, reported that the crop of *O.basilicum* was severely affected by *M. incognita*. Since then, in the past two decades there is no report available on the association/damage caused by plant-parasitic nematodes on Ocimum species. Recently *M. incognita* and *M. javanica* are found as main problems in the cultivation of *O basilicum*, *O. canum*, *O. sanctum*, *O. gratissimum* and *O. kilimandescharicum* (Anonymous, 1984; Haseeb & Pandey, 1987). The pathogenicity of *M. incognita* on *O. basilicum*, *O. sanctum* and *O. canum* has also been established in glasshouse condition (Haseeb *et al.*, 1988a; Haseeb & Butool, 1989; Haseeb & Pandey, 1987). Pathogenicity experiments demonstrated the damaging potential of the parasite on these crops in terms of its detrimental effect on all growth characters of the plant including oil yield. They further reported that with an increase in initial population densities of the nematode, there was a corresponding decrease in root-shoot length and fresh/dry weight. In general, the maximum reduction in all the parameters was obtained at highest inoculum level.

Tylenchus sp., *Tylenchorhynchus* sp., *Hoplolaimus* sp., *Helicotylenchus* sp., *Pratylenchus* sp., *Rotylenchulus reniformis*, *Xiphinema* sp. and larvae of *Meloidogyne* sp. were consistently isolated from the soil around different species of *Ocimum* (Anonymous, 1984).

Not much work is, however, done on the control of plant-parasitic nematodes affecting this crop. Recently, Haseeb *et al.* (1988b) have studied the comparative efficacy of aldicarb, carbofuran, bavistin and neem and mahua cakes for the control of *M. incognita* on *O. basilicum*. Studies indicated that the soil application of oil-cakes and nematicides significantly suppressed root-knot development and increased plant growth and oil yield of *O. basilicum*. Neem cake provided considerable improvement in the plant growth and oil content, thus being most effective treatment for reducing *M. incognita* population.

HENBANE (*Hyoscyamus* spp.)

Henbane (*Hyoscyamus* sp.), an important medicinal herb, is a rich source of tropane alkaloids particularly hyoscine, hyoscyamine and atropine. The hyoscine and its derivatives have been used in pharmaceutical preparations because of their mydriatic, antispasmodic and anticholingeric properties. About eleven species are distributed from the Canary Islands over Europe and North Africa to Asia. While *H. niger* is of indigenous origin, *H. muticus* has been introduced in India by the Central Institute of Medicinal and Aromatic Plants, and agro-technology for large scale cultivation has also been developed (Husain, 1983).

Previous records on the association of plant-parasitic nematodes indicate that *H. muticus* and *H. niger* are the hosts of *Meloidogyne* species (Ustinov, 1939; Minz, 1956).

Studies conducted during 1983-1988 at the experimental plots of CIMAP, Lucknow revealed that the crops of *H. niger* and *H. muticus* were severely infested with *M. incognita* and *M. javanica* with 60-70% plants showing chlorotic and stunted growth showing patchy appearance in the plots with reduced plant growth. The infested plants bear fewer leaves and flowers with varying sizes of galls on the root system. Pathogenicity of *M. incognita* and *M. javanica* was established separately on *H. muticus* and *H. niger* in glasshouse conditions (Haseeb, 1992, 1994; Haseeb & Pandey, 1989a). Further, Haseeb *et al.* (1990) determined the influence of different inoculum levels of *M. incognita* on the root-knot development, nematode multiplication, total alkaloid yield, physiological responses (total leaf chlorophyll, CO_2 exchange rate and concentration of sodium, potassium, iron, manganese, copper and zinc).

They reported that with an increase in inoculum level, there was a corresponding decrease in fresh and dry plant weight, total alkaloid yield, total chlorophyll content, photosynthetic rate, sodium, potassium, iron, manganese, copper and zinc in root and shoot. However, sodium and potassium increased in the shoot. The greatest reduction in all the characters was obtained when plants were inoculated with highest nematode inoculum. The relationship between initial and final nematode population showed a maximum reproduction rate of *M. incognita* at initial population of 50 second stage juveniles which decreased as the initial population increased up to 15,000 second stage juveniles.

Efficacy of different nematicides and oil-cakes has been tested against *M. incognita* infesting *H. muticus* and *H. albus* (Haseeb & Pandey, unpublished).

DAVANA (*Artemisia pallens*)

Davana (*Artemisia pallens* Wall). an annual aromatic herb belonging to the family *Asteraceae*, is considered to be of immense importance for its high quality of essential oil in U.S.A., Europe and Japan. The crop is cultivated in India in Andhra Pradesh, Tamil Nadu and Kerala. According to a general estimate, in India about 1-2 tonnes of Davana oil is produced annually. The main constituents of the oil are cinnamoyl, cinnamate, fenchyl alcohol, cadinene, 10-14 phenols or acids and several sesquiterpenes, linalool, eugenol and geraniol.

Before the 80s, nothing was known about the association of plant-parasitic nematodes with this crop. Haseeb & Pandey (1989c) reported that the real problem in the cultivation of this crop is the root-knot nematode. The main symptoms of the plant were stunted growth with less number of flower buds showing patchy appearance in the field. For the first time, they have established the pathogenicity of *M. incognita* on Davana. Results indicate that an increase in initial inoculum level of the nematode causes corresponding decrease in shoot height, root length, fresh/dry root and shoot weight and oil yield. With the increase in initial nematode population, there was corresponding decrease in nematode multiplication.

So far no report is available on the control of root-knot nematode infesting Davana. Recently, Haseeb & Butool (1991) conducted an experiment to control *M. incognita* by using aldicarb (4 kg a.i./ha), bavistin (4 kg a.i./ha), carbofuran (4 kg a.i./ha,) oncol (4 kg a.i./ha), roughly (4 kg a.i./ha) and neem cake (1 g N/kg soil). Various treatments resulted in a significant increase in total fresh/dry weight, and oil yield of *A. pallens* in aldicarb treated plants followed by mocap, neem cake, carbofuran, roughby, bavistin and oncol respectively. Pesticides and oil-cakes also significantly suppressed nematode population development as evidenced by root-knot index, J2 population in soil and roots.

Several other plant pathogenic nematodes were isolated from the rhizosphere of this plant (Haseeb & Pandey, *unpublished*). However, pathogenicity of individual nematode species is yet to be established.

YAM (*Dioscorea* spp.)

Various species of *Dioscorea* are the sources of diosgenin, a steroidal sapogenin, most commonly used as precursor for the synthesis of drugs which include cortisones, sex hormones and oral contraceptives. At present more than 70 per cent of total requirement of these drugs is obtained from diosgenin. The present demand in the world is estimated to be over 3000 tonnes which is bound to increase in future depending upon the family planning programmes of the developing countries. The estimated requirement of diosgenin for Indian pharmaceutical industry is around 60 tonnes. The present production of diosgenin, mostly produced from *D. deltoidea* Wall. obtained from the forest of Jammu & Kashmir, H.P. and U.P., is approximately 15-20 tonnes annually.

Steiner in 1931 reported for the first time that the yam tuber was affected by plant-parasitic nematode, *Hoplolaimus* sp. and the disease was named as "Nematosis" and the nematode species was described as *Hoplolaimus bradys* Steiner & Lettew. Later Andrassy (1958) transferred the species to the genus *Scutellonema*. West (1934) reported that *S. bradys* was associated with yam in Nigeria. He described the symptoms of loosened tuber cortex and named the disease as "dry rot". Similar symptoms were observed by Steiner & Buhrer (1934) on yam from Puerto Rico. This nematode species was also found to be associated with *D. alata* L., *D. cayenensis* Lam., and *D. rotundata* Poir. Jenkins & Bird (1962) found *Pratylenchus brachyurus* (Godfrey) Filip. & Schuur. Stekh. and *Criconemoides* sp. associated with wild yam. Ayala (1966a, b) and Ayala & Acosta (1971) reported the association of *P. coffeae* (Zimmermann) Filipjev & Schuur. Stekh., *Rotylenchulus reniformis* Linford & Oliveira, *Helicotylenchus* sp., *Aphelenchoides* sp. and *Aphelenchus* sp. with *D. rotundata* in Puerto Rico. In Jamaica, several parasitic nematodes were associated with yam plants (Dixon & Latta, 1965; Hutton, 1978). Acosta & Ayala (1975) established the pathogenicity of *S. bradys* and *R. reniformis* on *D. rotundata*. They found that *P. coffeae* was responsible for the deterioration in quality of the yam tuber in Puerto Rico, a condition known as dry rot of yam. *P. coffeae* and *S. bradys* were pathogenic to yam cultivar Guinea, but *M. incognita* and *R. reniformis* did not cause necrosis or cracking of the tuber cortex.

P. coffee, *S. bradys* and *Hoplolaimus* sp. were involved in the etiology of a dry rot of yam tubers (Coates-Beckford, 1976; Hutton, 1977). Hutton (1978, 1979) reported that in Jamaica, *P. coffeae* is considered to be the most serious pest of yam, causing 'dry rot' or burning. This condition is characterized by cracking in the skin underlined by a brown, corky rot the storage tissues. Rot progresses deeper into the tissues following harvest and prior to planting or consumption and is generally more pronounced towards the stem. In yam tuber heavily "burnt" heads, stem primorida appear to be damaged or destroyed by the dry rot resulting in such heads not sprouting or veins growing from them being less thrifty than those from less affected heads.

For the first time Buhrer (1938) reported that *D. alata* is the host of *Meloidogyne* sp. Hawely (1956) and Schieber (1961) reported that *D. floribunda* is the host of *M. arenaria, M. javanica* and *M. incsognita*. Schieber (1961) and Schieber & Lassmann (1961) observed that tubers of *D. spiculiflora* Hemsl. and *D. floribunda* were severely infested by *Meloidogyne* species in Guatemala. Jenkins & Bird (1962) found that *M. incognita* damaged the wild yam with other nematodes.

Nadakal & Thomas (1967) reported that *S. bradys* was associated with dry rot of *D. alata*. Raveendran & Nadakal (1975) reported that *D. alata* and *D. spinosa*

Roxb. was severely infested with *M. incognita* in Kerala. In southern Nigeria, *M. incognita* was the most important pest of white guinea yam, *D. rotundata* (Acosta & Ayala, 1975; Adesiyan & Odihirin, 1978).

Atu *et al.* (1983) reported that in *D. rotundata* cv. gave at a soil temp. of 28°C, the economic threshold and economic injury levels were fixed at 250 and 1250 *M. incognita* larvae per plant respectively. Tubers from plots inoculated with more than 1250 nematodes per plant were so heavily galled that the market value was reduced by 40%.

Recently, Montalvo & Melendez (1980) studied the histopathology of interrelationships between *Fusarium oxysporum* f.sp. *dioscorea* and *M. incognita* and *P. coffeae* on *D. rotundata* cv. Habanero. Combined and separate inoculations were made with the fungus and the nematodes. Longitudinal and transverse section of root tissues collected at 14, 42 and 72 days after inoculation showed that colonization by the fungus occurred indiscriminately, either inter- or intra-cellularly. Colonization in the stem mainly in the nodal tissues was restricted to the vascular system, where abundant proliferation of hyphae was observed. Apparently, the nematode as well as the fungus colonized different sectors of the tissue invaded. However, abundant and vigorous hyphae were present in section of tissue modified by *M. incognita*. It may be concluded from the studies that a positive nematode-fungus interaction occurs in yam.

Coates-Beckford (1976) tested the efficacy of oxamyl, phenamiphos and diazinon for the control of *P. coffeae* infesting *D. rotundata*. Treatment with oxamyl at 1200 or 2400 ppm was the most effective since nematode populations were significantly low and plant growth was unaffected; phenamiphos and diazinon were ineffective treatments.

AROMATIC GRASSES

The Cymbopogons, possessing a large number of species (a total of 140 species of the grass family Gramineae, Sobti *et al.*, 1982), are characterized by plants bearing essential oil in all parts except root (except in *Cymbopogon jwarancusa* where roots also contain oil). Essential oils derived from these plants are widely used in perfumery and also in pharmaceutical industries. Further, these oil constitute the natural sources of valuable active principles, e.g., citral in lemon-grass oil, geraniol in palmarosa oil, citronellal in citronella oil and so on. Among different species of *Cymbopogon*, *C. winterianus* Jowitt (Citronella Java), *C. flexuosus* (Nees ex. Steud.) Wats. (Lemongrass) and *C. martinii* var. *motia* (Roxb.) Wats. (Palmarosa) have been recognised in perfumery and cosmetic trade throughout the world. These are cultivated in India on large scale in order to meet out the requirements of the industry. The current annual production of citronella, lemongrass and palmarosa oils is approximately 800, 400 and 70 tonnes respectively.

Survey conducted during 1983-1986 at CIMAP, Lucknow revealed that besides fungi, plant-parasitic nemtodes are also important disease causing agents associated with apparently diseased plants in citronella, palmarosa and lemongrass (Anonymous, 1984, 1985, 1986, 1987, 1988). Important pathogenic nematodes, viz. *Tylenchus* sp., *Tylenchorhynchus* sp., *Hoplolaimus* sp., *Helicotylenchus* sp., *Rotylenchulus reniformis*, *Pratylenchus* sp., *Longidorus* sp., *Xiphinema* sp. and the larvae of *Heterodera* and *Meloidoyne* species were consistently isolated from the rhizosphere of citronella and different cultivars of lemongrass, palmarosa and vetiver (Khus). The most dominant parasitic species of the nematodes were *Tylenchorhynchus* sp. and *Pratylenchus* sp. Generally very high populations of these nematodes were encountered from the rhizosphere of apparently diseased (stunted, chlorotic and reduced number of tillers) plants of citronella.

Important pathogenic fungi and nematodes were isolated and their pathogenicity was established separately. In case of nematode inoculated plants, symptoms were not much similar to those of field conditions. However, studies pertaining to the influence of different inoculum levels of *Tylenchorhynchus* sp. and *Pratylenchus* sp. individually on the growth of citronella, palmarosa and lemongrass clearly indicate that both the nematode species reduced the fresh and dry weight of herb and oil yield to a considerable extent. Since, in nature, *Tylenchorhynchus* sp. and *Pratylenchus* sp. cohabit the same rhizosphere of the plants, studies were also made on the combined effect of these nematode species on the disease development in citronella. Results indicate that both the nematode species together had more effect on the growth of the plant and yield than either of the nematode alone (Haseeb & Pandey, unpublished).

CONCLUSION

Scanning of the literature, reveals that research on economically useful medicinal and aromatic plants in relation to the plant-parasitic nematodes still remains neglected. The area assumes greater importance as the raw materials used in perfumery, cosmetics, flavouring and pharmaceutical industries are obtained from nature, more precisely, the plants up to the large extent. In fact these plants are also subjected to attack of several pathogens. Recently, interest has been generated in the pathology of medicinal and aromatic plants and our Institute has taken the lead. Based on the results of the preliminary work carried out in our lab, concerted efforts are being made in finding out the role of plant-parasitic nematodes in the causation of the diseases in medicinal and aromatic plants.

ACKNOWLEDGEMENTS

The authors are grateful to Dr. R.S. Thakur, Director, Central Institute of Medicinal & Aromatic Plants (CIMAP), Lucknow, India, for the encouragement and critical suggestions in preparing the manuscript.

REFERENCES

Acosta, N. & Ayala, A. (1975). Pathogenicity of *Pratylenchus coffeae*, *Scutellonema bradys*, *Meloidogyne incognita* and *Rotylenchulus reniformis* on *Dioscorea rotundata*. *J. Nematol*. **7:** 16.

Adesiyan, S.O. & Odihirin, R.A. (1978). Root-knot nematode as pests of yams (*Dioscorea* spp.) in Southern Nigeria. *Nematologica* **24:** 132-134.

Alam, M.M.; Saxena, S.K. & Khan, A.M. (1978). Suitabilty of crops to certain ecto-parasitic nematodes. *Acta Bot. Indiaca* **6:** (Supp.): 205-208.

Alam, M.M.; Saxena, S.K. & Khan, A.M. (1979). Suitability of crops to the reniform, stunt and root-knot nematodes. *In* : *Recent researches in plant sciences* (Ed.—S.S. Bir). Kalyani Publishers, New Delhi. pp. 630-634.

Andrassy, I. (1958). *Hoplolaimus tylenchiformis* Daday, 1905 (Syn. *H. coronatus* Cobb. 1923) und die Tottungen der unterfamilie Hoplolaiminae Filipjev., 1936. *Nematologica* **3:** 44-46.

Anonymous (1984). *Annual report* (1983-84). Central Institute of Medicinal and Aromatic Plants, Lucknow, India.

Anonymous (1985). *Annual report* (1984-85). Central Institute of Medicinal and Aromatic Plants, Lucknow, India.

Anonymous (1986). *Annual report* (1985-86). Central Institute of Medicinal and Aromatic Plants, Lucknow, India.

Anonymous (1987). *Annual report* (1986-87). Central Institute of Medicinal and Aromatic Plants, Lucknow, India.

Anonymous (1988). *Annual report* (1987-88). Central Institute of Medicinal and Aromatic Plants, Lucknow, India.

Atu, U.G.; Odurukwe, S.O. & Ogbuji, R.O. (1983). Root-knot nematode damage to *Dioscorea rotundata*. Plant Dis. **67** : 814-815.

Ayala, A. (1966a). *Annual report of the Department of Entomology*. Agric. Exp. Stn., Rio Piedras P.R.

Ayala, A. (1966b). Nematode problems in Puerto Rican Agriculture. *In* : *Proc. Symp. Trop. Nematol.* Univ. P.R. Agric. Exp. Stn., Rio Piedras, Puerto Rico. pp. 135-145.

Ayala, A. & Acosta, N. (1971). Observations on yam (*Dioscorea alata*) nematodes. *Nematropica* **1:** 39-40.

Bergeson, G.B. (1963). Influence of *Pratylenchus penetrans* alone and in combination with *Verticillium albo-atrum* on growth of peppermint. *Phytopathology* **53:** 1164-1166.

Bergeson, G.B. & Green, R.J. Jr. (1979). Damage to cultivars of peppermint, *Mentha piperita*, by the lesion nematode, Pratylenchus penetrans in Indiana. Plant Dis. Reptr. 63: 91-94.

Breda De Haan, J. Van (1899). Levensgeschiedenis en bestrijding vanhet tabaksaaltje (*Heterodera radicicola*) in Deli. *Meded PITuin*. Batavia: 35.

Buhrer, E.M. (1938). Additions to the list of plants attacked by the root-knot nematode (*Heterodera marioni*). *Plant Dis. Reptr.* **22:** 216-234.

Butool, F. and Haseeb, A. (1993). Relationship between root-knot nematode, *Meloidogyne incognita* and growth/yield of *Trachyspermum ammi*. *New Agricult.* **2:** (2)

Coates-Beckford, P.L. (1976). *Nematological Investigations*. Bull. No. 66 (New Series). Min. Agric.; Jamaica. pp. 52-59.

Dixon, W.B. & Latta, R. (1965). *Nematological Investigations* 1958-1961. Bull. No. 59 (New Series). Min. Agric. & Lands, Jamaica. p. 35.

Faulkner, L.R. (1962). Pathogenicity and population dynamics of *Pratylenchus hamatus* Thorne and Allen on *Mentha* spp. *Phytopathology* **52:** 731.

Faulkner, L.R. (1964). Pathogenicity and population dynamics of *Paratylenchus hamatus* on *Mentha* spp. *Phytopathology* **54:** 344-348.

Faulkner, L.R. & Bolander, W.J. (1969). Interaction of *Verticillium dehliae* and *Pratylenchus minyus* in *Verticillium* wilt of peppermint: Effects of soil temperature. *Phytopathology* **59:** 868-870.

Faulkner, L.R. & Skotland, C.B. (1963). Control of *Verticillium* wilt of mint and plant-parasitic nematodes associated with mint by soil fumigation. *Plant Dis. Reptr.* **47:** 662-665.

Faulkner, L.R. & Skotland, C.B. (1965). Interactions of *Verticillium dehliae* and *Pratylenchus minyus* in Verticillium wilt of peppermint. *Phytpathology* **55:** 583-586.

Goffart, H. (1931). Beobachtungen uber Pflanzenschadliche Nematoden an Fartenpflanzen. *Gartenbauwiss* **5:** 353-359.

Goodey, T. (1940). *The nematode parasites of plants catalogued under their hosts*. Imp. Bur. Agric. Parasit. (Helminth). St. Albans, England. p. 80.

Haseeb, A. (1992). Nematode pests of medicinal and aromatic plants. *In* : Nematode Pests of Crops (Eds.: D.S. Bhatti & R.K. Walia). CBS Publ. Dist., Delhi. pp. 239-249.

Haseeb, A. (1994). Plant parasitic nematodes of medicinal and aromatic plants. *In* : *Vistas in seed Biology* (Eds. — T. Singh & P.C. Trivedi). Printwell, Jaipur. pp 98-119.

Haseeb, A. & Butool, F. (1989). Pathogenicity and reproduction of *Meloidogyne incognita* on *Ocimum sanctum*. *Int. Nematol. Network Newsl.* **6:** 27-30.

Haseeb, A., Butool, F. (1991). Evaluation of different pesticides and oil-cake on the control of root-knot nematode (*Meloidodyne incognita*) infesting on Davana (*Artemisia pallens*). *Curr. Nematol.* **2:** 201-204.

Haseeb, A. & Butool, F. & Pandey, R. (1988a). Influence of different initial population densities of *Meloidogyne incognita* on the growth and oil yield of *Ocimum basilicum* cv. French. *Indian J. Plant Path.* **6:** 176-179.

Haseeb, A. & Pandey, R. (1987). Incidence of root-knot nematodes in medicinal and aromatic plants — new host records. *Nematropica* **17:** 209-212.

Haseeb, A. & Pandey, R. (1989a). Root-knot disease of Henbane, *Hyoscyamus* — a new disease record. *Tropical Pest Manag.* **35:** 212-213.

Haseeb, A. & Pandey, R. (1989b). Observations of *Meloidogyne* species affecting Japanese mint — new disease record. *Nematropica* **19:** 93-97.

Haseeb, A & Pandey, R. (1989c). Root-knot nematode — a constraint to cultivation of Davana, *Artemisia pallens*. *Trop. Pest Manag.* **35:** 317-319.

Haseeb, A.; Pandey, R. & Husain, A. (1984). Some new host records of root-knot nematode,

Meloidogyne incognita (Kofoid & White, 1919) Chitwood, 1949. *Curr. Sci*. **35:** 1151-1152.

Haseeb, A.; Pandey, R. & Husain, A. (1985). New host records the root-knot nematode, *Meloidogyne javanica*. *FAO Plant Prot. Bull*. **33:** 123.

Haseeb, A.; Pandey, R. & Husain, A. (1988b). A comparison of nematicides and oil seed cakes for control of *Meloidogyne incognita* on Ocimum basilicum. *Nematropica* **18:** 65-69.

Haseeb, A. & Shukla, P.K. (1994). Effect of *Pratylenchus thornier* on growth, physiology and oil yield of *Mentha spicata. Afro-Asian J. Nematol*. **4:** 51-53.

Haseeb, A. & Shukla, P.K. (1995). Influence of *Pratylenchus thornie* on growth, photosynthetic rate, chlorophyll, total sugar and oil yield of Mentha citrata. Nematol. medit. 23: 89-91.

Haseeb. A.; Srivastava, N.K. & Pandey, R. (1990). The influence of *Meloidogyne incognita* on the growth, physiology, nutrient concentration and alkaloid yield of *Hyoscyamus niger, Nematol. Medit*. **18:** 127-129.

Hawely, W.O. (1956). Hot water treatment for the control of root-knot nematodes on *Dioscorea floribunda. Plant Dis. Reptr*. **40:** 1045-1046.

Horner, C.E. & Jensen, H.J. (1954). Nematodes associated with mints in Oregan. *Plant Dis. Reptr*. **38:** 39-41.

Husain, A. (1983). Present and future of medicinal plants in India — A scientific appraisal. *In* : *Report of the seminar on Medicinal Plants, Phytochemicals* and *Bulk Drugs* (23-25 March) New Delhi: pp. 33-46.

Hutton, D.G. (1977). *Nematological investigations*. Bull. No. 67 (New Series), Min. Agric. Jamaica. pp. 23-48.

Hutton, D.G. (1978). *Nematological investigations*. Bull. No. 68 (New Series), Min. Agric., Jamaica.

Hutton, D.G. (1979). *Nematological investigations*. Bull. No. 69 (New Series), Min. Agric., Jamaica.

Jatala, P. & Jensen, H.J. (1974). Oxamyl controls *Longidorus elongatus* on peppermint in greenhouse experiments. *Plant Dis. Reptr*. **58:** 591-593.

Jenkins, W.R. & Bird, C.W. (1962). Nematodes associated with wild yam *Dioscorea* sp. with special reference to the pathogenicity of *Meloidogyne incognita. Plant Dis. Reptr*. **46:** 858-860.

Jensen, H.J. & Horner, C.E. (1956). A decline of peppermint caused by an ectoparasitic nematode, *Longidorus sylphus. Phytopathology* **46** : 637.

Krishnamurthy, G.V.G. & Elias, N.A. (1967). Host-range of *Meloidogyne incognita* causing root-knot on tobacco in Hunsur, Mysore State. *Indian Phytopath*. **20:** 374-377.

Minz, G. (1956). The root-knot nematode, *Meloidogyne* spp., in Israel. *Plant Dis. Reptr*. **40** : 798-801.

Montalvo, A.E. & Melendez, P.L. (1980). Histopathology of interrelations between *Fusarium oxysporium* f. sp. *dioscoreae* and two nematode species on yam. *J. Agric. Univ. Puerto Rico* **70:** 245-254.

Nadakal, A.M. & Thomas, N. (1967). Observations of nematodes associated with dry rot of *Dioscorea alata L. Sci. & Cult*. **33:** 142-143.

Pinkerton, J.N. & Jensen, H.J. (1983). Chemical control of *Longidorus elongatus* on peppermint with non-volatile nematicides. *Plant Dis*. **67:** 201-203.

Pinkerton, J.N.; Jensen, H.J.; Newcomb, G.B. & Ingham, R.E. (1988). Management of *Pratylenchus penetrans* damage to peppermint with selected nematicides. *Plant Dis.* **72:** 167-170.

Rangaswami, G.; Balasubramanian, M. & Vasantarajan, V.N. (1961). The host range of sugarcane root-knot nematode, *Meloidogyne janvanica* (Treub) Chitwood. Curr. Sci. 30: 149-150.

Raveendran, V. & Nadakal, A.M. (1975). An additional list of plants infected by the root-knot nematode, *Meloidogyne incognita* (Kofoid & White, 1919) Chitwood, 1949 in Kerala. Indian *J. Nematol.* **5:** 126-127.

Rhoades, H.L. (1983). Effect of *Belonolaimus longicaudatus, Dolichodorus heterocephalus* and *Pratylenchus scribneri* on growth of peppermint. *Phytopathology* **53:**1164-1166.

Rhoades, H.L. (1984). Control of *Pratylenchus scribneri* on spearmint, *Mentha spicata*, with non-fumigant nematicides. *Nematropica* **14:** 85-89.

Schieber, E. (1961). Parasitic nematodes on Dioscorea in Guatemala. *Plant Dis. Reptr.* **45:** 425.

Schieber, R. & Lassmann, D.K. (1961). Root-knot nematode on *Dioscorea* in Guatemala. *Plant Dis. Reptr.* **45:** 981.

Skotland, C.B. & Menzies, J.D. (1957). Two peppermint diseases found in Yakima Valley of Washington. *Plant. Dis. Reptr.* **41:** 493.

Sobti, S.N.; Verma, V. & Rao, B.L. (1982). Scope for development of new cultivars of Cymbopogons as a source of terpene chemicals. *In : Cultivation and utilization of aromatic plants.* (Eds.—C.K. Atal & B.M. Kapoor). Reg. Res. Lab., Jammu. pp. 302-307.

Steiner, G. (1931). A nematode of yam caused by a new species of *Hoplolaimus. Plant Dis. Reptr.* **15:** 121.

Steiner, G. & Buhrer, E.M. (1934). Observations of interest on nematode diseases of plants. *Plant Dis. Reptr.* **18:** 100.

Ustinov, A.A. (1939). "The root-knot nematode, *Heterodera marioni* (Cornu) on nematodes of Agricultural Crops" (Ed.—Kirjanova, E.S.). Leningrad Acad. Sci., Moscow & Leningrad. p. 274.

West, J. (1934). Dry rot of yams. *Bull. Imp. Inst. London* **32:** 448-450.

Chapter 18

Control of Nematode Diseases of Tobacco

S.S. Hussaini

CTRI Research Station, Hunsur-571 105, India

TOBACCO is a high value crop which is in wide demand throughout the world for the production of cigarettes, cigars and other products and is the most widely grown commercial non-food crop in the world. The total world production in 1989 was estimated to be about 7.1 million tonnes, of which one third was produced in China, followed by the United States, India and Brazil. Out of the 64 species, only two species, viz. *Nicotiana tabacum* and *N. rustica* are cultivated extensively. India grows both the species but by far the largest area is under *N. tabacum* which is grown all over the country. Tobacco originated as a natural hybrid in Central America and has been under cultivation for many centuries. By the time the explorers from Europe came to America, tobacco cultivation was widespread in North, Central and South America and since then has spread to all over the world.

MAJOR NEMATODE PROBLEMS, SYMPTOMS AND CROP LOSSES

Plant-parasitic nematodes are widely distributed throughout the world wherever tobacco is grown. The severity of attack and damage depends on the soil and climatic conditions. The dominant nematodes are *Meloidogyne* spp. (Fig. 1) particularly *M. incognita, M. javanica, M. arenaria* and *M. hapla* (Table 1). Others like *Globodera* sp., *Pratylenchus* sp., *Tylenchorhynchus* sp., *Ditylenchus* sp., *Rotylenchulus reniformis* are restricted in distribution causing losses in specific areas. Some species of *Xiphinema*, *Longidorus* and *Trichodorus* are reported as vectors transmitting virus diseases.

Infection of root-knot nematode commences in nurseries. Infected seedlings are stunted unthrifty with sparse and shallow root system. In the field, nematode infestation is patchy. However, in light soil areas of Karnataka it is not uncommon to final all plants in a field affected with root-knot nematode. Affected plants show wilting, deficiency, stunting and premature yellowing. Leaves are thin, scorch

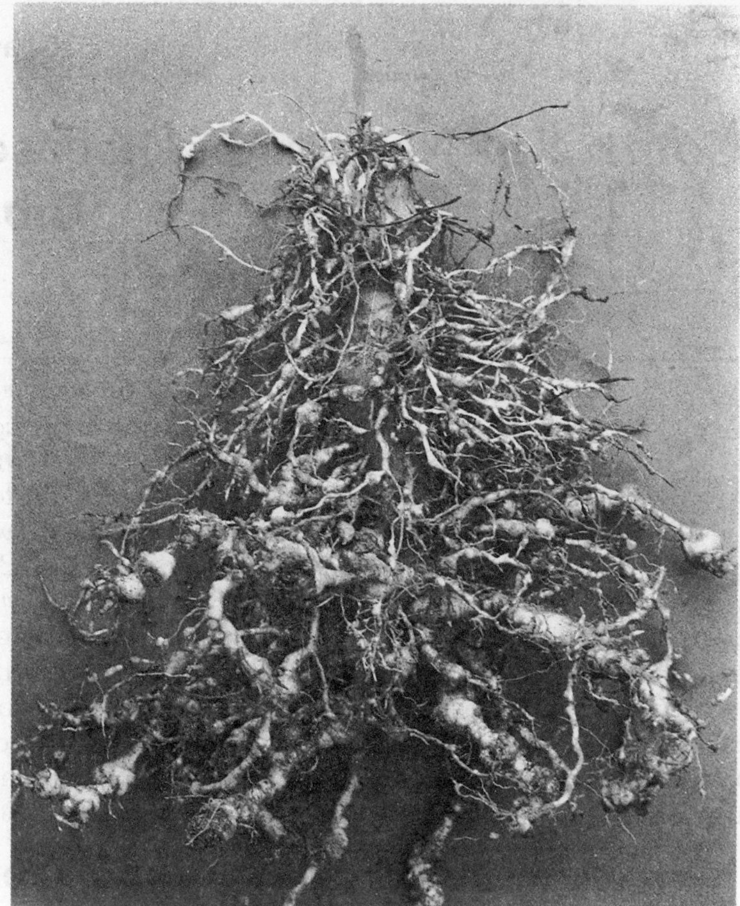

Fig. 1. Root system of FCV tobacco infected with root-knot nematode.

easily and eventually turn brown and on curing produce low grade tobacco. Quantification of damage and loss caused to tobacco was assessed by Hussaini (1983). It was found that when nematode-free seedlings obtained from fumigated beds were planted in infested field, the green and cured yields were reduced by 29.6% and 15.1%, respectively compared to seedlings from fumigated beds planted in fumigated fields. Contrarily planting of infected seedlings in fumigated plots resulted in 36.7% and 34.7% reduction in green and cured leaf yields, respectively. However, the reduction was 54.9% and 52.9% as a result of infection in nursery and field. Similarly, bright grade out-turn was reduced with infection in both nursery and field up to 43% and when infected seedlings were planted in fumigated plots the reduction was 37.2%. No significant reduction was noticed when root-knot free seedlings were planted in non-fumigated plots. Leaf length and breadth of cured leaf were also adversely affected due to nematode damage. Delayed maturity of leaves was observed as a result of infection, particularly when carried from nursery. Clayton *et al.* (1988) observed that infected plants supported very little growth of suckers and delayed flowering was found (Rich & Barker, 1984). Qualitatively widespread potash deficiency, in field crop was found associated with infection (Ramakrishnayya & Hussaini, 1987). Fumigation and subsequent elimination of infection resulted in increased plant yield, per cent nicotine sugars and potassium while only nicotine and potash increased with potash application. Barker & Weeks (1981) also reported reduced nicotine and sugar content in cured leaf due to *M. incognita* infection. Moderate to high infection may reduce yield in excess of 600 kg/ha (Todd, 1981). Influence of different levels of root-knot infection in seedlings on the cured leaf yields was studied (Hussaini & Moses, 1989). Absolutely healthy and apparently gall-free seedlings showed difference to the extent of 41-46 per cent in green and cured leaf yields. As infection in seedlings increased, yield as well as quality of final produce deteriorated further (Table 2).

Lamberti (1979) reported 30-75% yield reduction caused by *M. incognita*. Root-knot is probably the major disease of tobacco throughout the world (Milne, 1972; Lucas, 1975). The initial density required to cause significant damage and yield losses varies with nematode species. Low numbers of *M. arenaria* can cause severe damage to tobacco (Ferris, 1972, 1974; Hanounik *et al.*, 1975; Hanounik

Table 1 : Distribution of *Meloidogyne* spp. on tobacco in different countries.

Meloidogyne spp.	Country
M. incognita	Zimbabwe, Malawi, Greece, Libya, El Salvador, Argentina, Yugoslavia, Cuba, U.S.A., Bulgaria, Costa Rica, Tanzania, Ghana, Portugal, Turkey, South Africa, Brazil, Philippines, Sri Lanka, Malaysia, India.
M. incognita acrita	Zimbabwe, Zambia, Greece, India.
M. javanica	Zimbabwe, Malawi, Zambia, Sudan, Greece, Libya, Mexico, Argentina, Nigeria, Tanzania, Ghana, Portugal, Turkey, South Africa, Brazil, Philippines, Sri Lanka, Malaysia, India.
M. arenaria	Zimbabwe, Libya, Nicaragua, Argentina, Cuba, U.S.A., Bulgaria, Tanzania, Ghana, Portugal, Turkey, South Africa, India.
M. hapla	Libya, Portugal, Turkey, South Africa, Canada.
M. thamesi	Turkey, India.
M. microcephala	Thailand
M. salasi	Costa Rica
M. platani	U.S.A.

Table 2 : Yield in kg/ha of green, cured and bright grade leaf and root-knot index (RKI) of different categories of crop.

RKI in seedl-ings	Seedling category	RKI in crop	Yield(kg)/ha			
			Green leaf	Total cured leaf	Bright grades	
					X position	L position
0	Absolutely gall-free	2.7	13750	2013	325	416
1	Apparently gall-free	3.5	8133	1097	163	393
2	Light infection	4.6	6833	1021	143	350
3	Medium infection	4.7	6133	683	64	195
4	Heavy infection	4.8	5100	711	72	285
5	Very heavy infection	5.0	5616	771	78	214

(Source : Hussaini & Moses, 1989)

& Osborne, 1977). *M. hapla* causes less galling with a maximum of 50-60% roots being galled and little necrosis. Thus this species appears much less important in predisposing tobacco roots. Hence, root gall indices combined with root necrosis indices is essential for yield loss assessment. Yield losses for each 10 fold increase in initial population per species were *M. javanica* 19%, *M. arenaria* 16.5% *M. incognita* 8.9%, and *M. hapla* 3.4% (Barker *et al.*, 1981). Various investigators have obtained conflicting results on the potential of nematode assays and root galling as predictors of tobacco yield losses. Daulton (1964) showed a striking relationship between the magnitude of root galling and yield of tobacco. Ferris (1974) obtained poor correlations between yield and root galls. Mid-season root-gall rating was found closely related to yield (Johnson *et al.*, 1973; Johnson, 1974). However, Barker *et al.* (1981) found rate of reproduction to be a primary key in relative pathogenicity of four species studied. Several workers have worked on the relationship of nematodes to yield and growth of tobacco (Table 3).

Meloidogyne javanica was more aggressive than *M. incognita* and caused approximately 2 fold more yield suppression than *M. incognita*. A 7% yield suppression was caused by 4 eggs or 2nd stage larvae/100 cm^3 of soil of *M. incognita* and 1 egg or 2nd stage larva/100 cm^3 of soil of *M. javanica* (Arens & Rich, 1981).

The cyst nematodes have been important on shade grown tobacco. Yield losses can be very high. Komm *et al.* (1983) observed that some farmers in Virginia reported complete crop failure due to the attack. A sub-species *Globodera tabacum*

Table 3 : Relationship of certain nematodes of yield or growth of tobacco.

Nematode spp.	Type of experiment	Approx. threshold density 100 cm^3 soil	Corr. Coef.	Reference
Meloidogyne incognita	Pot	100	-	Sheikh *et al.*, 1975
Meloidogyne incognita	Field	5-10	- 0.83 to NS	Barker & Nusbaum, 1971; Ferris, 1974
Meloidogyne incognita	Field	2	- 0.67	Rickard, 1973
Heterodera tabacum	Pot	5000 larvae	-	Rickard, 1973
Pratylenchus brachyurus	Field	40	-	Barker & Nusbaum, 1971
Pratylenchus penetrans	Field Microplot	100-200	- 0.88	Lange *et al.*, 1964; Olthof *et al.*, 1973
Tylenchorhynchus claytoni	Field	not damaged	-	Aycock *et al.*, 1976

solanacearum is known to predispose FCV tobacco to granville wilt. However, yield losses up to 57% can normally be expected.

In green house trials, *Pratylenchus brachyurus* invaded tobacco roots and reproduced causing a reduction in yield, while *P. zeae* did not reproduce and had no effect (Southards, 1965; Shepherd, 1982). Low to moderate infection levels of *P. brachyurus* and *P. scribneri* failed to have any significant effect on yield in North Carolina.

CONTROL STRATEGIES

The methods adopted for control of nematodes in seed beds and field varies in different countries. Root-knot nematode infection in light soil areas is of universal occurrence. Routine nematode control schedules based on population estimation are being adopted in countries like U.S.A., Zimbabwe and Brazil. In India, particularly in Karnataka, nursery fumigation is being carried out every year under subsidy from Tobacco Board for production of healthy infection-free seedlings. In the field, use of nematicides is still restricted and root-knot tolerant varieties, where available, are in cultivation.

RABBING

Rabbing with paddy husk at 20 kg/sq m was found effective in controlling *M. incognita* in FCV tobacco nurseries and the control was absolute resulting in increased seedling growth and numbers (Krishnamurthy & Elias, 1969). However, quick growth of seedlings induced succulence, a disadvantage in establishment. Infection of nematode occurred after 55 days (Hussaini, 1985). A slow burning material with deep penetration and long period of heat is required for effective treatment. Surface burning with petroleum products is hence ineffective.

SOIL SOLARIZATION

Katan (1981) presented a comprehensive review on solar heating by mulching with white transparent polyethylene sheets for control of soil borne pests. Preliminary trials during 1990 nursery season at Hunsur, India revealed the efficacy of solarization, using white transparent polyethylene sheets, against root-knot nematodes and weeds resulting in increased number of tobacco transplants.

CHEMICAL CONTROL (SEED BEDS)

In seed beds the present trend is to use as overall multipurpose chemical to take care of nematodes, weeds and soil borne fungi (Fig. 2, 3). Fumigation with Durofume (MB:EDB 1:1) was reported very effective for the control of root-knot nematodes and weeds, particularly *Cyperus* sp. and *Cynodon* sp., and increasing number of transplants. Methyl bromide application was also equally effective. EDB also showed partial weedicidal action and caused reduction of weeds up to 49-63%. Root-knot infection was absolutely reduced in Durofume and MB treated beds and up to 15% in EDB compared to untreated control. An increase of 238-248 kg/ha of cured leaf in field was obtained (Table 4) using seedlings from Durofume and MB treated beds over non-fumigated seedlings category (Hussaini, 1985, 1989).

MB (98% + 2% CP) at 50 g/m^2 or *Thielaviopsis basicola* and DD (1-3 D 50% + 1,2D 50%) at 100 g/m^2 was effective against *Meloidogyne* spp. in Italy (Della & Cappelli, 1988). Fumigation in seed beds is a routine practice in many countries like U.S.A., Zimbabwe, Brazil, Argentina and Philippines. In Japan, chloropicrin is commonly used against nematodes and damping-off soil organisms especially in sandy soils. In microplot experiment at Anand DD (500 l/ha), DBCP (30 l a.i./ha and 15 l a.i./ha at surface or 15 l a.i./ha 30 cm deep), and fensulfothion (4 kg a.i./ha) remained effective up to 10th week of sowing (Patel *et al.*, 1981). Krishna Prasad (1985) reported application of 4 and 8 kg quinalphos and phorate were as effective as DD application at 300 l/ha. Mephospholan and bendiocarb were not

Table 4 : Impact of seed bed fumigation on the incidence of weeds, seedling number, root-knot infestation of tobacco seedlings and subsequent yield of crop.

Fumigant	Rate (ml/m^2)	Weeds per m^2		Tobacco seedlings m^2	Root-knot Index (0-5)	Cured yield of tobacco in field kg/ha
		nut-grass	others			
Durofume	18	1.05	5.8	403	1.0	1476
Methyl Bromide	15	0.08	1.3	420	1.0	1471
EDB	18	22.9	10.3	213	2.8	1418
Untreated	-	44.6	27.7	198	3.3	1233

(Source : Hussaini, 1989)

Fig.2. Healthy nursery of FCV tobacco after fumigation and weed control.

effective against *M. incognita* in FCV nurseries at Shimoga. Ramjilal *et al.* (1981) found DD 360 l/ha and DBCP 50 l/ha to control the nematode up to 60 per cent. Patel *et al.* (1981) found that seed germination was higher in 'bidi' tobacco nursery beds treated with DBCP at 15 l/ha, ethoprop at 4 kg a.i./ha and metham sodium at 80 I a.i./ha. DD at 500 I/ha and DBCP at 15 I/ha gave the best control of the nematode up to 8 weeks. Carbofuran at 4 kg a.i./ha was better than aldicarb at 4 kg a.i./ha but both were inferior to ethoprop, fensulfothion and metham sodium. In FCV tobacco nurseries in sandy loam soils at Hunsur, Karnataka, Phenamiphos at 1 kg a.i./ha in two split doses controlled the root-knot nematode up to 52 per cent till 45 days which was at par with aldicarb 8 kg a.i./ha (4 + 4) in efficacy (Anon., 1983). However, aldicarb at 8 kg a.i./ha in single dose at sowing controlled *M. incognita* up to 50 per cent (Hussaini, 1985). Phenamiphos 400 EC at 1 ml/m^2 offered 91 per cent disease control, while phenamiphos 5G at 8 g/m^2 offered 50 per cent control in the first pulling and the same treatments gave 61 and 25 per cent control respectively in the second pulling resulting in increased number of transplants (Hussaini, 1986). Krishnamurthy & Elias (1969) found DD at 56 ml/m^2, metham sodium at 125 ml/m^2 and phenamiphos at 5 g/m^2 to be effective for control of *M. incognita* in light soil nurseries. Abdul Wajid & Elias (1978) found decrease in nematode population in soil as well as least disease incidence on seedlings with application of methomyl at 9 kg a.i./ha in tobacco nurseries. It was effective up to 30 days after which there was a gradual nematode build-up. In black tobacco, seed bed treatment with methyl bromide, 1,3-D + MIT, dazomet and phenamiphos resulted in

Fig.3. Root systems of FCV tobacco from non-fumigated (left) and fumigated (right) plots.

effective control of *M. incognita* with increased number of seedlings up to 50%. Best economic effect was obtained with dozomet and 1,3-D + MIT treatments (Teran *et al.*, 1986). Similarly, Teran (1982) reported that phenamiphos 10G at 30 g/m^2 treatment showed phenamiphos to be the most effective and economical and DBCP the least. Basamid was also effective against weeds and possibly other pathogens.

CHEMICAL CONTROL (FIELD)

DD is used at the rate of 112-125 l/ha in overall treatment or at 90-169 l/ha in row placement, whereas EDB is applied at 56-112 l/ha for total land treatment. In Zimbabwe EDB at 9-60 l a.i./ha is recommended costing Z $ 40-50 which can be recovered by production of 22-28 kg/ha of flue cured tobacco. It is effective for control of *Pratylenchus* spp. Also, besides *Meloidogyne* spp. approximately 90% of tobacco acreage in Ontario, Canada is fumigated for control of root-lesion nematodes. Major fumigants are Telone 11, Telone C17, Vorlex or Vorlex CP applied at 6-7" depth and covered with a ridge 2-3 weeks before planting. The hills are opened shortly before or at the time of planting. In North Carolina, Telone C17 is injected at a depth of 12" behind each subsoiler shank at a cost of approximately $ 100/ac with excellent control of root-knot, gainville wilt and black shank. Broadley (1979) reported that in Queensland EDB at 36 kg a.i./ha in row treatment was consistently superior to non-volatile nematicides, ethoprop, phenamiphos, fensulfothion, oxamyl, aldicarb, UC 21865, CGA 12223, AC 64475, triazophos and chloropyriphos on tobacco against *M. javanica*. Patel *et al.* (1978) reported effectiveness of DBCP as preplant treatment on bidi tobacco in Gujarat. Aldicarb or carbofuran at 1-2 kg/ha significantly decreased *M. incognita* populations up to 45 days increasing the fresh leaf yield up to 80% at higher dose (Naganathan, 1986). At the dosage at 1.5-4.5 kg a.i./ha, aldicarb enhanced tobacco growth and yield but the rates above 4.5 kg resulted in a neutral to negative effect in the growth of tobacco 'NC 82', 'SpG 70' and 'McNair 944' but excessive moisture (Ca 7-8 cms) limited cured leaf yields (Barker & Powell, 1988). In polyspecific communities with root-knot and blank-shank complex, metalaxyl (2.24 kg a.i./ha) through irrigation water was very effective and economical (Csinos *et al.*, 1988). Reddy & Seshadri (1971) reported control of root-knot by application of methomyl (5G 90 WP) at 18, 26 and 32 kg a.i./ha before transplanting tobacco, however, it was phytotoxic at higher doses. Fensulfothion, aldicarb and carbofuran at 4 kg a.i./ha (row treatment) and 1 kg a.i./ha (spot treatment) controlled the nematode significantly (Reddy, 1975). Aldicarb, aldicarb sulfone and fensulfothion at 0.5 and 1.0 kg a.i./ha as spot application reduced the incidence of *M. incognita* on FCV tobacco increasing cured leaf yields. Phenamiphos 40 EC as root dips at 0.2 per cent for 1/2 hour and soil application in planting hole at 0.1, 0.2 an 0.4 per cent gave effective nematode control increasing the yield of tobacco (Hussaini, 1986).

Carrillo & Lopez (1979) tested aldicarb (5 kg a.i./ha), fensulfothion (4 kg), phenamiphos (5 kg), carbofuran (6 kg), ethoprop (liquid/granular formulation, 7 kg) and oxamyl (foliar application of 0.65 kg or a bare-root dip at 1500 ppm on tobacco cv 'SpG 28' in Costa Rica against *M. incognita*. Garcia & Rich (1983) found poorest control with carbofuran and aldicarb. Ethoprop and phenamiphos varied in effectiveness. Post-plant application of ethoprop was inferior to preplant application. Avigliano & Sannino (1983) tests AC 641475, a systemic nematicide-insecticide in liquid and granular formulations and found it to be effective at 2.4 and 3.4 kg/ha on burley tobacco against *Meloidogyne* spp. Rich *et al.* (1986) compared FMC 67825 20G and Isazophos 4E for three years for control of *M. javanica* on *N. tabacum* alongwith phenamiphos 3SC applied 1-3 days prior to planting. All the three nematicides were effective in lowering root-knot index. Johnson & Wright (1987) found that phenamiphos in combination with other nematicides controlled root-knot with yield increase of FCV tobacco. Between tobacco yield and root-knot index there was negative correlation. Similarly, phenamiphos and aldicarb at 50 and 100 kg a.i./ha were found effective, whereas dazomet was ineffective (Castelli & Mecheli, 1986). DS 38697 at 3 lb a.i./ac, DS 38297, DS 46995 at 6 lb a.i./ac and F 3843-4 EC gave best control of *M. incognita* in Georgia (Johnson & Wright, 1984, 1985). Katudat (1978) preported efficacy of Vydate 1% applied 3 times at two weeks intervals against *M. javanica* in FCV tobacco in Thailand. Nordemeyer *et al.* (1982) tested effect of ethoprop, carbofuran, and aldicarb at 6.7 kg a.i./ha on three species of *Meloidogyne*, viz. *M. incognita*, *M. javanica* and *M. javanica* dn *M. arenaria*, which increased plant height with the exception of carbofuran for *M. javanica* infested plants. Aldicarb significantly increased yields of plants infected with *M. javanica*. Nematode numbers were reduced by aldicarb or ethoprop 11 and 15 weeks after planting. Root-gall indices at final harvest were high and did not differ among different treatments. Fortnum & Currin (1982) found that fensulfothion-phenamiphos applied to tobacco infected with *Phytophthora parasitica nicotianae-M. incognita* complex in South Carolina significantly increased yields. Metalazyl also reduced the root galling over control. The experimental nematicide SD 386997 at 120 lb a.i./ac gave the greatest reduction in number of *M. incognita* larvae per 100 cm^3 of soil. The chemical was found to be at par with phenamiphos at similar rate (Fortnum & Currin, 1985). Nordemeyer & Dickson (1985) found ethoprop, aldicarb, phenamiphos, oxamyl and carbofuran at 6.5 kg a.i./ha to be effective in increasing yields of tobacco infected with *M. incognita*. Carbofuran did not increase yields of *M. javanica*-and *M. arenaria*-infected plants. Root-gall index and soil populations were reduced with all the nematicides. Similarly ethoprop was found effective at 27, 18 or 9 kg a.i./ha in reducing root-galling due to *M. javanica* (Rich *et al.*, 1984). Foliar application of carbofuran at 2400 ppm reduced root-galling due to *M. incognita* and 2400 ppm of FMC 35001 controlled *Tylenchorhynchus claytoni* (Di Sanzo, 1981). Application of ethoprop at 7 kg a.i./ha to tobacco cv. 'NC 3555' infected with *Rotylenchulus reniformis* resulted in an increase in yield by 852 kg/ha over untreated control (Hutton, 1983). Aldicarb 10G at 10-15 kg a.i./ha ploughed in one week before planting was found effective in the control of *Pratylenchus, Aphelenchoides, Aphelenchus, Dorylaimus* and *Helicotylenchus* on FCV tobacco in Romania (Romascu & Dragoescu, 1979).

Control of *Globodera tabacum solanacearum* by the application of Mocap + Nemacur, Nemacur-Dasanit + Nemacur 3 at 1.3 gal/ac, Nemacur 3 + Furadan 4F at 0.75 gal/ac and Nemacur 3 + Larsban 4E resulted in higher yield of FCV tobacco compared to Furadan 4F alone and Terr-O-Cide II at 12 gal/ac (Johnson, 1987). Post-harvest population densities of cyst nematodes were reduced with phenamiphos at 7.47 kg a.i./ha using a resistant cultivar 'NC 567' relative to a susceptible cultivar 'K 326' or 'Mc Nair 944' (Johnson *et al.*, 1989). Application of Soil-brom-90 (32 l/ha) to cyst nematode-infested plots increased yields. Cultivars 'Clemson', 'PD-4', 'Va 81' and 'NC 567' significantly decreased second stage larvae, cyst and egg densities (Komm & Terrill, 1984).

Efficacy of avermectin Bia, Av-B2a and Av B2a 2-3 ketone, for control of *M. incognita* in tobacco in field was studied in U.S.A. Application of 0.05-0.50 kg a.i./ha suppressed root-galling and egg production. Eggs/plant were suppressed by 21-86% depending upon dosage and comparable to those of ethoprop and phenamiphos at 6.73 kg a.i./ha (Sasser *et al.*, 1982).

OILCAKES

Application of oilcakes to soil results in reduction of plant-parasitic nematodes, besides being of manurival value. Several workers have obtained effective reduction of *Meloidogyne* spp., *Globodera rostochiensis, Heterodera schachtii, T. semipenetrans, Rotylenchulus reniformis, Tylenchorhynchus* spp., *Hoplolaimus indicus*, etc. in different crops by the application of deoiled cakes of castor, mahua, neem, mustard, peanut, linseed, sesamum, safflower, and karanj (Singh & Sitaramaiah, 1970; Muller & Gooch, 1982; Alam, 1993). In case of tobacco the application of oilcakes has not become popular since excess of nitrogen in nurseries results in succulent seedlings and in field affects the quality of cured leaf. *In vitro* studies with punnakkai cake *Calophyllum inophyllum* revealed its nematicidal property. Application of the cake 30 g/plot checked the root-knot (*M. javanica*) infection on tobacco. Decomposition for 15 days before application was more effective (Desai *et al.*, 1973).

CULTURAL PRACTICES

Patel *et al.* (1985) explored the possibility of using castor plants in management of *M. incognita* and *M. javanica*. Anand 39/1' a castor genotype caused maximum trapping of larvae in a pot culture study. Hence the authors felt that castor 'Anand 39/1' could be successfully used preceding tobacco in nurseries.

CROPPING SYSTEMS

Construction of a realistic cropping system model requires quantitative information about the dynamic interaction between the host plant and the pest. Besides, the crop in a rotation scheme should be of economic value to the user of the system.

Peanut, corn, wheat and *Sorghum* sp. plant prior to tobacco reduced the incidence of *M. incognita* (Muro, 1975; Cortado & Davide, 1968). At Anand, tobacco followed by wheat and cotton had low root-knot index, A 2-year rotation with summer bajra-hybrid cotton minimised the root-knot infection (Anon., 1974).

Cotton (*Gossypium hirsutum*) is injured by *Pratylenchus* in South Carolina and is fairly resistant to *M. incognita* in North Carolina whereas it was found resistant to *M. javanica* and *G. barbadense* to *M. incognita*. At Tobacco Research Station, Hunsur, India during 1976-79, rotations with cotton, groundnut and gingelly either alone or in combination with summer ploughings were tried against *M. incognita*. It was found that any rotation with summer ploughings was better than the one with it; cotton rotation was the best among the three crops without ploughing in summer. However, groundnut and gingelly rotations alone did not increase yields of following crop of tobacco. All the rotation crops reduced root-knot infection on tobacco when combined with summer ploughings (Anon., 1978).

In Zimbabwe, Daulton (1962) extensively investigated the inclusion of pasture grasses in rotation with tobacco resulting in reduction of root-knot incidence and increased tobacco yields. A three-year rotation with *Eragrostis curvula* cv. 'Ermelo', *Chloris gayana* cv. 'Katambora', and *Panicum maximum* cv. 'Sabi' was effective.

RESISTANCE

There was marked variability in the reactions of five accessions of *Nicotiana otophora* to race 4 of *M. incognita*. The most resistant was 'La Quinta' which did not develop egg masses or root necrosis (Arcia & Wernsman, 1980). The hypersensitive reaction of *N. repanda* to *M. incognita acrita* has been attributed to the rapid accumulation of oxidation products of chlorogenic acid (Milne *et al.*, 1965).

Patel *et al.*, (1986) found 'PCT 4' ('Mc Nair 135') to be highly resistant to *M. incognita* and *M. javanica*. Gwynn *et al.* (1986) found a croos of *N. repanda* x *N. sylvestris* backcrossed with N. tabacum to have resistance to tobacco mosaic virus (TMV), 'cyst nematodes and *M. incognita*. Kadotani *et al.* (1985) transferred *M. javanica* resistance from 'Okinawa' to 'Coker 319' and 'Coker 354' by backcross. The F1 of 'MjR Coker 319' x 'Coker 319' was similar to 'Coker 319' in resistance, growth and yield characteristics. Resistance was also obtained by segmental substitution from *N. repanda* and *N. longiflora* and from crosses between them. Two breeding lines resistant to either race 2 or 4 of *M. incognita* and 14 lines resistant to 2 and 4 and those moderately resistant to *M. javanica* were evolved by Cornilissen & Wyk (1987). Quality of the breeding lines with crosses with accepted cvs. 'Hicks', 'Hicks 76', Delcrest 202' and 'TL 33' approximated with that of Hicks. Burley tobacco lines KY 8651, 8654, 8635 were found resistant to *M. incognita* race 1 (Nielsen & Collins, 1988). These were derived by backcrossing NC 95 and burley tobacco cv. KY 10, KY 14 and KY 17 as recurrent parents. The lines performed well.

Resistance to *M. incognita* is conferred by a single dominant gene transferred through interspecific hybridization with *N. tomentosiformis* (Stavely, 1979). In *N. repanda* and *N. longiflora* x *N. tabacum* numbers of *M. incognita* and *M. javanica* egg masses were reduced compared to 'NC 2326', a susceptible cultivar of *N. tabacum*. However, fewer differences in egg mass numbers were found with the inoculation of *M. arenaria*. Similarly in *N. tabacum* x *N. repanda* number of egg masses of *M. arenaria* or *M. javanica* were not reduced significantly. Among different species, *N. knightiana*, *N. sandrae* and *N. velutma* showed reduced egg mass production with M. arenaria; N. megalosiphon exhibited egg mass production when inoculated with *M. javanica* ; *N. glauca*, *N. longiflora*, *N. plumbaginifolia* and *N. repanda* showed long egg mass production when inoculated with either *M. arenaria* or *M. javanica*. Several selection of *N. longiflora* and *N. repanda* cv. '46-G' had the lowest egg mass production when inoculated with either *M. arenaria* or *M. javanica* (Davis *et al.*, 1989).

The resistance to *M. javanica* and *M. incognita* can break down at temperatures between 30°C and 35°C (Fukudome & Kamigana, 1982; Rufty *et al.*, 1983).

In the U.S.A. many tobacco cultivars have been bred with resistance to *M. incognita* races 1 and 3 (Clayton *et al.*, 1988) and these cultivars can be grown in any part of the world which has a problem with these races of *M. incognita* and where the type of tobacco produced by these cultivars is acceptable. These cultivars are grown in China, Mexico and the Philippines where they are also breeding their own cultivars with resistance to *M. incognita* and also in Brazil. The gene for this source of resistance apparently came from *N. tomentosa* (Slana & Stavely, 1981). Other sources of resistance to *M. incognita* race 3 are available from a *N. repanda* x *N. tabacum* cross and for races 1 and 4 from selections of *N. otophora* (Arcia & Warnsmann, 1980).

In Zimbabwe where *M. javanica* is the most important nematode species attempts to breed resistance into commercial cultivars started in 1941. Sources of resistance to *M. javanica* have been found in *N. repanda*, some strains of *N. longiflora* and some locally grown primitive tobacco types. These lines were later crossed to *M. incognita*-resistant tobacco obtained from America. Recently hybrids between local susceptible cultivars and the advanced resistant breeding lines have been tested and found to have a high level of resistance to *M. javanica* and produced good quality tobacco. However, when grown in fields with a high nematode population, the resistant cultivars, though producing lower yields than the susceptible cultivars and having much less root galling do benefit from the extra protection provided by nematicides. These cultivars are not yet ready for general use but are in the final stages before release. Other sources of resistance to *M. javanica* have been reported by Slana (1978) and Kadotani *et al.* (1985). Cultivars with tolerance can be effective in fields with low to moderate infestation.

The amount of invasion and gall damage by *M. javanica* in both susceptible and resistant cultivars was found to increase as the soil temperature was raised from 25°C to 30°C and fell as the temperature increased to 35°C. The effect on the resistant cultivars was always much less than on the susceptible cultivars. *M. incognita* races 2 and 4, *M. javanica*, *M. arenaria* and *M. grahami* can break the resistance of the *M. incognita* races 1 and 3 resistant tobaccos. Some of the *M. javanica*-resistant breeding lines were susceptible to *M. arenaria* and *M. incognita* race 3.

The biochemical basis for resistance to *Meloidogyne* in tobacco is not known, though as a hypersensitive reaction it is usually found in the nematode infested resistant plants; phytoalexin may be involved.

X10 cultivars with resistance specific to *Pratylenchus* have been developed. Nevertheless, several *M. incognita*-resistant cultivars have some resistance and some burley cultivars have resistance to *P. penetrans*. Source of resistance to the tobacco cyst nematode have also been established and breeding lines are being developed. There has been no attempt to bread resistance to other nematode parasites of tobacco.

Search for sources of resistance to *M. javanica* in available wild species at CTRI, Rajahmundry, India revealed that *N. benthamiana*, *N. nesophila* and *N. solanifolia* were moderately resistant and *N. plumbaginifolia* and *N. nudicaulis* were resistant (Gopalachari, 1984). Varieties like 'Diebel Basma' 'Dixie Bright' and 'Mandrene Koseme' showed 1.40, 1.80 and 2.45 root-knot index, 'SpG 28' and 'SpG 41' were completely free of galls, while 'Va 770', 'Coker 254' and SC

66 PCT 4' had very low infection under field conditions (Anon., 1982, 1983). At Hunsur, *N. repanda* and *N. nesophila* were found highly resistant to *M. incognita*, *M. javanica* and *M. arenaria*.

Fukudome & Yamaguchi (1976) found 'Okinawa 6' and 'Okinawa 12' to be resistant to *M. javanica* Japan. However, they were susceptible under Gujarat conditions to the same species. Perhaps Gujarat population is a different pathotype (Shah *et al.*, 1985). Ohashi & Kodato (1981) reported low root-knot index and slow development of *M. javanica* in 'Hirac' and 'Okinawa' varieties. 'Florida 22' and 'RK 70' were found resistant to *M. incognita* (Mine, 1972). 'NC 95' was released in U.S.A. in 1962 at the first flue cured commercial variety resistant to *M. incognita* (Moore *et al.*, 1962). In Yugoslavia, Klinoic *et al.*, (1978) reported 'NC 95' to be resistant to local population of *M. incognita*. It has added advantage of being resistant to black shank, *Fusarium* wilt and to some extent bacterial wilt and brown spot diseases. It is however susceptible to *M. javanica* and *M. arenaria* (Graham, 1964; Milne, 1966; Miller, 1968) and where *M. javanica* is present blank shank is a problem with this variety (Miller, 1968). A new race of *M. incognita* attacking 'NC 95' was later found (Graham, 1969; Anon., 1963; Krishnamurthy & Elias, 1968; Reddy *et al.*, 1972). Loss of resistance in 'NC 95' to *M. incognita* was found due to infection of *M. arenaria* or *M. hapla* (Eisenbeck, 1983).

Under field conditions with predominant population of *M. incognita* 213 tobacco selections were screened. 'SpG 28' was the only var. to have to egg masses after 60 days and 'Ga 1469', 'NCPY 10' and 'Coker 148' to have low root-knot index (Hussaini & Moses, 1981). Subsequently 'Va 145' was found to have the lowest root-knot index. Under nurseries and controlled conditions in glasshouse with *M. incognita* vars. 'Va 145', 'FCH 6005' ('Bhavya') and 'Coker 254' showed low RKI at 60 days while 'Va 770' is the only var. found to be promising against *M. arenaria*. 'FCH 6005 ('Bhavya') has RKI against *M. javanica* also. 'Bhavya' has been released for commercial cultivation in Karnataka, India.

Lagana *et al.* (1986) found 'NC 95', 'SpG 28', 'Cv. 227', Cv. 347', 'Golden Green', 'Coker 258', 'Cv 116', 'Cv 563', 'Mammoth Gold', 'Coker 86', 'Coker 51', 'Mammoth delcrest' and 'Coker 347' to be resistant to *M. incognita*.

Gwynn (1986) found that 'NC 60' a cross between 'Mc Nair 944' and SpG 28' was highly resistant to *P. parasitica* var. *nicotianae*, *Pseudomonas solanacearum* and *Thielviopsis basicola* and resistant to race 1 and 3 of *M. incognita* yielding higher good quality tobacco. *N. repanda* crossed with *N. tabacum* was resistant to *M. incognita*, *Globodera tabacum solanacearum* and *Psudomonas syringae pvatabaci* (Gwynn *et al.*, 1986).

Burley tobacco lines 'UcvL 2', 'UcvL3' and UcvL16' were found highly resistant to *Meloidogyne incognita* and *Fusarium oxysporum* f.sp. *nictotianae* with high yield and quality characteristics in Venezuela.

Among *N. rustica* cultivars 'Motihari', 'Bitri', and 'Hepti' showed moderate resistance to both *M. incognita* and *M. javanica*. Patel & Patel (1986) reported cultivars 'Bitri', 'Gharrama' and 'Mandahata small leaf' to be resistant and 'Daiya', 'Kharwar', 'Kampuri Local', 'Maharka', 'NP 18', 'NP 219', 'P7', 'GC1' and 'T26' moderately resistant to *M. incognita*. Against *M. javanica* 'Motihari', 'Hepti' and 'Pakar' were resistant.

'Va 81 and 'PD 4' were tested for resistance to *Globodera tabacum*. Less number of cysts were noticed on these two cultivars compared to 'CT 86-4'. Subsequent reproduction was also much less and development of nematode in roots stained after 5 weeks was found to be less. During a growing season population densities of the cyst nematode in naturally infested field soil declined by 72-80% under 'Va 8' and 'PD 81', and increased by 144% under 'CT 86-4' (La Mondia, 1988).

Patel *et al.* (1986) reported that '*bidi*' tobacco varieties 'Anand 2' and 'GT 4' and chewing tobacco variety 'Anand 145' were consistently less preferred by *Rotylenchulus reniformis* compared to 'Va SCR' which was highly preferred.

BIOLOGICAL CONTROL

The major genera of predaceous fungi *Arthrobotrys, Dactylaria, Dactylella* and *Monacrosporum* probably contain over 100 species. Many species form traps spontaneously and consistently even in pure cultures while others produce traps feebly even in the presence of nematodes (Mankau, 1980). In France, use of *A. irregularis* as commercial product 'R 350' is effective at 140 g/m^2 applied one month before planting. An annual dose of 50 mg/m^2 is suggested (Cayrol, 1972). At the CTRI Research Station, Hunsur *Arthrobotrys, Dactylella* and *Monacrosporium* sp. were isolated from tobacco soils. Attempts at mass culturing were done and it was found that *Arthrobotrys* and *Monacrosporium* developed well on *ragi* husk medium while the growth of *Dactylella* was poor. *Arthrobotrys* was chosen for testing in nurseries against root-knot nematodes. The fungus culture was applied in dried powder form at 150, 300, and 600g incorporated in 10 m^2 bed. An untreated check was maintained. There was no improvement in any of the growth characters or yield of transplants. The number of nematodes in development stages in roots were found to be at par with untreated check.

Miller & Anagnostakis (1977) found that *Trichoderma viride* in soil was partly responsible for the poor survival of *Pratylenchus penetrans* and *Tylenchorhynchus dubious* and hatching of eggs of *Heterodera tabacum*.

Smart & Khuong (1987) tested efficacy of *Paecilamyces lilacinus* in green house microplots and field against *M. incognita*. Yields were higher in plots receiving the fungus and were at par with nematicide treated plots. In a 2-year trial application of a combination of *P. lilacinus* with phenamiphos at 4.5-9.0 kg a.i/ha and ethoprophos at 3.4-6.8 kg/ha or individually showed that the fungus did not control *M. javanica* in either year of the experiments (Hewlett *et al.*, 1988). Plants with nematode alone or in combination of *P. lilacinus* had galling indices of 5.0 in both the years in a 0-5 scale. The density of fungus propagules in the soil remained high up to harvest. Dickson & Mitchell (1985) evaluated *P. lilacinus* alongwith and in combination with ethoprop and phenamiphos against *M. javanica* on tobacco in Florida. The root-knot index was 2.7-3.9 in nematode-infested plots treated with nematicides. However, plots with *M. javanica* alone or in combination with *P. lilacinus* had a galling index of 5 in 1-5 scale.

Increased yield of field tobacco up to 24% was reported by Brown *et al.* (1985) following application of *Pasteuria penetrans* to soil infested with *M. incognita*. One of the major problems in utilizing this organism as an effective biological agent is our inability to culture it *in vitro* on any standard bacteriological media. Neither vegetative cells nor spores are harvested in sufficient numbers for large scale field tests. Large scale use still depends on commercial *in vitro* cultivation.

CONCLUSION

From the foregoing discussion it is clear that root-knot nematode infection is universal and serious on tobacco grown for various purposes. Other plant parasitic nematodes are *Rotylenchulus reniformis*, *Globodera* sp., *Pratylenchus* sp. and *Tylenchorhynchus* sp. It is practically impossible to eliminate the nematodes from soils once established, by any single method. Some of the guidelines for management of populations are suggested :

1. Use of fumigants in nurseries to be taken up to tackle high populations of nematodes besides weeds and soil borne pathogens.
2. Nematicides (non-volatile) to be selected and used when moderate levels of nematode populations exist.
3. Economical dosages of non-volatile nematicides to be used in field crop in conjunction with a less susceptible/tolerant cultivar of tobacco.
4. Weed control reduces the multiplication of nematode in nurseries and field.
5. It is to be obligatory on the part of tobacco farmers to take up stalk and root destruction immediately after final harvests to reduce the inoculum potential.
6. Dry tillage in summer reduces the nematode population.
7. Rotation of non-host crops like cereals, cotton, gingelly, and groundnut and grasses help in reducing populations.

8. Soil solarization in seed beds to be combined with low dose of deoiled cake or other organic amendments or a nematicide for a possible synergistic effect.
9. Resistance factors as available in wild species of *Nicotiana* to be transferred to cultivated varieties of *N. tabacum* by genetic engineering and advanced breeding techniques.

Although several plant-parasitic nematodes attack tobacco and inflict considerable damage, most of the work on different aspects has been confined to *Meloidogyne* spp. in India. Detailed studies on the other nematodes like *Rotylenchulus* sp., *Tylenchorhynchus* sp. and *Pratylenchus* sp. would provide information on their role in crop damage, biology, pathogenicity and control.

ACKNOWLEDGEMENTS

Author is grateful to Dr. M.S. Chari, Director, Tobacco Research, Central Tobacco Research Institute, Rajahmundry and Mr. J.S.L. Moses, Officer Incharge, CTRI Research Station, Hunsur for critically going through the manuscript and offering valuable suggestions. Thanks are due to Mr. Jaimon Thomas for help in preparation and to Mr. M.B. Gurraja for typing the manuscript.

REFERENCES

Abdul Wajid, S.M. & Elias, N.A. (1978). Effects of nematicides on the control of root-knot nematode (*M. incognita*) in tobacco nurseries. *Tob. Res.* **4:** 7-9.

Alam, M.M. (1983). Bioactivity against phytonematodes. *In* : *Neem Research & Development* (Eds.-N.S. Randhawa & B.S. Parmar). Soc. of Pesticide Sci. India. pp. 123-143.

Anon. (1963). *Ann. Rep. Bidi. Tob. Res. Stn., Anand, India.*

Anon. (1974). *Ann. Rep. Bidi. Tob. Res. Stn., Anand, India.*

Anon. (1978). Ann. Rep. Bidi. Tob. Res. Stn., Anand, India.

Anon. (1982). *Ann. Rep. Bidi. Tob. Res. Stn., Anand, India.*

Anon. (1983). *Ann. Rep. Cent. Tob. Res. Instt., Rajahmundry, India.*

Arcia, M.A. & Wernsman, E.A. (1980). Reaction of *Nicotiana otophora* accessions to the root-knot nematode *M. incognita* race 4. *Tob. Sci.* **24:** 15-16.

Arens, M.L. & Rich, J.R. (1981). Yield response and injury levels of *Meloidogyne incognita* and *M. javanica* on the susceptible tobacco Mc Nair 944. *J. Nematol.* **13**(2): 196-201.

Avigliano, M. & Sannino, L. (1983). A new product for control of root-knot nematode (*Meloidogyne*) on tobacco. *Helm. Abstr.* **53**(1): 214.

Aycock, R.; Barker, K.R. & Benson, D.M. (1976). Susceptibility of Japanese holly to *Criconemoides* xenoplax, *Tylenchorhynchus claytoni* and certain other plant-parasitic nematodes. *J. Nematol.* **8:** 26-31.

Barker, K.R. & Nusbaum, C.J. (1971). Diagnostic and advisory programmes. *In* : *Plant Parasitic Nematodes* Vol. I (Eds.-B.M. Zuckerman, W.F. Mai & R.A. Rohde) Academic Press, N.Y. pp. 281-301.

Barker, K.R. & Powell, N.T. (1988). Influence of aldicarb on the growth and yield of tobacco. *J. Nematol.* **20**(3): 432-438.

Barker, K.R.; Todd, F.A.; Shane, N.W. & Nelson, L.A. (1981). Inter-relationships between *Meloidogyne* spp. with flue cured tobacco. *J. Nematol.* **13**(1):67-69.

Barker, K.R. & Weeks, W.W. (1981). Influence of soil type and *M. incognita* on yield and quality of tobacco. *J. Nematol.* **13**(4): 432

Broadly, R.A. (1979). Comparison of FAB with non-volatile nematicides for control of root-knot of tobacco in North Queensland and Australia *J. Expt. Agri. & Animal Husbandry* **19**(100): 620-623.

Brown, S.M.; Kepner, J.L. & Smart, Jr. G.C. (1985). Increased crop yields following application of *B. penetrans* to field plot infested with *M. incognita. Soil Biol. Biochem.* **17**(4): 483-486.

Carrillo, A.M.A. & Lopez, C.R. (1979). Response of FCV tobacco to application of nematicides. *Nematropica* **9**(2): 129-134.

Castelli, F. & Micheli, F. (1986). Fumigant and systemic nematicides for root-knot nematodes. *Annali - dell Instituto Sperimentilale per tobacco* **12**: 195-207.

Cayrol, J.C. (1972). Possible biological control of *Meloidogyne* with the nematophagous fungus Arthrobotrys irregularis R. 350. Defence des Vegeteanusn **35** (210): 279-282.

Clayton, E.E.; Gaines, J.G.; Smith, T.E.; Shaw, K.J. & Graham, T.W. (1988). Control of the flue cured tobacco root diseases by crop rotation. *USDA Agric. Farm Bull.* **1:** 12.

Cornelissen, A.P.F. & Wyk, R.J. Van (1987). *Nicotiana tabacum* breeding lines resistant to various root-knot nematodes. *Nematologica* **33**(3): 316-321.

Cortado, R.G. & Davide, R. (1968). Survey, identification and pathogenicity test of nematodes associated with tobacco in Cagayan valley and Illoco region. *Philip. Agric.* **51:** 779-801.

Csinos, A.S.; Johnson, A.W. & Golden, A.M. (1988). Metalaxyl and phenamiphos applied thro 'irrign. water to control black root-knot complex on tobacco. *Pl.Dis.* **70**(3): 210-213.

Daulton, R.A.C. (1962). Survey of land use and plant parasitic nematode control practices in the south western flue cured tobacco areas of Rhodesia. *Rhod. J. agric. Res.* **59:** 216-217.

Daulton, R.A.C. (1964). Effect of soil fumigation in tobacco in Southern Rhodesia. *Biokemia* No. **5:** 10-15.

Davis. E.L.; Rich, J.R.; Gwynn, G.R. & Sisson, V. (1989). Greenhouse evaluation of *Nicotiana* spp. for resistance to root-knot nematodes. *Nematropica* **18**(2): 99-107.

Delta, T.G. & Cappelli, C. (1988). Results of 2 years trials on root parasites of tobacco in Umbria. *Helm. Abstr.* **57**(1): 181.

Desai, M.V.; Shah, M.M. & Pillai, S.N. (1973). Nematicidal property of some plant species. Indian *J. Nematol.* **3:** 77.

Dickson, D.W. & Mitchell, D.J. (1985). Evaluation of *Paecilomyces lilacinus* as a biocontrol agent of *M. javanica* on tobacco. *J. Nematol.* **17**(4): 519.

Di Sanzo, C.P. (1981). Effect of foliar application of carbofuran and a related compound on plant-parasitic nematodes under greenhouse and growth chamber. *J. Nematol.* **13**(1): 20-24.

Eisenbeck, J.D. (1983). Loss of resistance in tobacco cv. NC 95 by infection of *M. arenaria* or *M. arenaria* of *M. arenaria* or *M. hapla. J. Nematol.* **15** : 478.

Ferris, H. (1972). Population dynamics of *Meloidogyne* spp. in relation to epidemiology and control of root-knot of tobacco. Ph. D. Thesis, N.C. State Univ., Raleigh. p. 95.

Ferris, H. (1974). Correlation of tobacco yield value and root-knot index with early to mid season and post-harvest *Meloidogyne* population densities. *J. Nematol.* **6:** 75-80.

Fortnum, B.A. & Currin, R.E. (1982). Control of black shank, root-knot complex on tobacco with fumigant nematicies and Ridomil. *Fungicide Nematicide Tests.* **38:** 13.

Fortnum, B.A. & Currin, R.E. (1985). Evaluation of selected nematicides for control of root-knot nematode on flue cured tobacco. *Fungicide Nematicide Tests* **40:** 114.

Fukudome, N. & Kamigana, K. (1982). Resistance of tobacco to *M. javanica*. II. Effect of soil temperature on manifestation of resistance. *Jap. J. Nematol.* **11:** 13-18.

Fukudome, N. & Yamaguchi, Y. (1976). Studies on the tobacco varieties resistant to *M. javanica*. I. On the resistance of Okinawa native tobacco lines collected in Okinawa to *M. javanica*. *Kagoshima Jap. Exptl. Stn. Bull.* **20:** 135-144.

Garcia, M.R. & Rich, J. (1983). Efficacy of selected fumigant and non-fumigant nematicides to control *M. javanica* in Florida Tobacco. *Nematropica* **13:** (2): 125-134.

Gopalachari, N.C. (1984). *Tobacco*. Indian Council of Agri. Research. p. 328.

Graham, T.W. (1964). Response of selected tobacco varieties to *M. javanica* and *M. incognita acrita* in field pathogenicity trial. *Phytopathology* **54:** 623.

Graham, T.W. (1969). A new pathogenic race of *M. incognita* on flue cured tobacco. *Phytopathology* **59:** 114.

Gwynn, G.R. (1986). Registration of NC 60 tobacco. *Crop Sci.* **26**(5): 1085-1086.

Gwynn, G.R.; Barker, K.R.; Reilly, J.J.; Komm, D.A.; Burk, L.G. & Read, S.M. (1986). Genetic resistance to TMV, cyst nematode, root-knot nematode and wildfire incorporated into *N. tabacum*. *Pl. Dis.* **70**(10): 958-962.

Hanounik, S.B. & Osborne, N.W. (1977). The relationship between population density of *Meloidogyne incognita* and nicotine content of tobacco. *Nematologica* **23:** 147-152.

Hanounik S.B.; Osborne, N.W. & Pirie, W.R. (1975). Relationships between the population density of *Meloidogyne incognita* and growth of tobacco. *J. Nematol.* **7:** 352-356.

Hewlett, T.E.; Dickson, D.W.; Mitchell, D.J. & Kannwischer-Mitchell, M.E. (1988). Evaluation of *P. lilacinus* as a biocontrol agent of *M. javanica* on tobacco. *J. Nematol.* **20**(4): 578-584.

Hussaini, S.S. (1983). Quantification of root-knot nematode *Meloidogyne* spp. damage of FCV tobacco. *Tob. Res.* **9** (2): 61-65.

Hussaini, S.S. (1985). Efficacy of Durofume rabbing and aldicarb against root-knot nematodes and weeds in tobacco nurseries. *Indian J. Nematol.* **15**(1): 88-92.

Hussaini, S.S. (1986). Efficacy of non-volatile nematicides for control of *Meloidogyne incognita* in FCV tobacco. *Indian J. Nematol.* **16**(1): 100-101.

Hussaini, S.S. (1989). Comparative efficacy of durofume and other fumigants against root-knot nematodes and weeds in FCV tobacco nurseries. *Indian J. Nematol.* **19**(1): 14-17.

Hussaini, S.S. & Moses, J.S.L. (1981). Evaluation of tobacco germplasm for resistance to root-knot nematode. *Tob. Res.* **7**(2): 176-177.

Hussaini, S.S. & Moses, J.S.L. (1989). Effect of intensity of root-knot galling in seedling on yields of FCV tobacco. *Indian J. Nematol.* **19**(1): 18-20.

Hutton, D.G. (1983). Effect of *Rotylenchulus reniformis* on yield of tobacco. *Helm. Abstr.* 51(4).

Johnson, A.W. (1974). Relative efficacy of selected non-volatile nematodes for control of root-knot nematodes on flue-cured tobacco. *Tob. Sci.* **18:** 132-133.

Johnson, A.W.; Dukes, P.D. & Flowers, R.A. (1973). Organic pesticides for control of root-knot nematodes in flue cured tobacco. *Tob. Sci.* **17:** 153-154.

Johnson, A.W. & Wright, W.C. (1984). Evaluation of root-knot control on tobacco. *Fungicide Nematicide Tests.* **39:** 99-100.

Johnson, A.W. & Wright, W.C. (1985). Effect of nematicides on nematode control, plant growth and yield of FCV tobacco. *Fungicide Nematicide Tests* **40:** 112-113.

Johnson, A.W. & Wright, W.C. (1987). Evaluation of nematicides on root-knot control and yield of FCV tobacco No. 2326. *Fungicide Nematicide Tests* **43:** 117.

Johnson, C.S. (1987). Effects of tobacco cyst nematode (*Globodera tabacum solanacearum*). control on agronomic characteristics of FCV tobacco var. C 319: *Fungicide Nematicide Tests* **43:** 173.

Johnson, C.S.; Komm, D.A. & Jones, J.E. (1989). Control of *Globodera tabacum solanacearum* by alternating host resistance and nematicide. *J. Nematol.* **21**(1): 16-23.

Kadotani, N.; Hokaguchi, N. & Fukudome, N. (1985). Breeding FCV tobacco resistant to *M. javanica. Bull. Kogoshima Tob. Exp. Stn.* **26:** 19-33.

Katan, J. (1981). Soil solarization for control of soil borne pests. *Ann. Rev. Phytopathol.* **19:** 211-236.

Katudat, U. (1978). Concise Progress Report. *Kasetsart Journal* **12**(1): 67.

Kilinoic, O.; Petrovic, O.; Bumic, K. & Beljo, J. (1978). Investigation of the resistance of some imported varieties of alfalfa to the domestic population of stem nematode and of tobacco to root-knot nematode. *Zastita Bilja* **29**(4): 381-385.

Komm, D.A.; Reilly, J. & Elliot, A.P. (1983). Epidemiology of a cyst nematode in Virginia. *Pl. Dis.* **67:** 1249-1251.

Komm, D.A. & Terrill, T.R. (1984). Effect of resistant cultivars and a fumigant nematicide on the control of the tobacco cyst nematode. *Phytopathology* **74**(7): 830.

Krishnamurthy, G.V.G. & Elias, N.A. (1968). Additions to host range of *M. incognita* causing root-knot on tobacco in Hunsur. *Indian Phytopath.* **21**(3): 351-352.

Krishnamurthy, G.V.G.& Elias, N.A. (1969). Note on effect of nematicide and heat treatment on the control of root-knot in tobacco nursery. *Indian J. Agric. Sci.* **39:** 263-265.

Krishna Prasad, K.S. (1985). Relative efficacy of few pesticides on root-knot nematodes in FCV tobacco nurseries. *Indian J. Nematol.* **15**(1): 93-133.

Lagana, A.C.; Revellame, L.A. & Ruguian, V.C. (1986). Field screening of FCV tobacco CVS for resistance to *M.incognita. J. Tob. Sci. Techol.* **1**(1): 89-90.

Lamberti, F. (1979). Economic importance of *Meloidogyne* spp. in sub tropical and Mediterranean climates. *In* : *Root Knot Nematodes: Systematics, Biology & Control*. Academic Press. pp. 341-357.

La Mondia, J.A. (1988). Tobacco resistance to *Globodera tabacum. Ann. appl. Nematol.* **2:** 77-80.

Lange, B.; Meyer, J. & Bururcister, P. (1964). Voulanfige Erwillungen Uber Schadeu on Tabak durch Aeilebende Nematodeu and Versuche Au rhrer Bekampflug Mitt Bial Bundesan St hand Fortsiaji rtchsch. Berlin **111:** 85-95.

Lucas, G.B. (1975). *Diseases of Tobacco*. Scareerow Press, N.Y.

Mankau, R. (1980). Bio-control of nematode pests by natural enemies. *Ann. Rev. Phytopathol.* **18:** 415-440.

Miller, G.R. (1968). Interaction of *Meloidogyne javanica* and *Phytopthora parasitica* var. *Nicotinae* on *Nicotiana tobacum* NC 95. *Phytopathology* **58:** 553.

Miller, P.M. & Anagnostakis (1977). Suppression of *Pratylenchus penetrans* and *Tylenchorhynchus dubius* by *Trichoderma viride. J. Nematol.* **9:** 182-183.

Milne, D.L. (1966). Screening of Nicotiana plants for resistance to *Meloidogyne* spp. by use hyper-sensitive root reaction. *South Africa J. Agric. Sci.* **39:** 59-60.

Milne, D.L. (1972). Nematodes of tobacco. *In* : *Economic Nematology* (Ed.-J.M. Webster). Acad Press, N.Y. pp. 142-186.

Milne, D.L.; Boshoff, D.N. & Buchen, P.W.W. (1965). The nature of resistance of *Nicotiana repanda* to the root-knot nematode, *Meloidogyne javanica*. *J. Agric. Sci*. **8:** 557-567.

Moore, E.L.; Powell, N.T.; Jones, G.L. & Gwynn, G.R. (1962). Flue-cured tobacco NC 95. *Bull. N. Agric. Exp. Stn*. 419.

Muller, R. & Gooch, P.S. (1982). Organic amendments in nematode control. An examination of literature. *Nematropica* **12:** 319-326.

Muro, A.D. (1975). Root-knot nematode (*Meloidogyne incognita*). Control by two and three years crop rotation. *Anu. Ist Sper. Tab*. (Rome) **2:** 89-91.

Naganathan. T.G. (1986). Control of root-knot nematode on tobacco with non-volatile nematicides. *Nematol. medit*. **14:** 285-286.

Nielsen, M.T. & Collins, G.B. (1988). Registration of three root-knot resistant Burley tobacco germplasm lines. *Crop Sci*. **28**(1): 206.

Nordemeyer, D. & Dickson, D.W. (1985). Management of *Meloidogyne javanica*, *M. arenaria* and *M. incognita* on FCV tobacco with organophosphate carbamate and avermictin nematicides. *Plant Dis*. **69:** 67-69.

Nordemeyer, D.; Rich, J.R. & Dickson, D.W. (1982). Effect of ethoprop, carbofuran and aldicarb on FCV tobacco infected with three species of *Meloidogyne*. *Nematropica* **12**(2): 199-204.

Ohashi, Y. & Kodato, Y. (1981). Varietal differences of resistance to root-knot nematode, *M. javanica* of Japanese domestic tobacco cultivars. *Bulletin of the Utsunomiya Tobacco Experimental Station* No. **16:** 1-7.

Olthof, T.H.A.; Marks, C.F. & Elliot, J.M. (1973). Relationship between population densities of *Pratylenchus penetrans* and crop loss in FCV tobacco in Ontario. *J. Nematol*, **5:** 158-162.

Patel, D.J. & Patel, S.K. (1986). Relative susceptibility of *Nicotiana rustica* germplasm to root-knot nematodes. *Int. Nematol. Network Newsl*. **3**(4): 18-20.

Patel, D.J.; Desai, M.V. & Patel, D.B. (1986). Susceptibility of tobacco varieties to *Rotylenchulus reniformis*. *Tob. Res*. **12**(2): 202-203.

Patel, G.J.; Desai, M.V.; Naik, M.N. & Shah, H.M. (1978). Effect of applying Nemagon to the crop preceding tobacco in control of root-knot nematodes. *Tob. Res*. **4**(1): 59-61.

Patel, G.J.; Shah, H.M.; Patel, D.J. & Patel, S.K. (1985). Use of castor in management of root-knot nematodes. *Tob. Res*. **11**(1): 78-79.

Patel, G.J.; Shah, H.M.; Patel, D.J. & Valand, G.B. (1981). Effective duration of some proven nematicides in control of root-knot disease in bidi tobacco. *Indian J. Nematol*. **11**(1): 74-76.

Ramakrishnayya, B.V. & Hussaini, S.S. (1987). Preliminary observations on interaction of root-knot nematodes and potash in FCV tobacco. *Indian J. Nematol*. **17**(1): 27-29.

Ramjilal; Nagarajan, K.& Krishnamurthy, G.V.G. (1981). Evaluation of systemic nematicides for root-knot control in tobacco seed beds. *Pesticides* **17**(9): 20-21.

Reddy, D.D.R. (1975). Comparison of methods of application and rate of grannular nematicides for control of *Meloidogyne* spp. on tobacco *Pl. Dis. Reptr*. **59:** 83-85.

Reddy, D.D.R. & Seshadri, A.R. (1971). Studies on some systemic nematicides. I. Evaluation for systemic and contact action against root-knot nematodes. *Indian J. Nematol*. **1:** 199-208.

Reddy, P.P.; Shetty, K.G.H. & Govindu, H.C. (1972). Susceptibility of FCV tobacco var. NC 95 to southern root-knot nematodes. *Mys. J. Agric. Sci*. **6:** 192.

Rich, J.R. & Barker, K.R. (1984). Flowering delay in FCV tobacco infected with *Meloidogyne* spp. *J. Nematol.* **16**(4): 402-404.

Rich, J.R.; Hodge, C. & Harris, J. (1986). Experimental nematicides for control of *M. javanica* in tobacco. *Soil Crop Sci. Soc. Fla.* **45:** 192.

Rich, J.R.; Hodge, C. & Johnson, J.T. (1984). Population development and pathogenicity of *M. javanica* on FCV tobacco as influenced by Ethoprop and DD. *J. Nematol.* **16**(3): 240-245.

Richard, O.A. (1973). Responses of certain plant-parasitic nematodes to selected chemical soil treatments. Ph. D. Thesis, N.C. State Univ. Raleigh. p. 62.

Romascu, E. & Dragoescu, M. (1979). Efficacy of the preparation Temik 10G in control of nematodes on tobacco plantation Analele Institutum Ceretari Pentruprotection Plantelor. **15:** 219-221.

Rufty, R.C.; Powell, N.T. & Gooding, G.V. Jr. (1983). Relationship between resistance to *M. incognita* and necrotic response to infection by a strain of Potato Virus Y in tobacco. *Phytopathology* **73**(10): 1418.

Sasser, J.N.; Kirpatrick, T.L. & Dybas, R.A. (1982). Efficacy of avermictins for root-knot control in tobacco. *Pl. Dis.* **66**(8): 691-693.

Shah, H.M.; Patel, D.J. & Jaisani, B.G. (1985). Reaction of two Japanese tobacco varieties Okinawa 6 and Okinawa 12 to *Meloidogyne javanica* under Gujarat conditions. *Indian J. Nematol.* **15**(1): 108.

Sheikh, A.G.; Hussaini, S.S. & Nirula, K.K. (1975). Pathogenicity of root-knot nematode *Meloidogyne incognita* to tobacco. *Indian J. Pl. Protec.* **3**(2): 165-169.

Shepherd, J. (1982). Nematode pests of tobacco. *In : Nematology in South Africa*. Dept. of Agric and Fisheries.

Singh, R.S. & Sitaramaiah, K. (1970): Control of plant-parasitic nematodes with organic soil amendments. *PANS* **16:** 287-297.

Slana, L.J. (1978). Studies on resistance to *Meloidogyne* in *Nicotiana* species. *Diss. Abstr. Intl.* **40B**(4): 1450-1451.

Slana, L.J. & Stavely, J.R. (1981). Identification of the chromosome carrying the factor for resistance to *M. incognita* in tobacco. *J. Nematol.* **13:** 61-66.

Smart, G.C. Jr. & Khuong, N.B. (1987). Biological control of root-knot nematodes with the fungus *P. lilacinus. Tob. Abstr.* **31**(4): 825.

Stavely, J.R. (1979). Disease resistance in *Nicotiana*. Procedures for Exptl. Use. *Techl. Bull* No. 1586.

Southards, C.J. (1965). Host-parasite relation of the lesion nematodes, *Pratylenchus brachyurus, P. zeae* and *P. scribneri* and flue cured tobacco. Ph. D. Thesis, North Carolina State Univ., Raleigh, U.S.A.

Teran, O.G. (1982). Chemical control of *Meloidogyne incognita* in tobacco seedlings. *Ciencia Technica en la Agricultura Proteccion de Plantas Suppl.* **5:** 37-43.

Teran, O.G.; Perez, M. & Perez, J.A. (1986). Chemical control of *M. incognita* in Semilleros de tabaco var. C 30. *Gienciay Technicala Agricultura protection de plantas* **9:** 93-102.

Todd, F.A. (1981). *Flue cured tobacco: Producing a healthy crop*. Tobacco Consultants Inc., Route 3, Box No. 234, NC 27591. p. 173.

Chapter 19

Nematode Problems of Forest Trees and their Control

Bao-jun Yang

Research Institute of Forestry, Chinese Academy & Forestry, Wan Shou Shan, Beijing - 100091, China.

FOREST trees are frequently attacked by various pathogens including fungi, bacteria, nematodes, viruses, mycoplasma, etc. The diseases not only inhibit the tree growth but also decrease the quality and output of forest production. As for example, the annual losses of pines in 1981 amounted to 2 million cubic metres with the death of nearly 10 million trees which is equivalent to 130,000 houses made of pine wood in Japan. Owing to the death of trees due to various diseases, forest biological system as well as human living environment are greatly damaged. So, forest diseases are very important to us. To date several nematodes have been recognised as serious pests of forest trees, such as *Bursaphelenchus xylophilus, Meloidogyne*, spp. *Tylenchulus semipenetrans*, etc.

BURSAPHELENCHUS XYLOPHILUS

DISTRIBUTION AND DAMAGE

In 1972 Mamiya and Kiyohara reported that in Japan many pine trees wilted due to a nematode and named the nematode *Bursaphelenchus lignicolus* (Mamiya, 1972, 1973). Later studies of Nicle *et al.* (1981) showed that *B. lignicolus* is a synonym of *B. xylophilus*. Actually the earliest occurrence of pine wilt disease in Japan dates back to 1905 when Yani reported an outbreak of the disease on pine trees in Nagasaki, Kyushu. He did not determine the causal agent, but the disease symptoms described by him were similar to those recently noted by workers. Since then, infested areas have increased progressively along the coastal areas of southwestern Japan, and inland areas of low elevation. The infestation now extends to almost all of Japan except Aomori and Hokkaido.

In 1979 pine wood nematode was found in a dying Austrian pine in Missouri (Dropkin & Foudin, 1979). Soon after an extensive survey was made in North

America which revealed the widespread occurrence of pine wood nematode from South Canada to Louisiana and from the Atlantic to the Pacific (Bekder *et al.*, 1984; Dropkin, 1984; Dropkin *et al.*, 1981; Malek, 1984; Myers, 1984; Nickle, 1984; Wingfield & Balanchette, 1984; Wingfield *et al.*, 1982, 1984). The disease is probably native to the United States. Because of the widespread distribution of the nematode, it is found in many pine species and also transmitted through more than one species of *Monochamus*. But the disease has not achieved great importance in the pine forests of the continental United States and Canada as the native pine species might be resistant to the nematode or the vectors such as beetles, etc. are not very effective like the vectors present in Japan. Moreover the natural environment is protected very well in North America and the beetle population is also lowered due to the presence of large number of birds in the region.

In 1982 the pine wood nematode was first found in Nanjing, China. Till date, the nematode has spread in four provinces and killed about 1 million pine trees (Yang, 1985, 1987; Yang & Wang, 1988, 1989). In the first several years Japanese black pine, *Pinus thunbergii* was mainly damaged and later the native pine species, *P. massoniana* got infected which was earlier found to be resistant to the nematode by artificial inoculation. In 1985 the disease was reported in Taiwan and Hong Kong.

The disease was found in Korea in 1989 (Yi *et al.*, 1989). The Korean government has paid great attention to it and promulgated emergency measures, such as restricting the transportation of diseased wood out of the diseased area, felling and burning of infested trees and spraying of chemicals 4 times a year for controlling the vector beetles. Thus the disease was eliminated completely from Korea in 1990.

Pine wood nematode has been found on *P. pinaster* in France in 1978 (Baujard, 1980; de Guiran & Boulbria, 1981). This nematode is morphologically similar to *B. mucronatus* and different in several biological features. So, the nematode occurring in France may not be *B. xylophilus*. Various pine species common in Europe such as *P. sylvestris, P. pinaster, P. nigra*, are known to be highly susceptible to Japanese isolates of *B. xylophilus*. The fact is that death of these species does not occur in Europe and also suggests that *B. xylophilus* can not be endemic in Europe.

Pine trees after getting infected by pine wood nematode, reduction in oleoresin flow from the wound made artificially on the stem is characteristic and no external symptoms are observed. Then, two-year-old leaves began to change their colour to yellow and oleoresin flow is progressively reduced than that of the previous one. Subsequently, foliage of these trees have changed their colour to abnormal yellow and apparently showed the wilting symptom. The diseased trees do not yield any appreciable amount of resin. Thus the decline of the diseased tree starts. Leaves have been shrivelled and changed their colour to brown completely. Pine trees found to be healthy in early summer die in late summer after showing yellowish foliage. Therefore, this nematode is very serious pest of pine and the disease is called "Pine's Cancer" by many foresters.

Many pine species can be attacked by the nematode, e.g. *Pinus densiflora, P. sylvestris, P. sylvestris var. mongolica, P. mugo, P. thunbergii, P. nigra, P. luchuensis, P. bungeana, P. armandii, P. parviflora, P. echinata, P. koraiensis, P. massoniana, P. taiwanensis, P. elliottii, P. taeda, P. tabulaeformis, P. engelmannii, P. leiophylla, P. lambertiana, P. monticola, P. muricata, P. occarpa, P. pentaphylla, P. pinaster, P. ponderosa, P. radiata, P. rudis, P. strobiformis, P. yunnanensis, etc. Besides, Cedrus atlantica, Cedrus deodara, Larix decidua, Larix laricina, Picea glauca* and *Picea excelsa*.

DISEASE CYCLE

The pine wood nematodes have two cycles of reproduction. The propagative cycle begins in the spring and continues throughout the growing season. Under optimum environmental conditions, development of nematodes from eggs to reproductive maturity is completed in 4-5 days. During this period the nematodes develop very rapidly in newly infected trees and also on fungi in dying or recently killed trees. In addition to the propagative cycle, nematodes may also develop through a dispersal phase. This phase takes place during the time of stress such as when the wood becomes dry after tree death or in late fall. Here, as the nematodes moult from the second larval stage to the third larval dispersal stage, where they contain high levels of lipids and are extremely resistant to adverse environmental conditions such as freezing and desiccation.

The pine wood nematodes are carried by cerambycid beetles from infected to non-infected host trees. The beetles provide wounds on the host trees which serve as entry points for the nematodes. In Japan the principal vector is *Monochamus alternatus*, while in North America *M. carolinensis, M. scutellatus, M. titillatus, M. obtusus* have been reported to be the vectors where the first one is most important. These beetles are best known as wood borers in their larval stage. They oviposit on the bark of dead and dying trees, leaving eggs in the bark-wood interface. Larvae penetrate the wood and feed on the wood cells. The larvae are inactive during the winter, pupate and emerge in spring. After emergence, the beetles feed on young pine branches for maturation feeding.

At the end of winter, nematodes moult from the dispersal stage to the fourth larval stage which is called dauer-larvae. Dauer-larvae develop in response to the presence of maturing cerambycid beetle pupae. Dauer-larvae enter the spiracles of the beetles and are carried from the dead trees with emerging beetles in spring. Beetles carrying nematodes emerge from trees in spring and begin maturation feeding (Kobayashi, 1984a). Dauer-larvae leave the tracheae through the spiracles and enter host pines through feeding wounds made by the beetles. The nematodes moult to the adult stage, mate, and reproduce rapidly in the resin canals of the host pines. The pine trees become chloritoc and die (Mamiya, 1980, 1984). Cerambycid beetles carrying the nematodes are attracted to dying trees and oviposit on them. Therefore, cerambycid larvae develop in trees containing millions of nematodes that later colonize the adult insects before their emergence as adults. The beetles thus transmit the nematodes to healthy trees. Primary infection occurs when nematodes enter an apparently healthy tree through the maturation feeding wounds of beetle vectors. Alternately, a secondary infection occurs when nematodes colonize a tree already stressed by abiotic factors or which is under attack by other diseases or insects. Such a tree is an attractive host for *Monochamus* oviposition, and pine wood nematode enters the tree through oviposition wounds (Okamoto, 1984).

Toshiya *et al.* (1981) reported that the disease cycle including beetles, pine wood nematodes and pine trees was formed because of the chemical ecology. The noteworthy aggregation behaviour of the nematodes to the pupal cell of the beetle was defined chemically. Three unsaturated fatty acids, linoleic, oleic and palmitoleic acid, and their glycerides were isolated from both the body extracts or excretions of the beetle and the pupal cell wall as the attractive substance. The behaviour of nematode dauer-larvae moving into the tracheae was also characterized by the existence of carbondioxide as the attractive substance. Newly emerged adults of the cerambycid beetle preferred the current or one-year-old pine twigs, from which β-sitosterol, campesterol, several sugars, sugar alcohols and several glycosides were found as the feeding stimulants. Both sexes of matured beetles aggregate to the freshly felled or the nematode-infected pine trees. The attractive substances vaporized from there were mixtures of pine-derived monoterpene hydrocarbons and ethanol. Among them α-pinene as well as ethanol played a significant role for the attraction of the beetles.

STRATEGIES FOR THE CONTROL OF PINE WILT DISEASE

The strategy for the management of pine wilt disease should be to reduce tree mortality to tolerable levels in polluted areas and to stop the invasion of nematodes into virgin areas (Kobayashi, 1984b).

The occurrence of pine infestation is based on the combined effect of the three components : pine trees, pine wood nematodes, and beetle vectors. Therefore, the counterplan for the control should be aimed at the each component or blockage of linkages of two components. Recently, most of control methods are aimed at the destruction of beetle populations. In order to exterminate the hot bed of disease, felling the infested trees and killing the beetles in them by burning, insecticide spraying, or removing them from the forest for their utilization are widely in practice. For preventing the maturation feeding of beetles and reducing their population, chemical spraying onto the crown of standing trees gave excellent results. The main components involved were fenitrothion, carbaryl, fenthion, or prothiofos. The disease can be controlled by using systemic nematicides injected into tree trunk or given as soil treatment. But it is too expensive to be used as a practice.

Different pine species have different degrees of resistance to pine wood nematode (Yang *et al.*, 1987). The resistance of each pine, however, is not stable in different areas. For example, *Pinus tabulaeformis* is resistant in Japan, but it is very susceptible in China. Some resistant pine species, such as *P. massoniana* and *P. taeda*, are infected by the nematodes in serious diseased areas. Some pine species are resistant to pine wood nematode but susceptible to beetles. Moreover, the environmental conditions, other diseases, and pests on the pines extremely influence the resistance to pine trees to the nematodes. Therefore, using pine resistance to control this disease has been doubtful.

TYLENCHULUS SEMIPENETRANS

Tylenchulus semipenetrans was initially reported to damage citrus. Cai *et al.* (1990a,b,c) found that this nematode can also infect chinafir, *Cunnighamia lanceolata*, in Fujian Province of China. The needles of diseased chinafir trees become yellow and the growth in length of diseased trees is reduced by 20-30%. The roots of diseased trees become short and enlarge as cock-foot. The disease is more serious in pure forest of chinafir than in mixed forest. Using nematicides in soil is effective to control this nematode. In order to obtain non-diseased seedlings, the roots of seedlings should be treated by 40-50°C hot water.

MELOIDOGYNE spp.

The root-knot nematodes belonging to the genus *Meloidogyne* includes about 80 species. Among them many species attack forest trees (Baldwin & Sasser, 1979; Eisenback *et al.*, 1985; Feng *et al.*, 1989; Golden, 1979; Golden & Kaplan, 1986; Sagalina *et al.*, 1985; He & Duan, 1981, 1984; Riffle, 1963; Shen, 1989; Shen *et al.*, 1990; Wu *et al.*, 1988; Yang, 1984; Yang & Eisenback, 1983; Yang & Zhong, 1980; Yin *et al.*, 1989; Yu *et al.*, 1990; Zhao & Yang, 1990; Zao *et al.*, 1989) such as *Meloidogyne platani* on *Platanus occidentalis* and *P. hispanica; M. enterolobii* on *Enterolobium contortisiliquum*; *M. turkestanica* on *Tamarix* sp.; *M. pini* on *Pinus clausa*; *M. ovalis* on Acer saccharum; *M. megatyla* on *Pinus teada*; *M. quercian* on *Quercus palustris*, *M.*

christiei on *Quercus laevis*, *M. econincki* on *Fraxinus excelsior*, *M. litoralis* on *Ligustrum* sp. But the most important species are *M.incognita*, *M. arenaria*, *M. javanica* and *M. hapla*. These four common *Meloidogyne* species damage many trees such as *Catalpa* spp., *Morus alba*, *Acacia* spp., *Buxis* spp., *Trachycarpus fortunei*, *Michelia alba*, *Rhapis humilis*, *Salix babylonica*, *Olea europaea*, *Ceioa pentantra*, *Paulownia fortunei*, *Rauvolfia verticillata*, *Maytenus hookerii*, *Enterolobium* spp. *Albizzia* spp., *Leucaena* spp., *Gmelina arborea*, *Tectona grandis*, *Heteropqnax fragrans*, *Schefflera octophylla*, *Itoa orientolis*, *Anacardium occidentale*, *Fraxinus americana*, *Lindera communis*, *Podocarpus fleuryi*, *Melia azedarach*, *Cunninghamia lanceolata*, *Cercis chinensis*, *Gleditsia sinensis*, *Robinia pseudoacacia*, *Sophora japonica*, *Ligustrum lucidum*, *Pterocarya stenoptera*, *Celtis sinensis*, *Ulmus pumila*, *Broussoretia papyrifera*, *Ficus carica*, *Firmicana platanifolia*, *Koelreuteria paniculata*, *Populus tomentosa*, etc. and cause economic losses. The most serious damage, however, is in nursery. For example, in Hainan Island over 90% of seedlings of pacara earpod tree, *Enterolobium contrortisiliquum* died because of the presence of root-knot nematodes in heavily infested area.

The strategy of control of root-knot nematodes on trees should be through reduction in soil population of eggs and infective larvae. Recently most common method is the use of nematicides to reduce the soil population of root-knot nematodes in nursery or the valuable trees of forest. Hot water treatment of roots can be used for the diseased seedlings.

REFERENCES

Baldwin, J.G. & Sasser, J.N. (1979). *Meloidogyne megatyla n.* sp. a root-knot nematode from loblolly pine. *J. Nematol.* **11:** 47-56.

Baujard, P. (1980). Trois nouvelles especes de *Bursaphelenchus* (Nematoda : Tylenchida) at remarques sur le genre. *Revue Nematol.* **3:** 167-177.

Bedker, M.J.; Wingfield, M. J. & Blanchette, R.A. (1984). Host specificity of the pine wood nematode in North Central United States and Canada. *Proc. US-Japan Seminar, The Resistance Mechanisms of Pines Against Pine Wilt Disease*. pp. 46-53.

Cai, Q.J. *et al.* (1990a). Studies not the nematode disease (*Tylenchulus*) of *Cunnighamia lanceolata*. I. Identification of the pathogen nematode. *J. Fujian Coll. Porestry* **10** (1): 1-6.

Cai, Q.J. *et al.* (1990b). Studies on the nematode disease of *Cynnighamia lanceolata*. II. The life history and the population change in natural condition, and the biological characteristics. *J. Fajjan Coll. Forestry* **10** (2): 89-95.

Cai, Q.J. *et al.* (1990c). Studies on *Tylenchulus semipentrans* Cobb attacking *Cunninghamia lanceolata. Scientia Silvae Sinica* **26:** 506-514.

de Guiran, G. & Boulbria, A. (1981). Importance and characters of *Bursaphelenchus lignicolus* attacks on *Pinus pinaster* in France. *Proc. 27th IUFRO World Congr., Div. 2. Kukizaki-Japanese IUFRO Congress Committee*. p. 273.

Dropkin, V.H. (1984). Pine wilt in the United States. *Proc. US-Japan Seminar, The Resistance*

Mechanisms of Pines Against Pine Wilt Disease. pp. 3-5.

Dropkin, V.H. & Foudin, A.S. (1979). Report of the occurrence of *Bursaphelenchus lignicolus* induced pine wilt disease in Missouri. *Plant Dis. Reptr*. **63:** 904-905.

Dropkin, V.H. *et al*. (1981). Pinewood nematode: a threat to U.S. forests. *Plant Disease* **65:** 1022-1027.

Eisenback, J.D.; Yang, B.J. & Hartman, K.M. (1985). Description of *Meloidogyne pini n*. sp., a root-knot nematode parasitic on Sand pine (*Pinus clausa*), with additional notes on the morphology of *M. megatyla. J. Nematol*. **17:** 206-219.

Feng, R.Z.; Yang, B.J.; Wang, Q.L. & Hu, K.J. (1989). The identification of root-knot nematode populations in Guangxi. *J. Guangxi Agri. Coll*. **8** (2): 9-14.

Golden, A.M. (1979). Description of *Meloidogyne camelliae n*. sp. and *M. querciana* n. sp. (Nematoda : Meloidogynidae), with SEM and host-range observations. *J. Nematol*. **11:** 175-188.

Golden, A.M. & Kaplan, D.T. (1986). Description of *Meloidogyne christiei* n. sp. (Nematoda : Meloidogynidae) from oak with SEM and host-range observations. *J. Nematol*. **18:** 533-540.

He, H.Z. & Duan, D.R. (1981). The investigation of root-knot nematodes on tropical forest plants in Hainan Island. *Forest, Pest & Disease* **1:** 6-10.

He, H.Z. & Duan, D.R. (1984). Root-knot nematode of forests in Hainan Island. *Chinese J. Zool*. **6:** 1-4.

Kobayashi, F. (1984a). Biology of insects associated with pine mortality. *Proc. US-Japan Seminar, The Resistance Mechanisms of Pines Against Pine Wilt Disease*. pp. 64-75.

Kobayashi, F. (1984b). Strategies for the control of pine wilt disease *Proc. US-Japan Seminar, The Resistance Mechanisms of Pines Against Pine Wilt Disease*. pp. 171-177.

Malek, R.B. (1984). Current status of pine wilt and recent research on the disease complex in Illinois-Center of an epidemic in scotch pine. *Proc. US-Japan Seminar. The Resistance Mechanisms of Pines Against Pine Wilt Disease*. pp. 164-169.

Mamiya, Y. (1972). Pine wood nematode, *Bursaphelenchus lignicolus* Mamiya and Koyohara, as a causal agent of pine wilting disease. *Review of Plant Protc. Res*. **5:** 46-60.

Mamiya, Y. (1980). Pine wilt and pine wood nematode : hisohtopathological aspects of disease development. *Proc. Third Int. Workshop Genetics of Host-Parasite Interactions in Forestry*. Wageningen, the Netherlands. pp. 153-160.

Mamiya, Y. (1983). Pathology of the pine wilt disease caused by *Bursaphelenchus xylophilus. Ann. Rev. Phytopathol*. **21:** 201-220.

Mamiya, Y. (1984). Behaviour of the pine wood nematode, *Bursaphelenchus xylophilus*, associated with the disease development of pine wilt. *Proc. US-Japan Seminar, The Resistance Mechanisms of Pines Against Pine Wilt Disease*. pp. 14-25.

Mamiya, Y. & Kiyohara, T. (1972). Description of *Bursaphelenchus lignicolus n*. sp. (Nematoda: Aphelenchoididae) from pine wood and histopathology of nematode-infested trees. *Nematologica* **18:** 120-124.

Myers, R.F. (1984). Comparative histology and pathology in conifers infected with pine wood nematodes, *Bursaphelenchus xylophilus. Proc. US-Japan Seminar, The Resistance Mechanisms of Pines Against Pine Wilt Disease*. pp. 91-95.

Nickle, W.R. (1984). Pine wood nematode (*Bursaphelenchus xylophilus*) in Japanese black pine (*Pinus thunbergii*) and red pine (*P. resinosa*) in Delaware and Maryland. *Proc. US-Japan Seminar, The Resistance Mechanisms of Pines Against Pine Wilt Disease*. pp. 26-30.

Nickle, W.R.; Golden, A.M.; Mamiya, Y. & Wergin, W.P. (1981). On the taxonomy and morphology of the pine wood nematode, *Bursaphelenchus xylophilus* (Steiner & Buhrer, 1934) Nickle, 1970, *J. Nematol.* **13:** 385-392.

Okamoto, H. (1984). Behaviour of the adults of Japanese pine sawyer *Monochamus alternatus. Proc. US-Japan Seminar, The Resistance Mechanisms of Pines Against Pine Wilt Disease.* pp. 82-89.

Riffle, J.W. (1963). *Meloidogyne ovalis* (Nematoda : Heteroderidae), a new species of root-knot nematode. *Proc. Helm. Soc.* **30:** 287-296.

Sagalina, L.T.; Ivanova, T. & Krall, E. (1985). Two new root-knot nematode species of the genus *Meloidogyne* (Nematoda : Meloidogynidae) — parasites of shrubs and trees. *Proc. Acad. Sci. Estonian SSR. Biology* **34** (4): 279-287.

Shen, B.K. (1989). Identification of root-knot nematode on London planetree. *J. Nanjing Forestry Univ.* **13** (4): 65-68.

Shen, B.K.; Sun, S.Y. & Dai, Y.S. (1990). Studies on root-knot nematodes of trees, ornamental plants and forest weeds. *J. Nanjing Forestry Univ.* **14** (3): 37-42.

Toshiya, Keda *et al.* (1981). The chemical ecology of *Monochamus alternatus* on the relationship with pine wood nematodes and host trees. Proc. *XVII IUFRO World Congress*. pp. 297-303.

Wingfield, M.J. & Blanchette, R.A. (1984). Pathogenicity and insect associates of the pine wood nematode in the North Central States. *Proc. US-Japan Seminar, The Resistance Mechanisms of Pines Against Pine Wilt Disease.* pp. 32-43.

Wingfiled, M.J.; Blanchette, R.A. & Nicholls, T.H. (1984). Is the pine wood nematode an important pathogen in the United States ? *J. Forestry* **82:** 232-235.

Wingfield, M.J. *et al.* (1982). The pine wood nematode: a comparison of the situation in the United States and Japan. *Can. J. Res.* **12:** 71-75.

Wu, Q. *et al.* (1988). The investigation of root-knot nematode in Henan Province. *Henan Agric. Sci.* **1:** 17-19.

Yang, B.J. (1984). The identification of 15 root-knot nematode populations. *Acta Phytopathologica Sinica* **14:** 107-112.

Yang, B.J. (1985). The identification of nematode on pine tree. *Scientia Silvae Sinica* **21:** 305-309.

Yang, B.J. (1987). Pine wilt disease. *World Agric.* **6:** 38-39.

Yang, B.J. & Eisenback, J.D. (1983). *Meloidogyne enterolobii n.* sp. (Meloidogynidae), a root-knot nematode parasitizing pacara earpod tree in China. *J. Nematol.* **15:** 381-391.

Yang, B.J. & Wang, Q.L. (1988). Distribution of pine wood nematode in China. *Forest Res.* **1:** 450-452.

Yang, B.J. & Wang, Q.L. (1989). Distribution of the pine wood nematode in China and susceptibility of some Chinese and exotic pines to the nematode. *Can. J. Forestry Res.* **19:** 1527-1530.

Yang, B.J. & Zhong, X.W. (1980). The identification of root-knot nematodes in *Olea europea. Scientia Silvae Sinica* **16:** 264-265.

Yang, B.J. *et al.* (1987). The resistance of pine species to pine wood nematode, *Bursaphelenchus xylophilus. Acta Phytopathologica Sinica* **17:** 211-214.

Yi, C.K. et. al. (1989). First finding of the pine wood nematode, *Bursaphelenchus xylophilus* (Steiner et Burhrer) Nickle and its insect vector in Korea. *Res. Rep. For. Res. Inst.* No. **38:** 141-149.

Yin, Y.Q.; Yang, B.J. & Wang, Q.L. (1989). Identification of root-knot nematodes in Guangdong. *Acta Phytophylacica Sinica* **16:** 152.

Yu. S.F.; Yang, B.J. & Wang, Q.L. (1990). Investigation and identification of root-knot nematodes on plants in Yunnan. *J. Yunnan Agric Univ.* **5:** 212-217.

Zhao, L.S. & Yang, B.J. (1990). A preliminary survey of *Meloidogyne hapla* Chitwood on cultivated chinese medicinal herbs. *Acta Phytopathologica Sinica* **20:** 159-160.

Zhao, L.S. *et al.* (1989). The investigation on Chinese medicinal herbs hosts of *Meloidogyne hapla* Chitwood. *J. Hebei Agric. Univ.* **12** (4): 149-152.

Chapter 20

Control of Insect-Transmitted Plant-parasitic Nematodes

P.S. Grewal* & P.N. Richardson
Microbiology and Crop Protection Department, Horticulture Research International, Worthing Road, Littlehampton, West Sussex, BN17 6 LP, UK.

THREE types of symbiotic association between organisms are commonly recognised: commensalism, mutualism and parasitism (Giblin, 1987). Although no example of classical mutualism between insects and nematodes has yet been reported, commensal and parasitic associations are common. With the exception of *Fergusobia*, a nematode that parasitises its vector, most insect-transmitted plant-parasitic nematodes exhibit a commensal relationship with their vectors. Commensal associations take the form of phoresy in which nematodes benefit by being transported to new habitats that provide, at least temporarily, abundant food resources. Most nematode phoretics are dispersed by the host as dauer juveniles which are generally regarded as dispersal or survival stage juveniles. The dauer juvenile is a unique, morphologically distinct, non-feeding, development variant, produced as an off-shoot of the normal, reproductive cycle. This stage is capable of tolerating stresses that would prove fatal to other developmental forms. It is, for example, usually more resistant to desiccation, starvation and exposure to chemicals (Riddle, 1988). In rhabditoid nematodes the third developmental larval stage is facultatively replaced, and in aphelenchoid and tylenchoid nematodes the third or fourth stage is obligatorily replaced, by the dauer juvenile (Foder *et al.*, 1990).

In phoretic associations dauers may be transported externally on the legs, for example, or underneath the elytra of Coleoptera. Some phoretic nematodes are transported internally by the insect host, in the reproductive tracts of both sexes, in the tracheal system, gut or accessory glands (Giblin, 1987). Where economically important nematodes are vectored by benign or pest insect species there is a need to understand the relationship fully in order to approach the combined pest problem in the most effective way. In this chapter, the bionomics and control

* Present address : Department of Entomology, Rutgers University, New Brunswick, New Jersey, 08903, U.S.A.

of the following most significant plant diseases caused by insect-borne nematodes are discussed : (a) Pine wilt disease, (b) 'Red-ring' disease of coconut and oil palms, (c) Nematode diseases of mushrooms, and (d) the *Fergusonina-Fergusobia* association.

PINE WILT DISEASE

Pine wilt is the most serious disease of native pines in Japan (Mamiya, 1984) and, potentially, is the most important nematode disease of conifers worldwide. Although, in Japan, the recognition of this lethal condition can be traced back to the early twentieth century (Yano, 1913), it was in 1969 that the association of nematodes with wilted pines was first discovered. Soon afterwards, the pine-wood nematode was described as *Bursaphelenchus lignicolous* (Mamiya & Kiyohara, 1972); it was later synonymised with *B. xylophilus* (Steiner & Buhrer, 1934) Nickle, 1970; Nickle *et al.* 1981). Since its discovery in Kyushu, this parasite has been found in many other regions of Japan (Mamiya, 1984), and in the U.S.A. (Dropkin & Foudin, 1979; Robbins, 1982; Malek & Appleby, 1984), France (Scotto La Massese *et al.*, 1978), Hong Kong (Riddell-Swan, 1984), China (Cheng *et al.*, 1986) and Canada (Moodey, 1988).

ETIOLOGY

Nematodes found in Long Leaf Pine, *Pinus palustris* Mill, at a sawmill in Louisiana, U.S.A. were first described as *Aphelenchoides xylophilus* by Steiner & Buhrer (1934). Nickle (1970) transferred this species to the genus *Bursaphelenchus*, a group that now includes 44 species most of which have phoretic relationships with insects especially Coleoptera such as bark beetles and wood-borers (Yin *et al.*, 1988). The species are predominantly mycophagous and readily reproduce on fungi associated either with their vector insects or with the wood habitats of their vectors (Wingfield, 1987).

Dispersal stage dauer larvae of *B. xylophilus* are transported to new host trees on the body surface and within the tracheal system of several beetle species. Worldwide, 21 species of Cerambycidae, one of Buprestidae, and two of Curculionidae are known to carry pine wood nematode dauer larvae upon emerging from nematode-infested trees (Table 1). Based on the frequency of occurrence of beetles and the number of nematodes that they carry only two species have been considered as major vectors: *Monochamus alternatus* Hope (the Japanese pine sawyer) in Japan and *M. carolinensis* (Oliver) in the U.S.A. (Kobayashi *et al.*, 1984). Primary transmission to healthy trees occurs through wounds caused by beetles feeding on young branches. Secondary transmission to stressed trees or to recently cut logs occurs though insects' oviposition sites.

Table 1 : Insect species known to carry *Bursaphlenchus xylophilus* dauer larvae in Japan and North America.

Country	Family	Species
Japan		
	Cerambycidae	
		Acaloptera fraudatrix Bates
		Acanthocinus griseus F.
		Arthopalus rusticus L.
		Corymbia succedanea Lewis
		Monochamus alternatus Hope
		Monochamus nitens Bates
		Monochamus saltuaricus Gebler
		Spondylis buprestoides L.
		Uraecha bimaculata Thompson
North America		
	Cerambycidae	
		Amniscus sexguttatus Say
		Arhopalus rusticus obsoletus Randall
		Asemum striatum L.
		Monochamus carolinensis Oliver
		Monochamus marmorator Kirby
		Monochamus mutator Le Conte
		Monochamus obtusus Casey
		Monochamus scutellatus Say
		Monochamus titillator Fabricius
		Neacanthocinus obsoletus Oliver
		Neacanthocinus pusillus Kirby
		Xylotrechus saggitatus Germe
	Buprestidae	
		Chrysobothris spp.
	Corculionidae	
		Hylobius pales Herbst
		Pissodes approximatus Hopkins

The feeding habits of *B. xylophilus* allow its life cycle, summarised in Fig. 1, to be conveniently divided into phytophagous and mycophagous phases. The nematode is carried by cerambycid beetles that feed on healthy pine twigs and oviposit in dead and dying trees (Linit, 1988). Nematodes migrate from beetle tracheae and enter wounds made during beetle feeding. In the phytophagous phase, the

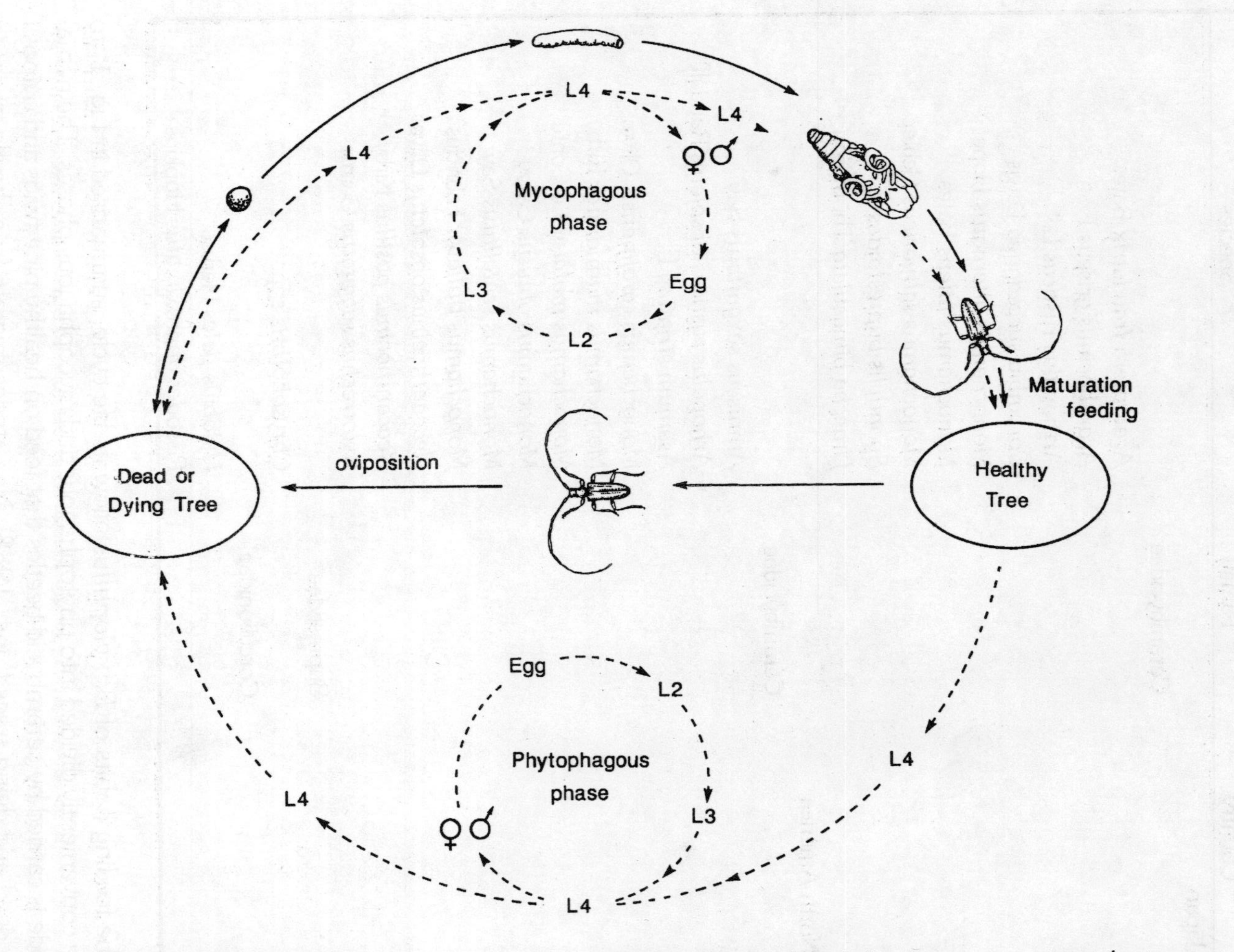

Fig. 1. Disease cycle of *Bursaphelenchus xylophilus*. Vector (—), nematode (——), L_2 - L_4= juvenile stages.

nematodes colonise resin canals, feed on the surrounding thin-walled epithelial cells, multiply rapidly and cause death of the pine.

After the completion of maturation feeding, cerambycid beetles are attracted for oviposition to dead and dying pine trees. The nematode that leave the beetles at this stage enter the mycophagous phase of their life cycle feeding on fungi including those transmitted to the dead trees by scolytid bark beetles. After hatching cerambycid larvae feed on the surface of the timber and then penetrate the wood to mature. Here both the beetles and *B. xylophilus* overwinter. In spring, the beetles pupate and the nematodes moult to dauer (dispersal stage) larvae that invade pupal chambers prior to eclosion of the adult beetles. Dauers enter the tracheae of the adult beetle through its abdominal spiracles (Mamiya, 1984). They are subsequently transported to healthy pine twigs where they initiate another phytophagous phase.

When nematodes colonize resin canals and destroy epithelial cells they cause the oleoresin flow to cease and the transpiration mechanism to be disrupted. Cessation of oleoresin exudation may occur within two weeks of infestation; the transpiration rate may be affected within 20-30 days. Death of the tree is inexorable and follows a characteristic sequence of wilting, yellowing of foliage and rapid destruction of the wood. Rapid wilting has been linked to phytotoxin production (Shaheen *et al.*, 1984).

CONTROL STRATEGIES

Control of pine wilt disease is complicated by the endophytic nature of the nematode and by the sudden death of infected trees. The synchronization of the nematode's life cycle with that of its vector and the production of dispersal stage juveniles, ensures efficient dissemination of nematodes from dead to healthy plants. Furthermore, the ability of nematodes to feed on a wide range of wood-inhabiting fungi, in the absence of host trees, aids their survival and seasonal 'carry-over'. Various control strategies have been devised; they are mainly preventive and are based on the direct control of nematodes, the destruction of infested plant material, the breeding of resistant trees and the control of insect vectors.

A. CONTROL OF THE NEMATODE

Chemical Control

Systemic nematicides have been shown to be the most effective chemicals for use against *B. xylophilus*. Prophylactic treatment of pine trunks, or their surrounding soil, made up to three months before nematode invasion, has demonstrated that fensulfothion, disulfoton, mesulfenfos and thionazin can all effectively protect pine trees (Takase *et al.*, 1982; Mastuura, 1984). Successful 'spot-treatment' of particularly valuable park and garden trees has been achieved (Mamiya, 1984) with a vermicide, morantal-tartrate, widely used for veterinary purposes.

The mode of action of specific compounds has been studied. Matsurra & Hosono (1987) demonstrated that fensulfothion, mesulfenfos and thionazin strongly inhibit acetyl chlonestrase activity in *B. xylophilus*; anthelmintic agents and insecticides, such as methomyl, had similar effects. After screening a large number of alkylamines, Nagase *et al.* (1982, 1983) identified didecylamine as potentially the most active nematicidal compound (LC_{50}; 4.19 μm, 1.24 ppm). Some metal ions are comparable in effect to organic nematicides (Kobamoto & Izumi, 1984) and some benzimidazole derivatives inhibit the hatching of *B. xylophilus* eggs more than other derivatives (Shuto *et al.*, 1989).

Nematicidal Plant Products

Certain plants produce allelochemicals with nematicidal properties that affect *B. xylophilus*. Kawazu *et al.* (1980a) tested plant extracts and found the allelochemicals from 11 species to be particularly effective against the pine wood nematode. Amongst them were the roots of *Cirsium japonicum* and *Erigeron annus*, the leaves of *Prunus verecunda*, the leaves and stems of *Artemisia capillaris* and the flowers of *Coreopsis lanceolata*. Further work led to the characterisation of nine nematicidal compounds from five plants species (Fig. 2).

Biological Control

Although a range of potentially useful biological control agents are known, and some field trials have taken place, it is not yet possible to use biological control to eradicate *B. xylophilus* populations. So cryptic is the life cycle of this nematode that there are only limited opportunities for contact with natural enemies and diseases. However, at least seven species of mites having a phoretic association with cerambycid beetles are commonly found (Ikeda, 1987; Kinn, 1987). Of these, three species of *Dendrolaelaps* and two of *Trichouropoda* prey on nematodes and have a close association with vector insects (Ikeda, 1987). The mites inhabit the pupal chambers of the beetles and become attached to them after they emerge from pupation. Although these predators can suppress *B. xylophilus* populations before they invade the trees, their performance in the field remains untested (Ikeda, 1987).

Nematophagous fungi are capable of capturing and killing nematodes using trapping devices as diverse as hyphal networks, sticky knobs or hyphae and constricting or non-constricting rings. Several species of fungi from the genera *Arthrobotrys*, *Dactylaria, Dactylella* and *Harposporium* have been frequently

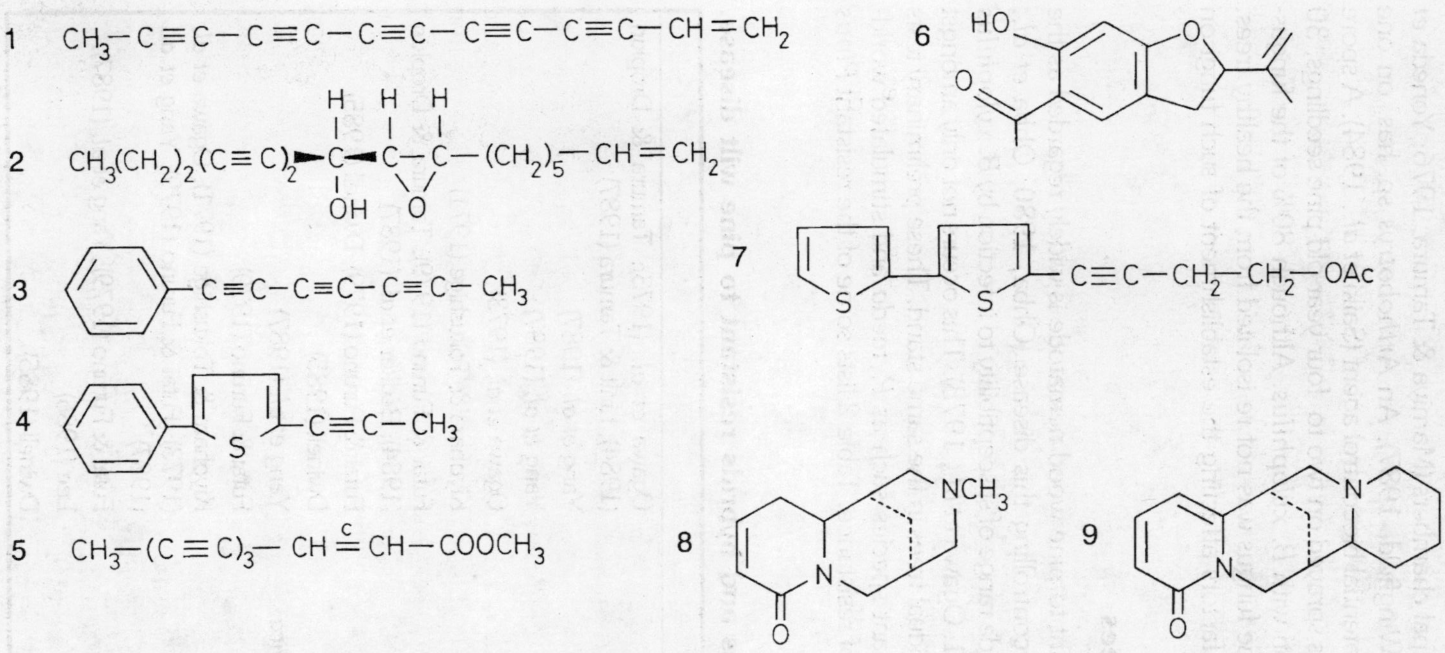

1 $CH_3-C\equiv C-C\equiv C-C\equiv C-C\equiv C-C\equiv C-CH=CH_2$

5 $CH_3-(C\equiv C)_3-CH \overset{c}{=} CH-COOCH_3$

Fig.2. *Bursaphelenchus xylophilus* nematicidal compounds isolated from plants. **1** = tridec + ene-3, 5, 7, 9, 11-pentayne and **2** = 9, 10-epoxy-heptadec-16-ene-4, 6-diyn-8-ol from roots of *Cirsium japonicum*; **3** = 1-phenylhepta-1, 3, 5-triyne and **4** = 2-phenyl-5-(1'-propynl)-thiophene from flowers of *Coreopsis lanceola*; **5** = cis-dehydromatricaria ester from *Solidago altissima* (from Kawazu *et al.*, 1980b). **6** = hydroxytremetone ((-)-2-Isopropenyl-5-acetyl-6-hydroxy-2, 3-dihydrobenzofuran) and **7** = bithienyl derivative (5-(4-acetoxybut-1-ynyl)-2, 2'-bithienyl) from *Tagetes patula* (from Hatakeda *et al.*, 1985). **8** = (-)-n-methylcytisine and **9** = (-)-anagryrine from roots of *Sophora flavescens* (from Mastuda *et al.*, 1989).

isolated from cerambycid pupal chambers (Mamiya & Tamura, 1976; Yoneda *et al.*, 1980; Tamura, 1980; Wingfield, 1987). An *Arthobotrys* sp. has, on one occasion, been tested as a potential biocontrol agent (Saiki *et al.*, 1984). A spore suspension of the fungus was sprayed on two to four-year-old pine seedlings, 30 to 60 days before inoculation with *B. xylophilus*. Although 80% of the fungus-treated seedlings survived, the fungus was not re-isolated from the healthy trees. Further researches into the factors affecting the establishment of such fungi on trees is needed.

Breeding Resistant Trees

The breeding of trees resistant to pine wood nematode is widely regarded as the most promising method of controlling this disease (Ohba, 1980; Ohba *et al.*, 1984). Pine trees exhibit a wide range of susceptibility to infection by *B. xylophilus* (Kiyohara & Tokushige, 1971; Ogawa *et al.*, 1973). This occurs not only amongst species but also among individual trees in the same stand. These preliminary tests identified some highly resistant species such as *P. taeda* and stimulated world-wide search for the sources of resistance. Table 2 lists some of the resistant *Pinus* spp. that have been found.

Table 2 : *Pinus* species and hybrids resistant to pine wilt disease.

Species	
P. echinata	Ogawa *et al.* (1973); Tamura & Dropkin (1984); Linit & Tamura (1987)
P. fenzeliana	Yang *et al.* (1987)
P. griffithi	Yang *et al.* (1987)
P. palustris	Ogawa *et al.* (1973)
P. pungens	Kiyohara & Tokushige (1971)
P. resinosa	Futai & Furuno (1979); Tamura & Dropkin (1984); Bedker *et al.* (1987)
P. rigida	Futai & Furuno (1979); Dwinell (1985)
P. serotina	Dwinell (1985)
P. sylvestris var mangolica	Yang *et al.* (1987)
P. tabulaeformis	Futai & Furuno (1979)
P. taeda	Kiyohara & Tokushige (1971); Ogawa *et al.* (1973); Futai & Furuno (1979); Yang *et al.* (1987)
P. taiwanensis	Futai & Furuno (1979); Yang *et al.* (1987). Lee (1986)
P. virginiana	Dwinell (1985)

Species	
Hybrids	
P. densiflora x P. mugo	Furukoshi & Sasaki (1984)
P. densiflora x P. nigra var corsicana	Furukoshi & Sasaki (1984)
P. thunbergii x. P. massoniana	Furukoshi & Sasaki (1984); Futai & Furuno (1979); Ishii *et al.* (1982)
P. thunbergii x P. tablaeformis	Furukoshi & Sasaki (1984)
P. thunbergii x P. nigra var. corsicana	Furukoshi & Sasaki (1984)

The degree of resistance or susceptibility of pines is affected by environmental conditions such as water status (Suzuki, 1987), temperature (Rutherford & Webster, 1987) and level of pollution (Bolla & Fitzsimmons, 1988; Huber *et al.*, 1989). Recently β-myrcene, a hormone-like substance that stimulates the moulting and multiplication of *B. xylophilus* (Ishikawa *et al.*, 1986; Hinode *et al.*, 1987) has been isolated from pine trees. As the content of β-myrcene is greater in susceptible (e.g., *P. sylvestris*) than in resistant pines (e.g., *P. taeda*), its use as an indicator for the screening of resistant Japanese red and black pines has been suggested (Ishikawa *et al.*, 1987).

In 1978, a project involving 17 research organizations and field stations in Japan was started, with the aim of breeding resistant pines (Ikeda, 1984). Scions were taken from the trees that has survived in heavily-infested natural stands and 20 grafts/tree were prepared for the first nematode inoculation tests. Each organisation annually conducted grafting of 3,000-20,000 trees. The nematodes that were inoculated had been pre-tested for pathogenicity to pines and two-year-old seedlings of *P. taeda* were used as controls to compare degree of resistance. Over 23,000 candidate trees were tested; 940 of these had passed the first inoculation tests by 1982. The second inoculation tests, completed by 1984, were performed with strains of nematodes that were more pathogenic. The propagation of strictly selected resistant clones, the preparation of seed orchards and the mass production of seeds of resistant trees for distribution continues.

Cross-protection

Cross-protection is a method of inducing resistance in susceptible plants by prior inoculation with avirulent isolates/strains of the causal organism. A species-specific resistance to pine wilt has been induced in normally susceptible *P. thunbergii* by previous inoculation with avirulent isolates of *B. xylophilus* (Kiyohara, 1984). A high level of resistance was obtained when 10-30 days were left between inducer and challenger inoculations and the protection lasted for at least 120 days after inoculation. The degree of induced resistance was proportional to inoculum concentration. Pathotypes of *B. xylophilus* with varying degrees of virulence are being genomically and physiologically characterised (Bolla *et al.*, 1987, 1988; Bolla & Tamura, 1989) and might form the basis for induced resistance.

Quarantine

Plant quarantine not only involves legal embargo on importation of disease or pest organisms into a country where they are not endemic, but can also include decontamination or destruction of infested plant materials at the site of detection. Pine wood nematode is reported from many parts of China, Taiwan, Japan, Canada and North America. Closely-related nematode species are known to occur in the temperate forests of Europe and Siberia, but there are not reports of *B. xylophilus* in these areas. There is concern that pine wilt could become more widespread in North America (Wingfield *et al.*, 1982), Europe (Braash, 1983; Magnusson, 1986; Kinn, 1986; McNamara & Stoen, 1988) and Asia (Mamiya, 1987) as well as in the pine plantations of the southern hemisphere (Rutherford & Webster, 1987). The European and Mediterranean Plant Protection Organisation (EPPO) has placed *B. xylophilus* on its A1 list of quarantine pest (Anon., 1986) because: (i) it is a perceived threat in view of the devastation it causes in Japan and (ii) it is presumed still to be absent in Europe. This action has profound implications for the marketing of softwoods.

For quarantine purposes, precise identification of the suspected pests or pathogens is of paramount importance. Difficulties have arisen with respect to the precise identity of some isolate of *B. xylophilus* and *B. mucronatus*. Variability, and overlap in the range of several definitive taxonomic characters of these, and similar species, is such that accurate identification of the nematodes is often difficult. In order to overcome this problem, De Guiran *et al.* (1985) used electrophoresis of extracted enzymes to differentiate some *Bursaphelenchus* spp. More recently, two species-specific probes based on restriction fragment analysis following hybridization with heterologous probes, have been developed to differentiate isolates of *Bursaphelenchus* spp. within the pine wood nematode species complex (Webster *et al.*, 1990).

During the last decade, *B. xylophilus* has been detected in Finland, Norway and Sweden in softwoods imported from Canada, Japan and the U.S.A.; heat-treatment of wood chips is being evaluated as a feasible method of nematode decontamination (Kinn, 1986; Kinn & Deister, 1987).

B. CONTROL OF THE VECTOR

Chemical Control

Prevention of maturation feeding by beetles on twigs is an effective means of controlling pine wilt. Just prior to the feeding period, the crowns of healthy pine trees are sprayed with insecticides (Yamane, 1981). In Japan, from 1972, aerial spraying of insecticides has reduced pine mortality. Bioassays of the efficacy of various chemicals conducted during 1972-1975 resulted in standard treatments being established. These were 60 I of 3% fenitrothion/ha (= 1.8 kg fenitrothion/ha) or 7-8 1 of 40% carbaryl/ha (= 3 kg carbaryl/ha), applied twice to coincide both the beginning and with the peak of beetle emergence (Ikeda, 1984).

Ground sprays were used when aerial sprays were not possible. The standard practice involved 2-3 I of 0.25-0.5% fenitrothion/fenthion or 1.5-2% carbaryl per tree (or 600-1200 I/ha) sprayed once at the beginning of the season (Ikeda, 1984). This method proved particularly successful in protecting valuable trees in public places but was less effective in dense pine stands.

Several important considerations are known to effect the efficacy of spraying strategies. Spraying protects pine trees by preventing vectors from feeding on twigs, but it does not control the source of outbreaks. Helicopters are not always available at the optimum time to spray emerging beetles. The persistence of chemicals on twigs may be affected by rain, and the impact of chemicals on non-target arthropods and their birds predators is largely unknown. Phytotoxic damage to Japanese Cypress *Chamaecyparis obtusa* (one of the most valuable tree species in Japan) has been encountered and populations of vectors can appear to recover quickly after spraying (Ikeda, 1984).

Poisoned Baits

The characterisation, from infected trees, of volatiles such as α-pinene and ethanol, as important attractants for both sexes of *Monochamus alternatus* (Ikeda *et al.*, 1980; Ikeda & Oda, 1980), has led to the development of a mass-trapping and surveying technique involving commercial lures. A bait tree technique using either paraquet, a herbicide which increases pine oleoresin production and thus attracts *M. alternatus* (Yamasaki *et al.*, 1980; Yamasaki & Suzuki, 1982), or a solvent treatment combined with chemical sprays (Ikeda *et al.*, 1984) have resulted.

Crop Husbandry

The maintenance of a high level of crop husbandry is also an effective method of controlling vectors and reducing the incidence of pine wilt (Mamiya, 1984). Infected trees should, wherever possible, be felled and removed from the forest burnt, or treated with insecticides to kill beetle larvae. In Japan, standard practices include 0.4-0.6 l of insecticide/m^2 of the log surface. Eighteen chemicals including fenitrothion, fenthion, carbaryl and diazinon have been registered (Ikeda, 1984) for this purpose.

Biological Control

Natural enemies of *Monochamus* spp. include microbial pathogens (bacteria and fungi), invertebrates (parasitic and predatory arthropods and parasitic nematodes) and birds (Ikeda, 1987). Intensive surveys in many localities revealed yeasts, eight genera of fungi and a few bacteria as microorganisms associated with the pine sawyer (Katagiri & Shimazu, 1980). Inoculation tests determined the pathogenicity of *Beauveria bassiana, Verticillium* spp. and *Serratia marcescens. B. bassiana* showed promise as a control agent (Shimazu & Katagiri, 1981; Shimazu *et al.*, 1982), while a combined application of *S. marcescens* and *B. bassiana* resulted in higher mortality than that caused by the fungus alone (Ochi *et al.*, 1982). *Beauveria tenella*, from cultures grown on *Anomala* (Coleopetra : Scarabaeidae), also proved pathogenic to the beetles. Although the fungi were easily mass-produced, results were inconsistent mainly because of the difficulties encountered in delivering fungal spores to the sites of infection beneath bark (Ikeda, 1987).

Over 30 insect species are reported as natural enemies of *M. alterntus* (Nobuchi, 1980). Among the *Temnochila japonica* Reiter (Coleoptera : Ostomidae) and *Thanasiinus lewisi* Jacobson (Coleoptera : Cleriidae) are frequently found beneath the bark and may cause 60-80% beetle mortality during the early larval stages (Ochi & Katagiri, 1979).

Susceptibility of *Monochamus* larvae to parasitism by the rhabditid nematode *Steinernema feltiae* Filipjev (= *Neoaplectana carpocapsae* Weiser) has also been demonstrated (Mamiya & Tamura, 1983; Katagiri *et al.*, 1984; Mamiya & Shoji, 1985). Several isolates of *S. feltiae* have been introduced into Japan from France, the U.S.A. and Australia and the potential of the nematodes of parasitise *M. alternatus* in logs has been investigated. When 600 million *S. feltiae* infectives/m^2 were sprayed onto logs, 19-75% mortality of the beetles was recorded. Injection of a suspension of 1000 infectives/0.1 ml of water into larval tunnels, just before beetle emergence, resulted in 87% mortality (Mamiya & Tamua, 1983). However, a whole range of factors affecting nematode activity, including their persistence on logs and in larval tunnels, needs to be evaluated and the time and method of their application standardised. The comparative efficacy of nematodes of the genus *Heterohabitis* should also be studied.

With regard to avian predation, Igarashi (1980) observed that some woodpeckers could cause high mortality of larvae; all the overwintering larvae in pupal cells were devoured but larvae under bark usually avoided detection.

RED RING DISEASE OF COCONUT AND OIL PALMS

The disease was first reported in Trinidad by Hart (1905); its etiology was established by Nowell (1919). It is now widely distributed throughout the West Indies, Latin America and Guatemala (Dean, 1979). The causal nematode was first described as *Aphelenchoides cocophilus* by Cobb (1919) and was later transferred to the genus *Rhadinaphelenchus* by Goodey (1960).

ETIOLOGY

The nematodes are mainly transmitted by female palm weevils (*Rhynchophorus palmarum*); *R. cruentus* may also act as a vector. The biology and life cycle of *Rhadinaphelenchus cocophilus* is intimately associated with that of its vector. *R. palmarum* is a pest of palms in it own right; its larvae feed by burrowing through coconut stems. In diseased trees, nematodes enter the haemocoel of weevil larvae through the gut; in adult weevils, they can be found in the gut, the body cavity, the ovipositor and sometimes in tracheal sacs (Griffith & Koshy, 1990). Weevils may become infested with many thousands of nematodes, but only a certain proportion of the insect population is so affected. Non-susceptible weevil larvae are capable of immobilizing the nematodes in their haemocoel. Weevils that become infested are generally smaller than those that are non-susceptible (Griffith, 1968). Nematodes are injected into the tissue of the coconut tree during oviposition, normally in a leaf axil in the crown of the tree.

Newly-affected trees become chemically attractive to all palm weevils including potential vectors of nematodes. The rate of spread from the primary source of infection depends upon the development of vector palm weevils within the diseased tree. Parasitization by nematodes may limit the number of developing vectors and reduce their size, fecundity and longevity (Griffith, 1968). Within three months of infection, new trees can be infected by female weevils emerging from infected trees.

Coconut palms of three to ten-year-old are extremely susceptible. In newly-invaded tissue, the parasites occur intercellularly but during the later stages of disease development, they can also be found intracellularly. In many cases, lysigenous cavities are formed in which large numbers of nematodes are present (Blair & Darling, 1968; Blair, 1970a, b). The most characteristic symptoms are the internal lesions. In a cross-section of the stem, they appear as an orange to brick-red coloured ring, two to four cm wide, and at a distance of three to five cm in from the periphery. The external symptoms of red-ring disease include the appearance of chlorotic tips to older leaves. Nuts are shed prematurely, either simultaneously with the development of leaf symptoms, or slightly before. Four to six weeks after the symptoms first appear, the crown of the palm, weakened by internal weevil-induced damage, may topple over. In Surinam (West Indies). *R. cocophilus* has also been found associated with 'little leaf' syndrome of coconut palms (Van Hoof & Seinhorst, 1962).

Oil palms (*Elaies guineensis* Jacq.) in Venezuela are also susceptible to *R. cocophilus* (Webster & Gonzales, 1959). Despite some fundamental differences in pathogenesis, the symptoms of the disease in oil palms are similar to those in coconut palms. In coconut, three to ten-year-old palms die within three months of infection; in oil palms of the same age group, the process can take three to four years. One reason for this is that the nematode does not appear to colonize as rapidly in oil palm tissue as it does in coconut.

CONTROL STRATEGIES

Control measures are mainly preventive rather than curative and are based on the destruction of infested material, and on the trapping and killing of the weevil vectors. Since external symptoms of the disease are not easily recognisable, it is necessary to take a 'core sample' of the trunk (Griffith & Koshy, 1990) to determine the development of internal symptoms and to check for the presence of nematodes. Various measures taken to limit the spread of the disease are outlined below :

Destruction of Diseased Trees

Nematode multiplication and vector development proceed simultaneously in a newly-infested tree. Only three months after the primary infection, weevils emerging from infested palms spread the nematodes to new hosts. Since disease symptoms are apparent before the weevil progeny emerge, the trees should be destroyed as soon as symptoms are recognised. 'Cut and burn' is one of the oldest and most effective practices used to prevent further spread of the disease. More recent methods involve the destruction of palms using arboricides, herbicides or phytotoxic nematicides. Such chemicals are applied to holes made in tree-trunks. Based on a screen of large numbers of chemicals, Hoyle (1968) recommended the use of cacodylic acid ('Silvisar') to poison trees. He found that the chemical prevented weevil breeding, killed the diseased trees within two weeks, but had no nematicidal effects. Later, Victoria *et al.* (1970) used sodium and potassium arsenate to kill both diseased trees and nematodes. Trees in an advanced stage of disease, or in a 'broken neck' condition should not be treated with herbicides but cut down and their, stumps sprayed thoroughly with 0.1% methomyl solution to eradicate weevil larvae and pupae (Griffith & Koshy, 1990).

Chemical Control

Pesticides are largely ineffective for the control of *R. cocophilus* and *R. palmarum* mainly because of their poor rates of absorption and translocation within the coconut tree and their lack of stability within its tissues. Stem injection of chemicals has been found to be impracticable due to the damage that is caused to the trees (Hoyle, 1971). Despite the difficulties some success has been achieved. Fenwick (1969) reported good control of both nematodes and weevils with 'Cocosev', a formulation of ground coconut shells coated with carbaryl that is applied to leaf axils at two-monthly intervals. Fumigation of living trees has also been tried as a means of curing diseased trees. Hoyle (1971) reported that when applied to leaf axils, only aldicarb was absorbed by the plant tissues in significant amounts. He also reported that methomyl degraded too quickly to have any prolonged effect.

Traps or Guard Baskets

Adults weevils are attracted to anaerobic fermentation products such as ethyl alcohol, n-butyl alcohol and to certain volatile esters extracted from diseased palm tissue (Griffith, 1987). This behaviour can be exploited to attract weevils to poisoned baits. Traps or 'guard baskets' are made to 2-cm mesh wire and are 1 m high and 0.3 m in diameter (Griffith & Koshy, 1990). The baskets are filled with chunks of fresh tissue from diseased coconut trees and are thoroughly sprayed with 0.1% methomyl solution. Alternative materials especially those from palmiste or 'gru-gru' palm, pineapple or papaya can also be used. The treated baskets are distributed on the ground in plantations of young coconut trees, at a rate of 25 baskets/10 ha. The baskets are replenished every two weeks with fresh plant material and the previous plant materials are burnt. Morin *et al.* (1986) described the construction of traps made from cubes cut from trunks of oil palms and reported that such traps had been in use in Brazil since 1975. This procedure is especially effective in the cool nights of the dry season when weevils are most active.

Biological Control of the Vector

Griffith (1977) reported the pathogenicity of a bacterium *Micrococus agilis* to the palm weevil and suggested its potential use as a biocontrol agent. Biological control strategies are, however, currently concentrated more on the use of rhabditid nematodes of the genera *Steinernema* and *Heterorhabditis*. Since the vector can be heavily parasitised, the maintenance of a high population of nematodes in the environment introduces a selective pressure against the vectors. Such measures are being employed in Trinidad (Griffith & Koshy, 1990).

NEMATODE DISEASES OF MUSHROOMS

Worldwide over 18 species of *Aphelenchoides*, one species of *Ditylenchus* (Sharma & Seth, 1986; Seth & Sharma, 1986) and over 14 species of rhabditoid nematodes (Grewal, 1990) have been implicated as being potentially detrimental to the growth of the cultivated mushroom *Agaricus bisporus* (Lange) Imbach.

ETIOLOGY

Nematodes are adapted to a passive mode of dispersal and are introduced into mushroom growing houses in various ways. They can gain entry via infested mushroom substrates (compost or casing materials), contaminated water, infested trays, and can be spread by farm workers and by flies (Hesling, 1979a). In countries such as India, where unpasteurised substrates are commonly used in mushroom culture (Grewal, 1989b), all the above factors may contribute towards nematode dissemination. In Europe and the U.S.A., however, composts are thoroughly pasteurised before being spawned and a high standard of hygiene is maintained throughout the cropping cycle. In these situations, flies remain, the most important potential source of nematode contamination (Haglund & Milne, 1973; Rinker & Bloom, 1982). Three nematode species have been demonstrated to be transmitted through flies; *Aphelenchoides composticola*, *Caenorhabditis elegans* and *Cruznema tripartitum* (syn. *C. lambdiensis*).

The pathogenic potential of, and the extent of damage that can be caused by, *A. composticola* in mushroom production is long established (Arnold & Blake, 1968). The effects of saprophagous nematodes have, however, only recently been detailed (Ross & Burden, 1981; Kaufman *et al.*, 1984). In an intensive study, Grewal & Richardson (1991a) demonstrated that when mass-produced *C. elegans* were introduced into casing the effects were far-reaching. Not only was the onset of mushroom production delayed, the growth pattern of crops was disrupted, mushrooms were discoloured and physically distorted, and most importantly, significantly yield losses incurred. Steiner (1933) implicated *C. tripartitum* in the spread of the pathogenic bacterium, *Pseudomonas tolaasii* ('bacterial blotch') in mushroom beds. Both of these rhabditid nematodes are known to be disseminated by at least one mushroom sciarid fly, *Lycoriella mali* (Rinker & Bloom, 1982).

Flies, especially fungus gnats (Sciaridae), are adapted to mycophagy and generally colonize dead or decaying plant materials. Due to the abundance of bacteria and fungi in such environments, saprophagous and mycophagous nematodes are also common. Most of the rhabditids and a large number of aphelenchoid nematodes produce a morphologically distinct dispersal stage larva ('dauer') in the life cycle — a stage that is adapted to dissemination by emerging flies. Sudhaus *et al.* (1988) studied co-habitation of rhabditid nematodes with coleopteran and dipteran insects in dung and reported their synchronous development. Mushroom sciarid flies are attracted to odours emanating from compost cooling after pasteurization and also to the crop during the early stages of mycelial colonization (Binns, 1973). When sciarid flies that had invaded mushroom houses were trapped, large numbers of *A. composticola* and some free-living nematodes were found attached to their body surfaces (Haglund & Milne, 1973). There seems, however, to be no direct association between particular species of sciarids and particular species of nematodes.

A. CONTROL OF THE NEMATODES

Crop Husbandry

A high standard of hygiene is paramount if contamination and the spread of nematodes in mushroom houses is to be avoided. Cleanliness in the composting yard and out-buildings, before and after crops are filled, is essential. Hesling (1979b) discussed the role of fumigant and non-fumigant disinfectants for nematode decontamination. Apart from external sources, mushroom substrates (i.e., composts or casing materials) can also provide the initial nematode inoculum. In countries where composts are routinely pasteurized, nematodes are generally eliminated. However, casing materials such as moss-peat, spent compost or field-soil that are used to cover mushroom beds to initiate fructification, are generally not sterilized in order to preserve sporophore-inducing microflora (Eger, 1961). Casing can be fumigated with methyl bromide or formaldehyde, or steamed (Hesling, 1979b). In India, an effective method of casing pasteurization, based on the use of solar heat, was developed (Grewal & Grewal, 1988) wherein the thinly spread casing is exposed to the sun under a polythene cover (150 gauze) for 5-6 hours.

Chemical Control

Hesling (1979c) listed chemicals that have been used to control mushroom nematodes. Although nematicides such as thionazin (Cayrol & Ritter, 1972), aldicarb (Cayrol & Ritter, 1972; Chhabra & Kaul, 1982), fenamiphos (McLeod & Khair, 1978) have been shown to control *A. composticola*, some are toxic to mycelium and reduce mushroom yields (Grewal & Sohi, 1987; White, 1986; Richardson & Grewal, 1991); others leave residues in produce (Bhal & Agnihotri, 1989). At the present time, therefore, neither preventive nor curative use of nematicides is recommended.

Biological Control

Nematophagous fungi have been developed as commercial biological control agents. In France, a nematode trapping fungus, *Arthobotrys robusta*, is available for the control of mycophagous nematodes in mushrooms (Cayrol, 1978). When added to compost at spawning it traps nematodes by adhesive hyphal networks and has been shown, even in the absence of nematodes, to stimulate mycelial growth and improve cropping. *A. oligospora* and *A. superba* are also reported to reduce nematode populations and increase mushroom yields (Kermarrec, 1973). Grewal & Sohi (1988a) mass-produced *A. conoides* on wheat grain and, after incorporating it into compost at spawning, showed that the fungus reduced a mixed population of *A. composticola* and *Rhabditis* sp. by 42.5% in 25 days.

Nematicidal Plant Products

In India, the lack of bulk pasteurization facilities together with residue problems allied to the use of systemic nematicides, have resulted in nematicidal plant products being perceived as a safe and cheap alternative to chemicals for the control of mushroom nematodes (Grewal & Sohi, 1986; Sharma & Seth, 1986). A wide range of plant species was tested and leaf-extracts of *Leucaena leucocephala* (Grewal & Sohi, 1988b), *Bougainvillea spectabilis, Calotropis procera, Cedrella toona, Jacaranda acutifolia, Tageta patula* and leaf and seed-extracts of *Melia azedarach* and *Ricinus communis* (Grewal, 1989a) were found to be highly toxic to *A. composticola*. Furthermore, the incorporation of dried leaves of *Azadirachta indica, Canabis sativa* and *Ricinus communis* resulted in the suppression of inoculated populations of *A. composticola* below economic injury levels (Grewal, 1989b). Leaf incorporation also increased mushroom yields by 6.5-19%.

B. CONTROL OF THE VECTOR FLIES

Preventing Access

Effective exclusion of invading flies before they infest mushroom beds with nematodes is the most obvious method of alleviating the problem. This can be achieved by screening-off ventilation ducts (16 mesh/cm) and by maintaining efficient seals around the doors of mushroom houses. Newly-spawned crops should be protected with aerosols, smokes or fogs of pyrethrins, gamma-HCH or dichlorvos. The area in close proximity to mushroom houses should be kept clear of rotting organic matter, particularly spent compost.

Chemical Control of Fly Larvae

On a worldwide scale many different recommendations exist for the control of sciarid flies. Insecticides may either be incorporated into compost or mixed into casing. The recent extensive use of chemicals had led to some sciarid populations developing resistance to organophosphorus insecticides such as diazinon, dichlorvos and malathion (White & Gribben, 1989). At present resistant populations of *L. auripila* are controlled by the incorporation into casing of a chitin-synthesis inhibitor, diflubenzuron. Fletcher *et al*. (1989) list current chemical control measures for flies in mushroom houses.

Biological Control of Fly Larvae

Richardson (1987) reported the susceptibility of *L. auripila* larvae to parasitism by rhabditid nematodes of the genera *Steinernema* and *Heterorhabditis*. Due to their ability to search for, locate and invade a range of insect larvae, such nematodes are particularly suited to the mushroom environment where the larvae of Phoridae, Sciaridae and Cecidomyiidae are common. For sciarid fly control, a single application of *S. feltiae* in casing has been shown to provide results comparable to those achieved with the most effective chemical, diflubenzuron (Richardson & Grewal, 1991).

Rhabditid nematodes also offer other advantages over chemicals. They are persistent and capable of recycling (Grewal & Richardson, 1991b) whereas pesticides are effective over a more limited period and may produce phytotoxic breakdown products (Verloop *et al*., 1975; White, 1986) that reduce yields. The biocontrol approach also overcomes residue problems often associated with the use of chemicals.

Insect-pathogenic bacteria, *Bacillus thuringiensis* (White & Jarrett, 1990), and predatory mites (Al-Amidi *et al*., 1989) have also been demonstrated to have potential for the control of sciarids.

THE *FERGUSONINA-FERGUSOBIA* ASSOCIATION

Nematodes of the genus *Fergusobia* Currie are reported from flower-bud galls of many species of *Eucalyptus* in Australia (Currie, 1937) and *Syzygium cumini* (Myrtaceae) in India (Siddiqi, 1986). The nematodes not only infect plant shoot galls but also species of the dipteran genus *Fergusonina* Malloch (Table 3).

ETIOLOGY AND CONTROL

Nematode-infested flies emerge from galls, mate and lay eggs in young flower buds (Currie, 1937). With each egg, up to 50 juvenile nematodes are deposited inside the bud; these feed, as *Fergusonina* larvae do on plant tissues. Prior to pupation, fertilized female *Fergusobia* enter the body cavity of the host larva and, the time the female fly emerges, have already laid numerous eggs in its haemocoel. The nematode juveniles migrate to the host's ovary and are deposited during fly oviposition. The success of this nematode-insect association is in a balanced parasitism whereby the female fly is neither killed nor rendered infertile (Siddiqi, 1986).

Table 3 : Species of *Fergusobia* with their plant and insect hosts.

Nematode	Plant host	Insect host
F. indica (Jairajpuri) Siddiqi*	n.d.	n.d.
F. jambophilia Siddiqi	*Syzygium cumini*	*Fergusonina syzygii*
F. magna Siddiqi	*Eucalyptus* sp.	n.d.
F. tumifaciens (Currie) Wachek	Eucalyptus spp.	*F. carteri F. tillyardi*

* The species was originally described from soil around the roots of *Allium cepa* in India; n.d. = not determined.

In Australia, galling occurs on more than a dozen species of *Eucalyptus*, results in suppression of flower development; the sheer weight of the galls can cause branches to break. As *Eucalyptus* flowers are the main source of food for honey bees, honey production can be seriously affected (Currie, 1937). Galling of the flower buds of *E. maculata*, a tree valued for its timber, adversely affects its capacity to produce seeds. At present, no recommendations for the control of this insect-nematode complex are available.

CONCLUSIONS

Insect-transmitted nematodes can be the cause of devastating plant diseases and subsequent crop losses. Where synchronous development of nematodes and vector insects occurs, efficient dissemination of dispersal-stage juvenile nematodes is assured. Therefore, although the suppression of the primary foci of the disease forms the basis of any successful control measures for insect-borne plant nematodes, due to their complex etiology, no single method is likely to provide effective control. There is a continuing need to develop multi-facet control strategies that effectively blend together different aspects of nematode and vector control.

REFERENCES

Al-Amidi, A; Dunne, R. & Downe, M.J. (1989). Control of mushroom pests using a predatory mite. *The Mushroom J.* **194** : 65.

Anonymous (1986). EPPO data sheets on quarantine organisms No. 158, *OEPP/EPPO Bull.* **16:** 50-60.

Aronld, N.P. & Blake, C.D. (1968). Some effects of nematodes *Ditylenchus myceliophagus* and *Aphelenchoides composticola* on yield of cultivated mushroom. *Ann. Appl. Biol.* **61:** 161-166.

Bahl, N. & Agnihotri, N.P. (1989). Translocation of nematicide residues in mushrooms following their application at composting. *Mushroom Sci.* **12:** 883-888.

Bedker, R.J.; Wingfield, M.J. & Blanchette, R.A. (1987). Pathogenicity of *Bursaphelenchus xylophilus* on three species of pine. *Canadian J. Forest Res.* **17:** 51-57.

Binns, E.S. (1973). Laboratory rearing, biology and chemical control of the mushroom sciarid fly *Lycoriella auripila* (Diptera : Lycoriidae). *Ann. Appl. Biol.* **73:** 119-126.

Blair, G.P. (1970a). Studies on red-ring disease of the coconut palm. *Oleagineux* **25:** 19-22.

Blair, G.P. (1970b). Studies on red-ring disease of the coconut palm. *Oleagineux* **25:** 79-83.

Blair, G.P. & Darling, H.M. (1968). Red-ring disease of the coconut palm, inoculation studies and histopathology. *Nematologica* **14:** 395-403.

Bolla, R.I. & Fitzsimmons, K. (1988). Effect of simulated acid rain on *Bursaphelenchus xylophilus* infection of pine seedling. *J. Nematol.* **20** : 590-598.

Bolla, R.I. & Tamura, H. (1989). Mating potential and chromosome number of some Japanese and U.S. populations of the pine wood nematode, *Bursaphelenchus xylophilus. Japanese J. Nematol.* **19:** 7-11.

Bolla, R.I.; Weaver, C.; Koslowski, P.; Fitzsimmons, K. & Winter, R.E.K. (1987). Characterization of a non-parasitic isolate of *Bursaphelenchus xylophilus*. J.Nematol. **19**: 304-310.

Bolla, R.I.; Weaver, C. & Winter, R.E.K. (1988). Genomic differences among pathotypes of *Bursaphelenchus xylophilus. J. Nematol* **20:** 309-316.

Braasch, H. (1983). Der Kiefernholznematode, *Bursaphelenchus xylophilus* (Steiner and Buhrer, 1934) Nickel, 1970, im Blickpunkt der Pflanzenquarantane. *Nachrichtenblatt fur den Pflanzenschutzdient in der DDR* **37**: 227-230.

Cayrol, J.-C. (1978). Mise au point d'une methode de lutte biologique contre les nematodes des champignonnieres a l'aide d'un champignon predateur *Arthrobotrys robusta souche* "Antipolis" (Royal 300). *Bulletin de la Federation Nationale des Syndicates Agricoles des Cultivateurs de Champignons* **15:** 1445-1454.

Cayrol, J.-C. & Ritter, M. (1972). Nematodes and nematicides in mushroom culture. *Mushroom Sci.* **8:** 867-879.

Cheng, H.; Maosong, L. & Ruju, Q. (1986). A study on the morphological diagnosis and the pathogenicity of *Bursaphelenchus mucronatus. J. Nanjing Agric. Univ.* **2:** 55-61.

Chhabra, H.K. & Kaul, V.K. (1982). Occurrence and control of *Aphelenchoides composticola* on mushrooms. *Indian J. Nematol.* **12** : 191-192.

Cobb, N.A. (1919). A newly discovered nematode *Aphelenchus cocophilus n.* sp. connected with a serious disease of the coconut palm. *West Indian Bull.* **17:** 203-210.

Currie, G.A. (1937). Galls on *Eucalyptus* trees — A new type of association between flies and nematodes. *Proc. Linnaean Soc., New South Wales* **62:** 147-174.

Dean, C.G. (1979). Red-ring disease of *Cocos nucifera* L. caused by *Rhadinaphelenchus cocophilus* (Cobb, 1919) Goodey, 1960. *An annotated bibliography and review.* Technical Communication No. 47, CAB, St. Albans, U.K. pp. 70.

De Guiran, G.; Lee, M.J.; Dalmasso, A. & Bongiovanni, M. (1985). Preliminary attempt to differentiate pinewood nematodes (*Bursaphelenchus* spp.) by enzyme electrophoresis. *Revue Nematol.* **8:** 88-90.

Dropkin, V.H. & Foudin, A.S. (1979). Report of the occurrence of *Bursaphelenchus lignicolus*-induced pine wilt disease in Missouri. *Plant Dis. Reprt.* **63:** 904-905.

Dwinell, L.D. (1985). Relative susceptibilities of five pine species to three populations of the pinewood nematode. *Plant Dis.* **69:** 440-442.

Eger, G. (1961). Utersuchungen uber die Function der Deckschicht bei der Frucchtkorperbildung des Kulterchampignons, *Psalliota bispora* Lange. *Archiv fur Mikrobioligie* **39:** 313-334.

Fenwick, D.W. (1969). The present outlook on red-ring control. *J. Agric. Soc. Trinidad* **69:** 969-979.

Fletcher, J.T.; White, P.F. & Gaze, R.H. (1989). *Mushrooms : Pest and Disease Control* (Second Edition). Andover: Intercept Ltd. p. 174.

Foder, A.; Vecseri, G. & Farkas, T. (1990). *Caenorhabditis elegans* as a model for the study of entomiopathogenic nematodes. *In : Entomopathogenic Nematodes in Biological Control* (Eds.-R. Gaugler & H.K. Kaya). Boca Raton, Florida, CRC Press. pp. 249-269.

Furukoshi, T. & Sasaki, M. (1984). Hybridizations among species belong to sub sect. *Sylvestres* and their resistance to wood nematode, including future cross breeding strategy. *Annual Report (1982), Kanto Forest Tree Breeding Institute, Japan* **16:** 195-225.

Futai, K. & Furuno, T. (1979). The variety of resistances among pine-species to pine wood nematode, *Bursaphelenchus lignicolus. Bull. Kyoto Univ. Forests* **51:** 23-36.

Goblin, R.M. (1987). Culture of nematode associates and parasites of insects. *In: Vistas on Nematology: A commemoration of the Twenty-Fifth Anniversary of the Society of Nematologists* (Eds.—J.A. Veech & D.W. Dickson). Florida, Society of Nematologists. pp. 408-413.

Goodey, J.B. (1960). *Radinaphelenchus cocophilus* (Cobb, 1919). The nematode associated with red-ring disease of coconuts. *Nematologica* **5:** 98-102.

Grewal, P.S. (1989a). Nematicidal effects of some plant-extracts to *Aphelenchoides composticola* (Nematoda) infesting mushroom, *Agaricus bisporus. Revue Nematol.* **12:** 317-322.

Grewal, P.S. (1989b). Effects of leaf matter incorporation on *Aphelenchoides composticola* (Nematoda), mycofloral composition, mushroom compost quality and yield of *Agaricus bisporus. Ann. Appl. Biol.* **115:** 299-312.

Grewal, P.S. (1990). *Studies on saprobic rhabditid nematodes and their associated bacteria affecting mushroom culture.* Ph. D. Thesis, University of London. p. 239.

Grewal, P.S. & Grewal, S.K. (1988). Solar heat pasteurization of casing soil and chicken manure for mushroom cultivation. *Mushroom J. Tropics* **8** : 35-45.

Grewal, P.S. & Richardson, P.N. (1991a). Effects of *Caenorhabditis elegans* (Nematoda : Rhabditidae) on yield and quality of the cultivated mushroom *Agaricus bisporus. Ann. Appl. Biol.* **118:** (in press).

Grewal, P.S. & Richardson, P.N. (1991b). Effects of nematode application rates (*Steinernema feltiae*) and insecticides on sciarid fly (*Lycoriella auripila* populations and mushroom yields (*Agaricus bisporus*). *J. Invertebrate Pathol.* (in preparation).

Grewal, P.S. & Sohi, H.S. (1986). Bio-control of mushroom nematodes — a promising proposition. *Souvenir — inauguration of the National Centre for Mushroom Research and Training* (ICAR). pp. 39-40.

Grewal, P.S. & Sohi, H.S. (1987). Studies on the effects of different pesticides on the growth of *Agaricus bisporus* (Lange) Singer and *Pleurotus sajor-caju* (Fr.) Singer. *Mushroom J. Tropics* **7:** 25-29.

Grewal, P.S. & Sohi, H.S. (1988a). A new and cheaper technique for the rapid multiplication of *Arthorobotrys conoides* and its potential as a bio-nematicide in mushroom culture. *Curr. Sci.* **57:** 44-46.

Grewal, P.S. & Sohi, H.S. (1988b). Toxicity of some plant extracts to *Aphelenchoides composticola. Indian J. Nematol.* **18** : 354-355.

Griffith, R. (1968). The mechanism of transmission of the red-ring nematode. *J. Agric. Soc. Trinidad* **68** : 431-458.

Griffith, R. (1977). A species of bacterium pathogenic to the palm weevil, *Rhynchophorus palmarum*. *Annual Report, Red Ring Research Division*, *1976*. Ministry of Agriculture, Lands and Fisheries, Trinidad and Tobago. pp. 31-35.

Griffith, R. (1987). Red-ring disease of coconut palm. *Plant Dis*. **71:** 193-196.

Griffith, R. & Koshy, P.K. (1990). Nematode parasites of coconut and other palms. *In* : *Plant Parasitic Nematodes in Subtropical and Tropical Agriculture* (Eds.—M. Luc, R.A. Sikora & J. Bridge). Wallingford, Oxon, UK, CAB International. pp. 363-386.

Haglund, W.A. & Milne Jr., D.R. (1973). Nematode dissemination in commercial mushroom houses. *Phytopathology* **63:** 1455-1458.

Hart, J.M. (1905). Coconut disease. *Bulletin of Miscellaneous Information*, Royal Botanic Gardens, Trinidad **6:** 241-243.

Hatakeda, K.; Saito, N.; Ito, S.; Ikusima, Y. & Asano, T. (1985). A new nematicidal compound in French marigold. *Japanese J. Nematol*. **15** : 11-13.

Hesling, J.J. 1979a). The role of eelworms (nematodes) in the culture of mushrooms in the U.K. (Part I). *The Mushroom J*. **82:** 423-429.

Hesling, J.J. (1979b). The role of eelworms (nematodes) in the culture of mushrooms in the U.K. (Part II). *The Mushroom J*. **83:** 500-503.

Hesling, J.J. (1979c). The role of eelworms (nematodes) in the culture of mushrooms in the U.K. (Part III). *The Mushroom J*. **84:** 531-536.

Hinode, Y.; Shuto, Y. & Watanabe, H. (1987). Stimulating effects of β-myrcene on molting and multiplication of the pine wood nematode, *Bursaphelenchus xylophilus*. *Agricultural and Biological Chemistry* **51:** 1393-1396.

Hoyle, J.C. (1968). A simple method for poisoning coconut trees *J. Agric. Soc. Trinidad* **68:** 459-465.

Hoyle, J.C. (1971). Preliminary investigations into the use of systemics for control of red-ring disease of coconuts. *Experimental Agric*. **7:** 1-8.

Huber, M.C.; Winter, R.E.K. & Bolla, R.I. (1989). Effect of copper sulphate and lead acetate on infection of pines with *Bursaphelenchus xylophilus*. *J. Nematol*. **21:** 1-9.

Igarashi, M. (1980). Woodpecker predation on hibernating larvae of *Monochamus alternatus* Hope (Coleoptera : Cerambycidae). *Trans. Japanese Forestry Soc*. **91:** 363-364.

Ikeda, T. (1984). Integrated pest management of Japanese pine wilt disease. *European J. Forest Pathol*. **14:** 398-414.

Ikeda, T. (1987). Potential for biological control of pine wilt disease in Japan. *In* : *Pathogenicity of the Pine Wood Nematode* (Ed.-M.J. Wingfield). Minnesota, APS Press. pp. 74-80.

Ikeda, T.; Enda, N.; Yamane, A.; Oda, K. & Toyoda, T. (1980). Attractants for the Japanese pine sawyer, *Monochamus alternatus* Hope (Coleoptera: Cerambycidae). *Applied Entomol, and Zoology* **12** : 293-302.

Ikeda, T. & Oda, K. (1980). The occurrence of attractiveness for *Monochamus alternatus* Hope (Coleoptera : Cerambycidae) in nematode-infected pine trees. *J. Japanese Forestry Soc*. **63:** 201-207.

Ikeda, T.; Yamane, A.; Enda, N.; Makihara, H.; Okochi, I.; Ito, K. & Tabata, K. (1984). Attractiveness of ethanol-and paraquat-treated pine trees for *Monochamusalternatus* and other insects. (I). Attractiveness for *M. alternatus* and mortality dynamics of a pine stand. *J. Japanese Forestry Soc*. **66:** 386-390.

Ishii, K.; Kurinobu, S. & Furukoshi, T. (1982). Tree form of *Pinus thunbergii* X *Pinus massoniana* hybrid. *Annual Report, Kanto Forest Tree Breeding Institute, Japan* **16:** 235-240.

Ishikawa, M.; Kaneko, A.; Kashiwa, T. & Watanable, H. (1987). Participation of β-Myrcene in the susceptibility and/or resistance of pine trees to the pine wood nematode, *Bursaphelenchus xylophilus. Agric. Biol. Chem.* **51:** 3187-3190.

Ishikawa, M.; Shuto, Y. & Watanable, H. (1986). β-Myrcene, a potent attractant component of pine wood for the pinewood nematode, *Bursaphelenchus xylophilus*. *Agric. Biol. Chem.* **50:** 1863-1866.

Katagiri, K.; Mamiya, Y.; Shimazu, M.; Tamura, H. & Kushida, T. (1984). A spray application of *Steinernema feltiae* on pine logs infested with the pine sawyer, *Monochamus alternatus* and its mortality induced by the nematode. *Trans. Japanese Forestry Soc.* **95:** 479-480.

Katagiri, K. & Shimazu, M. (1980). Search for pathogenic bacteria and fungi to *Monochamus alternatus. Forest Pests* **29:** 28-33.

Kaufman, T.D.; Lukezic, F.L. & Bloom, J.R. (1984). The effects of free-living nematodes and compost moisture on growth and yield in *Agaricus brunnescens. Canadian J. Microbiol.* **30:** 503-506.

Kawazu, K.; Nishii, Y.; Ishii, K. & Tada, M. (1980a). A convenient screening method for nematicidal activity. *Agric. Biol. Chem.* **44:** 631-635.

Kawazu, K.; Nishii, Y. & Nakajima, S. (1980b). Two nematicidal substances from roots of *Cirsium japonicum. Agric. Biol. Chem.* **44:** 903-906.

Kermarrec, A.C. (1973). Investigations on the enemies of the cultivated mushroom. I. Study of the zoocoenoses of cultivated composts (*Agaricus bisporus* Lange). *Annales de Zoologie, Ecologie Animale* **5:** 425-464.

Kinn, D.N. (1986). Heat-treating wood chips: a possible solution of pinewood nematode contamination. *Tappi Journal* **69:** 97-98.

Kinn, D.N. (1987). Incidence of pinewood nematode dauer larvae and phoretic mites associated with long-horned beetles in Central Lousiana. *Canadian J. Forest Res.* **17:** 187-190.

Kinn, D.N. & Deister, C. 1987). Pinewood nematode-decontamination of North American wood chips for pulp and paper research in Europe. *Tappi. Journal* **70:** 131-132.

Kiyohara, T. (1984). Pine wilt resistance induced by prior inoculation with avirulent isolate of *Bursaphelenchus xylophilus. In : Proceedings of the United States-Japan Seminar : The resistance mechanisms of pines against pine wilt disease,* 7-11 May, 1984 (Ed.-V.H. Dropkin). Honolulu, Hawalii, U.S.A. pp. 178-186.

Kiyohara, T. & Tokushige, Y. (1971). Inoculation experiments of a nematode, *Bursaphelenchus* sp. onto pine trees. *J. Japanese Forestry Soc.* **53:** 210-218.

Kobamoto, M. & Izumi, Y. (1984). The nematicidal effect of metal ions on the pine wood nematode, *Bursaphelenchus xylophilus. J. Pesticide Sci.* **9:** 527-529.

Kobayashi, F.; Yamane, A. & Ikeda, T. (1984). The Japanese pine sawyer beetle as the vector of pine wilt disease. *Ann. Review Entomol.* **29:** 115-135.

Lee, M.J. (1986). Resistance of pine species in Taiwan to pinewood nematode. *Quarterly J. Chinese Forestry* **19:** 27-33.

Linit, M.J. (1988). Nematode-vector relationships in the pine wilt disease system. *J. Nematol.* **20:** 227-235.

Linit, M.J. & Tamura, H. (1987). Relative susceptibility of four pine species to infection of pinewood nematode. *J. Nematol.* **19:** 44-50.

Magnusson, C. (1986). Potential for establishment of *Bursaphelenchus xylophilus* and the pine wilt disease under Nordic conditions. *OEPP/EPPO Bull.* **16:** 465-471.

Malek, R.B. & Appleby, J.E. (1984). Epidemiology of pine wilt in Illinois. *Plant Dis.* **68:** 180-186.

Mamiya, Y. (1984). The pinewood nematode. *In : Plant and Insect Nematodes* (Ed.—W.R. Nickle). New York, Marcel Decker. pp. 589-626.

Mamiya, Y. (1987). Origin of the pinewood nematode and its distribution outside the United States. *In : Pathogenicity of the Pine Wood Nematode* (Ed.—M.J. Wingfield). Minnesota, APS Press. pp. 59-65.

Mamiya, Y. & Kiyohara, T. (1972). Description of *Bursaphelenchus lignicolus* n. sp. (Nematoda : Aphelenchoidide) from pinewood and histopathology of nematode-infested trees. *Nematologica* **18:** 120-126.

Mamiya, Y. Shoji, T. (1985). Application of *Neoaplectana carpocapsae* on pine logs infested with the pine sawyer, *Monochamus alternatus. Trans. (Kanto Branch) Japanese Forestry Soc.* **36:** 135-136.

Mamiya, Y. & Tamura, H. (1976). A kind of nematode trapping fungi *Dactylella leptospora* found in wood around pupal chambers of *Monochamus alternatus. Forest Pests.* **25:** 147-149.

Mamiya, Y. & Tamura, H. (1983). Mortality of the pine sawyer, *Monochamus alternatus*, induced by the entomogenous nematode *Steinernema feltiae. Trans. (Kanto Branch) Japanese Forestry Soc.* **35:** 163-164.

Matsuda, K.; Kimura, M.; Komai, K. & Hamada, M. (1989). Nematicidal activities of (-)-N-Methylcytisine and (-)-Anagyrine from *Sophora flavescens* against pinewood nematodes. *Agric. Biol. Chem.* **53:** 2287-2288.

Matsuura K. (1984). Preventive and therapeutic effects of certain systemic nematicides upon the pine wilt disease caused by pinewood nematodes, *Bursaphelenchus xylophilus*. (I). Chemotherapeutic effect of six systemic nematicides. *J. Japanese Forestry Soc.* **66:** 1-9.

Matsuura, K. & Hosono, R. (1987). Activity of acetylcholinesterase of *Bursaphelenchus xylophilus* and its inhibition by several control agents. *Japanese J. Nematol.* **17:** 17-22.

McLeod, R.W. & Khair, G.T. (1978). Control of *Aphelenchoides composticola* in mushroom compost with nematicides. *Ann. Appl. Biol.* **88:** 81-88.

McNamara, D.G. & Stoen, M. (1988). A survey for *Bursapehlenchus* spp. in pine forests in Norway. *OEPP/EPPO Bull.* **18:** 353-363.

Moodey, B.H. (1988). Forest insect and disease conditions in Canada 1987. *Forest Insect and Disease Survey, Forestry Canada*. p. 122.

Morin, J.P.; Lucchini, F.; Araujo, J.C.A. De; Ferreira, J.M.S. & Fraga, L.S. (1986). *Rhynchophorus palmarum* control using traps made from oil palm cubes. *Oleagineux* **41:** 57-62.

Nagase, A.; Kuwahara, Y.; Tominaga, Y. & Sugawara, R. (1982). Nematicidal activity of alkylamine against the pinewood nematode, *Bursaphelenchus lignicolus. Agric. Biol. Chem.* **46:** 167-172.

Nagase, A.; Kuwahara, Y.; Tominaga, Y. & Sugawara, R. (1983). Nematicidal activity of N-substituted and N, N-disubstituted alkylamines against the pinewood nematode, *Bursaphelenchus xylophilus. Agric. Biol. Chem.* **47:** 53-58.

Nickle, W.R. (1970). A taxonomic review of the genera of Aphelenchoidea (Fuchs, 1937) Thorne, 1949 (Nematoda : Tylenchida). *J. Nematol.* **2** : 375-392.

Nickle, W.R.; Golden, A.M.; Mamiya Y. & Wergin, W.P. (1981). On the taxonomy and morphology of the pinewood nematode *Bursaphelenchus xylophilus* (Steiner & Buhrer, 1934) Nickle, 1970. *J. Nematol*. **13:** 385-392.

Nobuchi, A. (1980). Natural insect enemies of the pine sawyer. *Forest Pests* **29:** 23-28.

Nowell, W. (1919). Red-ring or root disease of coconut palms. *West Indian Bull*. **17:** 189-192.

Ochi, K.; Igarashi, Y.; Katagiri, K. & Shimazu, M. (1982). Microbial control of the Japanese pine sawyer, *Monochamus alternatus* Hope. II. Mortality in the logs treated with *Beauvaria* and *Serratia*. *Trans. Japanese Forestry Soc*. **93:** 401-402.

Ochi, K. & Katagiri, K. (1979). Mortality factors in survival curves of *Monochamus alternatus* Hope (Coleoptera : Cerambycidae) population on the pine dead trees. *Bull. Forest & Forestry Products Res. Inst*. No. **302:** 125-152.

Ogawa, S.; Nakaajima, Y. & Hagihara, Y. (1973). A preliminary report on the distribution and biology of *Bursaphelenchus lignicokus* in Fukuola Prefecture. *Res. Note No. 2. Fukuoka Forest Exp. Sta*. pp. 3-27.

Ohba, K. (1980). Breeding of pines for resistance to the wood nematode, *Bursaphelenchus lignicolus* Mamiya *et* Kiyohara. *Proc. Workshop on Genetics of Host-Parasite Interactions in Forestry*, Wageningen, The Netherlands (Sept. 11-12).

Ohba, K.; Fukukoshi, T.; Kurinobu, S. & Ishii, K. (1984). Susceptibility of subtropical pine species and provenances to the pinewood nematode. *J. Japanese Forestry Soc*. **66:** 465-468.

Richardson, P.N. (1987). Susceptibility of mushroom pests to the insect-parasitic nematodes *Steinernema feltiae* and *Heterorhabditis heliothidis*. *Ann Appl. Biol*. **111:** 433-438.

Richardson, P.N. & Grewal, P.S. (1991). Comparative assessment of biological (Nematoda : *Steinernema feltiae*) and chemical methods of control for the mushroom fly *Lycoriella auripila* (Diptera : Sciaridae). *Biocontrol Sci. & Techno*,. (In press).

Riddell-Swan, J.M. (1984). *Annual departmental report by the Director of Agriculture and Fisheries for the Financial Year 1982-83*. Hong Kong Department of Agriculture & Fisheries. pp. 58-59.

Riddle, D.C. (1988). The dauer larva. *In*: *The nematode Caenorhabditis elegans* (Ed.—W.B. Wood), New York, Cold Spring Harbor Laboratory. pp. 393-412.

Rinker, D.L. & Bloom, J.R. (1982). Phoresy between a mushroom-infesting fly and two free living nematodes associated with mushroom culture. *J. Nematol*. **14:** 599-601.

Robbins, K. (1982). Distribution of the pinewood nematode in the United States. *In: Proc. Pine Wilt Dis. Workshop* (Eds.-J.E. Appleby & R.B. Malek). Illinois, Champaign. pp. 3-6.

Ross, R.C. & Burden, J.L. (1981). An unusual problem — saprophagous nematodes. *The Mushroom J*. **99**: 99-109.

Rutherford, T.A. & Webster, J.M. (1987). Distribution of pine wilt disease with respect to temperature in North America, Japan, and Europe. *Canadian J. Forest Res*. **17:** 1050-1059.

Saiki, H.; Saito, T.; Yoneda, K.; Kaijo, M.; Ichida, J. & Yamanka, K. (1984). Biological control of the pinewood nematode by spraying a nematode-trapping fungus. *J. Japanese Forestry Soc*. **66:** 30-32.

Scotto La Massese, C.; Boulbria, A. & Baujard, P. (1978). Epigeal nematofauna associated with *Pinus pinaster* decay. *3rd Int. Cong. Plant Pathol*. Munich (August 16-23).

Seth, A. & Sharma, N.K. (1986). Five new species of the genus *Aphelenchoides* infesting mushroom in Northern India. *Indian J. Nematol*. **16:** 205-215.

Shaheen, F.; Winter, R.E.K. & Bolla, R.I. (1984). Phytotoxin production in *Bursaphelenchus xylophilus*-infected *Pinus sylvestris*. *J. Nematol*. **16:** 57-61.

Sharma, N.K. & Seth, A. (1986). Nematode problems in mushroom : Losses, symptomatology and management. *In : Plant Parasitic Nematodes of India* (Eds.—G. Swarup & D.R. Dasgupta). New Delhi, ICAR. pp. 384-399.

Shimazu, M. & Katagiri, K. (1981). Pathogens of the pine sawyer *Monochamus alternatus* Hope, and possible utilization of them in a control program. *In: Proc. 17th, IUFRO World Cong.* Div. 2. Kukizaki Japanese IUFRO Congress Committee. pp. 292-295.

Shimazu, M.; Kushida, T. & Katagiri, K. (1982). Microbial control of *Monochamus alternatus*. Spray of the pathogens on the infested pine trees just before adult emergence. *Trans. Japanese Forestry Soc.* **93:** 399-400.

Shuto, Y.; Kuwano, E. & Watanabe, H. (1989). A new hatching test for the pinewood nematode and the inhibitory effect of benzimidazole derivatives. *Agric. Biol. Chem.* **53:** 1711-1712.

Siddiqi, M.R. (1986). A review of the nematode genus *Fergusobia* Currie (Hexatylina) with descriptions of *F. jambophilia* n. sp. and *F. magna* n. sp. *In: Plant Parasitic Nematodes of India* (Eds.-G. Swarup & D.R. Dasgupta). New Delhi, ICAR. pp. 264-278.

Steiner, G. (1933). *Rhabditis lambdiensis*, a nematode possibly acting as a disease agent in mushroom beds. *J. Agric. Res.* **46:** 427-435.

Steiner, G. & Buhrer, E.M. (1934). *Aphelenchoides xylophilus* n. sp., a nematode associated with blue-stain and other fungi in timber. *J. Agric. Res.* **48:** 949-951.

Sudhaus, W.; Rehfeld, K.; Schulter, D. & Schweiger, J. (1988). Beziehungen zwischen nematoden, Coleopteran and Dipteren in der sukzession beim abbau von kuhfladen. *Pedobiologia* **31:** 305-322.

Suzuki, K. (1987). Influence of water stress on disease development. *In : Pathogenicity of the Pine Wood Nematode* (Ed.—M.J. Wingfield). Minnoesota, APS Press. pp. 5-13.

Takase, I.; Oyama, H. & Fujishita, A. (1982). Chemical control of pine wilting disease caused by *Bursaphelenchus xylophilus*, and absorption and distribution of pesticides in the pine trees. *J. Pesticide Sci.* **7:** 473-480.

Tamura, H. (1980). Some nematode-trapping fungi in the pupal chambers of *Monochamus alternatus*. *Forest Pests* **29:** 20-23.

Tamura, H. & Dropkin, V. (1984). Resistance of pine tree to pine wilt caused by the nematode, *Bursaphelenchus xylophilus*. *J. Japanese Forestry Soc.* **66:** 305-312.

Van Hoof, H.A. & Seinhorst, J.W. (1962). *Rhadinaphelenchus cocophilus* associated with little leaf of coconut and oil palm. *Netherlands J. Plant Pathol.* **68:** 251-256.

Verloop, A.; Nimmo, W.B. & De Wilde, P.C. (1975). The fate of the insecticide diflubenzuron in the soil. *Proc. 11th Int. Plant Protec. Conf.* pp. 271-277.

Victoria, K.J.; Sanchez, P.A. & Barriga, O.R. (1970). Erradication de palmas de cocotero afectadas por el anillo rojo (*Rhadinaphelenchus cocophilus* Cobb, 1919, Goodey, 1960, Nematoda : Aphelenchoididae) mediante la utilization de sustancias quimicas. *Revista Inst. Colomb. Agropec.* **5:** 185-197.

Webster, J.M.; Anderson, R.V.; Baillie, D.L. Beckenbach, K.; Curran, J. & Rutherford, T.A. (1990). DNA probes for differentiating isolates of the pinewood nematode species complex. *Revue Nematol.* **13:** 255-263.

Webster, B.N. & Gonzalez, A. (1959). Infestigaciones sobre la enfermedad 'anillo rojo' en cocas y palma de aceite. *Ninth Convencion de la Associacion Venezolana el Avance de la Ciencia*, Servicio Shell para el Agricultor, Estado Araju, Venezuela.

White, P.F. (1986). Effects of bendiocarb and diflubenzuron on mushroom cropping. *Ann. Appl. Biol.* **108:** 11-20.

White, P.F. & Gribben, D.A. (1989). Variation in resistance to diazinon by the mushroom sciarid *Lycoriella auripila. Mushroom Sci.* **12:** 851-859.

White, P.F. & Jarrett, P. (1990). Laboratory and field tests with *Bacillus thuringiensis* for the control of the mushroom sciarid *Lycoriella auripila. Brighton Crop Protec. Conf. Pests & Diseases*. pp. 373-378.

Wingfield, M.J. (1987). Fungi associated with pinewood nematode, *Bursaphelenchus xylophilus*, and cerambycid beetles in Wisconsin. *Mycologia* **79:** 325-328.

Wingfield, M.J.; Blanchette, R.A.; Nicholls, T.H. & Robbins, K. (1982). The pinewood nematode: a comparison of the situation in the United States and Japan. *Canadian J. Forests. Res.* **12:** 71-75.

Yamane. A. (1981). The Japanese pine sawyer, *Monochamus alternatus* Hope (Coleoptera : Cerambycidae) : bionomics and control. *Review of Plant Protec. Res.* **14:** 1-25.

Yamasaki, T.; Hata, K. & Okamoto, H. (1980). Luring of Japanese pine sawyer *Monochamus alternatus* Hope by paraquat-treated pine trees. I. The effect of paraquat and ethephon on pine trees. *J. Japanese Forestry Soc.* **62:** 99-102.

Yamasaki, T.; & Suzuki, N. (1982). Luring of Japanese pine sawyer *Monochamus alternatus* Hope by paraquat-treated pine trees. II. The attractiveness of volatile oil from lightwood. *J. Japanese Forestry Soc.* **64:** 340-345.

Yang, B.J.; Wang, Q.; Zou, W. & Li, Y. (1987). The resistance of pine species to pinewood nematode, *Bursaphelenchus xylophilus. Acta Phytopathologica Sinica* **17:** 211-214.

Yano, N.J. (1913). Investigation on the cause of pine mortality in Nagasaki Prefecture. *Sanrinkoho* **4:** 1-14.

Yin, K.; Fang, Y. & Tarjan, A.C. (1988). A key to species in the genus *Bursaphelenchus* with a description of *Bursaphelenchus hunanensis* sp. n. (Nematoda : Aphelenchoididae) found in pinewood in Hunan Province, China. Proc. *Helminthol. Soc. Wash.* **55:** 1-11.

Yoneda, K.; Katsumata, J.; Saiki, H.; Tsubaki, K. & Tatsuma, S. (1980). A nematode-trapping fungus detected in pine sap. *J. Japanese Forestry Soc.* **62:** 227-229.

Chapter 21

Control of Major Species of the Root-Knot Nematode

B.K. Goswami

Division of Nematology, Indian Agricultural Research Institute, New Delhi-110 012, India.

AMONG the plant-parasitic nematodes which continue to threaten agricultural crops throughout the world, particularly the subtemperate, tropical and subtropical regions, the root-knot nematodes, a common name collectively given to the species of *Meloidogyne* Goeldi, causing knots (galls) on the roots of a wide variety of plants, are undoubtedly the most common. For centuries these microscopic organisms that feed mainly on the roots have caused pandemic to man's essential crop plants. The degree of damage to a particular crop by root-knot nematodes is influenced by the crop and cultivar, nematode species, level of soil infestation and the environment. Owing to the spectacular symptoms the root-knot is often the only or one of the few nematode diseases of crops known to the common farmers. Visible symptoms of the nematode attack often include reduced growth of individual plants, chlorosis, wilting of the foliage and sometimes death of the plants. These deleterious effects on plant growth results in reduced yield and poor quality of such crop as most vegetables, potato, tobacco, rice, cotton, peanut and pulse crops, etc. Although the genus *Meloidogyne* is polyphagous and have over 65 known species reported from wide variety of hosts, the present chapter would be dealt with three most common species, viz. *M. incognita, M. javanica* and *M. arenaria*.

The pioneer nematologists led by J.N. Sasser have done a major contribution on the importance of root-knot nematodes in crop production; the universal acclaim of which led to the formation of the International *Meloidogyne* Project (IMP) with its headquarters at the North Carolina State University, U.S.A. and collaborating centres at many developing nations.

Enormity of crop damage resulting from root-knot nematode invasion and the development and necessity for their effective control measures under field conditions have come about slowly since past several decades. In fields with constant soil types and climatic conditions, the type of crop grown and the intensity of culture are the chief determinants of nematode species and population. A field, thus, with a mild infestation of root-knot nematode if subjected to continuous vegetable production would lead to very high population of the nematode thus causing damage much above economic threshold level. The general persistance of nematodes, their relatively low reproductive rate as compared with bacteria, and their modest rate of locomotion as compared with bacteria, and their modest rate of locomotion as compared to insects influence population density. These characteristics combined with the non-specific nature of many disease symptoms caused by nematodes, may explain as to why the recognition of nematode damage often lags several years behind the initial infection. It is preferable, therefore, to determine the quantitative relationship between nematode population density and symptoms, and the potential for nematode increase, before initiating control measures.

The decision to take up one nematode management practice over another should ideally be based on an understanding of the crop production system and the ecology of its principal pests. The concept of economic threshold provides a foundation, using cost/benefit analysis, for such rational decisions.

Economic thresholds are variously defined (Ferris, 1978), but can be summarised as population density of a pest at which the value of the damage caused is equal to the cost of control. At densities up to the threshold there would be no economic incentive for pest control inputs since costs would exceed crop loss.

Control of nematodes and the damage they cause can be achieved by preventing entry by nematodes, by suppressing their population, by pacifying their effects, or by a combination of these principles. Control is usually difficult and often expensive. If nematode control fails, or it is impractical or uneconomic, then basic changes in the cropping pattern must occur. The essential element, therefore, in the economic control of root-knot nematodes is to be able to predict their importance under specified cropping and environmental conditions, and in order to achieve this, effective techniques are necessary to : 1) sample soil and plant tissue for nematodes, 2) separate nematodes from soil and plant tissue, 3) identify nematodes to race and species, 4) determine current population and reproduction potential of nematode species, 5) develop a model for the particular crop-nematode situation and 6) control the nematode infection by integrating a range of methods.

As nematode population rises above thresholds of economic loss, control becomes more difficult and more necessary. A grower with adequate financial and technical resources can protect crop from losses caused by nematodes. But poor farmers have limited means to combat with pests and need genetic resistance for their crops, especially on subsistence farms in tropical agriculture.

Control methods are either : 1) those that are highly effective and widely used, and 2) those that are more restrictive because they are either partially effective, or because they are used only under special circumstances.

Selection of a particular control method depends upon many factors, and the choice of tactics will vary with each field situation.

HISTORY OF ROOT-KNOT DISEASES OF CROP PLANTS

The root-knot disease was first reported on the glasshouse grown cucumbers in England by Berkeley (1855) who named the nematodes as "Vibrios". Later in 1872 Greef in Germany described the nematode including swellings in host roots as *Anguillula radicicola*. In 1884 Muller having observed some eggs within the females described them as *Heterodera radicicola* while Goeldi in 1887 named the root-knot nematode on coffee as *Meloidogyne exigua*. Goodey (1932) while reviewing the status of root-knot nematodes instead of giving a separate genus for them named them *Heterodera marioni*. Chitwood in 1949 considered all root-knot nematodes under the genus *Meloidogyne* giving the generic diagnosis differentiating them from *Heterodera*. He recognised four existing species, viz. *M. exigua* Goeldi, 1887; *M. javanica* Treub, 1885; *M. incognita* Kofoid & White, 1919 and *M. arenaria* Neal, 1889 and a new species *M. hapla* and a new variety, *M. incognita acrita*. Subsequently many new species were described.

Out of the developments showing the importance of root-knot nematodes in crop production, the first was the discovery of D-D soil fumigant, a relatively inexpensive nematicide by Carter (1943) which could be economically and effectively used on a field basis. The production and sale of nematicides thereafter increased rapidly.

The next key discovery was the characterization of "races" or the "physiological strains" of the root-knot nematodes by Christie & Albin (1944), at a time when all root-knot nematodes were considered to comprise a single species within the genus *Heterodera*, viz. *H. marioni*. This investigation by Christie and co-worker eventually resulted in Chitwood's assigning to root-knot nematodes to the genus *Meloidogyne* and describing several species in 1949. This landmark in the history of *Meloidogyne* made it possible to : 1) determine the host range of individual species, 2) design keys for their identification based on morphological differences, 3) develop crop cultivars resistant to certain species, and 4) study individual species cytologically and biochemically.

Another discovery which accelerated the development of nematology in U.S.A. was the demonstration of the role of root-knot nematodes in increasing the incidence of certain major diseases of tobacco, such as black-shank caused by *Phytophthora parasitica* var. *nicotianae* and Granville wilt caused by *Pseudomonas solanacearum* (Lucas *et al.*, 1955; Sasser *et al.*, 1955). These diseases were a serious threat to the tobacco industry in North Carolina and other southern states. Although cultivars resistant to these diseases had been developed, evidences indicated that they lost resistance in the presence of root-knot nematodes.

In 1975 the International *Meloidogyne* Project was initiated which was funded by the United States Agency for International Development through a contract with the North Carolina State University.

SYSTEMATICS AND IDENTIFICATION OF MOST COMMON SPECIES AND RACES

Since Filipjev (1934) included both root-knot and cyst nematodes in the family Heteroderidae, the taxonomic position of the genus *Meloidogyne* has been different by various workers, viz. Chitwood & Chitwood (1950), Skarvilovich (1959), Golden (1971), Wouts (1972), etc. Though there are some differences of opinion, the taxonomic position of *Meloidogyne* may be given as :

Phylum	:	Nematoda
Class	:	Secernentia
Order	:	Tylenchida
Family	:	Meloidogynidae
Genus	:	*Meloidogyne*

The species of *Meloidogyne* have been identified mainly on the basis of morphological features in addition to some other methods like host range (Sasser, 1954), biochemistry (Hussey, 1985) and cytogenetics (Triantaphyllou, 1985) which have also been used as supplementary tools. The three most common root-knot nematode species of subtropical and tropical parts of the world are *Meloidogyne incognita* (Kofoid & White) Chitwood, *M. javanica* (Treub) Chitwood, and *M. arenaria* (Neal) Chitwood. These are commonly known as southern root-knot nematode, Javanese root-knot nematode and peanut root-knot nematode, respectively. *M. hapla* Chitwood is commonly found in temperate countries and known as northern root-knot nematode.

Eisenback (1985) observed that the characters that are valuable for the identification of four most common species include perineal patterns, female stylets, male heads and stylets, and heads and stylets of second stage juveniles.

Supplemental parameters include response to a standard set of host differentials, cytological data (chromosome number and mode of reproduction) and, in some instances, biochemical data. In addition, recently scanning electron microscope studies (SEM) have detected significant differences in the morphology of larval, female and male heads that are proving useful in distinguishing species.

Under the IMP project in U.S.A. it has been investigated that the four common species mentioned above differ in ability to infect and reproduce on certain crop plants (Sasser, 1980). These differences may be with reference to a single crop or may involve two or more crops. The usual response of populations of the more common and widespread species on a particular set of differential indicates its identity (Table 1). Thus, a population that attacks the resistant tobacco, watermelon and tomato but not cotton, pepper and peanut would probably be *M. javanica*, while a population that reproduces well on resistant tobacco, pepper, peanut and tomato but fails to reproduce on cotton and watermelon would most likely be *M. hapla*. The most important species on a world basis is unquestionable *M. incognita* within which there is a physiologic variation. In this some populations attack cotton or resistant tobacco NC 95, while others do not. Differential host tests of this species showed a pattern of response to cotton and tobacco indicating at least four host races (Table 2). In view of the wide agroclimatic differences in the various regions of the world where the species of root-knot nematodes are distributed, there are possibilities of more races which need to be identified.

Table 1 : Usual response of plant species to attack by more common *Meloidogyne* spp.

Meloidogyne spp.	Differential host[a]					
	Tobacco	Cotton	Pepper	Watermelon	Peanut	Tomato
M. incognita	[b] (+)[c] [−]	[+](-)	+	+	-	+
M. javanica	+	-	[-]	+	-	+
M. arenaria	+	-	+	+	[+](-)	+
M. hapla	+	-	+	[-]	[+]	+

[a]Plant cultivars: tobacco 'NC 95'; Cotton 'Deltapine16'; Pepper 'Calif. Wonder'; Watermelon 'Charleston Grey'; peanut 'Florrunner'; tomato 'Rutgers'.
[b]Key differential hosts for that species.
[c]Some populations attack differential host but others do not.
(Source : Sasser, 1980)

SYMPTOMS

Root-knot nematodes are essentially parasites of underground stem, etc. affecting the nutrient and water uptake of the plants. The plants consequently become dwarfed, yellowish with smaller foliage with poorer and fewer fruits. In heavily infested soil, poor emergence and death of young seedlings may occur. Below ground symptoms on the roots are small galls which in case of multiple infections on the nearby tissue may coalesce to form large galls (Fig. 1). *M. incognita* and *M. javanica* normally produce larger galls, size of which depends upon the char-

Table 2 : Summary of differential host response to various *Meloidogyne* spp. and races

Meloidogyne spp.	Number of pathogenic populations						
	Cotton	Tobacco	Peanut	Pepper	Watermelon	Tomato	Number studied
M. incognita							
Race 1	0	0	0	197	199	199	199
Race 2	0	54	0	49	54	54	54
Race 3	35	0	0	34	35	35	35
Race 4	10	10	0	10	10	10	10
M. javanica	0	167	0	10	167	167	167
M. hapla	0	38	40	36	0	40	40
M. arenaria							
Race 1	0	8	8	8	8	8	8
Race 2	0	32	0	14[a]	32	32	32

[a]Variable on pepper.
(Source : Sasser, 1980)

acteristics of the plant since the gall is the product of the reaction of the hosts to the parasites. Thus cocurbits develop much larger galls while chillies produce smaller ones with the same species. The general appearance of the crop in a field is appearance of patchy growth due to uneven distribution of the nematodes. The symptoms advance with the age of the crop and are more spectacular under drought or low fertility levels affecting both quality and quantity of the produce.

LIFE CYCLE

These obligate endoparasites are most widely distributed in both tropical and temperate climates. Reproduction without male is the rule in many species, but in others both sexes are necessary. The role of male intersexes is unknown. Eggs are deposited into a gelatinous egg sac that protects them from desiccation and also microorganisms. In most host-parasite interactions a gall is formed from which an egg sac usually protrudes. Freshly deposited egg sacs are colourless and become brown upon aging. Eggs contain single celled zygote when deposited. The embryo develops into a juvenile that moults once within the egg. Second stage juveniles hatch under favourable temperature and moisture conditions and move freely through soil towards growing root tip. The juveniles have sufficient glycogen and glycolipid reserves for survival for several months in case the host is not available. These are also capable of surviving under adverse conditions. Towson & Apt (1983) have observed anhydrobiosis when the second stage juveniles shrink, coil tightly and survive for long. They penetrate usually in the region of root elongation, by breaking into a cell with repeated jabs of the stylet. Once within a

Fig. 1 : Root of eggplant, infected with *Meloidogyne incognita,* showing gall. Rotting, caused by secondary invaders, is also visible. (Courtesy : M.M. Alam, Aligarh, India).

root, a juvenile moves between cells until it locates a site close to the stele, often in the area of a developing side root. Here it becomes sedentary, and causes the transformation of cells upon which it feeds. The juvenile swells, moults rapidly a second and third time without feeding, and matures into an adult male or female. Adult males elongate within the fourth stage cuticle and emerge from roots. An adult female remains attached to its feeding side within the stele and with its posterior at the root surface. It continues to produce eggs throughout life, sometimes reaching a total of more than 1000 eggs. The life cycle varies in length according to hosts and temperature. It may be as short as 3 weeks and as long as several months. Under optimal condition total duration of life cycle is 3-4 weeks in most species. Tyler (1933) found that the duration from the effective juvenile to egg laying female was 17 days at optimum temperature of 27.5 to 30°C but longer up to 57 or more days at low and high temperature. The sex ratio is also influenced by the environment. Males are more numerous when roots are heavily attacked or nutrition is inadequate. Although root exudates enhance hatching they are not required for successful life cycles.

PLANT-NEMATODE RELATIONSHIP

In general, if root-knot larvae are present in soil and a favourable host is available, the population will increase as long as there are healthy roots on which to feed. Conversely population decreases when the root system of a host plant is destroyed or in poor condition. This phenomenon is undoubtedly related to food supply. Root galling induced by *Meloidogyne* spp. is well known host response which involve the production of giant cells (transfer cells) in the vascular tissues of susceptible plants. There may be three to six giant cells each with several nuclei. The stimulus from the digestive fluid from the nematodes apparently enzymatic and hormonal in nature, also causes an increase in number and size of other cells in the vicinity of feeding site.

The effect of *Meloidogyne* on plant growth and yield is varied and complex. The damage caused by this nematode on susceptible plants probably involves removal of plant nutrients, alter nutrient flow patterns in plant tissues and retard root growth, all of which contribute to suppressed plant yield. Adult female nematodes require considerable amount of nutrients for egg production and compete with the host for the pool of nutrients in the roots. The increased metabolic activity of giant cells stimulates mobilization of photosynthates from shoots to roots for utilization by the feeding nematodes, reaching its maximum when adult female commences egg laying and declines thereafter. Disturbance caused by *Meloidogyne* on plant roots obviously have significant impact on the physiology of the whole plant, ultimately affecting the yield.

ECONOMIC IMPORTANCE IN CROP PRODUCTION

It would be difficult to determine accurately the economic importance of nematodes in crop production on a worldwide or individual country basis since for many countries few or no studies have been made to determine the prevalence and extent of damage caused by root-knot nematodes and where surveys have been made, presence of nematodes have not always been correlated with crop performance. Studies made in Europe, United States and other countries, however, leave little doubt concerning the destructing nature of plant-parasitic nematodes and importance of their control for successful crop production.

In countries where highly effective control programmes are developed, efforts have been made directed primarily toward crops which the grower depends upon for his income, viz. tobacco, cotton, banana, pineapple, citrus, etc. for which he is spending millions of dollars annually for nematicides and as a result in most cases are receiving several fold returns. For some of these crops there are reliable estimates of damage caused each year, but where damage is not spectacular and control measures are not taken, there is little information. The range in crop loss can be from slight, perhaps less than five per cent to total loss.

CONTROL STRATEGIES

Control of nematodes in the agricultural soil is difficult and sometimes expensive, but nevertheless necessary for profitable crop production. Control is usually meant to reduce nematode population, by one or more methods, to a low level so that the damage is negligible or do not cause economic losses. The economic threshold is the density of nematodes where the losses incurred exceed the cost of nematode management.

The obvious and measurable effects of nematode control are increased yield and quality for the particular crop. In addition there are other important benefits as :

1. Nematode control may result in control of several other diseases caused by soil inhabiting organisms present in the soil, viz. the pathogenic fungi and bacteria which express themselves only in the presence of nematodes.
2. Nematode control assures greater protection of the overall investment in the crop. Growers although normally fertilize, cultivate, spray for insect and disease control and irrigate, money spent on these practices may to a large extent be wasted, since the crop yield and quality may be reduced, if nematodes are not controlled. On the other hand, if nematodes are controlled, a healthy and fibrous root system develops, and the crop has an excellent chance to grow to maturity.

It is not advisable to depend on a single method to control nematodes. Efficient management requires the carefully integrated combinations of several practices, all of which help reduce nematode populations.

There are two broad categories for management practices: non-chemical and chemical; the former constituting several options like physical, cultural and biological control methods. Using the best combination of available management strategies and tactics for the pest complex at hand (nematodes, insects, pathogens and weeds), including economic and environmental aspects, constitute an integrated pest management system.

NON-CHEMICAL CONTROL

Managing nematodes by non-chemical practices is very effective when the right combinations can be employed for combating the pest problems (Jenkin & Taylor, 1967; Mai, 1971; Johnson, 1985). Basically cultural management involves depriving nematodes of suitable host and thereby reducing the nematode populations by starvation. Out of many cultural practices used, crop rotation is very common in which most important factors to be considered are the hosts and the root-knot species. Some of the root-knot nematodes attack only a limited number of crops. This permits a wider choice of unfavourable host crops for the rotation. For example, *M. arenaria*, a serious pest of peanut, soybean and other crops can be effectively controlled by rotation with corn or cotton. Similarly *M. javanica* can be controlled by rotation with peanut. However, for *M. incognita* which has a wide host range, It is difficult to select a non-host. The problem is more difficult when more than one nematode species are to be managed simultaneously in the same field. In a rotation programme, it must be economically feasible to the farmers providing marketable crops. When all these criteria for consideration are established, an effective rotation programme can be put into practice. However, owing to important limitations in the practice of crop rotation, it is usually adopted as a part of integrated system of control consisting of rotation together with the use of nematicides, resistant cultivars, moisture control, soil amendments and in some cases biological control.

The following practices control or reduce nematode populations to some extent but are employed only under special conditions. These practices are not reliable for economic control when used alone ; and when integrated with other more effective measures, further reduce nematode populations in field soil or in plant parts.

Heat and Solarization

Hot water treatment (HWT) of root crops or root stocks infected with root-knot nematodes is an effective disinfection technique in which metabolic system at 50°C in nematodes are inactivated and they die. Plant enzymes, at this temperature are not destroyed if the hot water treatments are properly applied. Another method of killing nematodes with heat is by soil solarization. This technique consists of covering moist soil with a polyethylene sheet during periods of intense sunshine and heat, thereby capturing the radiant heat energy from the sun, causing physical, chemical and biological changes in the soil. Soil temperature is increased to levels lethal to many soil borne fungi, bacteria, nematodes and weeds (Raymundo *et al.*, 1986). This method of nematode management has been used in U.S.A., Israel and many Asian countries including India where heat from the sun's rays is sufficient when trapped under covers to penetrate the soil and effect kill of plant pests economically (Cartia & Greco, 1987).

Traps crops and antagonistic plants

An effective approach to the use of trap crops is to plant crops that are highly susceptible to invasion by the nematode but resistant to development of larvae to adults. In this case the crop can be harvested or used as a cover crop and turned under as green manure. *Crotalaria* is a crop that has been successfully used for this purpose in reducing the population of certain species of root-knot nematodes.

In recent years considerable interest has been focussed upon plants with roots containing nematicidal substances. These toxic substances found in *Tagetes* species and *Asparagus officinalis* reduce the population levels of some nematode species. Salem & Osman (1988) in pot experiment showed that intercropping *Tagetes* sp. with tomato plants was very effective in reducing nematode population, egg masses and galling of *M. javanica*. Since little is understood concerning principles involved for this and is being uneconomical as well, such a practice has not yet been developed.

Soil organic amendments

Soil amendments may result in improvements in soil structure and plant nutrition, and in the production of nematicidal breakdown products which reduce nematode damage irrespective of any effects on nematophagous fungi. Extracts of a large number of cultivated and weed plants have been found toxic to root-knot nematodes, chopped parts of which when added to soil have resulted in reduced nematode population (Singh & Sitaramaiah, 1967, 1971; Goswami & Vijayalakshmi, 1981, 1986a; Siddiqui & Alam, 1987, 1988; Alam, 1989). Parts of neem (*Azadirachta indica*), *Melia azedarach, Datura metel* and several other plants have been found effective (Nandal & Bhatti, 1983; Goswami & Vijayalakshmi, 1986b). Amendments of soil with non-edible oilseed cakes of neem, karanj, mahua, undi, etc. have been found effective against the root-knot nematodes (Goswami & Swarup, 1971; Singh & Sitaramaiah, 1973; Bhattacharya & Goswami, 1987). Combination of oilseed cakes and a nematicide has also been proved to be promising in the management of root-knot nematode, *M. incognita* (Bhattacharya & Goswami, 1988). Verma & Gupta (1987) while studying the effect of NPK fertilizers alone and in combination with aldicarb and carbofuran on okra infected with *M. javanica* observed that number of galls and eggs/plant were minimum when carbofuran or aldicarb were amended with N.

Biological control demands a multidisciplinary approach and, since success depends upon a detailed knowledge of host-soil interactions, problems would not be solved rapidly. Given an enlightened financial support from government agencies for more than five years, serious approach for the use of biological control practices could play an important part in the future management of root-knot nematode pests.

CHEMICAL CONTROL

For controlling a wide variety of nematodes, nematicides are one of the most important and reliable means. Nematicides fall into two groups: fumigants and non-fumigants. Out of available nematicides many are registered for use only on a few crops (Johnson, 1985). Nematicides can be applied as overall, row, strip, spot, side dress and irrigation treatments to crop.

Fumigants are volatile liquid that vaporize and dissolve in the soil solution. Their high vapour pressure distributes the gas in all directions through pores. Two types of fumigants are in use :

1. Halogenated aliphatic hydrocarbons such as EDB and 1, 3 dicholoropropene (1, 3 D), used at rates active against nematodes and soil arthropods.
2. Liberators of methylisothiocyanates such as metham sodium, dazomet and methylisothiocyanate itself. Together with methyl bromide and chloropicrin,

these are general biocides.

Nonfumigants do not fumigate the soil and instead, are distributed in the soil by incorporation and by percolation of water so that they come in contact with the nematodes. These chemicals are generally nontoxic to the plants on which they are recommended and can safely be used at planting or as a post-plant application. They are also of two types :

1. Organophosphates such as phenamiphos, thionazin, ethprop, fensulfothion, etc.

2. Carbamates such as oxamyl, aldicarb, carbofuran, etc.

Systemic nematicides are the newest types which can be taken up by the roots from the soil, or by foliage after being sprayed with the material, then translocated to roots where they kill nematodes that feed on the roots or prevent nematodes from feeding on the roots. Presumably they are taken into the nematodes during feeding, but they might also enter nematode bodies through the cuticle in contact with soil and plant tissues. Evaluation of various nematicides aldicarb, carbofuran, phanamiphos, ethoprophos and chloropyriphos against root-knot nematode, *M. javanica* on tomato (Javed *et al.*, 1989) and flue cured tobacco against *M. incognita* (Johnson, 1989) increased yield and control of nematode species. Naganathan (1986) also observed yield increase of tobacco and nematode reduction against *M. incognita* by aldicarb and carbofuran.

Species of plant nematodes differ in their sensitivity to nematicides. Nordmeyer *et al.* (1989) while studying the uptake, accumulation and metabolism of carbofuran and phenamiphos by *M. javanica* and *M. incognita* showed that carbofuran was extremely unstable while phenamiphos was rather stable in both nematode species. The data indicated that the difference in biological activity (fenamiphos about 10 times more active than carbofuran) is due largely to the rapid metabolization of carbofuran and the biological stability of fenamiphos in nematodes. Nordmeyer & Dickson (1989) investigated effects of carbamate, organophosphate and avermectin nematicies on oxygen consumption by *M. javanica*, *M. incognita* and *M. arenaria*. Second stage juveniles of *M. arenaria* consumed more oxygen than *M. incognita* juveniles which in turn consumed more than *M. javanica*; decrease in oxygen depends on nematicide. This finding, according to them may relate to differential sensitivity among *Meloidogyne* species in the field. Within a single species, some stages are more sensitive than others. Juveniles die at lower concentration of fumigants than adults while eggs are less sensitive than both.

Persistence of nematicides in soil after application is important in contamination of the environment. Some compounds are readily destroyed but others are more persistent. Aldicarb, a highly toxic pesticide, converts to aldicarb sulfoxide and aldicarb sulfone in soil and in plants all three of which are translocated through roots to all plant parts including fruits as has been shown in grapevines (Hafez & Raski, 1981). Carrots and other vegetables which mature more quickly than grapes, may contain higher level of these compounds. This nematicide showing environmental pollution was replaced by oxamyl and other compounds which does not accumulate in soil or ground water even after repeated treatments.

Use of nematicides in low rainfall areas is problematic. The organophosphate compounds like fensulphothion, fenamiphos, ethoprophos and carbamates like aldicarb, carbofuran, methomyl, oxamyl, etc. have been widely experimented. As the application due to high cost and their non-availability in developing countries have been greatest limitations which may partially be overcome through application to the nursery beds in transplanted crops, bare root-dip treatment and seed dressing with systemic nematicides. These treatments also protect the crop from nematode attack in early stage (Sivakumar *et al.*, 1976).

Safety and environmental risks can be overcome by careful handling of the nematicides which are poisonous; they should never come in contact with the skin or even with clothing, containers should never be used for any other purpose, etc.

In spite of above limitations of high cost, hazards, etc. primary advantage of chemical control over other methods is the quick reduction of nematode populations within days after the chemical is applied. Thus, a nematicide is to be evolved which would be : 1) more economical to use, 2) easier to use, and 3) safer to the environment. Avermectins, a new group of nematicides which are closely related to natural compounds isolated from the actinomycete fungus, *Streptomyces avermitilis* has been developed in the laboratory which have shown good control of *M. incognita* at very low concentration (Wright *et al.*, 1983). The avermectins are 10-30 times as potent as commercial nematicides (Putter *et al.*, 1981).

INTEGRATED PEST MANAGEMENT

Individual methods of nematode control having been proved inadequate, attempts have been made to manage the root-knot and other nematodes through utilization of best combination of available management strategies for the pest complex at hand (nematodes, pathogens, insect pests, etc.) constituting an Integrated Pest Management (IPM) system. Resistant cultivars, crop rotation, pesticides and sanitary and cultural practices can all be employed to the best possible advantage (Stephan *et al.*, 1988, 1989; Novaretti *et al.*, 1989). An integrated management strategy prevents the excessive build-up of any single nematode (Noe, 1988), insect or pathogen populations while the systems require flexibility and depend upon the specific pest problem and locally available management options. A fixed set of recommendations may keep a pest complex in check for a limited period of time, but as the pest population shifts, recommendations need to be changed. System development therefore, takes into account many factors, viz. species and race(s) of pest present, availability of resistant host plants, longevity of pests and the crops, cropping system and the ecological factors of the region. Seven components of integrated pest management system were proposed by Bird (1987).

Under tropical and subtropical conditions, use of nematicides can be limited to the nursery beds in transplanted crops and as seed dressers in transplanted as well as direct crops, bare root-dips and to main fields as spot or row application in perennial and other high value cash crops. These treatments may be combined with summer ploughing, organic manuring, supplementary fertilizers, resistant varieties wherever available, though weed control and crop rotation with commercial graminaceous or other non-host crops or other antagonistic crops.

Combination of nematicides with organic amendments (Bhattacharya & Goswami, 1988) or even farm yard manures (Verma & Gupta, 1987) have proved to be very effective management practice against root-knot nematodes, *M. incognita* and *M. javanica* respectively. Nematicidal seed treatments, organic amendments, solar drying of soil, bare root-dips, chemical treatment of nursery beds or growing a resistant variety are possible approaches for nursery in case of transplanted crops. Likewise crop rotation with non-host, deep summer ploughing preferably using polythene covers, spot or row treatments with nematicides and organic amendments, seed treatments with nematicides and organic amendments, seed treatments of direct seeded crops and through weed control are some of the possible approaches at field stage. Confining susceptible crops to cooler months when the temperature may be suitable for the crop but not so for the nematode, may be adopted.

FUTURE PROSPECTS

Root-knot nematodes having been intensively and extensively investigated worldwide, a number of practicable management approaches have emerged which need extension to the farmers, particularly in the tropical countries. Although a large number of *Meloidogyne* species have been described, most of the damages caused to the agricultural crops are due to *M. incognita, M. javanica* and *M. arenaria* being widely distributed throughout the world in agricultural soils. The frequency and relative importance of the major species have been determined by International *Meloidogyne* Project referring *M. incognita* and *M. javanica* to constitute large proportion of the populations in cultivated fields. Resistance in crops to these two species alone would help alleviate the damage caused by root-knot in the tropical and subtropical region of the world. As distinct host races occur in these two species, question of reaches has to be considered in any control programme including crop rotation and use of resistant cultivars. Although the available literature reflect advancement in research applicable to management of root-knot nematodes, the effort thus far have mostly been empirical and random. Following research and implementation strategy is, therefore, suggested in conclusion for a systematic approach for the management of root-knot nematodes :

1. Estimation of crop losses at different population levels under field conditions for correlation with crop health and determination of thresholds for important crops.
2. Collection and characterization of as many *Meloidogyne* populations as possible from different host plants in each state for identification of species and races and assessment of frequency and distribution.
3. Identification of sources of resistance among important food and fibre crops to *Meloidogyne* species/races common in various agroclimatic zones.
4. Development of integrated nematode management models for various crops.
5. Better utilization of indigenous methods like intercropping (Siddiqui & Alam, 1987; Tanda & Atwal, 1989) and indigenous materials like oil-cakes and other economical amendments (Alam, 1987; Siddiqui & Alam, 1989; Goswami *et al.*, 1989) as also to identify nematicides that could be safer, easy to apply and manufactured within the country, with provision of suitable subsidies.
6. To develop nematicides of biological origin and antimetabolites especially capable of stimulating or inhibiting egg hatch, or interfering with establishment of parasites in host tissue.
7. To identify and develop systemic nematicides effective as seed dressers, tissue baths or foliar sprays (Das & Sukul, 1988).
8. To develop cropping systems with susceptible crops grown in the low activity period of root-knot nematodes and economic crops in the active period alongwith weed control.
9. To study the compatibility and integration of nematode management methods with control of other concomitant pests, diseases and weeds, e.g. some oilcakes in addition their nematicidal effects also possess to fungicidal properties.

REFERENCES

Alam, M.M. (1987). Pollution free control of plant parasitic nematodes by soil amendments with plant wastes. *Biological Wastes* **22** : 75-79.

Alam, M.M. (1989). Control of root-knot and stunt nematodes with horn meal, bone meal and oil-seed cakes. *Indian J. Nematol.* **19** : 166-170.

Berkeley, M.J. (1855). Vibrios forming cysts on the roots of cucumbers. *Gardeners' Chronicle* **14**: 220.

Bhattacharya, D. & Goswami, B.K. (1987). Comparative efficacy of neem and groundnut oil-cakes with aldicarb against *Meloidogyne incognita* in tomato. *Revue Nematol.* **10** : 467-470.

Bhattacharya, D. & Goswami, B.K. (1988). Effect of oil-cakes used alone and in combination with aldicarb on *Meloidogyne incognita* infecting tomato. *Nematol. Medit.* **16** : 143-145.

Bird, G.W. (1987). Role of nematology in integrated pest management programs. *In : Vistas on Nematology* (Eds.- J.A. Veech & D.W. Dickson). SON, Maryland. pp. 114-121.

Carter, W. (1943). A promising new soil amendment and disinfectant. *Science* **97** : 383-384.

Cartia, G. & Greco, N. (1987). The effect of solar heating of soil on *Capsicum* in a glasshouse. *Colture Protetta* **16** : 61-65.

Chitwood, B.G. (1949). Root-knot nematodes. Part I. A revision of the genus *Meloidogyne* Goeldi, 1887. *Proc. Helminth. Soc. Wash* **16** : 90-104.

Chitwood, B.G. & Chitwood, M.B. (1950). *An introduction to Nematology*. Monumental Printing Co., Baltimore, Md. p. 213.

Christie, J.R. & Albin, F.E. (1944). Host-parasite relationship to the root-knot nematodes, *Heterodera marioni*. 1. The question of races. *Proc. Helminth. Soc. Wash.* **11** : 31-37.

Das, S. & Sukul, N. (1988). Nematicidal effect of the oil from the seeds of *Argemone mexicana. Environment & Ecology* **6** : 194-197.

Eisenback, J.D. (1985). Detailed morphology and anatomy of second stage juveniles, males and females of the genus *Meloidogyne. In: An Advance treatise on Meloidogyne Vol.* I (Eds.—J.N. Sasser & C.C. Carter). IMP, NC State Univ., pp. 47-77.

Filipjev, I.N. (1934). The classification of the free living nematodes and their relation to the parasitic nematodes. *Smithson. Misc. Collect.* **89** : 1-63.

Ferris, H. (1978). Development of nematode damage functions and economic thresholds using *Meloidogyne incognita. J. Nematol.* **10** : 341-350.

Goeldi, E.A. (1887). Relatoria sobre a molestia do cafeeiro na provincia de Rio de Janeiro. *Arch. Mus. Nac. Reo de Janeiro* **8** : 7-121.

Golden, A.M. (1971). Classification of the genera and the higher categories of the order Tylenchida (Nematoda) *In : Plant Parasitic Nematodes* Vol. I. (Eds.- B.M. Zuckerman, W.F. Mai & R.A. Rohde). Academic Press, New York. pp. 191-232.

Goodey, T. (1932). On the nomenclature of the root gall nematodes. *J. Helminth.* **10** : 21-28.

Goswami, B.K. & Swarup, G. (1971). Effect of oil-cake amended soil on the growth of tomato and root-knot nematode population. *Indian Phytopath.* **24** : 491-494.

Goswami, B.K. & Vijayalakshmi, K. (1981). Effect of some indigenous plant materials and oilcakes amended soil on the growth of tomato and root-knot nematodes. *Indian J. Nematol.* **11** : 121.

Goswami, B.K. & Vijayalakshmi, K. (1986a). Efficacy of some indigenous plant materials and non-edible oilseed cakes against root-knot nematodes on tomato. *Indian J. Nematol* **16** : 280-281.

Goswami, B.K. & Vijayalakshmi, K. (1986b). Effect of some indigenous plant materials and oilcake amended soil on the growth of tomato and root-knot nematode populations. *Ann. Agric. Res.* **7** : 363-366.

Goswami, B.K.; Haque, M.M.; Mishra, S.D. & Mojumder, V. (1989). Comparative efficacy of Hind-O-Meal, karanj and castor oilseed cakes and carbofuran against root-knot nematode in cowpea. *Ann. Agric. Res.* **10** : 316-318.

Greef, R. (1872). Uber Nematoden in Wurzelanschellungen (Gallen) vers-sciendener Pflanzen. *Sitzber. Gesetzt Naturwissens-chaften. Marburg.* **11** : 169-174.

Hafez, S.C & Raski, D.J. (1981). Residue dynamics and persistence of aldicarb and its biologically similar active metabolites in grapevines. *J. Nematol.* **13** : 29-36.

Hussey, R.S. (1985). Biochemistry as a tool in identification and its probable usefulness in understanding nature of parasitism. *In : An Advanced Treatise on Meloidogyne* Vol. I. (Eds.— J.N. Sasser & C.C. Carter). IMP, NC State Univ. pp. 127-134.

Javed, N.; Chohan, R.A. & Shahid, A.A. (1989). Evaluation of various nematicides against root-knot nematodes, *Meloidogyne javanica* on tomato. *Int. Nematol. Network Newsl.* **6** : 34-36.

Jenkins, W.R. & Taylor, D.P. (1967). *Plant Nematology*. Reinhold Pub. Corp., New York.

Johnson, A.W. (1985). Specific crop rotation effects combined with cultural practices and nematicides. *In : An Advanced Treatise on Meloidogyne* Vol. I. (Eds.— J.N. Sasser & C.C. Carter). IMP., NC State Univ. pp. 283-301.

Johnson, C.S. (1989). Evaluation of four nematicides for root-knot nematode control in flue cured tobacco. *Fungicide and Nematicide Tests* **44** : 145.

Lucas, G.B.; Sasser, J.N. & Kelman, A. (1955). The relationship of root-knot nematodes to granville wilt resistance in tobacco. *Phytopathology* **45** : 537-540.

Mai, W.F. (1971). Introduction. *In : Plant Parasitic Nematodes* Vol. I (Eds.- B.M. Zuckerman; W.F. Mai & R.A. Rohde) Academic Press, New York. pp. 1-8.

Muller, C. (1884). Nematoden als gallen in gemusse. *Landwirtschaftische Jahrberischt*. Scweiz. **13** : 1-42.

Naganathan. T.G. (1986). Control of root-knot nematode on tobacco with non-volatile nematicides. *Nematol. Medit.* **14:** 285-286.

Nandal, S.N. & Bhatti, D.S. (1983). Preliminary screening of some weed shrubs for their nematicidal activity against *Meloidogyne javanica. Indian J. Nematol.* **13:** 123-127.

Noe, J.P. (1988). Theory and practice of the cropping system approach to reducing nematode problems in the tropics. *J. Nematol.* **20:** 204-213.

Nordmeyer, D. & Dickson, D.W. (1989). Effect of carbamate, organophosphate and avermectin nematicides on oxygen consumption by three *Meloidogyne* species. *J. Nematol.* **21:** 472-476.

Nordmeyer, D.; Dickson, D.W.; On, L.T. & Cromroy, H.L. (1989). Uptake, accumulation and metabolism of carbofuran and phenamiphos by the phytoparasitic nematodes, *Meloidogyne javanica* and *M. incognita. Pesticides Biochem. Physiol.* **34:** 197-184.

Novaretti, W.R.T.; Carderan, J.O.; Strabelli, J. & Amorim, E. (1989). Effect of compost application alone or in combination with nematicides and mineral fertilizers, on nematode control and productivity of sugarcane. *Nematol. Brasileira* **13:** 93-107.

Putter, I.; Mac Connell, J.G.; Preiser, F.A.; Haidri, A.A.; Ristich, S.S. & Dybas, R.A. (1981). Avermectins: Novel insecticides, acaricides, and nematicides from a soil microorganism. *Experientia* **37** : 963-964.

Raymundo, S.A.; Alcazar, J. & Salas, R. (1986). Effectiveness of soil solarization and dazomet in controlling root-knot nematodes and weed in potato fields. *Philippine Agric.* **69** : 89-97.

Salem, F.M. & Osman, G.Y. (1988). Effectiveness of *Tagetes* natural exudates on *Meloidogyne javanica* nematode. Anzeiger fur Schadlingskunde *Pflanzenschutz umweltschutz* **61** : 17-19.

Sasser, J.N. (1954). Identification and host-parasite relationships of certain root-knot nematodes (*Meloidogyne* spp.). Md. Agric. Exp. Sta. Tech. Bull. A-77: p. 30.

Sasser, J.N. (1980). Root-knot nematodes: a global menace to crop production. *Plant Disease* **64**: 36-41.

Sasser, J.N.; Lucas, G.Ġ. & Powers, H.R. Jr. (1955). The relationship of root-knot nematodes to black-shank resistance in tobacco. *Phytopathology* **45** : 459-461.

Siddiqui, M.A. & Alam, M.M.(1987). Control of plant parasitic nematodes by intercropping with *Tagetes minuta. Nematol. Medit.* **15** : 205-211.

Siddiqui, M.A. & Alam M.M. (1988). Control of root-knot and reniform nematodes by bare root-dip treatment in leaf extracts of margosa and persian lilac. *Zeitschrift fur Planzenkrankheiten und Pflanzenschutz* **95** : 138-142.

Siddiqui, M.A. & Alam, M.M. (1989). Possible utilization of a noxious weed in nematode control. *Biological Wastes* **28** : 181-188.

Singh, R.S. & Sitaramaiah, K. (1967). Effect of decomposing green leaves, sawdust and urea on the incidence of root-knot of okra and tomato. *Indian Phytopath.* **20** : 349-355.

Singh, R.S. & Sitaramaiah, K.S. (1971). Control of root-knot through organic and inorganic amendments of soil. Effect of oilcake and sawdust. *Indian J. Mycol. Plant Pathol.* **1** : 20-29.

Singh, R.S. & Sitaramaiah, K.S. (1973). Control of plant parasitic nematodes and organic amendments of soil. *Res. Bull. No. 6, G.B. Pant University of Agric. & Technol.* Pantnagar, p. 289.

Sivakumar, C.V.; Palaniswamy, S. & Naganathan, T.V. (1976). Persistence of nematicidal activity in seed treatment of okra with carbofuran and aldicarb sulphone, *Indian J. Nematol* **6** : 106-108.

Skarvilovich, T.S. (1959). On the structure of the systematics of the nematode order Tylenchida. *Acta Parasitology Pol.* **7** : 117-132.

Stephan, Z.A.; Al-Mammoovry, I.K. & Antoon, B.G. (1988). The efficacy of nematicides, solar heating and the fungus, *Paeciliomyces lilacinus* in controlling root-knot nematode, *Meloidogyne javanica* in Iraq. *Iraq J. Agric. Sci. Zanco* **6** : 69-76.

Stephan, Z.A.; Michbas, A.H. & Shakir, I. (1989). Effect of organic amendments, nematicides and solar heating on root-knot nematodes infecting eggplant. *Intl. Nematol. Network Newsl.* **6** : 34-35.

Tanda, A.S. & Atwal, A.S. (1989). Effect of sesame intercropping against the root-knot nematode (*Meloidogyne incognita*) in okra. *Nematologica* **34** : 484-492.

Triantaphyllou, A.C. (1985). Cytogenetics, cytotaxonomy and phylogeny of root-knot nematodes. *In : An Advanced Treatise on Meloidogyne* Vol. I. (Eds.-J.N. Sasser & C.C. Carter). IMP, NC State Univ. pp. 113-126.

Towson, A.J. & Apt, W.J. (1983). Effect of soil water potential on survival of *Meloidogyne javanica* in fallow soil. *J. Nematol.* **15** : 110-114.

Tyler, J. (1933). Reproduction without males in aseptic root cultures of the root-knot nematode. *Hilgardia* **7** : 773-788.

Verma, M.K. & Gupta, D.C. (1987). Effect of NPK fertilizers alone and in combinations with aldicarb and carbofuran on okra infected with *Meloidogyne incognita. Indian J. Agric. Sci.* **57** : 609-611.

Wouts, W.M. (1972). A revision of the family Heteroderidae (Nematoda: Tylenchoidea). *Nematologica* **18** : 439-446.

Wright, D.J.; Birtle, A.J.; Corps, A.E. & Dybas, R.A. (1983). Efficacy of avermectins against a plant parasitic nematode. *Ann. Appl. Biol.* **103** : 465-470.

Chapter 22

Control of Cyst Nematodes

Gopal Swarup*

Division of Nematology, Indian Agricultural Research Institute, New Delhi-110 012, India.

THE cyst-forming nematodes constitute a highly specialised group of plant-parasitic nematodes. Economically, these rate as one of the most important plant-parasitic nematodes all over the world. Till 1950's, these were mainly recognised in the temperate regions but subsequently, with the availability of trained nematologists in other parts of the world, one or other species of these nematodes were recorded parasitising economically important crop plants in the tropical and sub-tropical regions. Agriculture in the tropics is quite diversified as well as complex with a range of cropping patterns as well as soil types. Such conditions coupled with moisture availability, permit a very favourable environment for the plant-parasitic nematodes including the cyst formers. Within the past 40 years, about 77 cyst-forming nematodes have been identified on various crops all over the world, and of these twenty or more species are recorded from tropical and sub-tropical regions only. Of the 19 cyst-forming nematodes recorded from India, 8 are new species. The earlier concept, therefore, that the cyst-forming nematodes were of importance in temperate regions only (for instance the earlier records of *Globodera* species and *Heterodera schachtii*) does not hold true any more with the present knowledge about their distribution. It is true that some of these (species) are more recorded with practically no other information available on their biology or host range. Perhaps, with more awareness than hitherto expressed in nematological research by the plant protection scientists, these obscure species may yet prove to be of economic value.

Typically, the cyst-forming nematodes are characterised by an expression of marked sexual dimorphism in the individuals. The female is sedentary and swollen while the male is vermiform. A great majority of the species are amphimictic but there are a few parthenogenetic species also. With maturity of the crops or because of adverse conditions, the female cuticle turns into hard walled brown cysts. Usually, the eggs are laid within the swollen body of the female (before turning itself into a cyst) but in some cases, the eggs may be laid also in a gelatinous matrix exuded

* Present address : 155C, DDA, Flats-MIG, Rajouri Gardens, New Delhi-110 027, India

at the posterior end of the female body. The eggs, and subsequently the hatched larvae, remain inside the cysts till suitable environmental conditions are available for emergence of larvae and this also protects them from the vagaries of the weather. Because of these features in the cyst-forming nematodes, any approach for management of the populations, under field conditions, requires a serious understanding of the biology and host range of the concerned species. An attempt is made here, therefore, to give an overall appraisal of the problem and the progress achieved with different practices followed in various regions of the world with a particular emphasis for Indian conditions.

The importance of cyst nematodes in agriculture was recognised as early as 1881 when the European sugar industry was threatened by infestation of sugarbeet fields by *Heterodera schachtii*. The impact was of such magnitude that for the first time a soil fumigant (carbon disulfide) was used for control of plant disease. However, even at this time, use of chemicals for control of plant diseases, caused by nematodes, was not considered an economically feasible proposition. In India, nematological research received a boost with the discovery of association of *H. avenae* with the 'molya' disease of wheat and barley in north-western regions of the country and subsequently the discovery of presence of the dreaded golden cyst nematode in the Nilgiri Hills of South India (Vasudeva, 1958; Jones, 1961). The latter nematode is also responsible for a global awareness of plant-parasitic nematodes as one of the most important limiting factors in crop production.

Although about 15 species (mainly belonging to the genera *Heterodera* and *Globodera*) are of economic importance, investigations on the quantitative estimation of field losses are very limited. Some estimates interpolated from experiments are presented in Table 1. In the United States the estimated crop loss, due to nematodes, is about 1.59 x 10^9 dollars (Feldmesser, 1971). Some similar esti-

Table 1: Cyst nematode problems of high economic importance.

Nematode	Year of first discovery	Distribution	Crops attacked	Economic damage (%)
Potato cyst nematodes (*Globodera rostochiensis, G. pallida*)	1988	48 potato growing countries	potato, tomato, eggplant	9-10
Cereal root eelworm (*Heterodera avenae*)	1874	31 wheat growing countries	wheat, barley, oat, rye, maize	73-89
Sugarbeet cyst nematode (*Heterodera schachtii*)	1859	40 countries	218 species belonging to 23 plant families (including sugarbeet)	10-25

mates have been put forward by others too. However, these estimates, by and large, are the results of a consensus of experts which may or may not have been supported by experimental evidences. This deficiency is being gradually rectified with the development of disease loss models which may incorporate synergistic interactions between a variety of pathogens and the crop. It is important to understand this aspect since the soil contains a wide variety of nematode species. In fields, with similar soil types and climatic conditions, the type of crop grown and intensity of culture are the main factors for nematode species and population dominance. It would be desirable to understand the quantitative relationship between nematode population density and symptoms (disease) and the potential of nematode population increase before formulating management schedules.

CONTROL STRATEGIES

Before the main topic of control of the cyst nematodes is elaborated, it would be useful to understand that, in general, control of nematodes and the damage to plants that may be caused by them in the process of parasitisation, can be achieved by either preventing the entry of the larvae in plant roots or by suppressing their population in soil or by a combination of both the aspects. Total eradication or even control of the nematode populations is very difficult proposition as well as expensive. Therefore, the objective has to be modified to management of the nematode populations so that economic returns are possible under a specified cropping and environmental conditions. Another important factor to be taken into consideration is the economic value of the crop before formulating the type of the management schedule.

As is generally true for all the plant-parasitic nematodes, management of the cyst-forming nematodes can be based also on the following facts : (a) Nematode damage is a function of nematode density and therefore, nematode management can be achieved by reducing the population below the damage threshold level, (b) The nematode damage could be mitigated, to some extent, by providing a clean environment, initially, for establishment of the plants in fields, (c) Exclusion or regulatory control to prevent introduction of cyst nematode species in a new territory.

Taking the above facts into consideration, the control of the cyst nematode populations is being discussed under the following headings : (a) cultural, (b) plant resistance, (c) chemical, (d) biological and (e) exclusion through regulatory measures.

CULTURAL METHODS

With the advent of modern high intensity cultivation, as characterised by monoculture and intensive cultivation, there has been a gradual accentuation of the nematode problems all over the cultivated areas. Although the importance of cultural practices like crop rotation, deep summer ploughing, fallowing, etc. in nematode management are well recognised, their successful implementation requires a thorough knowledge of specific and racial composition of the nematode community in the area, biology, host range, population dynamics, rate of decline in the absence of host or of increase under a host, host status of various crops, local agroclimatic conditions and economics. The efficacy of most cultural practices in nematode control vary with soil type, topography, climate and target species involved. The adaptability of any type of cultural control would also depend on the economics of land holdings as well as market conditions.

Summer ploughing

In the tropics, there is an abundance of sunshine and is a particularly severe during the summer months (May-June under Indian conditions). Two to three deep summer ploughings, at intervals of 15 days, bring down the *Heterodera avenae* population reasonably below the damage threshold level (Mathur *et al.*, 1987). However, in the present trend of high density cropping system, this practice is difficult to follow unless the farmer is prepared to keep his land idle for 2-3 months. Economically, it is quite feasible, even for the marginal farmers.

A slight variation of the principle has been suggested under the process known as solarisation. Exposure of moderately moist soil to solar heat has been recorded as more effective than the dry heat. It has been observed that population of *H. cajani* is considerably reduced if the infested field is covered with a polyethylene sheet for 58 days (Sharma & Nene, 1985). However, the cost is prohibitive and restricts its adoption to high value crops only. It is also not applicable for humid and temperate conditions.

Crop rotation

This is an ageold practice but still practicable as well as profitable under certain set of conditions. However, before adoption, host range of the species involved has to be fully documented. Cyst nematode species like *H. avenae, H. zeae, H. sorghi* and *H.oryzicola* are mainly confined to graminaceous plants; *Globodera rostochiensis* and *G. pallida* mainly to few solanaceous crops while *H. cajani* has a large host range. The difficulty arises in finding a suitable non-host crop which is not only acceptable to the farmer but is also economical and viable agronomically. Most of the species, particularly in the tropics and the sub-tropics suffer high annual rate of decline in absence of host (in comparison to the temperate regions). Therefore, relatively shorter intervals of rotation between the host crops may suffice under the tropical conditions. Under Indian conditions, a 1-3 years rotation (depending on the population density) with carrot, fenugreek, onion, etc. is normally effective in bringing down the nematode population (Gill & Swarup, 1971). Against *Globodera* species on potato, a 4-year rotation is a recommended practice in the temperate regions (Anon., 1977). The trend of continuous cropping pattern, however, precludes easy adoption of this practice.

Time of planting

In Europe, early plantings of spring sown crops have been observed to escape early attack of root-knot as well as cyst nematodes (Jones, 1956; Milne, 1972). Under Indian conditions, it is possible that October sown wheat or barley crop may escape much of the infection since the larvae are generally released in large numbers in soil from November onwards. Unfortunately, the timely sowing period for the presently accepted wheat cultivars is from middle November to late December when the larval population in soil is abundantly available. Similarly, *H. zeae* is not active during December to June (Srivastava & Sethi, 1986) and therefore, maize varieties suitable for growing, either before December or June may help escape from nematode damage.

Application of nitrogenous fertilizers

It is generally accepted that application of fertilizers including organic amendments to soil, directly or indirectly influence the nematode population as well as plant growth. Under such a situation, either the nutritional status of the soil is improved which boosts up the plant growth or the microbial flora (bio-control agents) of the soil is modified which may suppress the nematode population or there is an abundance of fibrous root system which may provide freely available sites for the nematode feeding and thus cause an increase in the population. Under Indian conditions, although application of nitrogenous fertilizer (urea) gives a boost to crop (wheat) growth, the nematode (*H. avenae*) population also goes up (Mukhopadhyay *et al.*, 1972). Similar effects have been reported for cereals in Europe (Gair *et al.*, 1969). Time of application of nitrogenous fertilizers has been considered also of importance. It has been suggested that application at seeding time gives a better effect on plant growth than when applied 6 weeks after seeding (McClelland & Griffiths, 1972). However, so far, there has not been any startling response which may suggest economic feasibility in adoption of this approach.

PLANT RESISTANCE

This is the most effective as well as economical method of management for any pest problem. However, so far as the cyst nematodes are concerned there has not been much headway in developing an active programme of breeding for resistant varieties. In the case of *H. avenae*, the first source of resistance in wheat was recorded in the cultivar, Loros (Nielsen, 1966) which was also found to be resistant to the Australian population (O'Brien & Fisher, 1974). In the recent years, much success has been reported from Australia where two resistant wheat cultivars, specifically bred for resistance, Katyil and Cocamba, have been released and which have been accepted commercially. These are the world's first wheat cultivars which have been bred specifically against *H. avenae* (Brown & Young, 1982; Archer *et al.*, 1984). In India, although some resistant sources in wheat have been identified, their reactions have not been very consistent in the different regions of the country. As against this, a marked success has been achieved with barley. Two varieties, Rajkiran and C 164, have been bred and identified as resistant to the cyst nematode (*H. avenae*) and these varieties have been accepted commercially also. The nematologists and the breeders in the state of Rajasthan in India have taken the lead in developing an organised breeding programme. In comparison to wheat there has been greater attention to the potato crop and the golden nematode problem. There is an active breeding programme in Europe and a number of resistant sources have been identified and the resistance incorporated in commercially acceptable cultivars. Resistant cultivars like CP 1422, 1781 and 1806 have been found to be quite effective in bringing down the soil population of the nematode. Incorporation of a resistant variety alone in a 4-year crop rotation programme against the potato cyst nematode can result in 67-68 per cent yield increase (Zawislak *et al.*, 1981). In India too, a limited breeding programme is underway against this nematode and an acceptable cultivar Kufri Swarna has been released. Several attempts have been made and are continuing also in identifying resistant sources against other cyst species but so far, there has not been much success except to a limited extent with *H. glycines* on soybean.

Although use of resistant varieties in the cropping programme is a very promising approach, there are inherent dangers in the continued use of resistant cultivars because of the existence of pathotypes in most of the cyst-forming species. Overcropping with resistant cultivars can lead to pathotype selection, as has occurred with both the potato and soybean cyst nematodes (Brown & Kerry, 1987). In England, Howard & Cotten (1978) have recorded increasing populations of *Globodera pallida* when *G. rostochiensis*-resistant cultivar, Maris Piper was grown continuously. The biology of the nematode has to be understood properly for maximum benefit from the use of resistant varieties.

CHEMICAL CONTROL

Although chemicals are the most effective in bringing down the soil nematode populations and are also the most effective method of highlighting the crop losses caused by the nematodes, their use is highly uneconomical for low value crops, apart from the deterrent factors of toxicity and environmental pollution. Nevertheless, their use can not be totally dispensed with and if used with caution, they can provide an effective management of the nematodes. The currently available nematicides can be put in 4 groups : halogenated aliphatic hydrocarbons, methyl isothiocyanate precursor compounds, organophosphates and carbamates. Initially, the fumigants have spectacular success in demonstrating the losses due to nematodes but because of the cumbersomeness of their application, hazards in handling and often cost prohibitive, these were soon replaced with the chemicals of the other groups. In experimental trials in India, use of DD resulted in 4 and 16 times yield increases in wheat and barley respectively from *H. avenae* infested fields (Swarup *et al.*, 1976). However, the cost of application was quite beyond the reach of the farmers. With the discovery of nematicidal properties in organophosphate and carbamate compounds, the fumigants have been gradually eased out. These compounds are granular in nature and are systemic in action. Therefore, these can be handled with ease and application quantities are quite low. Oximecarbamates like aldicarb, carbofuran, oxamyl, methomyl and organophosphates like fensulfothion, thionazin, diazinon, fenamiphos possess great potential as nematicides. The efficacy of these chemicals in the management of the cyst nematodes is dependent on a number of factors like property of the soil, age of the plant, distribution of the root system in soil, and the nematode species involved, etc.

Under a Indo-German collaborative programme, the golden nematode infested potato areas of Nilgiris (South India) were treated with dasanit which found to be quite effective in reducing the nematode population. Some other chemicals like mocap, aldicarb, carbofuran and fensulfothion have been reported also to be quite effective against the golden cyst nematode of potato under field conditions (Gill *et al.*, 1978; Raja *et al.*, 1978). Under the All India Co-ordinated Wheat Improvement Project, aldicarb as well as carbofuran proved to be the most effective chemicals against *H. avenae*, with a cost-benefit ratio as 2-6:1. With a general awareness now about the environmental pollution, the emphasis has now shifted towards reducing the amount of chemicals in field applications. Therefore, the approach is now towards using localised treatments instead of blanket treatment. For example, wheat seed treatment with oxamyl at 360 g or carbofuran at 630 g/100 kg seed has been found to be quite effective against *H. avenae* in Australia (Brown, 1984). Similarly, calcium hypochlorite solution, containing 9 per cent available chlorine, as a seed potato treatment, has been reported to be effective against *G. rostochiensis* (Manoharan *et al.*, 1978).

The other localised approach treatment is to limit the use of chemicals in seed furrows or rows. It can be done either by placing the chemicals alongwith seed in furrows or the chemicals are applied only to the bands or soil ridges. Aldicarb, oxamyl, carbofuran, fenamiphos and ethoprophos have proved to be quite effective against *G. rostochiensis* when these are applied to 10 cm wide bands over the seed furrows (Brodie, 1983).

Generally, the systemic chemicals, presently in vogue, are translocated from root to shoot. A few have been now demonstrated to exhibit basipetal translocation (Nelmes *et al.*, 1973). According to Srivastava & Sethi (1985), fenamiphos and oxamyl have also some basipetal translocation since foliar application of these chemicals on maize seedlings, infested with *H. zeae*, caused a reduction in the nematode population. Similar basipetal translocation has been recorded with aldicarb sulfone and also with an experimental eximecarbamate, UC 54229 (Meher *et al.*, 1984, 1985).

BIOLOGICAL CONTROL

Manipulation of biotic agents present in soil or introduction of such agents in soil, for reduction of nematode populations, has been one of the most important research areas engaging the attentions of the plant protection scientists all over the world. Fungi have been the foremost amongst such agents. In a series of field experiments, Duddington and his associates in U.K., employed some predacious fungi against *H. avenae* and demonstrated marginal efficiency in terms of increase in wheat yield and decrease in the nematode population. During the last 15 years, a major attention has been given to the parasites which attacked the nematode cysts and the contents. Kerry and his associates (1982a) observed that in U.K. despite the fact that cereal crops were grown continuously, year after year in the same field, *H. avenae* population did not exhibit any marked increase. A Entomophthora like fungus was later found to be infesting the cysts. The fungus, *Nematophthora gynophila*, was found to parasitise the cysts of *H. avenae* and high level of parasitism was associated with low rates of nematode multiplication (Kerry *et al.*, 1982b). Sharma & Swarup (1988) have recorded a number of fungi and other organisms parasitising the cysts. Amongst these, the most promising are : *Pasteuria penetrans, Glomus fasciculatum, Catenaria auxiliaris, C. vermicola* and *Allomyces anomalus. P. penetrans*, used as a seed treatment, has been recorded to be effective in reducing the cyst populations (Bhattacharya & Swarup, 1988). However, *P. penetrans* is an obligate parasite and there is a need to standardise the procedures for its multiplication and application for use under field conditions. Nevertheless, the concept of nematode population management by exploitation of 'microbial pesticides' appears to be within the realm of practical possibility.

EXCLUSION THROUGH REGULATORY MEASURES

With increasing trade and rapid exchange of germplasm for crop improvement programmes, all over the world and especially in the developing countries, the risk of introductions of new cyst nematode or even new pathotypes of the already existing species, has magnified greatly. For example, many dreaded cyst species like *H. schachtii, H. glycines, H. humuli, H. goettingiana*, etc. are not known to be present in India but some of these have been already intercepted in the introductions of the germplasm materials. Therefore, following the adage that prevention is better than cure, the regulatory control measures assume a great significance. Regulatory control, theoretically, is legal enforcement of measures planned to prevent the introduction of pests in a country or if present on a limited scale, from further spreading and multiplying. Thus, the three important facets of regulatory control are: prevention of introduction; prevention of spread; and prevention of multiplication. Apart from government legislations for restricting imports of plant materials carrying or suspected to be carrying destructive pests and diseases, a social awareness is also necessary amongst the farmers for achieving the maximum benefits of this important aspects of management of plant health problems. As is true of the developed countries, the Government of India has also legislated an act, known as Destructive Insects Pests Act which regulates the entry or movement of plant material into the country. Besides the central legislation, domestic regulatory measures are also possible as has been done in the case of golden cyst nematode of potato, restricting the movement of potato seed material from the infested fields of Tamil Nadu to other areas of the country.

In spite of the soundness of legislative actions, there are several other possible ways of spread of the nematode populations from one field to another. Adhering soil particles on farm implements or human feet or boots are some of the possible ways through which the nematode cysts can be, inadvertently, spread from one field to another. Chitwood (1951) put forward this point forcefully when he collected 32,000 cysts from tractor wheels which had been used in an infested field.

CONCLUSION

One of the most important considerations for obtaining satisfactory crop yield is to improve the health status of the plants. This is true for all types of plant problems and not necessarily confined to the problem of cyst nematodes only. Various types of inorganic and organic fertilizers play a vital role in the overall health of plants but again these have to be used judiciously for maximum benefit. Nitrogenous fertilizers improve the plant growth but at the same time the soil nematodes find an abundance of fibrous root system on which to feed and multiply. Green manuring as a nitrogen source has been advocated wherever feasible, since it not only improves the plant health and tolerance to diseases, but also suppresses nematode populations. Similarly, use of non-edibile oil-cakes like *neem*, castor, Karanj, *mahua*, etc. have been also demonstrated to be good for plant growth as well as harmful to nematode populations.

In the preceding paragraphs, an attempt has been made to enumerate the various measures that can be employed or have been employed in the management of nematode population especially the cyst nematode. Besides the cyst problems listed, it is quite possible that there may be several more cyst species which have not been recognised as yet because of the limited surveys. Further, the suspected or unrealised damage potential of many cyst species is still obscure and, therefore, cannot be evaluated with respect to management of their populations because of lack of knowledge about their biology. However, the fact remains that the cyst nematodes are one of the most important contributory factors in crop yield reductions. No single measure can provide absolute control of the populations. Adoption of an integrated approach (rotation, green manuring, resistant/tolerant varieties, nematicides, introduction of organisms attacking or parasitising cysts, etc.) can be used with success, in combinations of two or more, which will bring down the cyst populations to a level which will not be detrimental to plant growth.

REFERENCES

Anon. (1977). Cyst eelworms on potato. *Adv. Leaft. Minist. Agric. Fish. Fd. 284*. p. 8.

Archer, M.J.; Griffiths, J.B. & Pye, D.L. (1984). Cocamba — a new wheat variety resistant to cereal cyst nematode. *Agnote, Department of Agriculture, Victoria*. p. 2.

Bhattacharya, D. & Swarup, G. (1988). *Pasteuria penetrans*, a pathogen of the genus *Heterodera*, its effect on nematode biology and control. *Indian J. Nematol*, **18** : 61-70.

Brodie, B.B. (1983). Control of *Globodera rostochiensis* in relation to method of applying nematicides. *J. Nematol*. **15** : 491.

Brown, R.H. (1984). Ecology and control of cereal cyst nematode (*Heterodera avenae*) in southern Australia. *J. Nematol*. **16** : 216-222.

Brown, R.H. & Kerry, B.R., Editors (1987). *Principles and Practice of Nematode Control in Crops*. Academic Press, Australia. p. 447.

Brown, R.H. & Young, R.M. (1982). Katyil, a wheat resistant to cereal cyst nematode. *Agnote, Department of Agriculture, Victoria*. p. 2.

Chitwood, B.G. (1951). The golden nematode of potatoes. *U.S. Dept. Agric. Circ. 875*.

Feldmesser, H. (1971). Estimated crop losses due to plant parasitic nematodes in the United States. Special Publication No. 1, Society of Nematologists, Beltsville.

Gair, R.; Mathias, P.L. & Harvey, P.N. (1969). Studies of cereal nematode populations and cereal yields under continuous intensive culture. *Ann. appl. Biol*. **63** : 503-512.

Gill, J.S.; Krishnapanda, N. & Singh, N. (1978). Efficacy of some nematicides for the control of potato cyst forming nematodes. *J. Ind. Bot. Soc*. **57** : 55-56.

Gill, J.S. & Swarup, G. (1971). On the host range of the cereal cyst nematode, *Heterodera avenae* Woll. 1924, the causal organisms of 'Molya' disease of wheat and barley in Rajasthan, India. *Indian J. Nematol*. **1**: 63-67.

Howard, H.W. & Cotten, J. (1978). Nematode-resistant crop plants. *In : Plant Nematology* (Ed.— J.F. Southey). Her Majesty Stationery Office, London. pp. 313-325.

Jones, F.G.W. (1956). Soil populations of beet eelworm (*Heterodera schachtii* Schm.) in relation to cropping. II. Microplot and field plot results. *Ann. appl. Biol.* **44** : 25-26.

Jones, F.G.W. (1961). The potato root eelworm, *Heterodera rostochiensis* in India. *Curr. Sci.* **30**: 187.

Kerry, B.R.; Crump., D.H. & Mullen. L.A. (1982a). Studies on the cereal cyst nematode, *Heterodera avenae* under continuous cereals, 1974-1978. I. Plant growth and nematode multiplication. *Ann. appl. Biol.* **100** : 477-487.

Kerry, B.R.; Crump, D.H. & Mullen, L.A. (1982b). Studies of the cereal cyst nematode, *Heterodera avenae* under continuous cereals, 1975-1978. II. Fungal parasitism of nematode females and eggs. *Ann. appl. Biol.* **100** : 489-499.

Manoharan, A.S.; Vijayaraghavan & Samboorharaman, S. (1978). Effect of treatment of seed tubers of potato to make potato cyst nematodes nonviable, a new method of disinfestation. *J. Ind. Bot. Soc.* **57**: 57.

Mathur, B.N.; Handa, D.K. & Swarup, G. (1987). Effect of deep summer ploughing on the cereal cyst nematode, *Heterodera avenae* and yield of wheat in Rajasthan, India. *Indian J. Nematol.* **17** : 292-295.

McClelland, V.F. & Griffiths, J.B. (1972). Types and Times of application of nitrogen for wheat. *J. Aust. Inst. Agric. Sci.* **38** : 219-221.

Meher, H.C.; Agnihotri, N.P. & Sethi, C.L. (1984). Persistence of aldicarb sulfone residue in cowpea and soil following foliar application and its nematicidal activity. *Indian J. Nematol.* **14** : 160-165.

Meher, H.C.; Sethi, C.L. & Agnihotri, N.P. (1985). Persistence and basipetal translocation of nematicidal activity of UC 54229, a carbofuran analog in pea. *Indian J. Nematol.* **15** : 71-74.

Milne, O.L. (1972). Nematodes of tobacco. *In : Economic Nematology* (Ed.— J.M. Webster). Academic Press, London. pp. 159-186.

Mukhopadhyay, M.C.; Dalal, M.R.; Saran, S. & Kharub, S.S. (1972). Studies on the 'Molya' disease of wheat and barley. *Indian J. Nematol.* **2** : 11-20.

Nelmes, A.J.; Trudgill, D.L. & Corbett, D.C.M. (1973). Chemotherapy in the study of plant parasitic nematodes. *In : Chemotherapeutic Agents in Study of Parasites* (Eds. — A.E.R. Taylor & R. Muller). Blackwell Scientific Publication, Oxford, U.K. pp. 95-112.

Nielsen, C.H. (1966). Untersuchungen Uber die Vererbung der Resistenz gegen den Getreide — nematoden (*Heterodera avenae*) beim weizen. *Nematologica* **12** : 575-578.

O'Brien, P.C. & Fisher, J.M. (1974). Resistance within wheat, barley and oat cultivars of *Heterodera allenae* in South Australia. Aust. *J. Exp. Agric. Anim. Husb.* **14** : 399-404.

Raja, J.S.; Vijayaraghavan & Soundararajan, S.K. (1978). Control of potato cyst nematodes with aldicarb 10G in the Nilgiris. *J. Ind. Bot. Soc.* **57** : 56.

Sharma, R. & Swarup, G. (1988). *Pathology of cyst nematodes*. Malhotra Publishing House, New Delhi. p. 88.

Sharma, S.B. & Nene, Y.L. (1985). Effect of presowing solarization of plant parasitic nematodes in chickpea and pigeonpea fields. (Abstr.). *Indian J. Nematol.* **12** : 277-278.

Srivastava, A.N. & Sethi, C.L. (1985). *Annual report*. Division of Nematology, IARI, New Delhi for year 1985. pp. 9-14.

Srivastava, A.N. & Sethi, C.L. (1986). Seasonal population fluctuations of *Heterodera zeae, Rotylenchulus reniformis* and *Tylenchorhynchus vulgaris* in maize-cowpea wheat rotation. *Indian J. Nematol.* **16** : 16-18.

Swarup, G.; Mathur, R.L.; Seshadri, A.R.; Raski, D.J. & Mathur, B.N. (1976). Response of wheat and barley to soil fumigation by DD and DBCP against 'Molya' disease caused by *Heterodera avenae. Indian J. Nematol.* **6** : 150-155.

Vasudeva, R.G. (1958). *Scientific report*. Division of Mycology and Plant Pathology, IARI, New Delhi for the year ending 30th June, 1958.

Zawislak, K.; Sowa, S. & Gronowicz, Z. (1981). Wplyz matwikoodpornych odmiar Ziemniaka ha zenikanie w glabie populacji matwika ziemniaczanego. (*Globodera rostochiensis*). *Zeszyty Naukowe Akademi Rolniczo - Technicznoj w olsztynie, Rolnictwo, No. 31* : 49-58.

Chapter 23

Control Strategies for Root-Lesion Nematodes

R.K. Walia & R.S. Kanwar
Department of Nematology, Haryana Agricultural University, Hisar-125 004, India.

THE root-lesion nematodes (*Pratylenchus* spp.) rank second among the ten most important genera of plant-parasitic nematodes on a worldwide basis. This order of importance of various genera is fairly representative for most regions of the world (Sasser, 1987). About 77 species of *Pratylenchus* are known to be associated with plants world over. However, only a few have been studied to explore their role in crop production. In most cases, they have been implicated as limiting factors in crop yields but their recognition as key pests has relegated them to second position, next only to root-knot nematodes (*Meloidogyne* spp.)

Incidentally, the first nematode species to be proved as a plant pathogen to the satisfaction of Koch's postulates was *P. penetrans* — the causal organism of 'peach replant' problem in Canada (Mountain & Patrick, 1959). Various other species have since been recorded to cause economic damage to many crops, and the information has been reviewed by Thames (1982) and Walia (1986).

Development of control strategies for any particular nematode species warrants judicious extrapolation of various control options available to fulfil the specific objectives. The objectives could vary from eradication of the target species — if it is newly introduced, to management — in complex cases. An accurate and comprehensive knowledge of the nematode identification, biology, ecology, and host-parasite relationship is the backbone for the development of management skills. It is in this backdrop that a brief account of these aspects is presented in the beginning, before discussing the control strategies.

SYSTEMATICS

The first species of *Pratylenchus* was described by de Man (1880) under the name of *Tylenchus pratensis*. Filipjev erected the genus *Pratylenchus* in 1936 to include nematodes that had been placed in the genera *Anguillulina* and

Tylenchus. Sher & Allen (1953) revised the genus and recognised 10 species. Loof (1978) reviewed the taxonomic status of the genus and recognised 29 valid species. Several new species were added thereafter till Siddiqi (1986) again reviewed the genus and recognised 68 valid species. Since then, 9 more new species have been described.

Accurate identification is critical to proper management decisions. Though root-lesion nematodes are easy to recognize, but the group apparently is difficult as such to delineate into species. Studies on the intra-specific variations among populations from different geographical areas could perhaps lead to realignment of species. While the use of modern tools like SEM are likely to resolve the morphological details, yet the presence of sibling species, and the apparent effects of environment on the expression of taxonomic characters in the genus *Pratylenchus* require that investigations depending on accurate identifications should be preceded by careful morphometric studies of populations of the area.

DISTRIBUTION

Several species of root-lesion nematodes namely, *P. brachyurus, P. coffeae, P. crenatus, P. neglectus, P. penetrans, P. scribneri, P. thornei* and *P. zeae* are worldwide in distribution. The distribution of root-lesion nematodes is facilitated due to the endoparasitic mode of life, besides other reasons. Upon introduction to new geographical areas, the prevailing ecological factors (temperature, etc.) play important role in the establishment of species, which may lead to restricted distribution (Loof, 1978). Consequently, certain species, viz. *P. brachyurus, P. coffeae* and *P. zeae* are prevalent mainly in the tropical areas, while others like, *P. penetrans, P. pratensis* and *P. neglectus* are considered temperate zone species (Corbett, 1969; Loof 1978).

ECONOMICALLY IMPORTANT SPECIES AND THEIR HOSTS

Most species have wide host ranges and are frequently found associated with a number of plant hosts. However, their pathogenic potential is restricted to selected few crops. Visible damage in fields, followed by proof of their pathogenicity is not consistent with the hosts for a particular species. The pathogenic potential is undoubtedly, influenced by several edaphic and biological factors. A list of economically important species which have been observed/proved to be pathogenic to their listed hosts is given below. It can be inferred from the list that at least 15 species can be considered economically important and designated as 'limiting factors'. In several such cases, the nematode species is the 'key pest' and warrants the adoption of control measures for economic returns.

Pratylenchus spp. found pathogenic/associated with plant damage :

(i) *P. alleni* - Soybean.

(ii) *P. brachyurus* - Cotton, *Eucalyptus* sp., long leaf pine, maize, peach, peanut, pineapple, potato, soybean, sweet orange, tobacco, yam.

(iii) *P. coffeae* - Abaca, apple, cacao, citrus, coffee, peanut, plantain, temperate vegetables, yam.

(iv) *P. convallariae* - Lily of the valley.

(v) *P. crenatus* - Barley, maize, oats, rye.

(vi) *P. hexincisus* - Maize, soybean.

(vii) *P. indicus* - Rice.

(viii) *P. loosi* - Tea.

(ix) *P. neglectus* - Alfalfa, barley, maize, rye, wheat.

(x) *P. penetrans* - Alfalfa, apple, *Berberia juliana*, cherry, citrus, Easter lily, garden balsam, *Ilex* sp. leather leaf fern, maize, narcissus, peach, pear, peppermint, potato, raspberry, rose, soybean, strawberry, temperate vegetables, tomato, zinnia.

(xi) *P. pratensis* - Carrot, cherry, spinach, strawberry, tomato.

(xii) *P. scribneri* - Amaryllis, soybean, strawberry, white clover.

(xiii) *P. thornei* - Chickpea, maize, mint, wheat.

(xiv) P. vulnus - Apple, boxwood, cherry, *Citrus aurantium*, grape, peach, raspberry, rose, walnut.

(xv) *P. zeae* - Chickpea, maize, peach sugarcane.

NATURE OF DAMAGE

The nature of damage is similar for all species. Root-lesion nematodes are primarily cortical feeders, resulting in the cell death and breakdown. The symptom of attack first appears as a tiny, elongate, water-soaked spot, which soon turns brown to almost black. These lesions mainly appear on young feeder roots but may be found anywhere on the roots. The lesions enlarge, coalesce and ultimately girdle the entire root which appears constricted. This follows the secondary attack by fungi and bacteria and rotting sets in. The affected area may be sloughed off and the overall root system reduced. The amount of damage or the degree of pathogenicity is usually, but not always, related to the extent of discolouration or necrosis.

The extent of damage depends on the phenolic content of the cells, which are released as a result of interaction with nematode enzymes (Mountain, 1965; Pitcher *et al.*, 1960). The oxidation of these phenolic substances causes the characteristic browning of the lesion. The classical work of Mountain & Patrick (1959) has shown that *P. penetrans* hydrolyses amygdalin in peach roots and this reaction appears to be the main mechanism involved in necrosis. Differences in pathogenicity in different hosts have been attributed to variations in the phenolic substances in the affected tissues (Mountain, 1965; Pitcher *et al.*, 1960).

The above-ground symptoms resemble those due to malnutrition or water deficiency. In the initial stages, infestation can be noticed as tiny patches of stunted growth and yellowing of leaves. These patches increase year after year. Stunting becomes more pronounced as the growth advances. Leaves may turn yellowish brown and wilt during hot summer months. In severe cases the plants may be killed. The trees or shrubs are rarely killed, but they gradually become unthrifty and unproductive due to gradual decline.

BIOLOGY

Pratylenchus species are migratory endoparasites with all vermiform stages capable of infecting the host roots. Eggs are laid singly in the root cortical parenchyma, however, they reproduce and oviposit in soil also. Migration within the root may be inter- or intra-cellular. In the beginning of crop season, majority of the population is in the soil but as the growth advances, most of the nematodes invade the roots and multiply. Multiplication rates are low compared to sedentary endoparasites. On an average, a single female lays between 10-25 eggs. The reproduction may be sexual or parthenogenetic. At crop maturity, the lesion nematodes either leave the roots and come out into the soil; or may survive in the root stubbles mostly as adults or in egg stage.

The biology of lesion nematodes is influenced by a variety of edaphic factors. Temperature plays an important role in governing the life cycle duration. The information is presented in Table 1 for an overview. Several species of lesion nematodes are known to survive for long periods in the absence of host or unfavourable environmental conditions. Adults and 4th stage larvae are the main overwintering stages of *P. penetrans* (Kable & Mai, 1968; Miller, 1968), although many eggs are also capable of surviving winters in roots (Dunn, 1972). *P. vulnus* does not survive out of doors in cool temperate area. It is generally a pest of glasshouse crops and eggs are important overwintering form (Dickerson *et al.*, 1964). *P. coffeae* can survive in moist soil for 8 months in the absence of host (Colbran, 1954). *P. zeae* overwinters in dead roots of crabgrass, maize, cotton and to a lesser extent of tobacco in U.S.A. Nematodes are also capable of surviving the winter in soil without host (Graham, 1951).

CONTROL STRATEGIES

Identification of the key pest problem is the first and basic step towards pest management. Development of a management programme for a particular nematode species should be done in a systematic way with specific objectives in mind.

Table 1 : Effect of temperature on the biology of *Pratylenchus* species.

Nematode species	Plant host	Life cycle duration/ reproduction	Temp (°C)	Reference
P. alleni	Soybean	No reproduction below	27.5	Dickerson, 1979
	Tomato	Reproduced at	20.0	
	"	Optimum for reproduction	35.0	
P. brachyurus	Maize	14 weeks at	15.0	Olowe & Corbett, 1976
	"	3 weeks at	30.0	
	"	No egg laying at	5.0	
	Citrus jambhiri	Optimum for reproduction	32.0	Radewald *et al.*, 1971
	"	Population lowered at	35.0	
P. coffeae	*Citrus jambhiri*	Reproduced at	26.0-32.0	Radewal *et al.*, 1971
	"	Optimum for reproduction	29.53	
	"	Population lowered at	35.0	
P. penetrns	Alfalfa	30 days at	30.0	Mai *et al.*, 1977
	"	37 days at	25.0	
		92 days at	15.0	
	Maize	Maximum population at	24.0	Dickerson *et al.*, 1964
	Cryptamaria	86 days at	15.0	Mamiya, 1971
	"	42-44 days at	20.0	
	"	35 days at	24.0	
	"	30-31 days at	30.0	
	"	Not completed at	35.0	
	Pea	35 days at	28.0	Dickerson, 1961
	Potato	Maximum population at	16.0	Dickerson, *et al.*, 1964
P. scribneri	Soybean	No reproduction below	27.5	Dickerson, 1979
	Tomato	Reproduced at	20.0	
	"	Optimum for reproduction	35.0	
P. alleni	Soybean	Optimum for reproduction	30.0	Acosta & Malek, 1981
P. coffeae	Soybean	Optimum for reproduction	30.0	
P. brachyurus	Soybean	Optimum for reproduction	30.0	
P. neglectus		Reproduced at	35.0	
P. scribneri	"			
P. zeae	"			
P. penetrans	Soybean	Optimum for reproduction	25.0	
P. vulnus	"	Reproduced at	15.0	

A sound management programme envisages the judicious utilization and blending of available control options to formulate a long term, economically viable and environmentally sound approach. The utilization of various control components will vary according to the objective, geographical area, ecology, economics, crop value and other factors. While the aim of this chapter is to discuss certain control strategies for the management of root-lesion nematodes, an overview of the various control options available in respect of economically important species could form the basis for the development of a pest management approach. The same has been highlighted later in this chapter through citing a couple of examples by tracing the development of management programmes in their historical perspective.

CHEMICAL METHOD

Chemicals have been used to control nematodes and other soilborne problems for a long time. In the period up to First World War, carbon bisulphide, formaldehyde, cyanides and other chemicals were investigated but proved to be too expensive for general nematode control. After the war, however, surplus chloropicrin became available and proved efficacious for the control of soil-borne diseases. The discovery of DD in 1943 led to the development of various halogenated hydrocarbons such as EDB and DBCP for nematode control. Since the early 1960's systemic organophosphates and carbamates have been developed which are active against both nematodes and insects, but their action against nematodes is largely nematoxic rather than systemic. A major disadvantage of all nematicides is their inherent toxicity coupled with their potential to be environmentally hazardous, particularly if their residue or breakdown products reach unacceptable levels in the produce or in soil water.

Despite the disadvantages of using chemicals, nematicides are still the most effective methods of controlling nematodes including root-lesion nematodes in modern crop production. The details of the effective nematicides, their doses against important *Pratylenchus* species infecting various crops have been summarised in Tables 2 and 3.

NON-CHEMICAL METHODS

Resistance tolerance

Growing crop cultivars resistant/tolerant to nematodes provide an ideal method for maintaining nematode population densities below damaging levels. Its advantages over other methods are : it can completely prevent nematode reproduction, allow rotations to be shortened, does not produce toxic residues, no special applicators required, no additional cost involved. Resistant cultivars offer the best method

Table 2 : Chemical control of economically important *Pratylenchus* species.

Nematode species	Crop(s)	Nematicide & Dose	Results	Reference
P. brachyurus	Peanut	DBCP	Demonstrable effect	Bosewell, 1968
	Peanut	Aldicarb, Carbofuran, Ethoprop, Phenamiphos, Fensulfothion	Excellent control of *P. brachyurus* and other nematodes	Minton & Morgan, 1974
	Peanut	MBr 24.5 mg/l for 24 h MBr 44.6 or 50.9 mg/l for 24 h	Killed all nematodes in empty shells Failed to kill all nemas in whole pods & resulted in 15% reduction in germination of nuts	Minton & Gillenwater, 1973
	Rough lemon	Dowco 275	Significant increase in top weight	Fredrick & Tarjan, 1975
	Tobacco	DD, EDB, 1,3-D Carbofuran Ethoprop	Significant yield increases, most treatments resulted in increased leaf quality also	Miller *et al.* 1970
" +*P.zeae*	Pearl millet Sorghum -Sudan grass hybrids	Aldicarb Phenamiphos at 11.2 a.i./ha	Increased plant height and yield	Johnson & Burton, 1973
" +Stem borer *(Sesamia calamistis)*	Maize	Carbofuran 1.5/2, 5 3.5, 4.5 kg a.i./ha at planting and tesselling	Significant nematode & insect control at 2.5-4.5 kg/ha. 2.5 kg a.i./ha cost effective	Atu & Duru, 1990
P. coffeae	Citrus	Paste with 4.5-18 mg Phenamiphos applied to bark of trees	Controlled nemas in roots, greater length & wt. of aerial parts greater root wts.	Tarjan, 1972
	Negro Yam	Oxamyl 1200 or 2400 ppm for 30 min.	Most successful control	Coastes-Beckford, 1977
P.hexincisus	Maize	Carbofuran or 1,3-D	12.9-26.4% increase in yield	Norton & Hinz, 1976
	Maize	Aldicarb, Ethoprop OGA 12223 Carbofuran	Nematode population reduced in maize roots	Zirakparvar 1979

Nematode species	Crop(s)	Nematicide & Dose	Result	Reference
P. loosi	Tea	Phenamiphos (Nemacur-P 15% G) 7 g/plant	Suppressed infestation up to 60% plant growth also stimulated	Gnanapragasam *et al.*, 1987
P. mediterraneus	Wheat	Promet 666 (seed dressing formulation of Furathiocarb) at 5 & 15 ml/kg seeds. Also same insecticide with soil application formulation	Best nematode kill (75%) by soil application, 50% kill by high dose of seed dressing. Highest yield with high dose of seed dressing	Orion & Glazer, 1987
P. penetrans	Apple	1,3-D 50 g/acre	90% increase in growth of trees	Mai & Parker 1970
	Apple	Nematicides, preplant treatment	Girth of trees increased by 81%	Mai *et al.*, 1977
	Apple	Oxamyl spray 5 weeks after inoculations on seedlings	Excellent nematode control	Abawi & Mai, 1975
	Cherry	1,3-D 50 g/acre	Yields twice over untreated trees	Mai & Parker, 1970
	Easter lily	Phorate at planting EDB preplanting	Phorate more effective	Jensen, 1966
	Easter lily	Aldicarb, Methomyl, Carbofuran	Increased bulb wt.	Maggenti *et al.*, 1970
		Phorate	Suppressed yields	
	Maize	DD 32 gal./acre	Controlled nemas no increase in yield	Edminds *et al.*, 1967
	Narcissus	Chloropicrin and dichloropropenes	Preplant treatment to bulbs most effective	Apt & Gould, 1961
	Narcissus	1,3-D - 10 ppm	100% control	Youngston & Goring, 1970
	Ornamentals	Aldicarb Phenamiphos	Controlled nematode for 10 months	Krusberg, 1971
		Chloropicrin DBCP	Failed to control nematodes	
	Potato	Metham Sodium 3.1 or 45. a.i/m row, and in combination of 1,3-D (8.2 g a.i./m row) or Aldicarb (0.21 kg a.i./m row)	Decreased soil population, tuber yield increased from 27-46% in 1984, and 23-37% in 1985.	Olthof, 1989

Nematode species	Crop(s)	Nematicide & Dose	Result	Reference
	Potato	Oxamyl (1000-32000 μg/ml) in aqueous sol. in pots. Soaking cut/uncut tubers in aqueous solutions	Plant growth greater; at 1000 μg nema population in soil reduced by 20% and in roots by 35%; at 32000 μg-73-86% in soil, 86-97% in roots	Townshend & Olthof, 1988
	Potato	Aldicarb 1.12 or 2.24 kg/ha Oxamyl 2.24 kg/ha at planting	Higher dose of Aldicarb most effective yield increased by 12-88%	Kimpinski, 1986
	Rasp-berry	Di-Trapex 300 & 500 l/ha Dazome 300 & 500 kg/ha	Di-Trapex most efficient practically nema free soil in following year. Yield 54-162% higher. Doses should not exceed 400 kg/ha for field treatment.	Rabandel & Szczygiel, 1989
	Tobacco	Oxamyl dichloropropenes	Best control Excellent nematode control	Elliot *et al.*, 1977
"+*Meloidogyne hapla*	Alfalfa	Oxamyl 10 or 50 mg/ml in water, acetone or methanol; seed coat with 2% aqueous cellulose; seed soaking for 20 min.	Oxamyl content in seeds did not decline after 26 months storage. No effect on germination. Seedling growth at 50 mg increased 62% nodulation 267%, nemas/g root reduced by 73%	Townshend & Chiba, 1987
Pratylenchus spp. + other nemas	Maize	Aldicarb, Ethoprop, Fensulfothion Phenamiphos Terbuphos CGA 12223	Major portion of maize yield loss attributed to *Pratylenchus* spp.	Norton *et al.*, 1978
Pratylenchus spp.	Maize	Carbofuran 2.25-4.5 kg/ha	Increased yield and reduced nema population	Bergeson, 1978
P. thornei	Chickpea	Aldicarb 5-10 kg a.i./ha	Greatly suppressed nematode root invasion & increased yield, seed treatment not effective	Greco *et al.*, 1988
	Chickpea	Seed coating with Carbofuran (Furadan 40 FP) 0.75% &1.0% a.i. of Fensulfothion (Dasanit 10G) 0.25% a.i. seed wt.	Significantly better plant growth and reduced nematode population	Walia & Seshadri, 1985

Nematode species	Crop(s)	Nematicide & Dose	Result	Reference
	Wheat	MBr	78% increase in yield controlled nemas up to 30-90 cm depth	Doyle *et al.*, 1987
		Aldicarb 4 kg/ha	Nemas eliminated in upper 10 cm, yield increased	
		EDB 2.7-10.8 l/ha	No effect on nemas or yield	
		Terbufos 0.6 & 1.2 kg a.i./ha	Small reduction in nema population and small increase in yield	
P. vulnus	Rose	Phenamiphos 100 & 200 ppm-10 min. 50 ppm - 30 min.	100% control, no toxicity concentrations higher than 200 ppm toxic at long dips	Veld *et al*, 1988
P. zeae	Greengram	Seed coating with Carbofuran (Furadan 40 FP) 0.5 % a.i. seed wt.	Best in enhancing plant growth and reducing nema population	Walia & Seshadri, 1988
P. zeae +*P. brachyurus*	Maize	Carbofuran (200 g a.i./100 m) Aminofuracarb (Benfuracarb) (300 g a.i./100 m, Aldicarb (100 g a.i./100 m), EDB (20 cm^3/100 m), BAS 263 (150 g a.i./100 m)	All nematicides significantly reduced nematode population. Aldicarb gave best overall control	Mc Donand *et al.*, 1987

Table 3 : Commercially recommended nematicide treatments against *Pratylenchus* species on some important crops.

Crop	Nematicide treatment			
	Common name	Formulation	Dose (kg a.i./ ha)	Application method
Banana	Phenamiphos	G/EC	2.5-3.0	applied as spot treatment around plants preferably before irrigation or rains
	Aldicarb	G	1.5-2.2	
	Isazophos	G	2.0-3.0	
Coffee	Aldicarb	G	4.5 or 0.7-1.5 g a.i./ plant	applied every 6 months
Cotton	Terbufos	G	2.0	applied in 15-30 cm band 8-15 cm below seed on 100 cm rows

Crop	Nematicide treatment			
	Common name	Formulation	Dose (kg a.i./ ha)	Application method
Grapevine	Aldicarb	G	5-10	applied in 200-400 l of water incorporated to 5-10 cm for nursery use (Europe)
Maize	Phenamiphos	G/EC	1.36-2.2	applied broadcast or in row and incorporated to 5-10 cm
Ornamentals	Oxamyl	EC	0.6-1.2	applied as foliar treatment in 200 l water
			7.0-9.34	applied broadcast in 200 l water incorporated to 10-20 cm
Peanut	Carbofuran	G	22-5.5	applied in 30-40 cm bands and incorporated to 10-20 cm
Potato	Ethoprophos	G	3.36	applied and incorporated in 30 cm bands
	Oxamyl	EC	2.24-4.67	applied in furrow at planting in 200 l water
			4.67-9.34	applied broadcast in 200 l water incorporated to 10-15 cm
Sugarcane	Aldicarb	G	2.35-3.36	applied over newly planted cane and covered to a depth of 15 cm
Tomato Cucurbits Vegetables	Oxamyl	EC	2.3-4.67	applied broadcast or in bands incorporated to 10 cm

Information gathered from technical information supplied by manufacturers
G = Granules; EC = Emulsifiable concentration
(Source : Hague & Gowen, 1987 in part)

for protection of low value crops against pest and diseases. Even for high value plantation crops resistant cultivars are truly advantageous. Resistant cultivars are not yet available for many crop-nematode combinations and even where available, their use is beset with certain difficulties, i.e. suitability of resistant cultivar to local agronomic practices, and existence of resistance-breaking races or pathotypes.

Pratylenchus spp. are implicated in *Prunus* replant diseases and there are reports of differences in root stock susceptibility. Cherry replants on Mahaleb stocks grew well and *P. vulnus* and *P. penetrans* caused no root lesions while Mazzard and Stockton Morello roots were susceptible (Edgerton & Parker, 1958; Serr &

Day, 1949). Layne (1974) found clones of peach breeding stocks from Russia relatively resistant to *P. penetrans*, as assessed by nematode population densities. An Okinawa and Nemaguard derivative resistant to *Meloidogyne* was also tolerant of *P. penetrans*. Peach hybrids with *P. myrobalan* and almond, and some wild Prunus species were poor hosts of *P. vulnus* (Scotto La Massese, 1975). Differences have been identified between seedling populations of peach root stocks in their ability to support *P. penetrans* populations increase (Potter *et al.*, 1984). These authors stressed the need for careful experimentation and consideration of the effects of plant vigour and size on nematode reproduction. Other records of resistant/tolerance in various crops against species of *Pratylenchus* are given in Table 4.

Some plants are also known to produce phytoalexins which inhibit nematode activity. Limabeans produce coumesterol in response to infection by *P. scribneri*, and production of phaseolin by red kidney bean inoculated with *P. penetrans* has also been reported (Rich *et al.*, 1977).

Crop Rotation

Any study of rotations in relation to root-lesion nematodes should provide for careful determinations of the per cent composition by species of the lesion nematode component of the soil fauna. Many otherwise valuable rotation studies are not as useful as they might be because the lesion nematodes present were deter-

Table 4 : Resistance/tolerance in crops to *Pratylenchus* species.

Crop	Nematode species	Cultivar	Reaction	Reference
Citrus	*P. coffeae*	*Poncirus trifoliata* *Microcitrus* selections	supported lowest populations	O'Bannon & Esser, 1975
Peanut	*P. brachyurus*	Virginia Bunch 67, Ga 186-28	shell tissues less sensitive	Minton *et. al.* 1970
		Florigiant, Early Runner, Argentine, Starr	shell tussues more sensitive	
Peanut	*P. brachyurus*	PI 290606, PJ 295233	lower pod damage and less nematodes/g shell tissues	Smith *et al.*, 1978
		Starr, Spancross	more pod damage	
Potato	*P. penetrans*	Katahdin, Kenebec, Superior	tuber yield suppressed	Bernard & Laughlin, 1976

Crop	Nematode species	Cultivar	Reaction	Reference
		Russett Burbank	yield not suppressed, highest nematode count	
Soybean	*P. brachyurus*	Lee	susceptible	Endo, 1967
	P. zeae	Peaking		
Soybean	*P. brachyurus*	Lee 74, Forest, Coker 74-478	useful in nematode management programme	Schmitt, 1976
Soybean	*P. scribneri*	Clark 63, Woodsworth, Williams	very susceptible	Acosta *et al.*,1979
		Cluster, Dyer, Forest	resistant	
Soybean	*P. hexincisus*	Clark 63, Cumberland, Lee 74, Bedford, Williams, Woodward, BSR 301, S 1346, S 1578	final population higher than Pi	Zirakparvar, 1982
Tobacco	*P. brachyurus*	Hicks	medium root decay, nematode population high in greenhouse	Graham & Heggestad, 1959
		Bell 549 E	severe root decay, nematode population low in greenhouse, tolerant to decay in field	
Tobacco	*P. brachyurus*	Hicks,	heavily damage	Graham, 1965
		N 95, Fla 22	suffered some damage	
		PD 406	least damage, resistant	

mined only to the level of genus. In case in *Pratylenchus* control, there is a need for determining local non-hosts in most areas before rotations can be designed and tested. In general, use of small grains, maize or sorghum results in an increase in population of lesion nematodes.

The influence of crops on 4 species of *Pratylenchus* over a five-year period in rotations of 16 fields revealed that *P. hexincisus*, *P. penetrans* and *P. scribneri* increased in the fields of maize. *P. neglectus*, *P. penetrans* and *P. scribneri* increased under soybean, but only *P. scribneri* increased under wheat as the current crop (Ferris & Bernard, 1967). *P.brachyurus*, *Paratrichodorus christiei* and *Meloidogyne* incognita were significantly affected by the choice of crop but not by the length of rotation. The density of *P. brachyurus* increased in rotations involving crotalaria, *Setaria italica* and sudangrass. Okra and fallow resulted in the lowest numbers of *P.brachyurus*, while a winter crop of rye had no appreciable effect on the population of *P.brachyurus* (Good *et al.*, 1973).

In tests of three 3-year rotation against *P. scribneri*, main season (summer) crops of maize, soybean or cotton were followed respectively by wheat, fallow or mixed common vetch (*Vicia sativa*) and crimson clover (*Trifolium incarnatum*). The rotations were imposed on plots receiving different combinations of N, P and K, minor elements and lime. Populations in summer crops were lowest in cotton, followed by soybean and maize. Winter wheat and vetch resulted in lower populations than did fallow. Nematodes were reduced as nutrient elements increased in rotations that included the legumes. Without the latter the number of *P. scribneri* increased in plots that received complete nutrition (Rodriguez-Kabana & Collins, 1980).

Miller (1978), based on host efficiency studies, recommended reed canary grass (*Phalaris arundinaceae*), smooth Saratoge brome grass (*Bromus inermis*), timothy (*Phleum pretense*), asparagus (*Asparagus officinalis*), pepper (*Capsicum annum* cvs. 'California Wonder' and 'Italian Sweet') for the control of *P. penetrans* populations.

Heat and Hot-water Treatment

A relatively new method of soil pasteurization by soil heating has been developed in Israel and successfully tried in many countries with a suitable climate. Moistened soil is covered with transparent polyethylene sheet for about one month during hot season. Such a practice has great potential in controlling pests and diseases in general, and nematodes in particular. In experiments where effective control of soil-borne parasites was obtained by soil solarization, maximum temperatures were within the ranges 45-50°C and 38-45°C at depths of 10 and 20 cm respectively, which is an increase of about 20% over untarped soil. Significant control of certain lesion nematodes has been obtained by solarization, e.g. *P. thornei* (Grinstein *et al.*, 1979) in Israel, and *P. penetrans* (Porter & Merriman, 1983) in Australia.

Hot-water treatment has been successfully used to control plant-parasitic nematodes since the beginning of this century. In Britain and the Netherlands, there is little doubt that the introduction of hot-water treatment of narcissus and other bulbs led to a spectacular improvement in their health (Southey, 1978). Hot-water treatment can be used as a preventive measure to control endoparasitic nematodes, including *Pratylenchus* species, that are disseminated on or in seeds, root stocks, corms, tubers, rhizomes and other planting material. A few examples are cited in Table 5.

Table 5 : Control of *Pratylenchus* species by hot-water treatment.

Nematode species	Plant	Temperature/Time	Reference
P. penetrans + *Meloidogyne hapla*	Dormat Strawberry plants	i) 49.4°C/17.5 min. ii) 52.8°C/3 min. iii) 54.4°C/1 min.	Christie, 1959
P. vonvallariae	Lily of the valley *Convallaris majalis*	43.3°C/1 h	Hauge, 1972
P. penetrans	Lily	46. °C/20 min.	D'Herde *et al.*, 1960
P. vulnus	Rose plants	i) 2.2°C/3 weeks (cold hardiness) ii) 37.8°C/24 h (heat hardiness) iii) 48.3°C/35 min. (hot-water treatment)	Towson & Lear, 1982

Biological Control

Very little is known about the parasites and predators of root-lesion nematodes. An endospore forming bacterial parasite, *Pasteuria penetrans* was shown to reduce the numbers of *P. scribneri* recovered from the soil and roots of beans. Small amounts of soil infested with spores of this bacterium could be used to transmit the organism to uninfested sites (Mankau, 1980). Recently, Starr & Sayre (1988) have renamed the group of this bacterium which is specific to lesion nematodes as *Pasteuria thornei* on the basis of spore morphology and host specificity.

Among the antagonistic plants, *Tagetes* spp. suppressed populations of *P. crenatus*, *P. neglectus*, *P. penetrans* and probably *P. thornei*. *T. patula* was more effective than *T. erecta* or *T. minuta*. *T. patula* cv. 'Harmony' grown in autumn after the main crop, or between rows of a main crop with plants spaced 10 cm apart, was advised for best results in Holland (Suatmadji, 1969). Gommers (1981) while discussing the biochemical interactions between nematodes and plants included the effects of naturally occurring nematicides, i.e. terthienyl and bithienyl from a number of species of Compositae, including *Tagetes* spp. on *P. penetrans*. He also listed the toxic effects of asparagusic acid and dihydroasparagusic acid from asparagus on *P. penetrans*, but listed no other toxins effective against *Pratylenchus* species.

Regulatory Measures

Legal enforcements to prevent pest and pathogens from spreading to new areas, or to prevent them from multiplying and spreading further in case they have already found entry, can be helpful particularly in case of endoparasitic nematodes like *Pratylenchus*. It is only a matter of speculation that how many lesion nematode species might have been introduced into new geographical areas by human efforts, unknowingly. With the increase in the international trade, exchange of plant germplasm material, etc. the risk of such hidden pest introductions has increased manifold. However, with the growing awareness about these, several countries have enforced quarantine measures and listed several nematode species as 'quarantine objects'. Sanwal (1990) surveyed plant quarantine requirements of 142 countries and listed 54 nematode species which were cited in the regulations of 63 countries. The list includes seven species of *Pratylenchus*, viz. *P. brachyurus, P. coffeae, P. convallariae, P. loosi, P. penetrans, P. pratensis* and *P. vulnus*.

INTEGRATED CONTROL

The development of pest management programme for the economic control of any pest implies the judicious selection of those available control techniques which will reduce the effects of the pest with minimum negative impact on the environment and with overall economic soundness. Two examples from among the root-lesion nematodes are cited below which highlight the development of a nematode management programme.

Pratylenchus thornei on wheat in Mexico — A case history

Wheat is grown in Sonora, Mexico as a winter irrigated crop, often in monoculture on 350,000 ha. Since 1967, serious yield reductions were attributed to a variety of soil problems such as wheat brown mite, fungi, bacteria, nematodes, and lack of proper fertilizers. Since then, studies have clearly identified *P. thornei* as one of the major contributors to the reduction in wheat yield (Perez *et al.*, 1970). Most of the samples contained 0-400 *Pratylenchus*/100 g of soil or 1 g of roots. Some samples had as high as 4800 nematodes/g root. The distribution and number of *P. thornei* were associated with fine-textured clay soils rather than coarse textured sandy soils (Cinco, 1970).

Symptoms of wheat decline in Mexican soils usually appear within the first 20 days as patchy, usually yellowish areas. Seldom is an entire field uniformly affected. Foliar symptoms consist of stunting, chlorosis, sometimes necrosis of leaf tips and flagging of young plants. Sometimes these young plants die and the stand is reduced, but more often tillering is reduced and only head is produced instead of 2-4 per plant. Head size is sometimes reduced. Nematode attack usually starts in the primary roots, causing the plant to be stunted and vulnerable to attack by soil fungi. Adventitious roots begin developing after 21 days and these permit the plants to recover and produce heads, although these are fewer in number than on plants growing in the absence of *P. thornei*. Pathogenicity experiments revealed that threshold density of *P. thornei* appeared to be about 42 nematodes/100 g soil, resulting in significant reduction in head, however, field observations indicated that the threshold level is somewhat higher 50-100/100 g soil) under Mexican conditions.

A soil temperature of 14°C during early portion of plant growth did not significantly affect them against nematodes. Maximum reduction in yield occurred in those plants grown at 25°C throughout their growth cycle. So, when the planting date of wheat on infested soil was delayed until the soil temperature had declined to 15°C, the soil temperature balance was shifted in favour of the host rather than the nematode pest. The application of a nitrogen fertilizer at planting time or shortly thereafter improved and extended the advantage of host over pest. Unfortunately, no host resistance appears to be available in the commercial varieties used in this region. Some of the new short-stemmed cultivars do show improved root vigour and many add host tolerance to the above practices. There is also some information to indicate that certain crop rotations favour the nematode pest while others are unfavourable. A wheat-fallow-wheat rotation should be avoided on fields known to be infested with the nematode. The use of nematicides, although effective, was not an economical solution for the relatively low value crop of wheat in this region.

A pest management approach using variety selection, nitrogen fertilizers, planting in cool soil (15°C) and a crop rotation avoiding wheat after wheat was the most practical solution to this problem on a commercial scale (Van Gundy *et al.*, 1974).

Pratylenchus coffeae on coffee in India — A case history

In India, coffee is grown on the hilly slopes of the traditional coffee tracts of Karnataka, Kerala, Tamil Nadu and Andhra Pradesh. Arabica (*Coffea arabica*) and robusta (*Coffea canephora*) are the two commercially grown species and are grown under a canopy of shade trees of various species along with other plantation crops like—black pepper, cardamom, banana, cacao, etc. Intensive survey for plant-parasitic nematodes associated with coffee revealed that out of about 50 nematode species, *P. coffeae* was more harmful to the crop. In India, it is serious problem in about 3000 ha.

The damage starts in the form of patches due to the death of a few plants. The patches extend slowly, and as a result, the land becomes useless. The loss is anywhere between 5-75%, and according to one estimate, a crop worth of little over Rs. 20 million is lost every year (Van Berkum & Seshadri, 1970). Both young and bearing plants are affected. The young plants become unhealthy with lean and lanky stem, older leaves become yellow and drop, leaving very few undersized chloritic and crinkled leaves at the tip of the main stem giving a 'tufted' appearance. Affected bearing plants have thinner main stem. Plants become prematurely old which calls for collar pruning or uprooting.

No single method of nematode management would offer effective solution to the nematode problem in coffee. Several soil fumigants and granular nematicides have been tested, of which phenamiphos (Nemacur 5G) was found effective. The chemical is required to be applied to young plant @ 3g per plant every three months at least for one year to protect them in the initial stages. The scope to manage nematodes through chemical means is limited in a perennial crop like coffee because of cost factor and also due to the interference of weeds and shade trees which normally harbour *P. coffeae*. Nematode control in coffee is achieved through cultural practices and use of nematode tolerant materials. Maintaining heavy shade, providing much to individual plants, application of farm yard manure and giving frequent irrigation or providing nutrients through foliar feeding help in renewing the vigour of the plants (Kumar, 1984).

Of several coffee species, varieties and inter-specific hybrids tested for their tolerance against *P. coffeae*, all the arabica varieties released by Central Coffee Research Institute were found susceptible to the nematode attack while, robusta and tree coffees like, excels and liberica (varieties of *C. dewevrei*) exhibited tolerance. Robusta was, however, found to be a more suitable material for replacing nematode affected arabica plants either as a pure material or as a root stock for arabica in grafting (Kumar, 1974, 1979). Kumar (1992) summarized the steps involved in nematode management as follows :

In nursery, soil should frequently be tested for the presence of nematodes. Care should be taken to keep the site free from weeds. Nursery site should be changed once every 7-8 years. Forest soil and farm yard manure should be dried well and sieved before use. Procuring soil from infested estate or preparation of secondary nursery on a vacant space in the estate should be avoided. Exchanging nursery plants among planters and procuring plants from commercial nurseries should be avoided.

In estate, soil samples from affected plants should be tested for the presence of *P. coffeae*. Affected plants should be pulled out and burnt. The planting site should be dug and the soil exposed to the sun for at least one summer season. Care should be taken to keep the site free from weeds. After taking up these precautionary measures, the site should be planted with either robusta or arabica-robusta grafted plants. Grafting of arabica scion onto robusta root stock in 40-day-old 'topee stage' plants by using a wedge-cleft method can be adopted (Kumar, 1974, 1988). Grafting gives 90-95% success.

CONCLUSION

Pratylenchus spp. have undoubtedly been recognised as one of the most important nematode pests of crops both in the temperate as well as tropical areas of the world. Several species have been recorded to be economically important in the annual and perennial crops. The adoption of control measures necessitates the determination of 'key pest' status of the species involved, and the extent of crop losses incurred. Thorough and complete information on the identification of species/race, biology and pathogenicity are pre-requisite to the formulation of control strategies.

The control strategies to be adopted will vary with the nematode species, crop value, nature of the crop, and the geographical area. Use of nematicides may be necessary in most cases, but should be avoided if possible. Crop rotations have little scope in the management of lesion nematodes attacking field crops, still avoidance of cereals and grasses may be useful in most cases. Summer solarization offers a promising means in reducing lesion nematode populations in tropical areas. Knowledge on the biocontrol organisms of lesion nematodes is incomplete.

Use of hot-water treatment among ornamental crops, which are propagated through bulbs, corms, rhizomes, runners, suckers and other such planting material, and in certain horticultural crops can be adopted successfully, but needs to be worked out carefully. Similarly, in horticultural crops, search for resistant/tolerant germplasm should be intensified. However, as already pointed out, no single method of control may be sufficient to develop a long term, environmentally safe and economically sound means of root-lesion nematode management. An integrated nematode management approach needs to be developed for individual problem, and for a particular area.

REFERENCES

Abawi, G.S. & Mai, W.F. (1975). Effect of foliar applications of ozamyl on movement of *Pratylenchus penetrans* in and outside host roots. *Plant Dis. Reptr.* **59:** 795-799.

Acosta, N. & Malek, R.B. (1981). Symptomatology and histopathology of soybean roots infected by *Pratylenchus scribneri* and *P. alleni*. *J. Nematol.* **13:** 6-12.

Acosta, N.; Malek, R.B. & Edwards, D.I. (1979). Susceptibility of soybean cultivars to *Pratylenchus scribneri*. *J. Agric. Univ. P.R.* **63** : 103-110.

Apt, W.J. & Gould, C.J. (1961). Control of root-lesion nematode, *Pratylenchus penetrans* on narcissus. *Plant Dis. Reptr.* **45** : 290-295.

Atu, U.G. & Duru, N. (1990). Control of stem borer (*Sesamia calamistis*) and lesion nematode (*Pratylenchus brachyurus* (Godfrey) Filipjev & Stekhoven) of late season sweet maize with carbofuran. *Beitrage zur Tropischen Landwirtschaff und Veterinarmedizin* **28:** 69-73.

Bergeson, G.B. (1978). Control of root-lesion nematode (*Pratylenchus* spp.) in corn with carbofuran. *Plant Dis. Reptr.* **62:** 295-297.

Bernard, E.C. & Laughlin, C.W. (1976). Relative susceptibility of selected cultivars of potato to *Pratylenchus penetrans. J. Nematol.* **8:** 239-242.

Boswell, T.E. (1968). Pathogenicity of *Pratylenchus brachyurus* to Spanish peanut. Ph. D. Diss. Texas A & M Univ. p. 156.

Christie, J.R. (1959). *Plant Nematodes. Their Bionomics and Control*. Fla. Agric. Exp. Stn. p. 256.

Cinco, G.L. (1970). Tesis (sin publicar) Esc. Agric. y Ganad. Universidad de Sonora, Mexico.

Coates-Beckford, P.L. (1977). Comparison of various treatments for the control of *Pratylenchus coffeae* in Yam. *Nematologica* **7** : 20-26.

Colbran, R.C.. (1954). Problems in tree replacement. 2. The effect of certain methods of management on the nematode fauna of an orchard soil. *Australian Jour. Agri. Sci.* **20** : 234-237.

Corbett, D.C.M. (1969). *Pratylenchus pinguicaudatus* n. sp. (Pratylenchinae : Nematoda) with a key to genus *Pratylenchus. Nematologica* **15:** 550-556.

de Man, J.G. (1880). Die einheimischen, frei in der rerinen Erde und im sussen-Wasser lebenden Nematoden. *Tijdschr. ned. dierk. Ver.* **5:** 1-104.

D'Herde, J.; Brande, J. van den & Gillard, A. (1960). Control of *Pratylenchus penetrans* (Cobb), causal agent of root rot in lilies. *Nematologica* (supplement). **2:** 64-67.

Dickerson, O.J. (1961). Pathogenicity, life cycle and morphology of certain *Pratylenchus* spp. on potato. *Diss. Abstr.* **22** : 700.

Dickerson, O.J. (1979). The effect of temperature on *Pratylenchus scribneri* and *P. alleni* populations on soybeans and tomatoes. *J. Nematol.* **11:** 23-26.

Dickerson, O.J.; Darling, H.M. & Griffin, G.D. (1964). Pathogenicity and population trends of *Pratylenchus penetrans* on potato and corn. *Phytopathology* **54:** 317-322.

Doyle, A.D.; McLeod, R.W.; Wong, P.T.W.; Hetherington, S.E. & Southwell, R.J. (1987). Evidence for the involvement of the root-lesion nematode, *Pratylenchus thornei* in wheat yield decline in northern New South Wales, *Australian J. expt. Agric.* **27** : 563-570.

Dunn, R.A. (1972). Importance of depth in soil, presence of host roots, and role of eggs as compared to vermiform stages in overwintering of *Pratylenchus penetrans* at Ithaca, New York. *J. Nematol.* **4:** 221-222.

Edgerton, L.J. & Parker, K.G. (1958). Effect of nematode infestation and root-stock on cold hardiness of Montmerency cherry trees. *Proc. Am. Sec. Hort. Sci.* **72** : 134-138.

Edmunds, J.E.; Boothroyd, C.W. & Mai, W.F. (1967). Soil fumigation with DD for control of *Pratylenchus penetrans* in corn. *Plant Dis. Reptr.* **51:** 15-19.

Elliott, J.M.; Marks, C.F. & Tu, C.M. (1977). Effects of certain nematicid on soil nitrogen, soil nitrifiers and populations of *Pratylenchus penetrans* in flue-cured tobacco. *Can. J. Plant Sci.* **57:** 143-154.

Endo, B.Y. (1967). Comparative population increase of *Pratylenchus brachyurus* and *P. zeae* in corn and soybean varieties Lee and Peking. *Phytopathology* **57** : 118-120.

Ferris, V.R. & Bernard, R.L. (1967). Population dynamics of nematodes in fields planted to soybeans and crops grown in rotation with soybeans. I. The genus *Pratylenchus* (Nemata : Tylenchida). *J. Econ. Entomol.* **60:** 405-410.

Filipjev, I.N. (1936). On the classification of the Tylenchinae. *Proc. Helminth. Soc. Wash.* **3:** 80-82.

Fredrick, J.J. & Tarjan, A.C. (1975). Control of *Pratylenchus brachyurus* in rough lemon seedlings with Dowco 275 (diethyl-fluoro-pyridyl-phosphorothioate). *Nematropica* **5:** 10-13.

Gnanapragasam, N.C.; Jebamalai, A.S. & Anapalangan. V.T. (1987). Control of root-lesion nematode, *Pratylenchus loosi* in contaminated tea nursery plants. *Sri Lanka J. Tea Sci.* **56:** 95-97.

Gommers, F.J. (1981). Biochemical interactions between nematodes and plants and their relevance to control. *Helminthol. Abstr.* **B 50:** 9-24.

Good, J.M.; Murphy, W.S. & Brodie, B.B. (1973). Population dynamics of plant nematodes in cultivated soil: length of rotation in newly cleared and old agricultural land. *J. Nematol.* **5:** 117-122.

Graham, T.W. (1951). Nematode root rot of tobacco and other plants. *Bull. S. Carol. Agric. Exp. Stn. No. 390.* p. 25.

Graham, T.W. (1965). Tobacco varieties used to show differential response to root-lesion and root-knot nematodes. *Plant Dis. Reptr.* **49:** 822-826.

Graham, T.W. & Heggested, H.E. (1959). Growth response and root decay development in certain tobacco varieties and breeding lines infected with root lesion nematodes. *Tobacco* **149:** 21-28.

Greco, N.; Vito, M. di; Saxena, M.C. & Reddy, M.V. (1988). Investigation on the root lesion nematode *Pratylenchus thornei* in Syria. *Nematol. medit* **16:** 101-105.

Grinstein, A.; Orion, D.; Greenberger, A. & Katan, J. (1979). Solar heating of the soil for the control of *Vercicillium dahliae* and *Pratylenchus thornei* in potatoes. *In : Soil-borne Plant Pathogens* (Eds.—B. Schippers & W. Gems). Academic Press, London. pp. 431-438.

Hague, N.G.M. (1972). Nematode diseases of flower bulbs, glasshouse crops and ornamentals. *In: Economic Nematology* (Ed.—J.M. Webster). Academic Press. New York, London. pp. 409-434.

Hague, N.G.M. & Gowen, S.R (1987). Chemical control of nematodes. *In: Principles & Practice of Nematode Control in Crops.* (Eds.— R.H. Brown & B.R. Kerry). Academic Press. pp. 131-178.

Jensen, H.J. (1966). Phosphate pesticides control root-lesion nematodes in Oregon Easter lily plantings. *Plant Dis. Reptr.* **50:** 923-927.

Johnson, A.W. & Burton, G.W. (1973). Comparison of millet and sorghum-sudan grass hybrids grown in untreated soil and soil treated with two nematicides. *J. Nematol.* **5:** 54-59.

Kable, P.F. & Mai, W.F. (1968). Influence of soil moisture on *Pratylenchus penetrans. Nematologica* **14:** 101-122.

Kimpinski, J. (1986). Effects of aldicarb and oxamyl on *Pratylenchus penetrans* and potato yields. *Canadian J. Pl. Pathol.* **8:** 189-192.

Krusberg, L.R. (1971). Control of *Pratylenchus penetrans* in heavy greenhouse soils. *Plant Dis. Reptr.* **55:** 276-279.

Kumar, A.C. (1974). Final Tech. Report, PL 480 Project No. 87-CR-260. p. 56.

Kumar, A.C. (1979). Relative tolerance or susceptibility of arabica, robusta and excelsa coffees to *Pratylenchus coffeae. Proc. PLACROSYM II.* 20-26.

Kumar, A.C. (1984). The symptoms and diagnosis of the disorder 'spreading decline' (Cannoncadoo dieback) with a note on spread and control of the causal agent, *Pratylenchus coffeae*. *J. Coffee Res.* **14:** 156-159.

Kumar, A.C. (1988). Nematode problem of coffee and its management. *Indian Coffee* **52:** 12-19.

Kumar, A.C. (1992). Nematode pests of coffee. *In* : *Nematode pests of Crops* (Eds.— D.S. Bhatti & R.K. Walia). CBS Publishers & Distributors Pvt. Ltd. Delhi. pp. 258-266.

Layne, R.E.C. (1974). Breeding peach root-stocks for Canada and the northern United States. *Hort. Sci*. **9:** 364-366.

Loof, P.A.A. (1978). The genus *Pratylenchus* Filipjev, 1936 (Nematoda: Pratylenchidae) — *a review of its anatomy, morphology, distribution, systematics and identification*. Swed. Univ. Agric. Sci. Uppsala. p. 50.

Maggenti, A.R.; Hart, W.H. & Lenz, J.V. (1970). Granular applications of phosphates and carbamates for control of *Pratylenchus penetrans* on Easter lily (*Lolium longiflorum* cv. Ace). *Plant Dis. Reptr*. **54:** 1012-1014.

Mai, W.F.; Bloom, J.R. & Chen, T.A., Eds. (1977). *Biology and ecology of plant-parasitic nematode, Pratylenchus penetrans*. Pa. Agric. Exp. Stn. Bull. No. 815.

Mai, W.F.; Parker, K.G. (1970). Controlling nematodes increases growth and yield of apples and cherries. *N.Y. State Hortic. Soc. Proc*. **115:** 207-209.

Mamiya, Y. (1971). Effect of temperature on the life cycle of *Pratylenchus penetrans on Cryptomeria* seedlings and observations on its reproduction *Nematologica* **17:** 82-92.

Mankau, R. (1980). Biological control of nematode pests by natural enemies. *Ann. Rev. Phytopathol*. **18:** 415-440.

McDonald, A.H.; Luis, J.H.; Loots, G.C. & Waele, D. de (1987). Chemical control of root-lesion nematodes (*Pratylenchus* spp.) on maize in South Africa. *Phytophylactica* **19:** 479-483.

Miller, C.R.; Clark, F. & Perry, V.G. (1970). Chemical control of nematodes on flue-cured tobacco. *Soil Crop Sci. Fla. Proc*. **29:** 369-372.

Miller, P.M. (1968). The susceptibility of parasitic nematodes to sub-freezing temperatures. *Plant Dis. Reptr*. **52:** 768-777.

Miller, P.M. (1978). Reproduction, penetration and pathogenicity of *Pratylenchus penetrans* on tobacco, vegetables and cover crops. *Phytopathology* **68:** 1502-1504.

Minton, N.A. & Gillenwater, H.B. (1973). Methyl bromide fumigation of *Pratylenchus brachyurus* in peanut shells. *J. Nematol*. **5:** 147-149.

Minton, N.A.; Hammons, R.O. & Parham, S.A. (1970). Infection of shell and peg tissues of six peanut cultivars by *Pratylenchus brachyurus. Phytopathology* **60:** 472-474.

Minton, N.A. & Morgan, L.W. (1974). Evaluation of systemic and non-systemic pesticides for insect and nematode control in peanuts. *Peanut Sci*. **1:** 91-98.

Mountain, W.B. (1965). Pathogenesis by soil nematodes. *In*: *Ecology of soil borne plant pathogens* (Eds.— K.F. Baker & C.S. Williams) Univ. of Calif Press. pp. 285-301.

Mountain, W.B. & Patrick, Z.A. (1959). The peach replant problem. VII. The pathogenicity of *Pratylenchus penetrans* (Cobb, 1917) Filip. & Stek. *Can. J. Bot*. **37:** 459-470.

Norton, D.C. & Hinz, P. (1976). Relationship of *Hoplolaimus galeatus*, *Pratylenchus hexincisus* in reduction of corn yields in sandy soils in Iowa. *Plant Dis. Reptr*. **60:** 197-200.

Norton, D.C..; Tollefson, J.; Hinz, P. & Thomas, S.H. (1978). Corn yield increases relative to nonfumigant chemical control of nematodes. *J. Nematol*. **10:** 160-166.

O'Bannon, J.H. & Esser, R.P. (1975). Evaluation of citrus, hybrids and relatives as hosts of

Pratylenchus coffeae, with comments on other hosts. *Nematol medit*. **3:** 113-122.

Olowe, T. & Corbett, D.C.M. (1976). Aspects of the biology of *Pratylenchus brachyurus* and *P. zeae*. *Nematologica* **22:** 202-211.

Olthof, T.H.A. (1989). Effects of fumigant and nonfumigant nematicides on *Pratylenchus penetrans* and yield of potato. *J. Nematol*. **21** (4 Suppl): 645-649.

Orion, D. & Glazer, I. (1987). Nematicide seed dressing for *Pratylenchus mediterraneous* control in wheat. *Phytoparasitica* **15:** 225-228.

Perez, B.L.G.; Van Gundy, S.D.; Stolzy, L.H.; Thomason, I.J. & Laird, R.J. (1970). *Pratylenchus thornei*, a nematode pest of wheat in Sonora, Mexico. *Phytopathology* **60:** 1307.

Pitcher, R.S.; Patrick, Z.A. & Mountain, W.B. (1960). Studies on the host-parasite relations of *Pratylenchus penetrans* (Cobb) to apple seedlings *Nematologica* **56:** 309-314.

Porter, I.J. & Merriman, P.R. (1983). Effects of soil solarization on nematodes and fungal pathogens at two sites in Victoria. *Soil Biol. Biochem*. **15** : 39-44.

Potter, J.W.; Dirks, V.A.; Johnson, P.W.; Olthof, T.H.A.; Layne, R.E.C. & McDonnell, M.M. (1984). Response of peach seedlings to infection by root lesion nematode, *Pratylenchus penetrans* under controlled conditions. *J. Nematol*. **16:** 317-322.

Radewald, J.D.; O'Bannon, J.H. & Tomerlin, A.T. (1971). Temperature effects on reproduction and pathogenicity of *Pratylenchus coffeae* and *P. brachyurus survival* of *P. coffeae* in roots of *Citrus jambhiri*. *J. Nematol*. **3:** 390-394.

Rebandel, Z. & Szczygiel, A. (1989). Effect of soil disinfection on *Pratylenchus penetrans* populations and on growth and yield of raspberries *Zeszyty Problemowe Postepow Nauk Rolniczych* **358:** 85-95.

Rich. J.R.; Keen, N.T. & Thomason, I.J. (1977). Association of coumestans with the hypersensitivity of lima bean roots of *Pratylenchus scribneri*. *Physiol. Pl. Pathol*. **10:** 105-116.

Rodriguez-Kabana, R. & Collins, R.J. (1980). Relation of fertilizer treatments and cropping sequence to populations of *Pratylenchus scribneri*. *Nematropica* **10:** 121-129.

Sanwal, K.C. (1990). Plant quarantine measures for nematodes — an overview. *In* : *Progress in Plant Nematology* (Eds.- S.K. Saxena *et al.*) CBS Publishers & Distributors Pvt. Ltd., Delhi. 349-365.

Sasser, J.N. (1987). A perspective on nematode problems worldwide. *In*: *Nematodes Parasitic to Cereals and Legumes in Temperate Semi-arid Regions* (Eds.— M.C. Saxena, R.A. Sikora & J.P. Srivastava) ICARDA. pp. 1-12.

Schmitt, D.P. (1976). Relative suitability of soybean cultivars to *Pratylenchus brachyurus*. *J. Nematol*. **8:** 302.

Scotto La Massese, C. (1975). Tests of some fruit root stocks and varieties as hosts for *Pratylenchus vulnus* Allen & Jenson. *C. rendus Seanc. Acad. Agric. Fr.* **61:** 1088-1095.

Serr, E.F. & Day, L.H. (1949). Lesion nematode injury to California fruit and nut trees, and comparative tolerance of various species of Jugalandacea. *Proc. Am. Soc. Hort. Sci.* **53:** 134-140.

Sher, S.A. & Allen, M.W. (1953). Revision of the genus *Pratylenchus* (Nematoda : Tylenchidae). *Univ. Calif. Berkeley Publ. Zool*. **57:** 441-470.

Siddiqi, M.R. (1986). *Tylenchida — Parasites of Plants and Insects*. Comm. Agric. Bureau. p. 645.

Smith, O.D.; Boswell, T.E. & Thames, W.H. (1978). Lesion nematode resistance. *Crop, Sci*. **18:** 1008-1011.

Southey, J.F. (1978). Physical methods of control. *In*: *Plant Nematology* (Ed.- J.F. Southey). Bull. 201, Min. Agric. Fish Food, London. pp. 302-312.

Starr, M.P. & Sayre, R.M. (1988). *Pasteuria thornei* sp. nov. and *Pasteuria penetrans sensu stricto* emend., mycelial and endospore forming bacteria parasitic, respectively, on plant-parasitic nematodes of the genera *Pratylenchus* and *Meloidogyne. Ann. Inst. Pasteur Microbiol.* **139**: 11-31.

Suatmadji R.W. (1969). Studies on the effects of *Tagetes* species on plant parasitic nematodes. Diss. Agric. Univ. Wageningen.

Tarjan, A.C. (1972). Response of *Pratylenchus*-infected citrus seedlings to stem applications of a paste of ethyl 4-(methyltio)-m-tolyl-isopropylphos phoramidate. Plant Dis. Reptr. **56**: 626-627.

Thames, W.H. (1982). The genus *Pratylenchus. In* : *Nematology in the Southern Region* of the United States. Southern Coop. Ser. Bull. **276**: 108-126.

Townshend, J.L. & Chiba, M. (1987). Control of *Pratylenchus penetrans* and *Meloidogyne hapla* and yield response of alfalfa due to oxamyl seed treatments. *J. Nematol.* **19**: 454-458.

Townshend, J.L. & Olthof, T.H.A. (1988). Growth of potato and control of *Paratylenchus penetrans* with oxamyl treated seed piece in green house studies. J. Nematol. **20**: 405-409.

Towson, A.J. & Lear, B. (1982). Control of nematodes in rose plants by hot water treatment preceded by heat hardening. *Nematologica* **28**: 339-353.

Van Berkum, J.A. & Seshadri, A.R. (1970). Some important nematode problems in India. *In : Proc. Int. Symp. (10th). European Soc. Nematologists*, Pescara (8-13th Sept.). pp. 135-137.

Van Gundy, S.D.; Perez, B.J.G.; Stolzy, L.H. & Thomason, I.J. (1974). A pest management approach to the control of *Pratylenchus thornei* on wheat in Mexico. *J. Nematol.* **6**: 107-116.

Veld, E.J.M.H. In't; Egberink, M.B.; Pol, P.A. Vande & Jacob, J.J.s' (1988). Fenamiphos as dip treatment for rose root stocks to control *Pratylenchus vulnus*. *Mededelingen van de Faculteit Landbouwwetenschappen*, Rijksuniversiteit Gent **53**(2b): 893-901.

Walia, R.K. (1986). The root-lesion nematode, *Pratylenchus. In* : *Plant Parasitic Nematodes of India—Problems and Progress* (Eds.—G. Swarup & D.R. Dasgupta). IARI, New Delhi. pp. 211-222.

Walia, R.K. & Seshadri, A.R. (1985). Chemical control of *Pratylenchus thornei* on chickpea through seed treatment. *Int. Chickpea Newsl.* No. 13: 32-34.

Walia, R.K. & Seshadri, A.R. (1986). Chemical control of root-lesion nematode on green gram. *Indian. J. Agric. Sci.* **56**: 735-737.

Youngston, C.R. & Goring, C.I.A. (1970). Nematicidal activity of 1,3-dichloropropene and 1,2-dichloropropane to three types of plant-parasitic nematodes. *Plant Dis. Reptr.* **54**: 196-199.

Zirakparvar, M.E. (1979). Population changes of *Pratylenchus hexincisus* as influenced by chemicals in fibrous and coarse roots of corn. *Plant Dis. Reptr.* **63**: 55-58.

Zirakparvar, M.E. (1982). Susceptibility of soybean cultivars and lines to *Pratylenchus hexincisus*. *J. Nematol.* **14**: 217-220.

Chapter 24

Control of the Reniform Nematode on Economically Important Crops

D.R. Dasgupta & A.K. Ganguly
Division of Nematology, Indian Agricultural Research Institute, New Delhi-110 012, India.

THE genus *Rotylenchulus* Linford & Oliveira, on a worldwide basis, is considered to be one of the ten most important genera of plant-parasitic nematodes (Sasser & Freckman, 1987). And, within this genus, the reniform nematode, *R. reniformis*, is by far the best studied and well researched species due to its widespread occurrence and potentiality to cause severe damage to many cultivated crops of economic importance. This nematode is essentially confined to tropical and sub-tropical areas of the world and reported to occur in more than 30 countries including Brazil, Sri Lanka, Cuba, India, Japan, Indonesia, Iraq, Jamaica, Liberia, Ghana, Mexico, Nigeria, Bangladesh, Philippines, U.S.A., Venezuela, Western Panama, Angola, Somalia, British Honduras, Fiji, Malawi, Spain, Greece, Malta, Israel, Egypt, Kenya, Lebanon, China, Pakistan, Dominican Republic and Puerto Rico (Dasgupta *et al.*, 1968). In India, the nematode has been reported from almost all the states. Since record indicates that the nematode survives in colder climate (Heald & Thames, 1982), there exists a possibility to suggest that the occurrence of *R. reniformis* may extend beyond tropical and sub-tropical areas of both the hemispheres. For example, the nematode was observed in a cotton growing area of Texas where the soil temperature at about 20 cm depth recorded to be below freezing point for prolonged period of winter. Nevertheless, since the thermal limits of reproduction of this nematode are 15°C and 36°C and optimum being 30°C, the likelihood of *R. reniformis* to cause serious problem is more under tropical and sub-tropical than temperate climatic condition.

R. reniformis is polyphagous, known to feed and multiply on a wide range of horticultural, vegetable and plantation crops. Its host range extends to more than 115 plant species. Some of the crops on which the nematode feeds and causes noticeable injury are cotton, papaya, pineapple, cowpea, castor bean, banana, tea, sweet potato, coffee, pigeonpea, tomato, tobacco, okra, eggplant, bean, soybean, cantaloupe and several pulse crops (Swarup *et al.*, 1989).

The widespread distribution, polyphagy, short life cycle and potentiality of the nematode to cause damage to variety of crops of economical importance make it imperative on the part of growers to contain the nematode to realize optimum crop yield potentiality. But contrary to this exception, paradoxically there is hardly any systematic effort chartered so far for the management of *R. reniformis* under field conditions.

In United States during mid 50's to 70's non-volatile and contact chemicals including aldicarb, fensulfothion, carbofuran and other toxicants have been tested for control of the reniform nematode, *R. reniformis* attacking variety of crops like grapes, banana, tobacco, cotton, soybean, cantaloupe, sweet potato and potato. Results obtained so far with regard to the nematode control and increased yield are rather encouraging. However, the withdrawal of most of the highly successful and efficient fumigants from commercial market coupled with high cost of pesticides and their detrimental effects on human health and environment and residue problems of toxicants prompted nematologists in recent years, to look for alternatives to chemical control. Under Indian conditions, so far as control of *R. reniformis* is concerned, most of the researchers experimented with organic amendments under pot conditions to evaluate their suppressive potentiality on the nematode. A compilation of information in this regard has been made in one of the recent publications (Varaprasad, 1986). Table 9 provides summarised updated additional information on soil amendments for the control of *R. reniformis*, in this country, infecting several crops. The present chapter primarily aims at presenting some of the achievements obtained in the area of management of *R. reniformis* infecting several important crops under field conditions.

CONTROL STRATEGIES

COTTON

Chemical control

Reports of occurrence and studies of the nature and scope of damage caused by the reniform nematode, *R. reniformis* to cotton in lower Rio Grande Valley, Texas (Birchfield, *et al.*, 1966; Heald & Heilman, 1971; Heald *et al.*, 1972; Lambe & Horne, 1963; Thames *et al.*, 1970) and elsewhere (Birchfield, 1962; Birchfield & Brister, 1963; Jones *et al.*, 1959; Khadr *et al.*, 1972; Oteifa, 1970; Taylor *et al.*, 1970; Heald & Thames, 1982) have shown it to be a serious pathogen sometimes causing losses of more than 60 per cent (Thames & Heald, 1974). Indeed, with regard to *R. reniformis*, cotton is one of the fewer crops which received utmost attention and researched extensively for the control of the nematode in United States. Long term field investigation were conducted to examine relation between reduction of the reniform nematode population through fumigation and increased yield of cotton. As early 1959, Jones *et al.*, showed the effect of large population of reniform nematode on the growth and development of cotton under greenhouse and field conditions for 4 years either in presence or absence of *Fusarium* wilt. Under field conditions, the nematode caused an appreciable reduction in yield for all varieties of cotton tested. In all the years under investigation, fumigated plots out yielded the checks by 22.4 to 79.6 per cent. The average significant increase in yield of cotton varieties tested in one year was about 88 per cent. The nematode also caused a delay in maturity, reduction in boll size and in some year a reduction in lint size per cent. Also the nematode increased the wilt development in wilt susceptible cotton varieties. Similar trials in the lower Rio Grande Valley of Texas usually reduced the nematode numbers but did not always increase yields, significantly (Thames *et al.*, 1969). Lamb & Horne (1963) observed that cotton following sorghum in these areas apparently made normal growth and suggested that a sorghum-cotton rotation would provide a practical control measure.

Long term field trials made during late 60s's to early 70's with fumigant, documented beneficial effect of chemical in reduction of the nematode population

Table 1: Effect of nematicides and rotation on *Rotylenchulus reniformis* and yield of cotton.

Treatment	Yields (lb seed cotton/A)	
	Cotton following cotton	Cotton following sorghum
DD 12 gal/A 1 at 20"	2752a	2516 a
DD 12 gal/A 2 at 10"	2540 ab	2328 a-c
DD 12 gal/A 10" & 20"	2529 ab	2491 a
DD 25 gal/A 10" & 20"	2483 a-c	2478 a
DD 12 gal/A 2 Dow bl. 10"	2292 b-d	2333 a-c
Aldicarb, 4 lb/A granular	2246 b-e	2202 a-d
DD 12 gal/A 1 at 10"	2164 c-f	2414 ab
DD 2.5 gal/A at planting	2099 d-g	2267 a-d
Aldicarb 2 lb/A granular	2039 d-g	2075 b-d
MBR 5667, 2 lb/A granular	2006 d-g	1977 d
MBR 6168, 4 lb/A granular	1998 d-g	1919 d
MBR 5667, 4 lb/A granular	1911 e-g	1944 d
MBR 6168, 2 lb/A granular	1834 fg	2014 cd
Check Untreated	1748 g	2251 a-d

Data without same letters are significantly different at 0.05% level.

(Source: Thames & Heald, 1974)

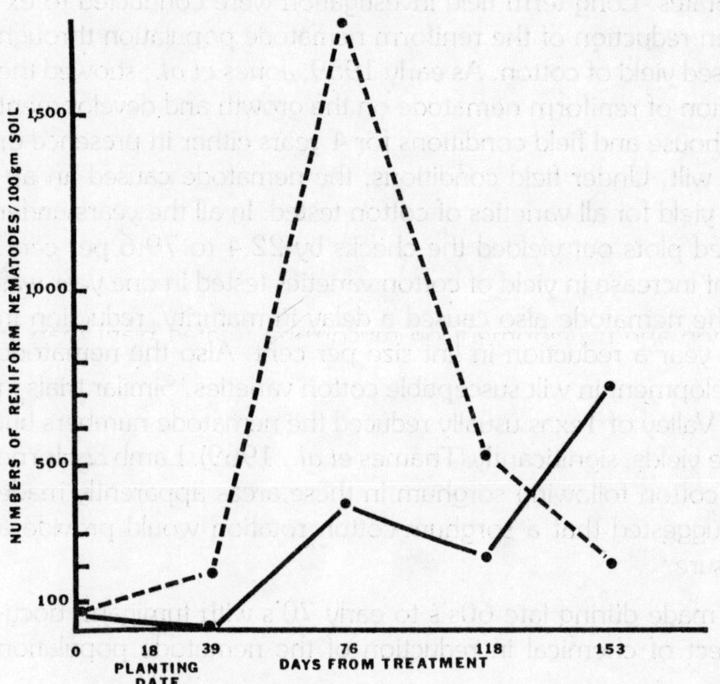

Fig. 1. Populations fo reniform nematode from chemcial treated (solid line) and untreated (broken line) seed cotton plots. (Thames & Heald, 1974).

(Fig. 1) and increase of crop yield (Table 1). Deep (20") and shallow (10") placement of soil fumigants were tested in paired trials for control of reniform nematode on cotton following cotton, and on cotton following sorghum. Seed cotton yields from chemical treatments were significantly higher than from checks when cotton followed 4 years of continuous cotton (Thames & Heald, 1974). Pretreatment nematode counts were 10 times greater when cotton followed cotton than when cotton followed grain sorghum in 1969. In 1970, there was no significant difference in yield between soil fumigants and checks when cotton followed one year with cotton. Counts of the nematode were negatively correlated with yields until near harvest, when counts indicated a change to positive correlation. These and similar other investigations consistently demonstrated that yield of cotton significantly increased with dischloropropene type fumigants (Thames *et al.*, 1970).

Generally, soil fumigation trials in other areas have successfully reduced number of reniform nematodes and increased yields on cotton on other crops (Birchfield, 1968, 1971; Birchfield & Pinckard, 1964; Holtzmann & Ishi, 1963; Jones *et al.*, 1959; Khadr *et al.*, 1972; Thames *et al.*, 1969).

Organic amendments

Various organic materials including agricultural and industrial byproducts, most of which are wastes, have been experimented for the control of *R. reniformis* infecting cotton. Under greenhouse conditions, fresh leaves with succulent stems when applied 2 weeks before planting reduced the nematode count in cotton *Gossypium hirsutum* cultivar 'Deltapine 16', 30 days after planting (de La Cruz, 1983). The most effective soil amendment was *Gliricidia* leaves, the least effective was *Anastacia* leaves (Table 2). Rate of application did not influence the degree of nematode control, which ranged 32.9% to 91.1% based on nematode recovery from control pot containing nematode infested soil only. All organic materials were equally effective as phenamiphos in controlling the nematode (Table 2). The effect of organic manuring on *R. reniformis* on cotton cultivar 'Deltapine 16' when evaluated under field condition exhibited almost identical trends; cow dung, sawdust and coir dust were also found effective against the nematode when used as soil amendments (Table 3). Except for rice straw and lower rate of cow dung of *Gliricidia* leaves, the soil amendments were equally effective as phenamiphos application (7.5 kg a.i./ha) in reducing *R. reniformis* population in 3 locations of Philippines.

Crop rotation

Inclusion of crop rotation as control measure seems to be feasible and economically viable practice. In most cases, a susceptible crop when rotated with the grass plants yielded excellent results as far as the nematode control is concerned. A two-year rotation of cotton with green sorghum was as effective as soil fumigant in keeping down the *R. reniformis* population in the fields where cotton was raised continuously (Thames & Heald, 1974). Likewise, two years rotation of cotton and resistant soybean cultivar 'Pickett 71' reduced the nematode population with increased yield of the susceptible crop (Gilman, *et al.*, 1978).

Resistance

In recent years, emphasis has been laid on the use of resistance as one of the major components for the management of the reniform nematode. Table 4 provides some important crops including cotton in which resistance against *R. reniformis* has been located.

In a field test 'Auburn 86-180', 'Auburn 634', 'La 434-810909' and 'La 434-1031-810910' were judged to be resistant on the basis of egg production, nematode population and yield components (Beasley, 1986). These 4 genotypes together with 'La-434-1031-4' were recommended as Parents in a breeding programme. The source, *Gossypium arborium* 'Nanking CB 1402' also possessed a high level of resistance to *R. reniformis* in U.S.A. (Carter, 1981). *Gossypium stocksii*, *G. somalense* and *G. barbadense* 'Texas 110' have potential value in breeding for resistance (Yik & Birchfield, 1984). *G. hirsutum* germplasm lines 'La RN 4-4', 'La RN 909', 'La RN 910' and 'La RN 1032' were developed from

Table 2 : *Rotylenchulus reniformis* population, 30 days after planting cotton in potted nematode-infested soil amended with plant leaves.

Treatment	Application Rate (g/23 cm diam. pot)	Nematode (soil+root)	Reduction (%)
Nematode-infested soil only (check)	-	671.5 d	-
Anastacia leaves	41	323.0 bc	51.9
	82	450.5 c	32.9
Manila grass leaves	41	161.5 ab	75.9
	82	238.0 abc	64.6
Gliricidia leaves	41	161.5 ab	75.9
	82	59.5 a	91.1
Leucaena leaves	41	229.5 abc	65.8
	82	128.0 ab	80.9

Similar letters indicate no significant differences at 5% level DMRT.

(Source : de La Cruz, 1983)

the selection within 'La 434 RKR' which has the parentage of 'Bajou 7769' x 'Deltapine 16'. These 4 lines have combined resistance against root-knot and reniform nematodes and *Fusarium* wilt. And, further these lines are of good yielding ability and fibre quality, equal to or greater than of 'Stenoville 825' and 'Deltapine 41' (Jones *et al.*, 1988).

PINEAPPLE

The reniform nematode did not become a significant agronomic problem in Hawaii until the mid 1950's. The tendency of the pineapple industry in Hawaii to use shorter and shorter fallow periods can be attributed to increasing problem of reniform nematodes (Rohrbach & Apt, 1986). In addition, the pH of pineapple soils steadily decreased from 1930 to 1950 due to the application of ammonium sulfate fertilizer. The pH in some fields in Hawaii was as low as 3.2 by 1950. The optimal pH for reproduction of the reniform nematode in Hawaii is approximately 4.8-5.2 (Rohrbach & Apt, 1986). Above mentioned experimental practices and monoculture created a soil environment ideal for massive build-up of the reniform nematode populations. Consequently, in 35-40 years, *R. reniformis* went from an initial limited occurrence to become a major limiting factor in Hawaiian pineapple culture.

The discovery of D-D as nematicide led almost the entire pineapple industry of Hawaii to adopt a control programme based on the use of this volatile soil fumi-

Table 3 : *Rotylenchulus reniformis* populations at harvest of cott+ 7o-n grown in three locations (I, II, III) that received different treatments.

Treatment/material applied in the plant row	Application rate/ha		Nematodes/300 cc soil and 1 g roots		
	I & II	III	I	II	III
Check (non-treated)			28 g	764 b	156 h
Phenamiphos	7.5 kg a.i.	2.0 kg a.i.	0 a	10 a	0 a
Rice straw mulch	2.5 t	0.625 t	17 f	360 ab	19 e
	5.0 t	1.25 t	11 ef	212 ab	0 a
Manuring with :					
Cow dung	2.5 t	2.5 t	8 de	456 ab	143 g
	5.0 t	5.0 t	2 abc	185 a	5 c
Saw dust	2.5 t	1.25 t	1 ab	436 ab	7 cd
	5.0 t	2.5 t	0 a	132 a	4 bc
Coir dust	2.5 t	1.25 t	0 a	183 a	27 f
	5.0 t	2.5 t	1 ab	145 a	4 bc
Gliricidia leaves	2.5 t	1.25 t	5 cde	376 ab	7 cd
	5.0 t	2.5 t	2 abc	136 a	1 ab
Leucaena leaves	2.5 t	1.25 t	2 abc	509 ab	10 d
	5.0 t	2.5 t	0 a	232 ab	7 cd

Means followed by a common letter are not significantly different at 5% level DMRT.
Location : I = Mnaoag II- MMSU Central Experiment Station and III = UPLB Central Experimen Station. All these locations are in Philippines.
(Source : de La Cruz 1983)

gant against *R. reniformis*. Initially the four halide soil fumigants D-D, EDB, DBCP and dichloropropene served as standard soil fumigants against the reniform nematode for the pineapple industry in Hawaii. With the withdrawal of most of these volatile halide soil fumigants from the commercial market in the United States, current interest in Hawaii has not shifted to pre-planting application of dichloropropene at the rate of 224-336 l/ha and other non-volatile halide soil chemicals for the control of the nematode infecting pineapple plants. In Puerto Rico, a two-year rotation of pineapple with sugarcane produced excellent results.

Table 4 : List of resistant varieties/sources against reniform nematode.

Crop/plant spp.	Resistant source/variety
Beans	Jack bean, Velvet bean
Coffee	*Coffea robusta, C. canefora*
Cotton	Auburn 56, Plains, Delfos 425-920, Delfos 9169, Deltapine 9169, Deltapine 6, Deltapine 6X, Delfos 6102, Cook 307-6X, Louisiana 33, Roxie, Cooker 100 wilt, Steno Ville 2B strains
Cowpea	V-16
Lycopersicon pimpinellifolium	P-1270453, PI-375937 Yellow strains
Tomato	Kalyanpur selection 1, Kalyanpur selection 111, LA 121, EC 118272, 11276, Peto 108, Peto 95, Nema 1400, ACE-55, Petopride
Mustard	Florida
Onion	Evergreen
Soybean	Peking, Dyer, Centennial, Pickett, Pickett 71, Cluster, D68-8, D 68-80, D 68-128, D 68-180, D 68-214, D 68-216, D 68-221, TK 5, Strain 99, Strain 12, PI 12, PI 200451, GC 40078-40, EGS Y 6, GC 40142-0-66, VPL SY 2
Sweet potato	Gold rush
Pigeonpea	UPAS 120, XPL-Prabhat, ICPL 87, ICPL 95, Pant A-10, Norman
Potato	La Rouge, Red Aboda, B7 642-2, B 7145-8, B 7680, PI 37741
Sorghum	Sweet
Pepper and related crops	Sweet Ball, California Wonder, *Capsicum annum, C. fasciculatum, C. frutescens* var. microcarpum
Sugarcane	C-P 44-101.

(Source : Swarup *et al.*, 1989)

SOYBEAN

Major efforts to control *R. reniformis* infecting soybean are currently centred around the use of host resistance. The ideal resistance rating against the nematode in soybean was found to be between 21-31 days after planting at juvenile populations of 1800-2800/500 ml of soil. Gilman *et al.* (1977) proposed that resistance in soybean cultivar 'Forrest' to be governed by one major and minor gene pairs and symbols 'Rm' and 'rm' were proposed for the gene pairs respectively. Resistance in 'Forrest' was recessive and controlled by alleles at one lucus. In a detailed study, resistance against the reniform nematode in soybean was found to be quantitative, controlled by two pairs of genes with unequal effects

(Harville *et al.*, 1985). Besides 'Forrest', 'Pickett' and 'Dyer' were reported to have shown varying degrees of resistance against the reniform nematode. Recently, one cultivar 'PADRE' has been reported to be highly resistant to *R. reniformis* (Hartwig *et al.*, 1988). Soybean cultivar 'Gregg', evolved from the crossing between 'Bragg' and 'Pickett 71', has combined resistance of *Heterodera glycines* race 3 and *R. reniformis* (Harville *et al.*, 1988).

TOMATO

Chemical control

Foliar application of systemic pesticide, oxamyl on tomato and cotton appears to be promising against the nematode (Rich & Bird, 1973). Single application of 600 and 2000 ppm of the chemical on tomato and cotton respectively blocked the entry of the nematode into roots. Tomato was more susceptible to *R. reniformis* than cotton and penetration of tomato roots by the nematode was inhibited with a lower rate of oxamyl. The time interval between multiple applications could be extremely important factor in the nematode control. Future role of systemic nematicides may be limited to their use in combination with other fumigant or contact chemicals to obtain residual nematicidal activity that will maintain population level below crop tolerance limit for more than one growing season. Fumigants containing methyl isothiocyanate and/or chlorinated C_3 hydrocarbon performed well as MBr-chloropicrin combination in suppressing pest populations and enhancing tomato yield (McSorley *et al.*, 1985).

Organic amendments

Growing of susceptible host for successive year often encourages build-up of population of mixed potentially pathogenic nematode species. Under such circumstances, application of organic soil amendments can successfully keep down polyspecific potential nematode pathogens at grower's advantage. Valuable information collected through series of experiments conducted at multilocated research centres, established conclusively their beneficial effects measured in terms of reduction in the nematode population and increased yields under field conditions (Castillo, 1985). The author cited an example of the results of the San Joaquin and San Pablo City experimental centres that had very high populations of *R. reniformis* and *M. incognita*. Phenamiphos and organic materials (rice straw, *Gliricidia* leaves, sawdust and chicken dung) were applied at 10 kg a.i./ha and 10 tons/ha, respectively. Nematode determination, 2 weeks after soil treatment, showed that phenamiphos application and soil amending with fresh *Giliricidia* leaves, and fresh and composted sawdust and chicken dung controlled the nematode population on tomato cultivar 'VC-11-1' (Castillo, 1985). The suppressive effect of fresh chicken dung, 2 weeks after treatment even surpassed that provided by phenamiphos (Table 5). Yield increases, statistically similar to that resulting from phenamiphos application were observed in soil amendment with fresh chicken dung.

Biocontrol

The endotrophic mycorrhizal fungus *Glomus fasciculatum* increased the resistance in tomato plant against *R. reniformis* infestation. Inoculation of tomato transplant on seed bed with the fungus significantly reduced the juvenile penetration and development on mycorrhiza treated plant than control (Sitaramaiah & Sikora, 1982). The development of gelatinous matrix was delayed and few eggs/egg masses were produced. Data showed *G. fasiculatum* adversely affected *R. reniformis* during several phases of its life cycle. Likewise *Paecilomyces lilacinus* has been reported to be pathogenic to reniform nematode infecting tomato and eggplant (Reddy & Khan, 1988, 1989).

Solarization

Compared to several summer management strategies studied so far for the control of the reniform nematode, solarization appears to be very promising. Soil solarization resulted in lower population of *R. reniformis* on *Cyperus* spp. than in control plants. The significant differences in *R. reniformis* population by summer management treatment were not maintained throughout at a subsequent tomato crop in autumn, but significant differences in *Cyperus* spp. and root-knot nematode galling were evident up to harvest of the tomato crop (McSorley & Parrado, 1986). Fumigation, following summer management resulted in significantly increased tomato yield when beds have been covered with opaque polyethylene. A second test revealed that efficacy of soil solarization was controlled by adding water to beds prior to covering or by changing the time during which beds were covered from 9 weeks during the summer to 6.5 weeks during the early or late summer (McSorley & Parrado, 1986).

CANTALOUPE

The reniform nematode causes significant reduction in the yield of cantaloupe primarily by reducing size of fruits and affecting quantity of melon in terms of reducing sucrose content of fruit (Heald, 1980). Fields having high population of the reniform nematodes when fumigated with nematicide yielded 105, 82 and 79 per cent higher yield of cantaloupe than non-treated plants for consecutive 3 years. These researchers also calculated gross increase of $ 965.42 for the investment of approx. $ 20 by soil fumigation (Tables 6-8). Likewise, fumigated plants had melon with increased sucrose than that of non-fumigated crops.

Table 5 : Plant-parasitic nematode populations and tomato yield in soil amended with fresh and composted organic materials (San Joaquin, San Pablo City, 1984 dry season). Nematodes were mostly *Rotylenchulus reniformis* and *Meloidogyne incognita.*

Treatment	No. of nematodes/300 ml soil and 1 g Roots			Yield (Kg/ha)
	No. of weeks after treatment			
	2	6	11	
Check (non-treated)	552 d	816 bc	2515 cd	22,200 a
Phenamiphos	96 b	366 b	465 a	31,100 c
Rice Straw :				
Fresh	215 bcd	500 bc	1921 bcd	29,100 bc
Composted	288 bcd	546 bc	1825 bcd	22,000 a
Gliricidia leaves :				
Fresh	201 b	533 bc	1885 abc	27,300 abc
Composted	517 cd	1074 c	4177 d	24,300 ab
Saw dust :				
Fresh	342 bc	859 bc	958 ab	25,300 ab
Composted	135 b	884 bc	928 ab	23,800 ab
Chicken dung :				
Fresh	17 a	169 a	1078 ab	32,600 c
Composted	216 b	655 bc	1406 bc	24,200 ab

Means followed by a common letter are not significantly different at 5% level DMRT.
Application rates: Phenamiphos, 10 kg a.i./ha; organic amendments 10 t/ha
(Source: Castillo, 1985)

SWEET POTATO

The reniform nematode, under field conditions, often acts as a limiting factor of sweet potato cultivation. *R. reniformis* caused significant reduction in yield of sweet potato by 60 bushels per acre. Application of chemicals, viz., organiphosphate and carbamate nematicides under field conditions resulted in significant increased in yield of U.S. No. 1 sweet potato cultivar 'Centennial grade' by 605, 615, 710 and 905 kg/ha when data was recorded for a number of years (Birchfield & Martin, 1976). Plot application of carbofuran checked the nematode population considerably, arrested the crop loss and increase sweet potato yield by 891-1643 kg/ha. Ethoprophos at 0.55, fensulfothion at 0.55, 0.82, 1.10 and 1.36 and carbofuran at 0.55 and 0.73 kg a.i./ha significantly reduced the reniform populations.

Table 6 : Effect of soil fumigation on reniform nematode population, cantaloupe yield and sucrose content (1975).

Parameter	Treatment	
	1, 3 D (46.7 l/ha)	Control
Nematode counts per 100 g of soil, 75 days after treatment	91	305
Yield (Kg/12 m row)	26.6[a]	12.9
Sucrose percentage	8.59[a]	5.93

[a] Significant increase at 1% level of probability.
(Source: Heald, 1980)

Table 7 : Effect of soil fumigation on reniform nematode population and cantaloupe yield (1976 and 1977).

Parameter	1976		1977	
	1, 3 D (46.7 l/ha)	Control	DBCP (9.3 l/ha)	Control
Nematode counts per 100 g of soil, before treatment	36	38	80	150
46 days after treatment	4	18	0	26
88 days after treatment	70	226	93	625
Yield (kg/row)	26.7[a] (12 m row)	14.7	96.6[a] (24 m row)	53.9
Marketable cantaloupe (kg)	10.4[a]	0.8	48.3[a]	18.0

[a] Significant increase at 1% level of probability.
(Source : Heald, 1980).

In addition to chemical control, other physical method like hot water treatment has been found to be useful and feasible for the control of *R. reniformis* infecting sweet potato. Hot water treatment of sweet potato, at 50°C for 3-5 minutes eliminated the nematode (Martin, 1970).

POTATO

Rebois *et al.* (1978) assessed losses in yield of four potato cultivars, 'Peconic', 'Katahdir', 'Red Pentoic' and 'Yampa' due to *R. reniformis* under greenhouse

Table 8 : Calculated cost benefit of soil fumigation to control reniform nematode infection of cantaloupe (1977).

Parameter	DBCP 9.3 l/ha	Control
Yield (Kg/24 m row)	48.3	18
Calculated yield/ha	10,819	4,636
Price per 100 g[a]	$ 32.19	$ 32.19
Calculated gross to grower	$ 3,457.75	$ 1,492.33

[a] Based on price quoted in 1977 Texas vegetable statistics.

(Source : Heald, 1980).

condition. The nematodes adversely affected marketable sized tuber to 'Red Pentioc' and 'Katahdir'. All the cultivars produced lower tuber yields in nematode-infested soil compared to uninfested soil and the average of marketable and total tuber weight per plant were significantly less. Cultivars 'Yampa' and 'Peconic' had significantly total tuber yield reduction. In the experiment having mixed population of reniform and root-knot nematodes, considerably damage to tuber was done and application of chemicals resulted in 55 per cent increase in yield of potato tuber.

COWPEA

Cowpea is one of the most favoured hosts of *R. reniformis*. And, despite the fact that this crop is severally damaged by the nematode, there is hardly any major work on the control of *R. reniformis* infecting cowpea. Major attempt to control the nematode on this crop involved organic amendment and search for source of resistance. Experiment conducted in the farmer's field in Philippines recorded significant yield increases in plots amended with chicken dung (15 tonnes and 30 tonnes/ha) over that of control plot only at 10 per cent level (Calinga, 1979). Although phenamiphos controlled the nematode population significantly but compared to control it did not influence increase in yield significantly (Calinga, 1979).

Under field conditions, soil solarization significantly reduced soil population of the nematode up to 0-15 cm deep and increased the yields of lettuce and cowpea. The length of time required for 90 per cent mortality of the nematode in soil heated under controlled conditions in the laboratory varied from 25 hours to less than 1 hour between 41 and 47°C (Heald & Robinson, 1987).

SUNFLOWER

In recent years, there has been an increased interest in the commercial sunflower production almost throughout the world. In cultivated fields, there is evidence that the reniform nematode causes injury to this crop (Heald & Stein, 1987). Soil fumigation with 1,3-dichloropropene at 47,5 56 or 74 l/ha by chisel injection 25 cm deep significantly reduced the nematode population in all the three years under investigation (1975, 1976 and 1981). The analysis of seeds from fumigated and non-fumigated plots showed significant increase in palmitic, stearic, oleic and linoleic acids in seed from fumigated plots (Heald & Stein, 1987).

EPILOGUE

The recognition of *R. reniformis* as one of the key pests of several crops of economic importance led to meaningful and useful investigation in many areas of the world with a major goal to contain the nematode. As it would be expected, in the earlier period, the application of nematicide had been the conventional method of the nematode control used by researchers, especially from affluent countries.

Table 9 : Organic materials used as soil amendment for the control of *R. reniformis* in India.

Source of organic material	Host plant	Reference
Zinnia elegans	Tomato	Tiyagi *et al.*, (1986a)
Organic extract from leaf	*In vitro*	Tiyagi *et al.* (1986b)
Neem and Karanj oil cakes	French bean	Padhi & Misra (1987)
Tagetes minuta	Cabbage, cauliflower, eggplant	Siddiqui & Alam (1987a)
Azadirachta indica and *Melia azedarach*	Tomato, okra, eggplant	Siddiqui & Alam (1987b)
M. azedarach and latex bearing plant	Vegetables	Siddiqui *et al.* (1987)
A. indica and *M. azedarach*	Tomato and eggplant	Siddiqui & Saxena (1987)
Persian lilac and neem leaf extract	Tomato and eggplant	Siddiqui & Alam (1988a)
Latex bearing plant	Vegetables	Siddiqui & Alam (1988b)
Chopped marigold waste	Tomato and eggplant	Siddiqui & Alam (1988c)
T. lucida, T minuta and *T. tenuifolia*	Vegetables and eggplant	Siddiqui & Alam (1988d)
Plant belonging to compositae	Eggplant	Tiyagi *et al.* (1988)
Water hyacinth	Tomato and eggplant	Siddiqui & Alam (1989)

However, in recent years, the range of acceptable management options for the nematode has increased with reduction in chemical alternatives. More and more emphasis currently is being laid on search for host resistance, and physical and biological control measures. The evidence for biotypes in this nematode (Dasgupta & Seshadri, 1971) warrants cautious approach while screening plant germplasm for resistance against *R. reniformis*. Since multispecific association of the nematodes under field condition is a rule rather than exception, plant cultivars having combined resistance against two or more potential pathogens including *R. reniformis* will be ideal as one of the major components of the nematode management strategies. The synthesis of soybean cultivar 'Gregg' equipped with combined resistance against *Heterodera glycines* race 3 and *R. reniformis* is illustrative example of such approach. The advent of biotechnology, advances in genetic engineering and plant breeding now offer unlimited opportunities for genotypic manipulation, conferral resistance and identification of nematode recognition factors in plant. These strategies are likely to be useful, practical and feasible for the management of the nematode especially infecting low value field crops.

Since *R. reniformis* is essentially a pathogen to be encountered under tropical conditions physical method like solarization appears to be promising as one of the control measures against the nematode. Results of explanatory investigations are encouraging enough to further this approach under tropical conditions in order to generate useful information on the nematode control and realization of increased yields (Abdel Rahim *et al.*, 1989). Experiences with organic soil amendments for the control of reniform nematode are worthy of greater scientific attention. But their availability in bulk for field application and economics remain a point of debate. Basic research that unfolds nematode suppressive mechanism(s) following soil amendments will be rewarding in this direction.

In summary, the success of the management of this or for any other plant-parasitic nematodes will depend upon many factors like choice of crop, crop value, environmental and cultural conditions. The final goal of *R. reniformis* or any other nematode control is to increase food and crop production. And the strategies employed to control nematode on land therefore should begin and end with growers.

REFERENCES

Abdel Rahim, M.F.; Satour, M.M.; Mickail, K.Y.; El Eraki, S.A.; Grinstein, A.; Chen, Y. & Katan J. (1989). Effectiveness of soil solarization in furrow irrigated Egyptian soils. *Plant* Dis. **72**: 143-146.

Beasley, J.P. (1986). Evaluation of cotton: *Gossypium hirsutum* L. genotypes for resistance to the reniform nematode, *Rotylenchulus reniformis* Linford & Oliveira. *Dissertation Abstr. Int.* B (Sciences & Engineering) **47** (3): 856B.

Birchfield, W. (1962). Host parasite relationship of *Rotylenchulus reniformis* on *Gossypium hirsutum. Phytopathology* **52:** 862-865.

Birchfield, W. (1968). Evaluation of nematicides for control of reniform nematodes on cotton. *Plant Dis. Reptr.* **52:** 786-789.

Birchfield, W. (1971). Systemic nematicides, control of *Rotylenchulus reniformis* of cotton. *Plant Dis. Reptr.* **55:** 362-365.

Birchfield, W. & Brister, L.R. (1963). Susceptibility of cotton and relatives to reniform nematodes in Louisiana. *Plant Dis. Reptr.* **47:** 990-992.

Birchfield, W. & Martin, W.J. (1976). Effect of some organophosphate and carbamate nematicides on reniform nematode and sweet potato yields. *J. Nematol* **8:** 279-280.

Birchfield, W. & Pinckard, J.A. (1964). New combinations of nematicides for control of reniform nematode on cotton. *Phytopathology* **54:** 393-394.

Birchfield, W.; Reynolds, H. & Orr, C. (1966). Plant parasitic nematodes of cotton in the Lower Rio Grande Valley of Texas. *Plant Dis. Reptr.* **50:** 149-150.

Calinga, R.H. (1979). *Field and laboratory studies on the effectiveness of dungs for the control of plant parasitic nematodes on cowpea.* M.Sc. Thesis, University of Philippines at Low Banos College, Lagume.

Carter, W.W. (1981). *Gossypium arboreum* 'Nanking CB 1402, possessed a high level of resistance to *Rotylenchulus reniformis. J. Nematol.* **13:** 368-374.

Castillo, M.B. (1985). Some studies on the use of organic soil amendments of nematode control. *Phill. Agr.* **68:** 76-93.

Dasgupta, D.R.; Raski, D.J. & Sher, S.A. (1968). A revision of the genus *Rotylenchulus reniformis* Linford & Oliveira, 1940 (Nematoda : Tylenchida). Proc. Helminth. *Soc. Wash.* **15:** 169-192.

Dasgupta, D.R. & Seshadri, A.R. (1971). Races of reniform nematode, *Rotylenchulus reniformis* Linford & Oliveira, 1940. *Indian J. Nematol.* **1:** 21-24.

De La Cruz, A. (1983). Control of nematodes on cotton (*Gossypium hirsutum* L.) *through organic manuring and other cultural practices.* M.S. Thesis, University of Philippines at Los Banos, College, Laguna.

Gilman, D.F.; Jones, J.E.; Williams, C. & Birchfield, W. (1978). Cotton-soybean forotation control of reniform nematode. *La Agric* **21:** 10-11.

Gilman, D.F.; Marshal, J.G.; Robb, J.L.; Lawrence, R.H.; Harbertz, R.; Allen, M.; Boquet, D. & Bartleson, J.L. (1977). Performance traits with soybean varieties in Louisiana, 1976. *La Agric. Expt. Stn. Agron. Res. Report No.* **449:** 24.

Hartwig, E.E.; Hinson, K. & Scott, A. (1988). Registration of 'PADRE' soybean. *Crop Sci.* **28** (6): 1028.

Harville, B.G.; Elius, W.M.; Hall, W. & Green, M.A. (1988). Registration of 'Gregg' soybean. *Crop. Sci.* **28:** 196.

Harville, B.G.; Green, A. & Birchfield, W. (1985). Genetic resistance to reniform nematodes in soybean. *Plant Dis.* **69:** 587-589.

Heald, C.M. (1980). Effect of reniform nematode on cantaloupe yield. *Plant Dis. Reptr.* **64:** 282-283.

Heald, C.M. & Heilman, M.D. (1971). Interaction of *Rotylenchulus reniformis*, soil salinity and cotton. *J. Nematol.* **3:** 179-182.

Heald, C.M. & Robinson, A.F. (1987). Effects of soil solarization of *Rotylenchulus reniformis*, on the lower Rio Grande Valley of Texas. *J. Nematol.* **9** (1): 93-103.

Heald, C. M. & Stein, E. (1987). Effect of *Rotylenchulus reniformis* on yield and oil quality of sunflowers. *Nematropica* **17:** 1-5.

Heald, C.M. & Thames, W.H. (1982). The reniform nematode, *Rotylenchulus reniformis. In* : *Nematology in the southern region of the United States. Sci. Crop Series Bull. No. 276.* p. 206.

Heald, C.M.; Thames, W.H. & Wiegand, C.L. (1972). Detection of *Rotylenchulus reniformis* infestation by aerial infra-red photography. *J. Nematol.* **4:** 298-300.

Holtzmann, O.V. & Ishi, M. (1963). Studies on the control of root-knot and reniform nematode with soil fumigation in Hawaii. *Hawaii Agric. Exp. Stn. Tech. Prog. Rep.* No. 139. p. 6.

Jones, J.E.; Beasley, J.P.; Dickson, J.I. & Caldwell, W.P. (1988). Registration of four cotton germplasm lines with resistance to reniform and root-knot nematodes. *Crop Sci.* **28**(1): 199-200.

Jones, J.E.; Newson, L.D. & Finley, E.L. (1959). Effect of reniform nematode on yield, plant characters and fibre properties of upland cotton. *Agron. J.* **51:** 353-356.

Khadr, A.S.; Salem, A.A. & Oteifa, B.A. (1972). Varietal susceptibility and significance of the reniform nematode, *Rotylenchulus reniformis*, in *Fusarium* wilt of cotton. *Plant Dis. Reptr.* **56:** 1040-1042.

Lambe, R.C. & Horne, W. (1963). The reniform nematode in cotton in the lower Rio Grande Valley of Texas. *Plant Dis. Reptr.* **47:** 941.

Martin, W.J. (1970). Elimination of root-knot and reniform nematodes and scurf infection from rootlets of sweet potato plants by hot water treatment. *Plant Dis. Reptr.* **54:** 1056-1058.

McSorley, R.; McMillan Jr., R.I. & Parrado, J.L. (1985). Soil fumigants for tomato production of Rockdale soils. *Proc. Florida St. Hort. Soc.* **98:** 232-237.

McSorley, R. & Parrado, J.L. (1986). Application of soil solarization to Rockdale soils in a subtropical environment. *Nematropica* **16** (2): 125-140.

Oteifa, B.A. (1970). The reniform nematode problem of Egyptian cotton production. *J. Parasitol.* **56** (Part 1): 255.

Padhi, N.N. & Misra, R.P. (1987). Control of *Rotylenchulus reniformis* on Frenchbean. *Indian J. Nematol.* **17**(1): 130-131.

Rebois, S.V.; Eldridge, B.J. & Webb, R.E. (1978). *Rotylenchulus reniformis* parasitism of potatoes and its effect on yield. *Plant Dis. Reptr.* **62:** 520-523.

Reddy, P.P. & Khan, R.M. (1988). Evaluation of *Paecilomyces lilacinus* for the biological control of *Rotylenchulus reniformis* infecting tomato compared with carbofuran. *Nematol. Medit.* **16:** 113-115.

Reddy, P.P. & Khan, R.M. (1989). Evaluation of biocontrol agent *Paecilomyces lilacinus* and carbofuran for the management of *Rotylenchulus reniformis* infecting brinjal. *Pak. J. Nematol.* **7** (1): 55-59.

Rich, J.R. & Bird, G.W. (1973). Inhibition of *Rotylenchulus reniformis* penetration of tomato and cotton root with foliar application of oxamyl. *J. Nematol.* **5:** 221-224.

Rohrbach, K.H. & Apt, W.J. (1986). Nematode and disease problems of pineapple. *Plant Dis.* **70:** 81-87.

Sasser, J.N. & Freckman, D.W. (1987). A world perspective on nematology. The role of the Society. *In* : *Vistas on Nematology*, (Eds.- J.A. Veech & D.W. Dickson). pp. 7-14.

Siddiqui, M.A. & Alam, M.M. (1987a). Control of plant parasitic nematodes by intercropping with *Tagetes minuta. Nematol. Medit.* **15** (2): 205-211.

Siddiqui, M.A. & Alam, M.M. (1987b). Efficacy of seed dressing with extracts of neem and Persian lilac against *Meloidogyne incognita* and *Rotylenchulus reniformis. Nematol. Medit.* **15** (2) 1: 399-403.

Siddiqui, M.A. & Alam, M.M. (1988a). Control of root-knot and reniform nematodes by bare root dip in leaf extracts of margosa and Persian lilac. *Zeitschrift fur Pflanzenkrankheiten und Pflanzenschutz* **95:** 138-142.

Siddiqui, M.A. & Alam, M.M. (1988b). Effect of latex seed dressing on *R. reniformis* and plant growth of some vegetables. *Nematol. Medit.* **16** (1): 129-130.

Siddiqui, M.A. & Alam, M.M. (1988c). Control of plant parasitic nematodes by soil amendment with marigold plant wastes. *Pak. J. Nematol*. **6:** 55-63.

Siddiqui, M.A. & Alam, M.M. (1988d). Control of plant parasitic nematodes by *Tagetes tenuifolia. Revue Nematol*. **4** (3): 369-370.

Siddiqui, M.A. & Alam, M.M. (1989). Possible utilization of a noxious weed on nematode control. *Biological Wastes* **28** (3): 181-188.

Siddiqui, M.A.; Haseeb, A & Alam, M.M. (1987). Evaluation of nematicidal properties in some latex bearing plants. *Indian J. Nematol*. **17:** 99-102.

Siddiqui, M.A. & Saxena, S.K. (1987). Effect of interculture of margosa and Persian lilac with tomato and eggplant on root-knot and reniform nematodes. *Int. Nematol. Network Newsl.* **4** (2): 5-8.

Sitaramaiah, K. & Sikora, R.A. (1982). Effect of the micorrhizal fungus *Glomus fasciculatus* on host parasite relationship of *Rotylenchulus reniformis Nematologica* **28:** 412-419.

Swarup, G.; Dasgupta, D.R. & Koshy, P.K. (1989). *Plant Diseases*. Anmol Publications, New Delhi. p. 395.

Taylor, D.P.; Schlosser, W.E. & Saad, A.T. (1970). First report of the reniform nematode, *Rotylenchulus reniformis*, from Lebanon. *Plant Dis. Reptr*. **54:** 435-436.

Thames, W.H. & Heald, C.M. (1974). Chemical and cultural control of *Rotylenchulus reniformis. Plant Dis. Reptr*. **58:** 337-341.

Thames, W.H.; Heald, C.M. & Anador, J. (1970). Populations of *Rotylenchulus reniformis* and yield of cotton following soil fumigation in Texas. *Phytopathology*. **60:** 1542.

Thames, W.H.; Sleeth, B. & Anador, J. (1969). Effect of soil fumigation on numbers of *Rotylenchulus reniformis* and yield of cotton in lower Rio Grande Valley of Texas. *Proc. Southern Agric. Workers Annual Convention*. p. 223.

Tiyagi, S.A.; Anver, S.; Bano, M. & Siddiqui, M.A. (1986a). Feasibility of growing zinnia as a mix crop alongwith tomato for control of root-knot and reniform nematodes. *Int. Nematol. Network Newsl*. **3:** 6-7.

Tiyagi, S.A.; Bano, M. & Alam, M.M. (1988). Evaluation of nematicidial potential in some plant species belonging to Family Composite. *Indian J. Nematol*. **18** (2): 228-231.

Tiyagi, S.A.; Siddiqui, M.A. & Alam, M.M. (1986b). Toxicity of an insect repellent plant to plant parasitic nematodes. *Int. Nematol. Network Newsl*. **3** (2): 16-17.

Varaprasad, K.S. (1986). *Rotylenchulus reniformis* Linford & Oliveira, 1940. A comparative account of systematics, biology and management. *In: Plant Parasitic nematodes of India: Problems and Progress* (Eds.—G. Swarup & D.R. Dasgupta). IARI Publication. pp. 194-210.

Yik, C.P. & Birchfield, W. (1984). Resistant germplasm in *Gossypium* species and related plants to *Rotylenchulus reniformis. J. Nematol*. **16:** 146-153.

Index